AF684961

Image Analysis in Biology

BB2
HAD

Image Analysis in Biology

INSTITUT PASTEUR
BIBLIOTHÈQUE CENTRALE
25, rue du Docteur-Roux
75724 PARIS CEDEX 15

23435
15.7.92

Editor

Donat-P. Häder, Dr.rer.nat.
Professor of Botany
Department of Botany and Pharmaceutical Biology
Friedrich-Alexander University
Erlangen, Germany

CRC

CRC Press
Boca Raton Ann Arbor London

Library of Congress Cataloging-in-Publication Data

Image analysis in biology / editor. Donat-P. Häder.
p. cm.
Includes bibliographical references and index.
ISBN 0-8493-6033-1
1. Microscope and microscopy. 2. Image processing. I. Häder, Donat-Peter.
QH205.2.I43 1991
578′.4—dc20 91-3778
CIP

This book represents information obtained from authentic and highly regarded sources. Reprinted material is quoted with permission, and sources are indicated. A wide variety of references are listed. Every reasonable effort has been made to give reliable data and information, but the author and the publisher cannot assume responsibility for the validity of all materials or for the consequences of their use.

All rights reserved. This book, or any parts thereof, may not be reproduced in any form without written consent from the publisher.

Direct all inquiries to CRC Press, Inc., 2000 Corporate Blvd., N.W., Boca Raton, Florida 33431.

© 1992 by CRC Press, Inc.

International Standard Book Number 0-8493-6033-1

Library of Congress Card Number 91-3778

Printed in the United States of America 1 2 3 4 5 6 7 8 9 0

PREFACE

Automatic image analysis has become an important tool in many fields in biology, medicine, and other sciences. The availability of powerful hardware and the development of specialized algorithms have dramatically improved the versatility of digital image analysis. Depending on the application, the techniques can be fully automatic or interactive with a human operator; they can operate in real time or require long calculation times dictated by the complexity of the task. The image can be either black and white or color, or a pseudocolor representation of gray values can be used.

The applications range from the submicroscopic (transmission and scanning electron microscopy) and even atomic level (scanning tunnel microscopy) to the macroscopic scale. Simple techniques include counting and quantitating objects such as leaf areas or sizes of microorganisms. Specific mathematical filters have been developed to enhance the quality of the original image or to extract specific features of interest. More complex programs attempt to analyze the form of objects in order to discriminate, e.g., cancer cells from normal tissue cells. Three-dimensional analysis of proteins, organelles, or macroscopic objects is even more involved since the number of calculations multiplies with the number of layers in the third dimension.

Real-time applications include cell and organism tracking used to analyze motility, velocity, and the direction of movement of motile organisms. Several techniques have been developed to detect the position of individuals and to follow these objects during a predetermined time interval.

As editor I very much appreciate the contributions of the authors to this book. Their specialized and diverse expertise with the various aspects of hardware, software, and biological application is reflected in each chapter, all of which meet a high quality standard.

Donat-P. Häder
Erlangen, Germany

THE EDITOR

Donat-P. Häder, Dr.rer.nat., is Professor of Botany, Department of Botany and Pharmaceutical Biology, at the Friedrich-Alexander University at Erlangen, Germany.

He received his doctoral degree and his habilitation from the University of Marburg. He has held a research associate position in the Department of Energy, Michigan State University, East Lansing, Michigan, and was visiting scientist in the Chemistry Department, Texas Tech University, Lubbock, Texas, CNR Institute of Biophysics, Pisa, Italy, and the National Research Laboratory, Okazaki, Japan.

Professor Häder has worked on the photomovement of microorganisms and the effect of solar ultraviolet radiation on phytoplankton and is involved in space biology, studying the effect of microgravity on motility in flagellates. He is a member of the Committee on Ecology for the German Ministry for Science and Technology, an expert for an Enquete Commission of the German Parliament, and a member of a United Nations Environment Program Commission on the Effects of Ozone Destruction.

One of the tools for his research activities is a real-time image analysis system developed over the last eight years. He has published over 130 original papers, coauthored 3 books, and translated 2 others.

CONTRIBUTORS

Shanti J. Aggarwal, Ph.D.
Department of Biomedical Engineering
Biomedical Engineering Program
The University of Texas at Austin
Austin, Texas

Norio Baba, Ph.D.
Department of Electrical Engineering
Kogakuin University
Tokyo, Japan

Egbert J. Boekema, Ph.D.
Biochemistry Laboratory
Rijksuniversiteit Groningen
Groningen, The Netherlands

Alan C. Bovik, Ph.D.
Department of Electrical and Computer Engineering
Biomedical Engineering Program
The University of Texas at Austin
Austin, Texas

Santos Carvajal-Gonzalez, M.Sc.
Laboratory of Quantitative Microscopy in Histopathology
INSERM U.263
University of Paris 7
Paris, France

Giuliano Colombetti, Ph.D.
CNR Institute of Biophysics
Pisa, Italy

Primo Coltelli, Ph.D.
CNR, CNUCE Institute
Pisa, Italy

Judith G. Croxdale, Ph.D.
Department of Botany
University of Wisconsin-Madison
Madison, Wisconsin

Kenneth R. Diller, Sc.D.
Department of Mechanical Engineering
Biomedical Engineering Program
The University of Texas at Austin
Austin, Texas

David B. Dusenbery, Ph.D.
School of Biology
Georgia Institute of Technology
Atlanta, Georgia

Günther Galfe, Dipl.-Ing.
Institut für Zoologie
Technische Universität München
Garching, Germany

David Gauthier, M.Eng.
McGill Research Centre for Intelligent Machines
McGill University
Montreal, Quebec, Canada

Paolo Gualtieri, Ph.D.
CNR Institute of Biophysics
Pisa, Italy

Donat-P. Häder, Dr.rer.nat.
Department of Botany and Pharmaceutical Biology
Friedrich-Alexander University
Erlangen, Germany

George Harauz, Ph.D.
Department of Molecular Biology and Genetics
University of Guelph
Guelph, Ontario, Canada

Tsuyoshi Hayakawa
Central Research Laboratory
Hamamatsu Photonics K. K.
Hamakita, Japan

Michael E. J. Holwill, Ph.D.
Physics Department
King's College
London, England

Nak H. Kim, Ph.D.
Department of Electrical and Computer Engineering
Biomedical Engineering Program
The University of Texas at Austin
Austin, Texas

Martin D. Levine, Ph.D.
Department of Electrical Engineering and McGill Research Centre for Intelligent Machines
McGill University
Montreal, Quebec, Canada

Roberto Marangoni, Ph.D.
Institute of Biophysics
Faculty of Medicine
University of Genova
Genova, Italy

Peter B. Noble, Ph.D.
Department of Oral Biology
Faculty of Dentistry
McGill University
Montreal, Quebec, Canada

Kenji Omasa, Ph.D.
Environmental Plant Science Department
The National Institute for Environmental Studies
Ibaraki, Japan

Michael Radermacher, Dr.rer.nat.
Wadsworth Center for Laboratories and Research
New York State Department of Health
Albany, New York

Jean Paul Rigaut, Sc.D., M.D., Ph.D.
Laboratory of Quantitative Microscopy in Histopathology
INSERM U.263
University of Paris 7
Paris, France

Peter Satir, Ph.D.
Department of Anatomy and Structural Biology
Albert Einstein College of Medicine
Bronx, New York

Nico A. M. Schellart, Ph.D.
Laboratory of Medical Physics and Informatics
University of Amsterdam
Amsterdam, The Netherlands

Tetsuo Takahashi, Ph.D.
Suntory Institute for Bioorganic Research
Osaka, Japan

Jany Vassy, Sc.D., Ph.D.
Laboratory of Developmental Biology and Cell Differentiation
University of Paris 13
Bobigny, France

Kurt Vogel, Dipl.Biol.
Department of Botany and Pharmaceutical Biology
Friedrich-Alexander University
Erlangen, Germany

Dieter G. Weiss, Dr.rer.nat.
Institut für Zoologie
Technische Universität München
Garching, Germany

TABLE OF CONTENTS

Introduction

Chapter 1

INTRODUCTION

Donat-P. Häder

Vision by man and machine differ in many respects and certainly the human visual apparatus and brain are still far superior to machine vision.[1] Both the spatial and gray or color shade resolutions of the human eye surpass those of high resolution video cameras and the image processing of the human brain far exceeds the speed and complexity of even the fastest computers.[2] However, using image technologies facilitates quantitative analysis and objective interpretation of selected parameters.

Since its first developmental stages in the 1970s, computerized image analysis has witnessed a rapid growth and enormous progress. Digitization of analog images lends itself to image enhancement and to the extraction of specific features of interest such as dimensions, areas of a distinct color or gray shade,[3,4] or movement of individual elements in a sequence of images.[5,6]

The advent of more powerful hardware and the development of robust and efficient algorithms[7,8] have laid the ground for highly specialized and automatic image analysis which has found applications in all fields of biology and medicine.[9-11] Early in the development of image analysis, researchers found interest in the digitization of cellular structures.[12-15] Both light microscopy[16,17] and electron microscopy[18,19] are important fields for image enhancement and analysis, and even holograms have been used as image sources.[20] Especially in medical samples, it is a declared goal to develop automated image analysis systems which allow the recognition and identification of specific features[21] in order to reliably determine quantitative distributions of, e.g., blood cells or sediments[22,23] and to identify abnormal cells such as cancer cells.[24-26] Reliable cell identification requires a high spatial resolution.[27] In addition to morphometric methods,[28] image analysis and reconstruction are used for three-dimensional (3-D) reconstruction of biological structures.[29-32]

This volume is divided into three sections, the first of which describes the methods of image analysis, including 3-D analysis, quantitation by laser scanning confocal microscopy and quantitative area determination. This section is intended to give a more general introduction into the techniques, includes a chapter on digital filters, and describes the principles of object detection in digital images as well as single photon imaging.

The second section focuses on the applications of image analysis in microscopy including contrast enhancement in video images, automatic detection and recognition of live cells, as well as image analysis of bending shapes of eukaryotic flagella and cilia. This section is also devoted to the analysis of subcellular structures in electron micrographs and protein analysis as well as the reconstruction of crystals and 3-D histological objects from serial section images.

The final section describes the techniques used for movement analysis of microorganisms, bacteria and animals, which require fast routines to track the organisms in real time in a sequence of video frames.

REFERENCES

1. **Poggio, T.,** Vision by man and machine, *Sci. Am.,* 250, 68, 1984.
2. **Blake, A.,** Real-time seeing machines?, *Nature,* 328, 759, 1987.
3. **McMillian, P. J., Yakush, A., Frykman, G., Nava, P. B., and Ras, V. R.,** Minima equalization: a useful strategy in automatic processing of microscopic images, *J. Microsc.,* 148, 253, 1987.
4. **Russ, J. C. and Russ, J. C.,** Automatic discrimination of features in grey-scale images, *J. Microsc.,* 148, 263, 1987.
5. **Turano, T. A., D'Arpa, P., Clark, W. L., and Williams, J. R.,** A time-lapse, image digitization videomicroscope system based on a mini computer with large peripheral memory, *Comput. Biol. Med.,* 15, 177, 1985.
6. **Thurston, G., Jaggi, B., and Palcic, B.,** Cell motility measurements with an automated microscope system, *Exp. Cell Res.,* 165, 380, 1986.
7. **Wampler, J. E. and Kutz, K.,** Quantitative fluorescence microscopy using photomultiplier tubes and imaging detectors, in *Methods in Cell Biology,* Wang, Y. L. and Taylor, D. L., Eds., Academic Press, San Diego, 1989, 239.
8. **Sasov, Y.,** An integrated PC-based image analysis system for microtomography and quantitative analysis of inner micro-object structure, *J. Microsc.,* 156, 91, 1989.
9. **Frank, J.,** Image analysis of single macromolecules, *Electron Microsc. Rev.,* 2, 53, 1989.
10. **Dhawan, A. P.,** A review on biomedical image processing and future trends, *Comput. Methods Prog. Biomed.,* 31, 141, 1990.
11. **Kenny, P. A., Dowsett, D. J., Vernon, D., and Ennis, J. T.,** A technique for digital image registration used prior to subtraction of lung images in nuclear medicine, *Phys. Med. Biol.,* 35, 679, 1990.
12. **Holmquist, J., Antonsson, D., Bengtsson, E., Danielsson, P. E., Eriksson, O., Hedblom, T., Martensson, A., Nordin, B., Olsson, T., and Stenkvist, B.,** TULIPS — The Uppsala-Linkoping image processing system, *Anal. Quant. Cytol.,* 3, 182, 1981.
13. **Smith, K. C. A.,** On-line digital computer techniques in electron microscopy: general introduction, *J. Microsc.,* 127, 3, 1982.
14. **Skarnulis, A. J.,** A computer system for on-line image capture and analysis, *J. Microsc.,* 127, 39, 1982.
15. **Spring, K. R.,** Application of video to light microscopy, in *Membrane Biophysics II, Physical Methods in the Study of Epithelia,* Alan R. Liss, New York, 1983, 15.
16. **Caldwell, D. E.,** New developments in computer-enhanced microscopy (CEM), *J. Microbiol. Methods,* 4, 117, 1985.
17. **Wollmer, W.,** Application of a small microcomputer to cell image analysis, *Anal. Quant. Cytol. Histol.,* 9, 535, 1987.
18. **Gronsky, R.,** Spectroscopic information from high resolution images, *Ultramicroscopy,* 24, 155, 1988.
19. **Stewart, M.,** Computer image processing of electron micrographs of biological structures with helical symmetry, *J. Electron Microsc. Tech.,* 9, 325, 1988.
20. **Lin, J. A. and Cowley, J. M.,** Reconstruction from in-line electron holograms by digital processing, *Ultramicroscopy,* 19, 179, 1986.
21. **De Paz, P., Barrio, J. P., and Renau-Piqueras, J.,** A basic program for determination of numerical density of cytoplasmic compartments. II. Analysis of ellipsoids and cylindrical particles, *Comput. Biol. Med.,* 16, 273, 1986.
22. **Koss, L. G., Sherman, A. B., and Adams, S. E.,** The use of hierarchic classification in the image analysis of a complex cell population. Experience with the sediment of voided urine, *Anal. Quant. Cytol.,* 5, 159, 1983.
23. **Preston, K., Jr.,** High-resolution image analysis, *J. Histochem. Cytochem.,* 34, 67, 1986.
24. **Preston, K., Jr., Fagan, L. M., Huang, H. K., and Pryor, T. A.,** Computing in medicine, *IEEE Trans. Biomed. Eng.,* 784, 294, 1984.
25. **Kaufman, A. G., Nathwani, B. N., and Preston, K., Jr.,** Subclassification of follicular lymphomas by computerized microscopy, *Hum. Pathol.,* 18, 226, 1987.
26. **Wittekind, C. and Schulte, E.,** Computerized morphometric image analysis of cytologic nuclear parameters in breast cancer, *Anal. Quant. Cytol. Histol.,* 9, 480, 1987.
27. **Gunzer, U., Aus, H. M., and Harms, H.,** Letter to the editor, *J. Histochem. Cytochem.,* 35, 705, 1987.
28. **Pradère, P. and Thomas, E. L.,** Image processing of partially periodic lattice images of polymers: the study of crystal defects, *Ultramicroscopy,* 32, 149, 1990.
29. **Nierzwicki-Bauer, S. A., Balkwill, D. L., and Stevens, S. E., Jr.,** Three-dimensional ultrastructure of a unicellular cyanobacterium, *J. Cell Biol.,* 97, 713, 1983.
30. **Gras, H. A.,** "Hidden line" algorithm for 3D-reconstruction from serial sections — An extension of the NEUREC program package for a microcomputer, *Comput. Programs Biomed.,* 18, 217, 1984.

31. **Jimenez, J., Santisteban, A., Carazo, J. M., and Carrascosa, J. L.,** Computer graphic display method for visualizing three-dimensional biological structures, *Science,* 232, 1113, 1986.
32. **Harauz, G. and Flannigan, D.,** Structure of ribosomes from *Thermomyces lanuginosus* by electron microscopy and image processing, *Biochim. Biophys. Acta,* 1038, 260, 1990.

Methods of Image Analysis

Chapter 2

PRINCIPLES OF OBJECT DETECTION FOR AN AUTOMATED CELL TRACKING SYSTEM

David Gauthier, Martin D. Levine, and Peter B. Noble

TABLE OF CONTENTS

I. INTRODUCTION

With recent developments in moderately priced video and computer technology, computer vision techniques, coupled with light microscopy, are assuming an increasingly important role in the examination of biological systems. The principles of object detection or segmentation in digital images, as outlined in this chapter, will deal with the problems of detecting and tracking cells as they migrate on either a two-dimensional (2-D) substratum or within a three-dimensional (3-D) collagen gel.

The study of cell movement and cell behavior is of fundamental importance in areas such as embryology, wound healing, host defense mechanisms, and mechanisms of tumor cell metastasis and invasion. Under both normal and pathological conditions, certain cell types exhibit extensive movements, often migrating to entirely different parts of the organism. Our particular research interests have evolved around the need to analyze the multitude of factors that control cell locomotory and replicative behavior on 2-D substrata and, more recently, in 3-D collagen gels.

Segmenting cells and tracking their patterns of migration are challenging problems owing to both the complex nature of cells and the problems inherent to video light microscopy. In both 2- and 3-D tissue culture environments, cells are nonrigid, irregularly shaped bodies. The cells' external environment influences their shape, which in turn affects their locomotory behavior and ultimately how they function. Cells that naturally migrate within organisms can take on a variety of different sizes and shapes and can migrate at different rates, depending on their current activities. For example, fibroblasts have been observed to take on round shapes when undergoing mitosis or not locomoting, stellate or bipolar shapes when they are moving slowly while examining their environment, and unipolar shapes when they are actively migrating. In addition, they can either attach to other cells or remain attached after mitosis to form cords, which locomote as a group. As the cell changes shape during locomotion, the contrast between the cell boundary and the background is continually changing. A round cell is uniformly thick, producing a high degree of contrast between the background and the cell, which makes it easy to detect. As it takes on unipolar, bipolar, and stellate shapes it becomes progressively thinner along its pseudopodia and filopodia, reducing the contrast between the boundary and background and making the cell more difficult to segment using computer vision techniques (Figure 1). In some instances, portions of the cell membrane can appear to be occluded. Artifacts in the environment, such as small pieces of membrane and organelles within the cell, can easily be mistaken for the cell's membrane. The problems of cell segmentation are further compounded when the cells are suspended in a 3-D environment, in that only portions of a cell are visible in any one focal plane. This requires deriving the 3-D boundary and shape information from several serial optical sections. Hardware-related factors which contribute to the quality of the image, such as uneven illumination, and electronic or optical noise, also play an important role in our ability to effectively segment an image. Finally, time constraints in real-time cell tracking call for compromises to be made between the efficacy and computational efficiency of whichever segmentation approach is adopted.

In this chapter, we will describe an *in vitro* imaging system based on a light microscope that can be used for the study of dynamic processes of cells. The system is called **TRACES**, which is an acronym for **TRA**cking **CE**lls **S**ystem. The approach used by TRACES for segmenting cells, a fast multistage boundary extraction algorithm based on a technique called component labeling, will be presented. Also, two other techniques that are under development in our laboratory will be briefly presented: a boundary extraction method based on active contour models called snakes, and a technique that segments moving cells by extracting the stationary background of a sequence of images.

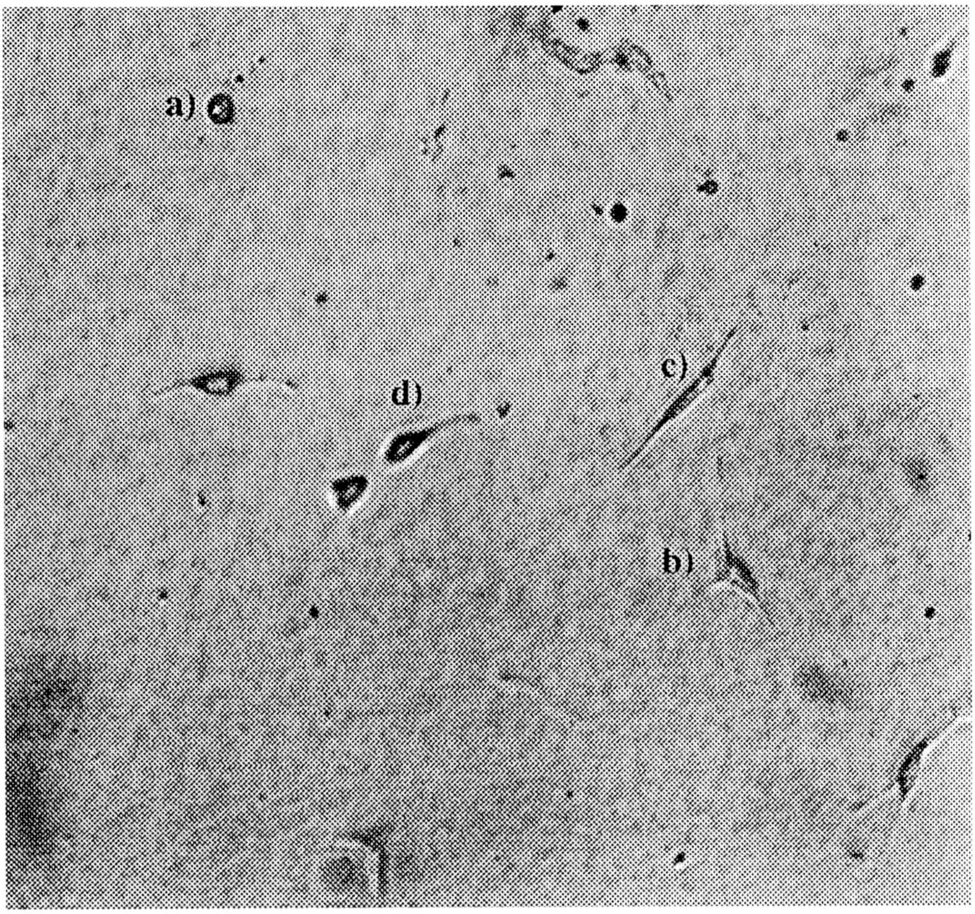

FIGURE 1. Example of (a) round, (b) stellate, (c) bipolar, and (d) unipolar cell shapes. The cells were plated onto a collagen gel, magnified 50×, and digitally acquired with 256 levels of quantization.

II. 3-D CELL TRACKING SYSTEM

In order to study cell movement and behavior *in vitro,* an automated 3-D cell tracking and optical sectioning system was developed. TRACES employs computer vision and artificial intelligence techniques to analyze and characterize sequences of digitized images. It is capable of concurrently tracking a number of cells using time-lapse techniques to follow their movements over long periods of time. The system records boundary information and the position of the cell's center at regular time intervals. The system hardware, which is illustrated in Figure 2, consists of four main subsystems: environmental control, positioning, image acquisition, and processing. We will describe each subsystem and discuss the appropriate design considerations.

A. Environmental Control Subsystem

The cell populations that we study are maintained in an environmental chamber, a custom-designed incubator with an optical window that allows light to pass through its contents. The environmental control system maintains the temperature of the chamber at 37°C in a humidified atmosphere of 5% CO_2 in air.

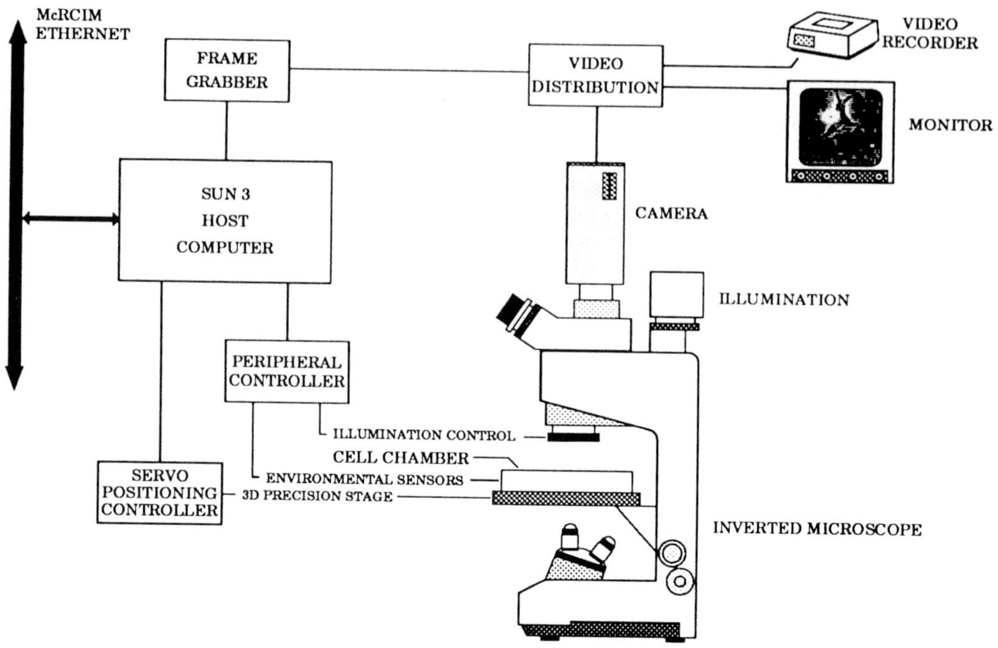

FIGURE 2. Three-dimensional cell tracking system.

B. Positioning Subsystem

The cell chamber is mounted in a locating fixture, which is part of a computer-controlled, linear, three-axis positioning system. The positioning system permits any point inside the chamber to be placed within the field of focus of the imaging system. Thus, the cells tracked are not limited to those that can be viewed in a single field and restrictions on the locations of the cells are eliminated. Also, the stages can be repositioned prior to the acquisition of each cell, ensuring that the given cell is within the optimum field of view of the camera at the time of digitization.

C. Imaging Subsystem

The imaging subsystem consists of an inverted light microscope, an illumination and shutter controller, a monochrome video camera, a digital frame grabber, and an analog time-lapse video recorder. A discussion of the relative merits of various microscope optical systems has already been presented.[6,7]

Under the constraints of our instrumentation,we have found that bright field optics are best suited to our computer vision techniques. Phase contrast microscopy produces higher contrast images, but the characteristic halos associated with these optical systems are inconsistent from one cell shape to the next. The halo is most pronounced with round objects and is less evident as the shape flattens. It also conceals details of the cell's boundary. Biochemical image enhancers such as fluorescing compounds cannot be used; cells are tracked over several generations and such compounds would be diluted with each mitotic cycle. They also could alter the physiology of the cells and their environment and consequently could affect the cells' locomotory behavior.

Choosing the microscope's magnification involves a tradeoff. The resolution is selected to provide a sufficient number of pixels so that the computer vision algorithms can extract accurate and detailed boundary information but not too high so as to require excessive processing. As the magnification increases, the working focal distance, and therefore the thickness of the gel that can be used, decreases. We have found that we obtain reasonable results with a magnification between 50× and 100×.

D. Processing Subsystem

Processing tasks such as high level sequencing of events, control of the subsystems, image analysis, and data analysis are performed on a SUN 3/160 computer workstation. The most computationally expensive tasks are those associated with image analysis, and these dictate the maximum number of cells that can be tracked in real time. Because cells locomote slowly, time-lapse techniques are employed to follow their movement over long periods. Images of the cells are acquired at regular sampling intervals and either analyzed in real-time or stored and processed at a later time. Several cells can be acquired and analyzed, in turn, within the sampling interval. Note that when we refer to real time we are not implying that the computer vision algorithms are executed at video rates, that is, 30 frames/s, but that all cells are analyzed within the specified sampling interval. On-line real-time processing of the images allows their contents to be interpreted and any necessary adjustments made prior to obtaining the next sample. For example, the vision system interprets the movement of the cell and directs the motorized positioning system to adjust the stages to ensure that the cell remains within the optimum field of view of the camera. The major challenge of real-time processing is the time constraint; the extraction of feature information must be completed for all cells before the next digitized image sequence commences. Constraints such as the number of samples being tracked, the sampling interval, and the generation time of the cells can markedly influence the computer vision algorithms that are employed. For quail-derived embryonic fibroblasts, a typical sampling interval is 2 min. Storing images and processing them off-line permits very elaborate computer vision algorithms to be executed that can extract more detailed information on each cell. The obvious drawback of off-line processing is that no position adjustments can be made during the course of the experiment. A cell that locomotes out of the field of view of the camera is lost. We have chosen to combine both on-line and off-line processing in our tracking system. The cells are tracked in real-time; but digitized images are also stored, both on disk and on a time-lapse video recorder, for more detailed analysis. Experiments that run over a period of several days can be replayed and any interesting events can be examined more closely. The image processing algorithms employed by TRACES are chosen to maximize the potential of the tracking system. Factors which have influenced their adoption include performance, processing throughput, flexibility, and reliability. The objective of our cell detection algorithms is to estimate either the two- or three-dimensional boundary and the center of each cell within some specified time interval. During the course of an experiment, this information is used by a tracking algorithm to predict where each cell will most likely be at the next sampling time. It also serves as the input for higher level feature- and object-based procedures that characterize the cell's shape.[30] Following an experiment, the information is used to analyze the cell's locomotory phenotypes and migratory patterns.[7]

E. Biological Methodology

Although we study a variety of different cell types, the majority of our experiments employ fibroblasts derived from the hearts of 7-day-old quail embryos using the explant technique. After 48 h, the beating cardiac cells, still within the explant proper, are removed using glass microneedles. The remaining cells are detached from the petri dish using a trypsin/EDTA solution. The embryos provide a source of cells having the same developmental age and properties for each experiment in lieu of having access to a liquid nitrogen storage facility. We have used these cells for our 2- and 3-D locomotory behavior studies with consistent results.[1-4]

For 3-D experiments, the cells are embedded in a volume of collagen gel, whose dimensions are approximately 16 cm^2 in the horizontal plane and 1500 μm in the vertical plane. The gels are made by polymerizing a solution of 1.5 mg/ml collagen fibers in cell growth media into a random lattice.[1] Just prior to polymerization, the cells are incorporated into the solution at a density that is much below that required to produce a contraction of the gel.[5] Embedding the cells in collagen gel affords the undertaking of experiments dealing with the complex biochemical and biophysical interactions that exist between cells and the polymer fibrils of the gel. In the case of 2-D experiments, the cells are plated onto glass or on top of a collagen gel. Tracking experiments are commenced 18 hours later.

III. IMAGE ACQUISITION

In many respects the image acquisition chain, which consists of the illumination source, the microscope, the video camera, and the input section of the digital frame grabber, can be considered as an optical analog processor. In order to obtain the best quality digital image, i.e., one which has good contrast between the cells being tracked and the background is low in noise and aberrations, and is evenly illuminated, each link in the acquisition chain must be optimized.

A major advantage of video-based microscopy is the ability to enhance image contrast. Structures that are not readily visible to the human eye can be dramatically enhanced with the application of image processing. Unfortunately, some imperfections resulting from poor-quality digital images also become more evident after contrast enhancement and these require extra processing to compensate for the problems introduced in the acquisition chain. The contrast of a specimen is a function of the illumination, the interaction of the illumination with the specimen, and how the microscope's optics project the image of the specimen into the video camera. It must therefore be ensured that the light source, collector, condenser, optics, filters, and camera are all properly matched and adjusted. Several innovative ways of doing this have been proposed.[6,38] When performing this task, one should bear in mind that the optimal adjustment of a microscope with a video camera does not necessarily correspond to that for visual adjustment through an ocular. That is, the optimal conditions for a digitized image are not the same as those of a scene viewed by a human through an ocular. Alignment and focusing of the illumination source and the condenser should therefore be made by looking at the output of a video camera connected to a monitor or a frame grabber. The final adjustments of the acquisition system alignment can be fine-tuned with the aid of simple processing techniques such as projections or contouring to evaluate the uniformity of the illumination and histograms to gauge the intensity of the illumination.

A horizontal or vertical projection is the plot of the sum of the gray-level intensities of each column or row of pixels, respectively, in an image.[7] If the intensity is uniform over the entire image, then the projection is obviously a line parallel to the horizontal or vertical axis. Otherwise, regions of high or low intensities would produce local maxima or minima (Figures 3b and 3c). Thus, we may use horizontal and vertical projections of an image of an unobstructed optical pathway to locate the positions of unevenly illuminated areas. This procedure is used to measure how well the illumination has been adjusted. A nonuniformly illuminated field can be effectively detected by reducing the number of gray levels displayed to produce an effect called contouring. Pixels are divided into groups by mapping gray-level intensities that are close to one another onto a single intensity. The resulting image is a contour map of bands of equal intensity. By eliminating differences in intensity levels within groups, differences between the groups are exaggerated. For example, a 256 gray-level image can be requantized into 8 levels by grouping each 32 adjacent intensity levels together as follows:

$$[(n - 1) \cdot 32, n \cdot 32 - 1] \rightarrow n, \quad \text{where} \quad n = 1 \ldots 8 \tag{1}$$

If the microscope's illumination is uneven, characteristic bands or concentric circles will be observed. If the center of the field of view is brightest and the illumination grows uniformly

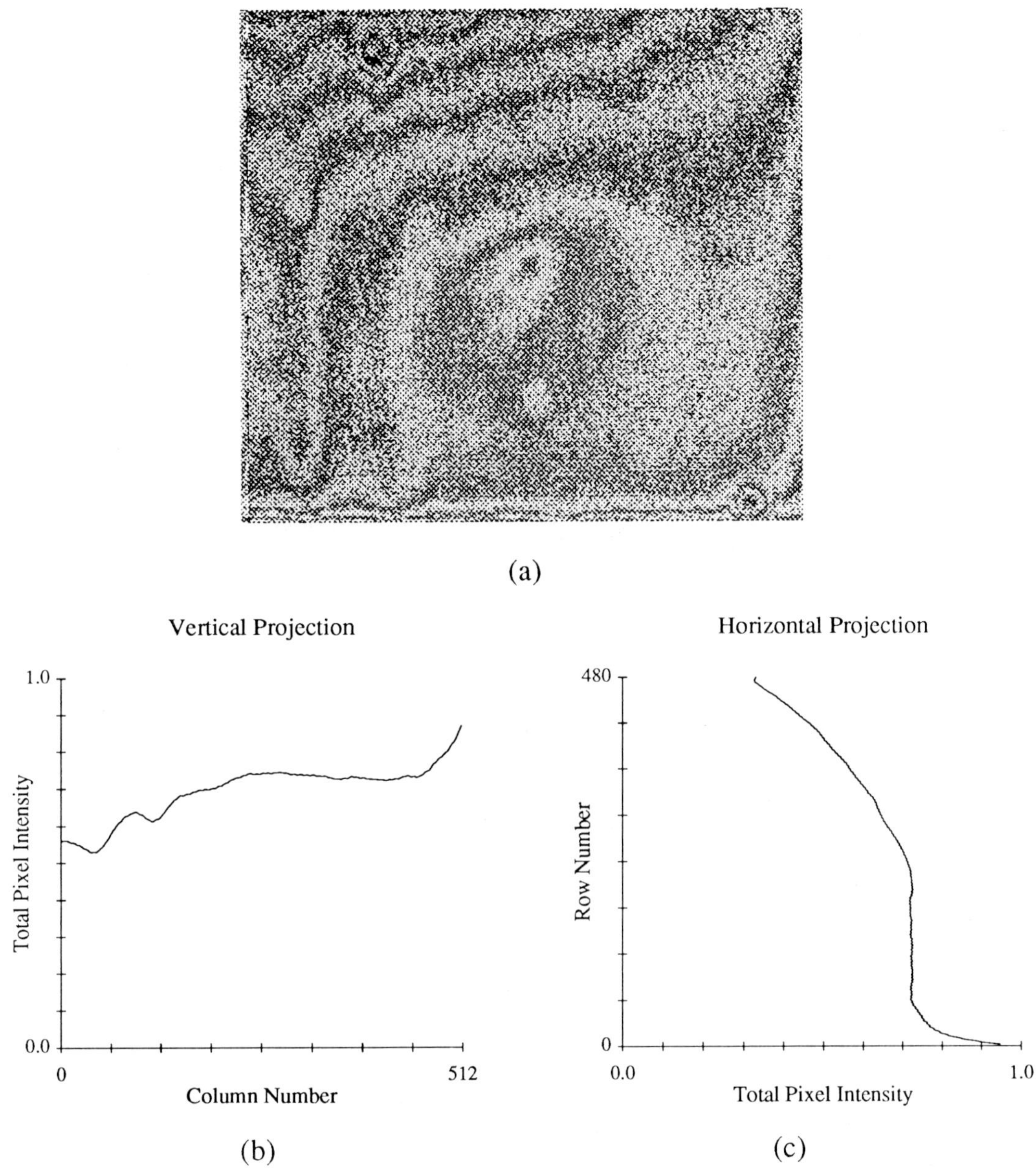

FIGURE 3. Evaluation of illumination uniformity. (a) Contour image, (b) vertical and (c) horizontal projections. The concentric rings in the contour image indicate that the illumination is concentrated near the center. The horizontal and vertical projections indicate the magnitude of the nonuniformity and that the image is darkest at the top left corner.

darker towards the periphery, the contoured image would be composed of a set of concentric circles (Figure 3a). If the illumination varies uniformly across the image, the contours would appear as bands. If the illumination were perfectly even, no contours would be observed. This latter case is not easily attainable, but through an iterative fine-tuning process, the optimal illumination uniformity can be achieved. In Section V.B, we will present a software-based method for further correcting the illumination uniformity, should it be necessary.

In comparison to the human visual system, a video camera can respond to only a very limited range of intensities. Below some critical brightness, called the cutoff level, the sensor

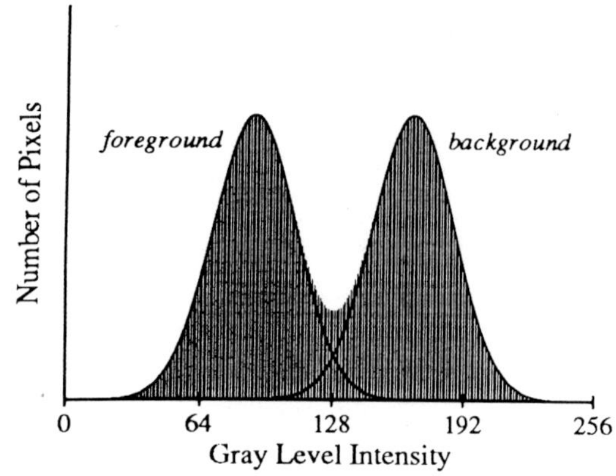

(a)

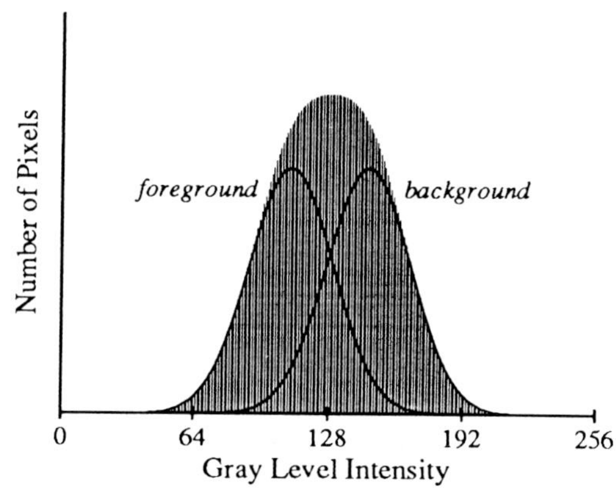

(b)

FIGURE 4. Examples of (a) bimodal and (b) unimodal intensity histograms.

response is low and the noise level high. Above a limiting intensity level, the sensor saturates and cannot respond to small intensity changes. The intensity of the illumination source is normally adjusted to expose the camera to the widest range without concentrating the intensity at either extreme of this range. The adjustment of the illumination intensity can be evaluated by examining an intensity histogram of an image.

With the environmental chamber containing cells placed under the microscope, an image is grabbed and an intensity histogram is plotted. The histogram measures the distribution of gray-level intensities. For each intensity level in the range of 0 (black) to the maximum gray-level (white), the histogram records the number of pixels at that intensity.[9] The intensity distribution of a multiple gray-level image is frequently modeled as the sum of two Gaussian distributions, one representing the foreground and the other, the background. In the ideal case, the image has a very distinct background and foreground, the two Gaussian distributions overlap only slightly, and the resulting histogram of the image is bimodal (Figure 4a). As the background and foreground distributions become less distinct, the overlap of the Gaussian distributions increases and the histogram of the image becomes unimodal (Figure 4b).

When the intensity of the illumination has been properly adjusted for uniform illumination, the measured intensities span the full range and there are no concentrations of

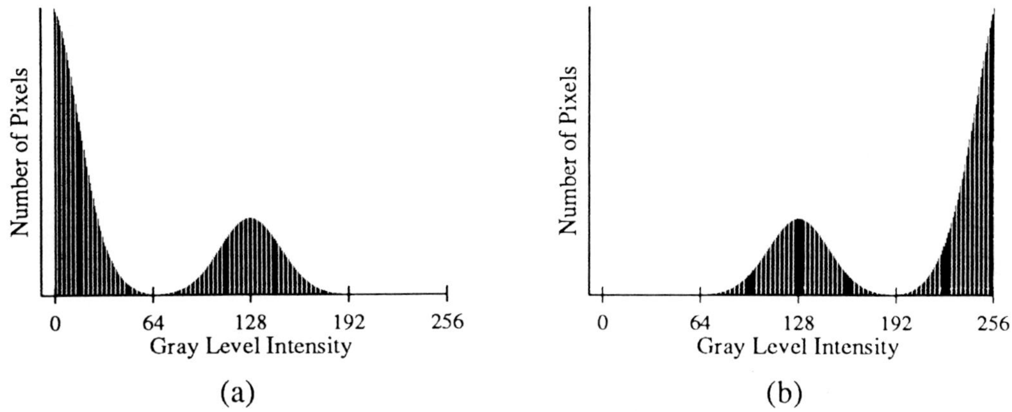

FIGURE 5. Histograms showing effect of illumination (a) cutoff and (b) saturation.

intensities about the black or white intensity levels, as shown in Figure 4. Examples of the undesirable cutoff and saturation situations are illustrated in Figure 5.

The intensity of the illumination source is normally adjusted to obtain uniform illumination across the range of the frame grabber. However, in some circumstances, it is advantageous to select specific types of nonuniform illumination. In these cases, it is important to be able to predict what type of cell images will be encountered over the course of an experiment. For example, when examining round cells, the illumination can be turned up to achieve a high degree of contrast between the cells and background. In this case there is very little intensity variation between the cells, which are almost completely black, and over the background regions, which are almost completely white. However, the same illumination is found to be unsuitable for examining cells that are not round, since the boundaries of pseudopodia and filopodia, which are thinner than round cells, are obscured.

As an alternative or an adjunct to the above, gray-level manipulation techniques can also be used. These techniques involve adjusting the input look-up table parameters of the frame grabber. A look-up table, also called an intensity transform table, maps the gray-level value of each pixel into another gray-level value. These mappings may be linear or nonlinear. Common transforms include scaling, which expands the dynamic range of gray-level intensities; compression, which contracts the dynamic range; and limiting, which limits high or low intensity values. The particular transform to be implemented in any given situation depends on the nature of the images to be acquired.

Finally, once an image has been acquired in digital format, there is a host of digital image preprocessing algorithms that can be applied to effect contrast and brightness, filter noise, and smooth or sharpen edges within the image.[9]

IV. IMAGE SEGMENTATION

Segmentation is the process of identifying regions of an image that have common properties while at the same time separating regions that are dissimilar. This assumes that the objects we are trying to segment within a scene can be characterized by some similar feature such as gray-level intensity, color, or texture. In this instance, we want to extract from the background the trace of a closed contour representing the cell membrane, using only the gray level intensity information.

Many techniques have been reported for image segmentation.[10-12] However, it is evident that no single method has yet been found that produces satisfactory results in difficult low contrast cases. One reason for this is that the segmentation techniques are often constrained by the instrumentation and experimental conditions. For example, in real-time applications, segmentation must be completed within short sample time intervals. Thus, only those techniques which can perform to a sufficient level of accuracy within the allocated time period can be considered. Also, techniques that rely on image enhancing stains or fluorescing compounds, which could alter the physiological activities of the cells or their environment are, for our purposes, excluded.

In the next section, we will present the approach used by TRACES for segmenting cells, a fast multistage boundary extraction algorithm based on a technique called component labeling.

V. CELL BOUNDARY EXTRACTION

A. The Window of Interest

When initializing the system, an operator selects which cells are to be tracked. The operator directs the computer to position a cell of interest within the field of view and selects the cell for tracking by placing a window around it. The center of this window provides an initial estimate of the location of the cell. This selection process is repeated until all cells to be tracked have been identified. When automated tracking is started, the system samples and analyzes each selected cell, in turn, within a specified time interval. The positioning system places the estimated location of the cell at the center of the camera's field of view. For the case of 2-D tracking, a working window is defined and a single image is acquired. For 3-D tracking, a working volume is defined and a series of vertical sections is acquired. Image processing algorithms are employed to determine the cell's boundary and center of mass. The center of mass of the cell provides an estimate of the location of the cell for the next sample. The purpose of the working window or volume is to impose a bound on the data size and therefore reduce the processing time. The size of the window is determined by the maximum expected cell motion between samples, which is based on *a priori* knowledge. When positioned at the estimated center of the cell in the previous sample, it must be large enough to ensure that the window will contain the entire cell in the present sample.

The image bounded by the working window is expected to contain the cell of interest, which appears as a region of pixels of darker intensity; other dark objects, which are smaller than the cell of interest; and the background, consisting of brighter pixels. The contrast between the intensity of the cell and the background is often poor and this is especially true for cells migrating within a 3-D gel. Both the background and foreground objects exhibit large variations in gray-level intensity. In part, this can be due to uneven illumination and can be compensated for through image processing.

B. Software-Based Lighting Compensation

It is exceedingly difficult to align and adjust the optics of a microscope system, as outlined in Section III, so that the illumination is perfectly uniform over the entire field when the digitized image is examined. Prior to segmenting the image, a procedure to compensate for these variations may need to be implemented.

The amount of variation introduced by the lighting system can be measured by examining the images of several blank scenes. The average intensity at each pixel of these images, b_{ij}, and the average intensity over all the pixels of the images, $\bar{b}$, are determined. The required intensity compensation at each pixel is determined by subtracting the intensity at the given pixel from the average intensity of the entire image, $\bar{b} - b_{ij}$. An arbitrary image, I, may then be compensated for imperfections in the acquisition system by adding to each pixel intensity the required compensations at that pixel:

$$I^c_{ij} = I_{ij} + (\bar{b} - b_{ij}) \tag{2}$$

where I^c_{ij} and I_{ij} are the corrected intensity and the measured intensity at pixel *ij*. The net result of this compensation procedure is a uniformly illuminated image.

C. Centering the Window of Interest

The working window is expected to contain the cell of interest. However, the cell will not necessarily be centered and, in some extreme cases, could extend slightly beyond the window. The position of the window is adjusted, using a converging squares algorithm, so that the cell will be approximately centered within the window.[13,14]

The purpose of the converging squares algorithm is to locate peaks in multidimensional data. Its advantages are that it executes extremely fast and is robust with respect to noise. Since, with bright field optics, the cell is on average darker than the background and in our configuration it comprises the largest dark object in the working window, this method is well suited for approximating the center of the cell.

At each step of the converging squares algorithm, the dimensions of the window are reduced by one pixel in each of the horizontal and vertical directions by eliminating border pixels. The total intensity along the left and right border columns and the top and bottom rows are determined; the row and column with the highest total intensity are eliminated from the window. Thus, at each step, the brighter boundary pixels are eliminated, retaining the darker pixels. This sequence of reducing windows ultimately converges towards the darkest region of the original window.

The converging squares algorithm is terminated once the original window has been reduced to a subwindow of specified size, in our case that approximated by the average diameter of the cell. The center of the dark region is estimated by the weighted average of the coordinates of all pixels within the final subwindow, with the weighting factor equal to pixel intensity. Assuming that the dark region of the window represents the cell of interest, repositioning the window to the center of the dark mass will better ensure that the cell will be approximately centered and that the entire cell will be within the window.

The converging squares algorithm can be easily adapted to a 3-D environment. The working window is defined by a cube formed by stacking the 2-D windows from each serial section. At each step, the dimensions of the window are reduced by one pixel in each of the horizontal, vertical, and depth directions, by eliminating boundary pixels. The total intensity is determined on each of the six boundary surfaces of the cube: left, right, front, back, top, and bottom. The combination of surfaces, left or right, top or bottom, and front or back, that yield the highest total intensity are eliminated. The converging squares algorithm is terminated once the original cube has been reduced to a size that is equal to the average volume of a cell.

The disadvantage of the converging squares technique is that it does not provide a highly accurate estimate of the cell center. The procedure simply finds the point of concentration of dark pixels, without attempting to segment the image. For irregularly shaped objects, or cases of high intensity variation of both the background and foreground with little differentiation between them, the estimated center can deviate significantly from the actual cell center. However, for the purpose of repositioning the cell window, converging squares is sufficiently accurate.

There are some situations where the converging squares algorithm alone has been shown to be sufficient for cell tracking. For instance, if the cells are similar in shape, such as with round cells, and the illumination is adjusted so that there is little or no variation within the cell, the algorithm has been shown to produce consistent results.[7]

D. Local vs. Global Thresholding

An image with multiple gray levels may be transformed into a binary image by classifying each pixel as either foreground or background. Each pixel is classified by comparing its intensity to a specified cutoff or threshold intensity. Normally, a pixel with intensity less than or equal to the threshold is classified as foreground; a pixel with intensity greater than the threshold is classified as background. The choice of scheme to achieve this classification is not obvious, particularly in the case of a histogram that is almost unimodal because there is very little difference in intensity between the foreground objects and the background. In addition, thresholding schemes can either be global or local, applying a constant threshold level to all pixels or varying the threshold for each individual pixel or region of pixels, respectively.

When there is a clear distinction between foreground and background, a global threshold procedure is appropriate. As the distinction decreases and there is overlap between the background and foreground intensity distributions, a global threshold procedure can result in a significant number of pixels being misclassified. This results in either large portions of the cell being classified as background or, as is most often the case, large portions of the background being classified as foreground. In cases where the thickness of the cell or gel varies uniformly or the illumination is nonuniform, a local threshold procedure is more appropriate. Such a procedure can adapt to variations in background intensity across regions in the image and vary the threshold to compensate for the nonhomogeneous variations in intensities.

The basis for local thresholding techniques is to partition the image into regions and derive a unique threshold for each region. The techniques vary as to how the regions are partitioned and how the thresholds are selected.[15-17] Local threshold procedures execute slower than global procedures and often require parallel processing in real-time applications.

E. Otsu's Global Thresholding Technique

Because of their simplicity and applicability, a large number of global thresholding techniques have appeared over the years.[10-12] We have chosen to adopt a global thresholding technique developed by Otsu.[18,19]

Once image pixels have been classified as background or foreground, the intensity of each group can be estimated by the average intensity of the pixels within that group. The suitability of the chosen threshold level can be evaluated by measuring how well the classification of pixels into foreground and background groups accounts for the variation in pixel intensity. The total variation of pixel intensity is comprised of the variation of intensity within each group plus the variation between groups. The Otsu method selects the best threshold level as that which maximizes the ratio of the variation between groups to the variation within groups.

More formally, assume that image pixels have been divided into two groups. Let n_i denote the number of pixels in group i and let P_{ij} denote the intensity of the jth pixel in group i, $i = 1,2$. A binary description of the image is provided by the one-way classification analysis of variance model:

$$P_{ij} = \mu + \alpha_i + \varepsilon_{ij} \tag{3}$$

where μ is the mean pixel intensity of the entire image; α_i measures the difference between the mean pixel intensity of group i and the mean of the entire image; ε_{ij} is the error which is distributed as a normal variate with mean zero. The mean intensities of groups 1 and 2 and of the entire image are estimated by the sample means:

$$\overline{P}_i = \Sigma_j \frac{P_{ij}}{n_i}$$

$$\overline{P} = \Sigma_i \Sigma_j \frac{P_{ij}}{n} \tag{4}$$

where $n = n_1 + n_2$.

The total variation of pixel intensity, the variation between groups and the variation within groups are measured, respectively, by the total sum of squares, SS_T, the sum of squares within groups or sum of squares model, SS_M, and the sum of squares between groups or sum of squares error, SS_E:

$$SS_T = \Sigma_i \Sigma_j (P_{ij} - \bar{P})^2$$

$$SS_M = \Sigma_i n_i (\bar{P}_i - \bar{P})^2$$

$$SS_E = \Sigma_i \Sigma_j (P_{ij} - \bar{P}_i)^2 \quad (5)$$

It can be shown that the total sum of squares is the sum of the sum of squares model and the sum of squares error:

$$SS_T = SS_M + SS_E \quad (6)$$

A model fits the data well if the sum of squares model is high and the sum of squares error is low, i.e., if most of the variability is described by the model.[20]

Clearly, the total sum of squares does not depend on the classification of pixels into group 1 or 2 and is constant. However, the sum of squares model and the sum of squares error vary depending on the particular classification of pixels into groups. For a classification scheme based on thresholding, the sums of squares model and error will depend on the threshold level. The optimal threshold is determined as the threshold that maximizes the sum of squares model and hence minimizes the sum of squares error.

In our implementation, the intensity histogram of the image is first determined. The sum of squares model is calculated from the histogram for each possible threshold. This algorithm is very fast and well suited to a large variety of cell images.

F. Component Labeling

Thresholding the image window transforms it to a binary image consisting of foreground and background pixels. In general, foreground pixels will not yet exactly correspond to the cell of interest. At this stage many pixels are misclassified due to noise or the truncating effect of the thresholding procedure itself. Also, larger regions of pixels may be misclassified due to variations in the background intensity that were not eliminated by the compensation for imperfections in the image acquisition system. In addition, the image may contain more than one distinct foreground object.

The region describing the cell is isolated from the other foreground regions using component labeling.[8] This technique assigns a unique label to each closed region by sequentially scanning the image, and examining each pixel and its connected neighbors to determine if the pixel in question (1) is part of the background, (2) identifies a new component, (3) is connected to a foreground component that has already been labeled, or (4) joins two regions that had been previously identified as separate foreground components. In the latter case, the two separate components are merged. This process is illustrated in Figure 6. A replica of the original image is constructed, with the background and each distinct foreground region described by different labels as shown in Figure 7.

Small isolated regions of background pixels that have been misclassified as foreground are easily identified following component labeling as those components with total size less than some specified amount.Foreground pixels that have been misclassified as background will result in holes in objects, within which may be nested other small foreground components. The largest foreground component is assumed to represent the cell. This assumption is considered valid since the initial density of cells placed in the gel is low enough to ensure that cells are not in close proximity to each other and that only one cell should be contained

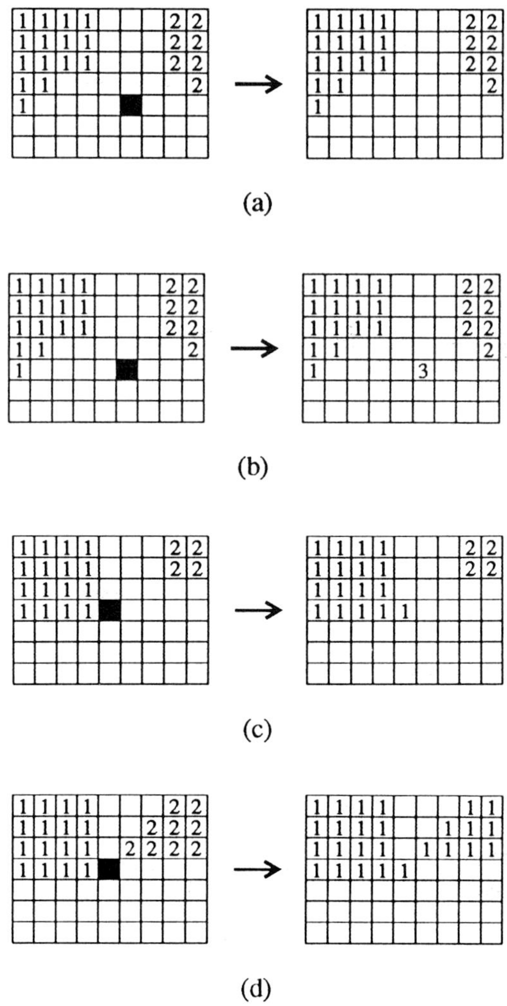

FIGURE 6. Components labeling. (a) Identification of a background pixel, (b) identification of a new component, (c) foreground pixel connected to an existing component, (d) foreground pixel merging two existing components.

within the working window. The exception to this is cell mitosis in which case special measures need to be taken.[21-23] In addition, the tracker examines various shape features of the extracted cell, such as length, width, area, perimeter, and number of pixels, and compares them to the previous samples to ensure that the correct component was selected. If there is a significant discrepancy, the system can backtrack and select the next largest component and repeat the process. If all else fails, the operator is notified and can manually intervene.

A single component is isolated by relabeling all other foreground components as background. A new binary image is produced, consisting of a single connected foreground region and a background, which may include several disconnected regions, appearing as holes in the foreground component.

G. Boundary Smoothing

The foreground component which has been isolated by the component labeling procedure may provide a somewhat inaccurate representation of the cell. Also, in addition to holes, the foreground component may be marked by an extremely rough boundary with sharp inlets.

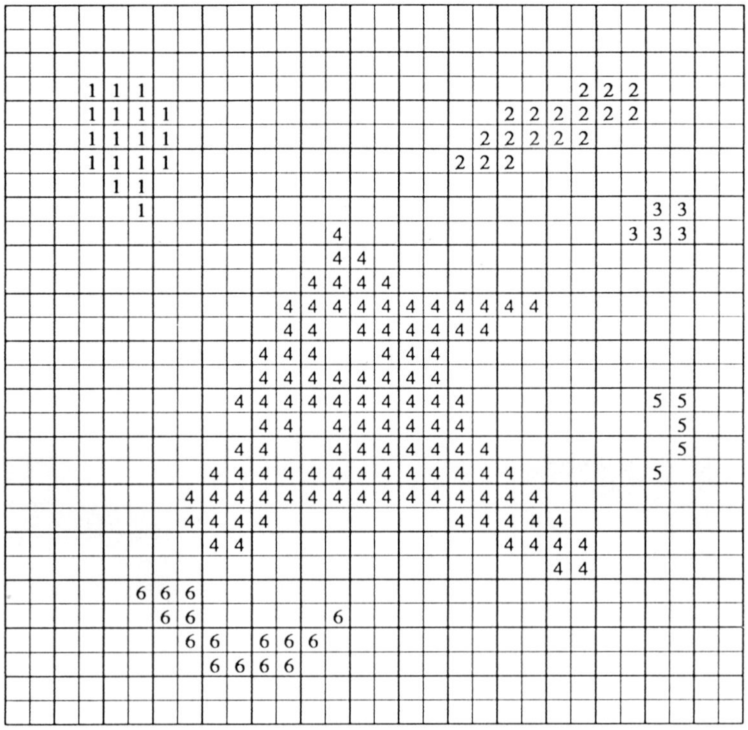

FIGURE 7. Labeled foreground components.

These would be uncharacteristic of the boundary of the type of cells under investigation and would affect the accuracy of any measurements used to determine the center of the segmented object. These problems are corrected with a dilation and erosion procedure.[24]

Dilating the image expands the foreground component. A pixel that is near a foreground pixel is classified as foreground. Erosion is complementary to dilation, reducing the foreground in the same way; any pixel that is near a background pixel is classified as background. The effect of dilation is to smooth the boundary, close narrow inlets, and fill small holes. Following dilation, erosion returns the component to its original size. The process of dilation followed by erosion is called closing. Dilation is defined as the morphological transformation which combines two sets using vector addition of set elements. Let *A* denote the set of all pixels that belong in the foreground and let *B* represent a second set, which is referred to as the structuring element. Dilation of the image by the structuring element B is defined as $A \oplus B = \{c | \ c = a+b \text{ for some } \alpha \ \varepsilon \ A \text{ and } b \ \varepsilon \ B\}$. Erosion of the image by structuring element *B* is defined as $A \ominus B = \{c | \ c + b \ \varepsilon \ A \text{ for every } b \ \varepsilon \ B\}$. Finally, the closing of image *A* by structuring element *B* is defined as $A \odot B = (A \oplus B) \ominus B$.

The amount of smoothing achieved by closing the image depends on the structuring element. We do not wish to smooth excessively and lose features, but would like to eliminate roughness in the boundary and close as many inlets as possible. The structuring element that defines a disk of radius 2,

$$B = \{(0,2), (-1,1), (0,1), (1,1), (-2,0), (-1,0), (0,0), (1,0), (2,0), (-1,-1), (0,-1), (1,-1), (0,-2)\} \quad (7)$$

was found to produce good results for our application (Figure 8h). The process of closing is illustrated in Figure 8. Note that small inlets closed by dilation are not reopened by erosion, the boundary is smoothed, and a number of holes remain in the object.

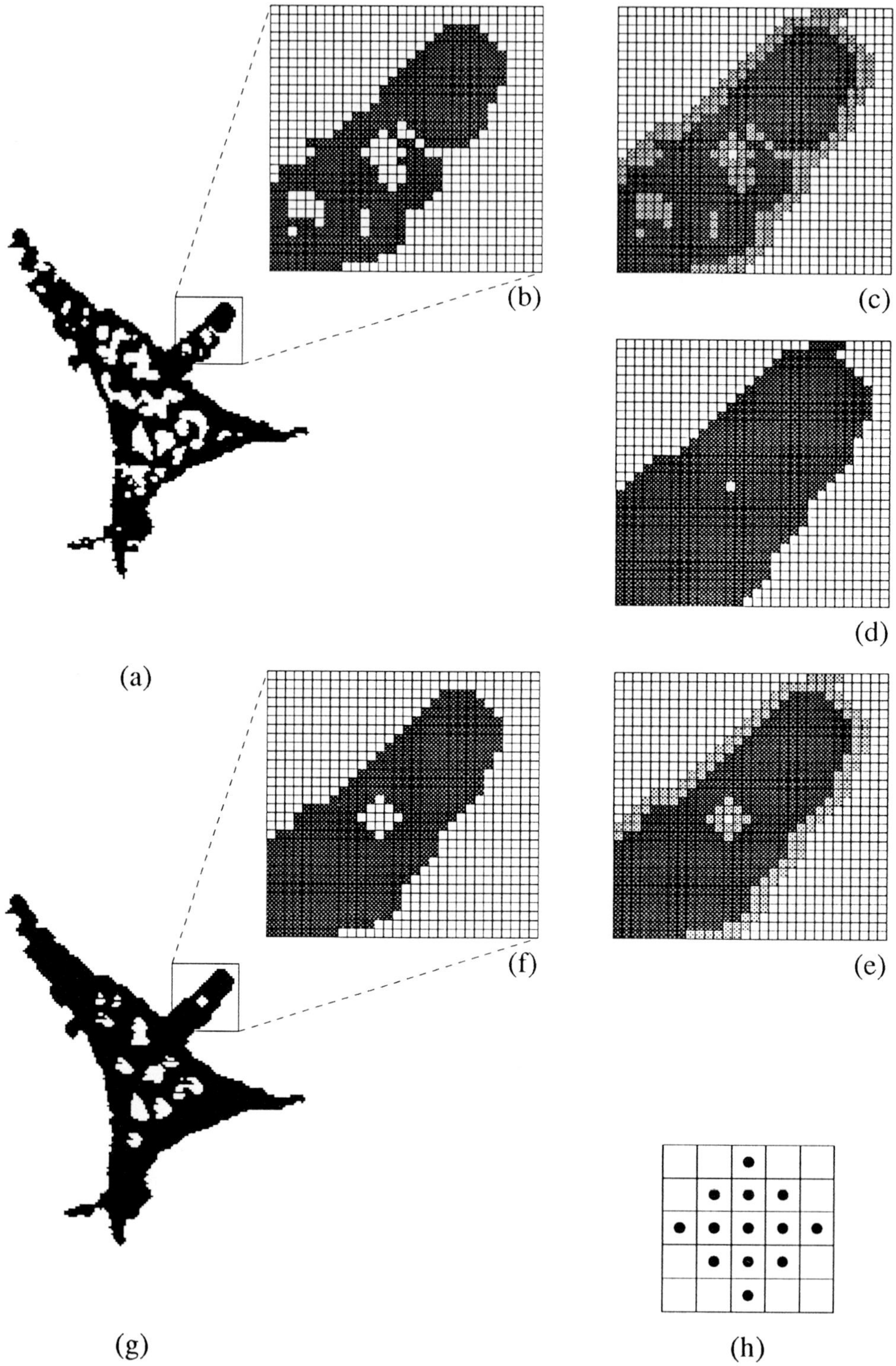

FIGURE 8. Magnified image closing process. (a) Binary image prior to closing, (b) magnification of a portion of (a), (c) identification of pixels to be dilated, (d) dilation of (b), (e) identification of pixels to be eroded, (f) erosion of (d), (g) binary image after closing, (h) structuring element.

We now have a binary image that consists of a white background and a black object with white holes in it. To eliminate all holes in the object, the component labeling routine is applied to the background. The true background is labeled as the largest component and the holes are each labeled as smaller distinct components. This distinguishes the holes in the object from the true background region. The isolated background components are then assimilated into the foreground, filling the holes.

H. Center of Mass and Boundary Extraction

Once we have obtained a smoothed representation of the cell of interest, the center and boundary can easily be determined. The center is simply computed as the center of mass of the foreground component:

$$c_x = \Sigma_j x_{f,j}/n_f$$

$$c_y = \Sigma_j y_{f,j}/n_f \qquad (8)$$

where $x_{f,j}$ and $y_{f,j}$ are the coordinates of the *jth* pixel of the foreground component and n_f is the size of the foreground component. The boundary is taken as the set of pixels contained in the foreground that touch the background.

I. Summary

The cell segmentation process is summarized in Figure 9. For 2-D cell tracking, this process is applied to a single image of the cell. For 3-D tracking, as in the case of converging squares, that data are defined by the solid formed by stacking the 2-D windows from each serial section. Each stage of the cell segmentation algorithm presented above has been extended to operate on the 3-D data set.

VI. OTHER SEGMENTATION APPROACHES

Another method that we are currently investigating is based on active contour models called snakes.[25] The snakes are used to solve both the image segmentation and object tracking problems.[26,27] Snakes are energy-minimizing splines guided by external constraints and image structures such as lines, edges, subjective contours, and region homogeneities, as well as internal spline forces that impose smoothness constraints on the modeled contours.

The cells are tracked by using the dynamic behavior of the snake. A snake *sticks* to the cell boundary from one image frame to the next by conforming to small deformations that occur when the cell moves. The snake provides an estimate of the cell boundary. A skeletonization process,[28,29] based on the grassfire transform[30-33] and on curvature extrema,[34,35] is then applied to the boundary to provide a description of the cell's shape. This approach is more robust than the previous segmentation algorithm but achieves this at the expense of execution time.

Another segmentation technique that we have studied differs from the previous two in that rather than segmenting the object from the background, it segments the background and stationary objects from the moving objects. Thus, it is not based on detecting an object or boundary.

The background generation algorithm integrates spatial and temporal information within a series of 2-D images acquired at regular time intervals.[36] The approach is an extension of a technique that was proposed by Yang and Long in work related to human body motion analysis.[37] Rather than eliminating background information, as is commonly done in motion analysis paradigms involving a stationary camera, a background generation process reconstructs the scene's background, effectively removing objects that are moving from the

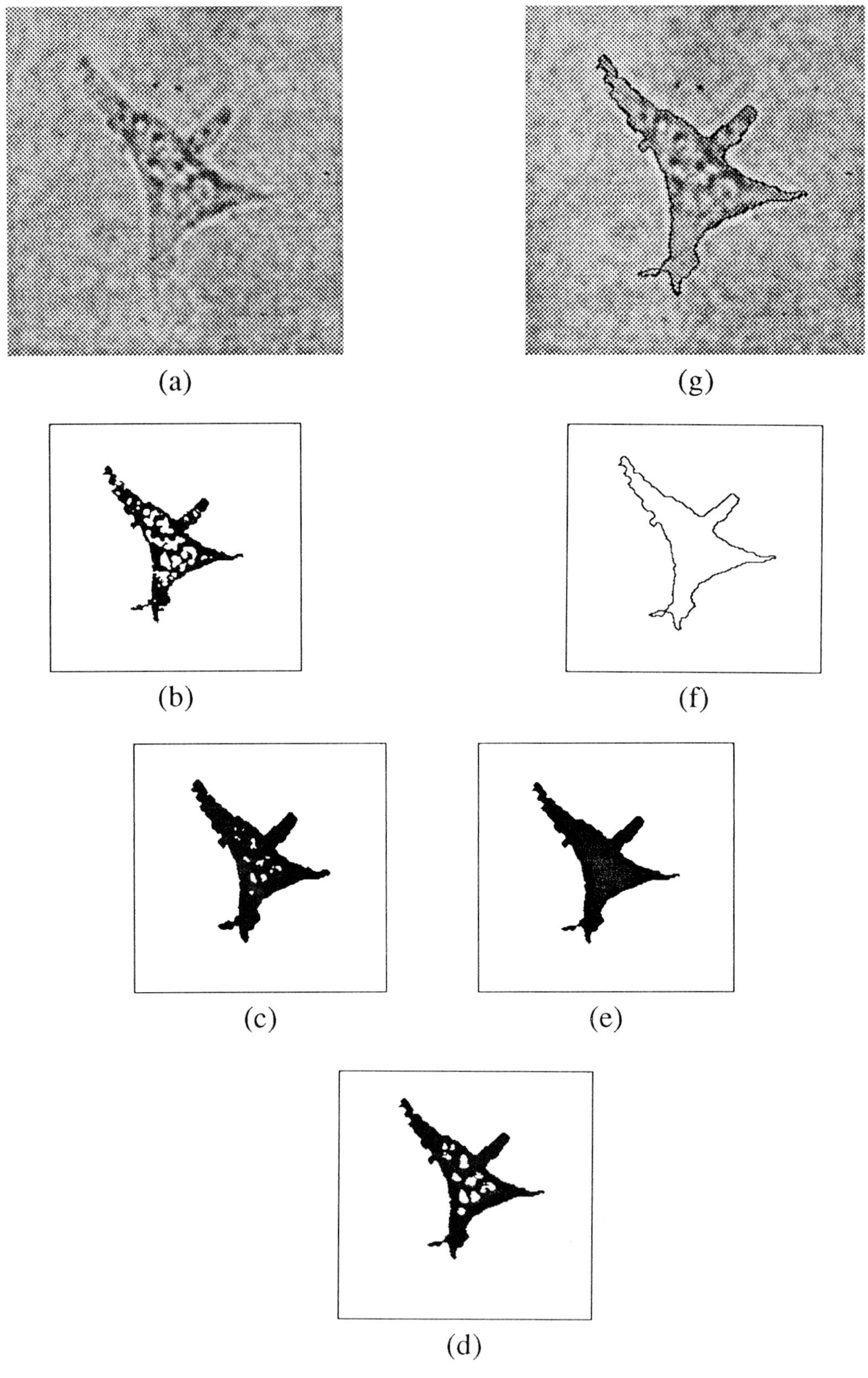

FIGURE 9. Summary of cell segmentation process. (a) Raw image, (b) largest component isolated, (c) dilation, (d) erosion, (e) internal holes filled, (f) extracted edge, (g) extracted edge superimposed on raw image.

sequence of images. Using the reconstructed background, pixels that are significantly different from the background values are assumed to belong to the moving edges.

The algorithm is computationally efficient and can concurrently track every moving object within a single field of view. This means that the operator does not have to select which cells to track. Limits on the size of the object being tracked can be imposed to ensure that only cells are followed and not smaller artifacts. This technique is especially well suited to cases where the cell boundaries are occluded since objects in close proximity to each other will move together and can be merged on that basis. The proposed approach has been tested on several sequences of moving cells with encouraging results.

VII. CONCLUSIONS

In this chapter, we have described the principles of image segmentation that are used by TRACES, an automated 2- and 3-D cell tracking system. Two other imaging techniques that are under development were also briefly mentioned: a boundary extraction method based on active contour models called snakes, and a technique that segments moving cells by extracting the stationary background of a sequence of images.

ACKNOWLEDGMENTS

This work was partially supported by the Medical Research Council of Canada, under Grant No. MT-3236, and the Fonds Pour la Formation de Chercheurs et L'Aide à la Recherge (FCAR), Québec, under Grant No. 91-ER-0528.

REFERENCES

1. **Noble, P. B.,** The extracellular matrix and cell migration: locomotory characteristics of Mos-11 cells within a three-dimensional gel, *J. Cell Sci.,* 87, 241, 1987.
2. **Shields, E. D. and Noble, P. B.,** A methodology for the detection of heterogeneity of locomotory phenotypes in three-dimensional gels, *Exp. Cell Biol.,* 55, 250, 1987.
3. **Noble, P. B. and Boyarsky, A.,** Analysis of cell three-dimensional locomotory vectors, *Exp. Cell Biol.,* 56, 289, 1988.
4. **Noble, P. B. and Shields, E. D.,** Time-based changes in fibroblast three-dimensional locomotory characteristics and phenotypes, *Exp. Cell Biol.,* 57, 238, 1989.
5. **Guidry, C. and Grinnell, F.,** Contraction of hydrated collagen gels by fibroblasts: evidence for two mechanisms by which collagen fibrils are stabilized, *Collagen Relat. Res.,* 6, 515, 1986.
6. **Inoué, S.,** *Video Microscopy,* Plenum Press, New York, 1986.
7. **Noble, P. B. and Levine, M. D.,** *Computer Assisted Analyses of Cell Locomotion and Chemotaxis,* CRC Press, Boca Raton, Fl., 1986.
8. **Rosenfeld, A. and Kak, A.,** *Digital Picture Processing,* Academic Press, New York, 1976, 347.
9. **Levine, M. D.,** *Vision in Man and Machine,* McGraw-Hill, New York, 1985.
10. **Sashoo, P. K., Soltani, Y. S., Wong, A. K. C., and Chen, Y. C.,** A survey of thresholding techniques, *Comput. Vision Graphics Image Process,* 41, 233, 1988.
11. **Haralick, R. M. and Shapiro, L. G.,** Survey on image segmentation techniques, *Comput. Vision Graphics Image Process.,* 29, 100, 1985.
12. **Weszka, J. S.,** A survey of threshold selection techniques, *Comput. Vision Graphics Image Process.,* 7, 259, 1978.
13. **O'Gorman, L. and Sanderson, A. C.,** The converging squares algorithm: an efficient method for locating peaks in multidimensions, *IEEE Trans. Pattern Anal. Mach. Intelligence,* 6, 280, 1984.
14. **O'Gorman, L. and Sanderson, A. C.,** Some extensions of the converging squares algorithm for image feature analysis, *IEEE Trans. Pattern Anal. Mach. Intelligence,* 8, 520, 1986.
15. **Bernsen, J.,** Dynamic thresholding of grey-level images, in *8th Int. Conf. on Pattern Recognition,* Paris, October 1986, 1251.

16. **Rodriguez, A. A.,** *Image Segmentation by Successive Background Extraction,* Research Rep. RC 15415 (#68603), T. J. Watson Research Center, IBM Corporation, New York, January 1990.
17. **Yanowitz, S. D. and Bruckstein, A. M.,** A new method for image segmentation, in *9th Int. Conf. on Pattern Recognition,* Rome, 1988, 270.
18. **Otsu, N.,** A threshold selection method from gray-level histograms, *IEEE Trans. Syst. Man Cybern.,* 9, 62, 1979.
19. **Otsu, N.,** Discriminant and least squares threshold selection, in *Proc. 4th Int. Jt. Conf. on Pattern Recognition,* Kyoto, 1978, 592.
20. **Draper, N. R. and Smith, H.,** *Applied Regression Analysis,* John Wiley & Sons, New York, 1966.
21. **Ferrie, F. and Levine, M. D.,** Cell Tracking by Inference, Rep. TR-81-11R, Department of Electrical Engineering, McGill University, Montreal, 1981.
22. **Ferrie, F., Levine, M. D., and Zucker, S. W.,** Cell Tracking: A Modeling and Minimization Approach, Rep. TR-80-4R, Department of Electrical Engineering, McGill University, Montreal, 1980.
23. **Ferrie, F. P.,** Experiments in Tracking the Morphologies of Proliferating Cell Cultures by Automatic Picture Processing, Master's thesis, Department of Electrical Engineering, McGill University, Montreal, 1979.
24. **Serra, J.,** *Image Analysis and Mathematical Morphology,* Academic Press, London, 1982.
25. **Kass, M., Witkin, A., and Terzopoulos, D.,** Snakes: active contour models, in *IEEE Proc. 1st Int. Conf. on Computer Vision,* June 1987, 259.
26. **Leymarie, F.,** Tracking and Describing Deformable Objects Using Active Contour Models, Master's thesis, Rep. CIM-90-9, McGill Research Centre for Intelligent Machines, McGill University, Montreal, 1990.
27. **Leymarie, F. and Levine, M. D.,** Tracking deformable objects in the plane using an active contour model, *IEEE Trans. Pattern Anal. Mach. Intelligence,* submitted.
28. **Dill, A. R., Levine, M. D. and Noble, P. B.,** Multiple solution skeletons, *IEEE Trans. Pattern Anal. Mach. Intelligence,* 9, 495, 1987.
29. **Dill, A. R. and Levine, M. D.,** Non-Rigid Body Motion, in *Computer-Generated Images, The State of the Art,* Magnenat-Thalmann, N. and Thalmann, D., Eds., Springer-Verlag, Tokyo, 1985, 79.
30. **Leymarie, F. and Levine, M. D.,** New method for shape description based on an active contour model, in *SPIE Conf. on Adv. Intelligent Robot Syst., Visual Commun. Image Process. IV,* Philadelphia, 1989, 1199.
31. **Levine, M. D. and Leymarie, F.,** Cells, skeletons and snakes, in 5th Israeli Symp. on Artif. Intelligence, Tel Aviv, 1988.
32. **Leymarie, F. and Levine, M. D.,** Skeletons from snakes, in *5th Int. Conf. on Image Anal. and Process.,* Positano, Italy, 1989.
33. **Leymarie, F. and Levine, M. D.,** Simulating the grassfire transform using an active contour model, *IEEE Trans. Pattern Anal. Mach. Intelligence,* in press.
34. **Leymarie, F. and Levine, M. D.,** Curvature morphology, in *Proc. Vision Interface '89,* London, Ontario, 1989, 102.
35. **Leymarie, F. and Levine, M. D.,** Shape features using curvature morphology, in *SPIE Conf. on Adv. Intelligent Robot Syst., Visual Commun. Image Process. IV,* Philadelphia, 1989, 1192.
36. **Yang, Y. H. and Levine, M. D.,** The Background Primal Sketch: An Approach for Tracking Moving Objects, Tech. Rep. 90-3, Department of Computational Science, University of Saskatchewan, 1990; also available as Tech. Rep. CIM-90-15, McGill Research Centre for Intelligent Machines, McGill University, Montreal, 1990.
37. **Long, W. and Yang, Y. H.,** Stationary Background Generation: An Alternative to the Difference of Two Pictures, Tech. Rep. 88-9, Department of Computational Science, University of Saskatchewan, 1988.
38. **Shotton, D. M.,** Review: video-enhanced light microscopy and its applications in cell biology, *J. Cell. Sci.,* 89, 129, 1988.

Chapter 3

QUANTITATIVE AREA DETERMINATION BY COMPUTER IMAGE ANALYSIS

Alan C. Bovik, Shanti J. Aggarwal, Nak H. Kim, and Kenneth R. Diller

TABLE OF CONTENTS

I. INTRODUCTION

Methods for obtaining numerical area data from two-dimensional (2-D) images of biological objects have been studied for over three decades. In the early 1960s and 1970s, various manual methods of area data collection were prevalent, of which the following were the most reliable. The areas of objects, such as cell images, in a positive print were measured by (1) projecting the outlines of the object using camera lucida or some similar microprojection apparatus onto millimeter ruled graph paper, and then counting the number of squares enclosed within each profile; (2) instead of counting the squares, measuring the areas of traced profiles with the aid of a planimeter; and (3) tracing the projected outlines of the object onto a uniform thickness paper and subsequently cutting out each profile and weighing it on a microbalance. By comparing the observed weights with samples of a precisely known area, it was possible to obtain a reasonable estimate of the cross-sectional area. For relatively large specimens such as ova, embryos, and even granulocytes, these techniques were straightforward and reasonably accurate, but they were of questionable use for area measurements of small organelles that must be magnified to the limits of optical resolution for viewing. Moreover, apart from the human errors inherent in the manual tracing and interpretation of cellular outline, these approaches were very slow, tedious, and impractical for the measurement of large numbers of cells.

Fortunately, rapid technological developments in video microscopy, video-camera technology, dedicated image-processing hardware and software, and the availability of powerful, inexpensive microcomputers or workstations have gone a long way towards ameliorating these difficulties. A variety of automated or semiautomated methods are now available that combine relatively simple camera, computer, and software techniques for the measurement of object areas. There are a fair number of systems that are highly effective for measuring objects that have images of high intrinsic contrast or that have well-defined and easily identified shapes, e.g., that are nearly circular in profile. Often these systems are designed for a specific area measurement application, where the incorporation of domain-specific information about the object and background can be used to simplify software algorithms and possibly lead to increased speed. Recently, however, more general techniques have been developed that effectively measure the areas of objects having more complex or tortuous boundaries or images of objects having low contrast, partial transparency, or diffuse detail.

The purpose of this chapter is to review the fundamental elements of microcomputer-based image area analysis techniques and their applications to various biological fields. Only representative examples will be given due to the very large variety of objects that are imaged in the various disciplines. Examples of rather generic image-processing software operations and general considerations involved in area computation in biomedical image analysis will be discussed and illustrated by a highly successful generic algorithm that has been used extensively for the quantitative analysis of wound healing.

II. BIOMEDICAL IMAGE ANALYSIS

In this section we review very briefly the types of information that can be used to identify image regions, the areas of which are to be measured, and more generally, give an overview of the kinds of preprocessing techniques that are currently available for determining the boundaries of objects or object regions. Video technology, and in particular video microscopy, coupled with computer image analysis presents an exciting potential and promise for the fast and accurate analysis of biological structures. The combination of the two technologies allows one to extract the useful information hidden in low contrast, dark, dynamically changing biological scenes by various well-known techniques such as noise reduction, contrast enhancement, feature extraction, and other manipulations that use the principles of classical image processing, image analysis, and mathematical morphology. Some of the most useful features that can automatically be measured from images of a biological scene are those that describe the size and shape of the individual object within the image, e.g., object area, object perimeter, degree of circularity, aspect ratio, and rectangularity. Local measurements such as boundary curvature can also be very useful. Of course, caution must be applied in the design and implementation of hardware and algorithms for image analysis, since it is possible to generate artifacts not present in the original scene by improper arrangements in the lighting conditions, imaging setup, or software.

Although image analyzers have gained considerable acceptance as analytical tools, their usefulness for quantitative image analysis in biological applications remains limited. Fully automated systems that allow the unsupervised measurement of a wide variety of numerical object parameters, for generic objects of variable shapes and reflectance/transmission properties, are not yet available. The main problem in realizing a fast, efficient, fully automated system resides first in the quality of the images obtained, which are often of low contrast (particularly micrographs) and can present highly-complex structures, and second, in the ability of the software system to decompose image data into information-rich primitives or on the basis of specific physical object properties. Many powerful image analysis and object recognition techniques, such as true color processing[1] and image texture analysis,[2,3] have remained relatively unexploited in the analysis of biomedical scenes. Color image analysis has obvious utility for the segregation or recognition of objects based on their chromatic emissive, reflectance, or transmission properties. Biomedical objects often have distinctive natural color or, in the case of stained slides, a truly differentiating color which could aid in better feature or object extraction. Texture analysis can provide important data about the local distribution of gray levels relative to those in surrounding areas, as illustrated by the following example. If one is interested in the therapeutic value of a drug as a wrinkle-preventive cream, then one viable approach for evaluating the results could proceed by determining the skin area that is occupied by discernible wrinkles before and after application of the cream. Traditional image-analyzing techniques often found in biomedical image-analysis systems will provide the ability to measure the relative local lightness and darkness of the sample, but likely will not be able to differentiate between the object classes (wrinkled vs. smooth) since the overall irradiance from the skin types may be the same. Automated texture analysis, on the other hand, can indicate changes in the texture pattern relative to the surrounding areas, and therefore, how the sample is responding to the cream.

The first stage in most biomedical image processing systems is an object isolation process in which a subject of interest, e.g., a cell, tumor, or bone, is separated from a background of varying optical complexity. Image-segmentation techniques devised for this purpose can be broadly divided into three categories:

1. Edge- or contour-based approaches, in which sustained intensity changes, or edges, are marked (usually with a discrete derivative operator)
2. Direct region segmentation approaches, in which the image is partitioned according to some homogeneity criterion
3. Gray-level thresholding, where bright (or dark) objects are separated from the background by the use of an appropriate threshold value[4,5]

Contour-based approaches are used mainly in applications where the object has a relatively smooth (low curvature) boundary[6-8] or when the object has a simple, invariant shape.[9] Region growing techniques, e.g., split-and-merge, can be useful when the local gray-level statistics of the object are dissimilar to those of the background.[10] Simple thresholding techniques are most adequate for cases where the object or objects of interest have a gray-level distribution that is markedly offset from that of the background. While these techniques each have been used successfully for many specific applications, they may be difficult to incorporate into a generic imaging system able to deal with a broad range of input images, since most of these methods have a limited range of applicability, i.e., they can be applied only to images with certain characteristics or types of intrinsic information that the algorithm assumes, and they also tend to not always be reliable, as they may misclassify a small portion of the region or boundary if the algorithm is not carefully defined.

In the following section, we focus primarily on the problem of measuring the area after some specific segmentation or boundary-finding technique has been applied. As a vehicle for demonstrating fundamental aspects of the shape correction and area calculation problem, a very simple, but often-effective, object extraction process is assumed, based on simple thresholding. This is done since the emphasis of this chapter is on automated area measurement rather than on preparing the images (or the specimens) obtained for the area measurement process. However, the principles described are highly generic and feature, in particular, methods for ameliorating the types of difficulties or errors that commonly occur with any kind of region segregation technique.

III. APPLICATION EXAMPLE: ANALYSIS OF WOUND HEALING

A demonstratively successful, generic, automated area analysis system which isolates an unknown object from its background and measures the area of the object will be used to illustrate the main themes given in this chapter.[11] The algorithm has been applied quite successfully to determine the progression of healing in standard burn wounds on experimental animals, i.e., the goal of the system is to measure the unhealed area of the inflicted skin wounds. The system described affords significant improvement over earlier attempts,[12] which limited the analysis to the use of simple gray-level thresholding techniques.

The skin wound presents a difficult image to analyze because its surface tends to have large spatial variations in both gray-level and light reflectance, and the wound area is often not well distinguished from the surrounding tissue. Also, the boundary tends to be quite irregular, and there may be considerable variation in wound shape within a population of experimental subjects or over time for a single subject. Therefore, standard object segmentation strategies often provide results that contain errors in object/background classification or in the accurate identification of object boundary location. Errors of these types can render area measurement calculations highly inaccurate. Thus, approaches for correcting the segmentations obtained are of interest; in later sections a new approach is described that has been applied with considerable success not only to the wound-analysis problem, but also, in turn, in the analysis of other challenging biomedical subjects, such as determining the projected cross-section of living cells and multicellular tissues such as pancreas islets imaged in light micrographs.

First, we will briefly discuss the characteristics of the wound image and discuss a simple segmentation technique that illustrates well the problems encountered by an object area analysis system. The most conspicuous features of the wound image are the diffuse boundary and the nonhomogeneous interior. Although the wound region appears, on the whole, brighter than the background, its boundary is not very well-defined in most cases, and its interior exhibits a considerable variation in intensity. In addition, the variations in brightness and in size among wounds at different times during the progressive healing process over the course of an experimental trial are large. Typically, the size of the wound area can vary by an order of magnitude. During the development stage of this system, it was observed that these image input characteristics were not adequate for any totally automated object segmentation system. These observations led to the following design strategy: design a highly-reliable imaging system with an inexpensive hardware configuration that requires a minimum amount of human intervention and runs in a reasonable amount of computation time.

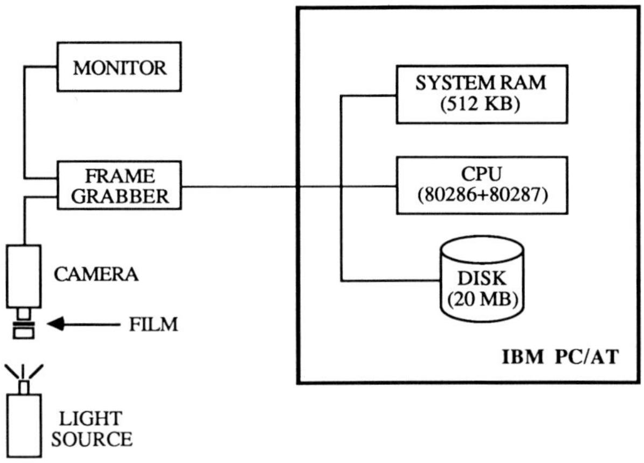

FIGURE 1. Physical configuration of the example system.

The system described is composed of two main stages. In the first, the image is segmented into object region and background using simple gray-level thresholding, clearly illustrating the types of classification errors that can occur. In the second, misclassified regions and errors in the object boundary estimates are corrected by a technique known as "blob coloring" and by methods of an important, generic branch of image-transformations known collectively as mathematical morphology. It is shown that, while the determination of an appropriate threshold is difficult, and an ideal threshold is impossible because of the input image characteristics, the gray-level thresholding technique combined with the object and shape-correcting devices produce measurements of area that are highly in accordance with accurate manual techniques.

IV. PHYSICAL STRUCTURE OF THE EXAMPLE SYSTEM

The area measurement system described here has been implemented on an IBM PC/AT microcomputer system equipped with an Imaging Technology video digitizer (frame-grab board) that converts a single video frame into a 512 × 512 × 8 digital image in 1/30 s. The frame-grab board also reconfigures the digitized signal into analog form for display on a video monitor. The microcomputer has 512 kbytes semiconductor memory on board and a hard disk unit of 20 Mbytes capacity. In the examples given later, the input wound images were taken directly from film negatives mounted in a pin-registered holder on a bellows apparatus and transilluminated from behind so that they can be converted into an electrical signal by the video camera. The original images were recorded on film because of the absence of a video camera and recording system at the remote site where the animal experiments were conducted. This arrangement simplifed the measurement of the true wound area, since it avoided the problem of determining the physical parameters of a combined video camera/optical adaptor/macroscopic system. The measurement of these parameters is discussed in Section IX. Figure 1 shows the physical configuration of the system.

V. SEPARATION OF OBJECT FROM BACKGROUND

Once the image has been digitized, it is processed through software in several stages. In the first stage, the image is segmented into object and background using a simple gray-level thresholding technique.[4] This stage is schematically shown in Figure 2 and is implemented as follows. First, the digitized image is reduced to 256 × 256 spatial resolution by averaging together every four adjacent original pixels, forming a single pixel in the condensed

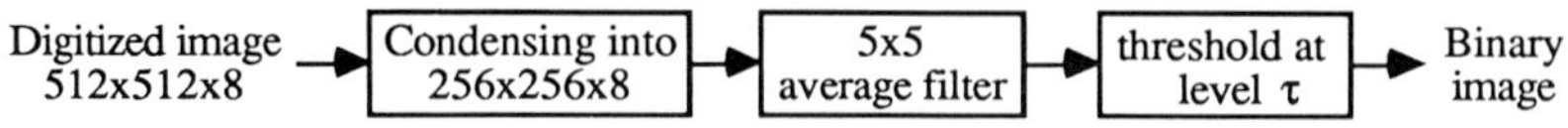

FIGURE 2. Functional diagram of the condensing, smoothing, and thresholding stages.

image. Thus, if the stored image is given by the array $\mathbf{I}^{orig} = [I^{orig}(i,j); 0 \leq i,j \leq N - 1]$, where in this instance $N = 512$, then the condensed image $\mathbf{I}^{cond} = [I^{cond}(i,j); 0 \leq i,j \leq (N/2) - 1]$ has elements given by

$$I^{cond}(i, j) = {}^1/_4[I^{orig}(2i, 2j) + I^{orig}(2i + 1, 2j) + I^{orig}(2i, 2j + 1) + I^{orig}(2i + 1, 2j + 1)]$$

for $0 \leq i,j \leq (N/2) - 1$. This stage is introduced mainly for increasing computational efficiency (speed) in later processing, although it is certainly not a necessary step if adequate computational resources are available. However, since the principal concern of the present system is to calculate the area of the segmented image, there is very little loss of information resulting from the condensing operation. For example, since the average area of the object in the application considered here is typically 20,000 pixels (in a 256 × 256 image), the loss of resolution due to the condensing is less than $4/(2 \times 10^4) = 0.02\%$. The second stage of the algorithm involves the smoothing of the image using a 5 × 5 (square) moving average filter. In general, the averaging operation results in a distribution of the image gray-levels that is more distinctive.[4] In simple terms, the intensity value of the center pixel of the window is replaced by the average intensity of its neighbors within the window. In mathematical terms that can be easily converted into computer code, the smoothed image $\mathbf{I}^{ave} = [I^{ave}(i,j); 0 \leq i,j \leq (N/2) - 1]$ has elements given by

$$I^{ave}(i,j) = \frac{1}{25} \sum_{n=-2}^{n=+2} \sum_{m=-2}^{m=+2} I^{cond}(i + n, j + m)$$

for $0 \leq i,j \leq (N/2) - 1$. It may be observed that the averaging operation at the boundaries of the image, i.e., wherever $\min(i,j) < 2$ or $\max(i,j) > (N/2) - 2$, is not well defined, since the indexes over which the average is computed will not all fall within the prescribed range $0 \leq i,j \leq (N/2) - 1$. This may be easily ameliorated by replicating pixels, i.e., assigning gray levels from the closest pixels to the undefined pixel elements.

A simple gray-level thresholding operation is then applied to the averaged image, in order to obtain a first attempt at separating the wound region from the background. This is a simple example of the general technique of segmentation that is appropriate for the example application considered here, since the image objects of interest generally have an average intensity that is significantly different from the average background intensity. Of course, in other applications, the segmentation may proceed on the basis of color separation, textural differences, or some other physically meaningful criteria. We assume, however, that the result of the segmentation is a binary image or map indicating the likely locations of object pixels vs. background pixels. This binary map will, of course, generally contain errors resulting from local perturbations of the image from the model assumed. For a determined threshold level τ that falls within the range of allowable gray-level values (usually coded as integers in the range 0 to 255 corresponding to 8 bits of resolution), the thresholded binary image $\mathbf{J} = [J(i,j); 0 \leq i,j \leq (N/2) - 1]$ is defined

$$J(i,j) = \begin{cases} \text{'1'}, & \text{if} \quad I^{ave}(i,j) \geq \tau \\ \text{'0'}, & \text{if} \quad I^{ave}(i,j) < \tau \end{cases}$$

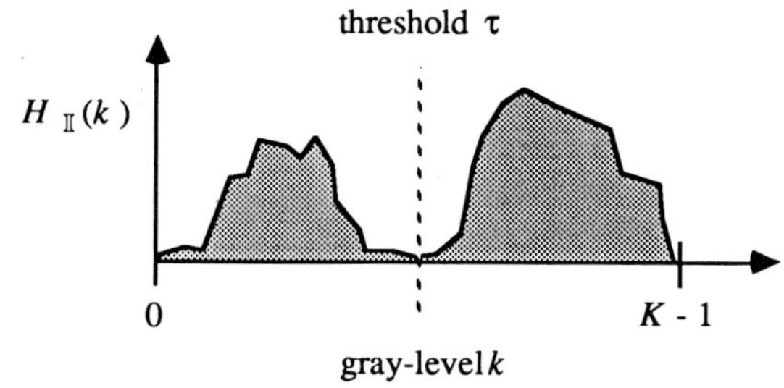

FIGURE 3. Appearance of typical bimodal histogram and one possible threshold location.

where '1' denotes a binary value indicating that the brightness exceeds the threshold, hence is likely an object pixel, and the binary digit '0' denotes low-intensity background pixels. For display purposes, it will be assumed that the binary value '1' corresponds to BLACK, the binary value '0' to WHITE. Of course, if the application is such that the objects appear as dark regions on a light background, these assignments may be reversed. Many of these assignments may be erroneous; hence the need for the region and shape correction procedures outlined next. Since the image to be operated on is now binary- or logical-valued, the correction procedures may be conveniently defined as highly efficient logical operations.

In order to determine an appropriate threshold τ, the gray-level distribution or histogram of the image must be measured. The histogram $H_{\mathbf{I}}(k)$; $0 \leq k \leq K - 1$ ($K = 256$ in most cases) of an arbitrary gray-level image **I** is a one-dimensional plot or graph of the frequency of occurrence of each image gray-level k in **I**:

$$H_{\mathbf{I}}(k) = n \quad \text{if} \quad I \text{ contains exactly } n \text{ occurrences of gray-level } k$$

for each $k = 0, \ldots K - 1$. The threshold operation is most efficacious if the histogram $H_{\mathbf{I}}$ of the image **I** contains two well-separated peaks or modes, corresponding to object and background pixel gray levels. Such a "bimodal" histogram has the appearance depicted in Figure 3.

Since the area of the object may vary over a wide range (typically 2000 to 30,000 pixels in the example application), and since there tends to be a large variation in the distribution of gray levels within a typical input image, automatic determination of the threshold is generally quite difficult; although automated threshold selection criteria have been proposed,[4] some manual intervention is generally required in order to select the best possible threshold. In the current implementation, the gray-level histogram of the image is displayed on the monitor, and a threshold is set manually. The output at this stage is the thresholded binary image. To this image, a simple region correction procedure is applied in order to determine the principal object regions of interest; a smoothing or shape correction operation based on concepts of mathematical morphology is then applied prior to measuring the area of the object of interest.

VI. REGION CORRECTION IN THE THRESHOLDED IMAGE

A binary image resulting from thresholding a gray-level image typically has an appearance similar to that shown in Figure 4; that is, there will usually be misclassifications of both object pixels and background pixels due to random variations in surface reflectance/absorption, extraneous objects, dust, or image noise. In this section, an algorithm is described that effectively corrects (eliminates) misclassified small regions. The result of the algorithm is a new binary image, where all small misclassified regions lying entirely outside the object

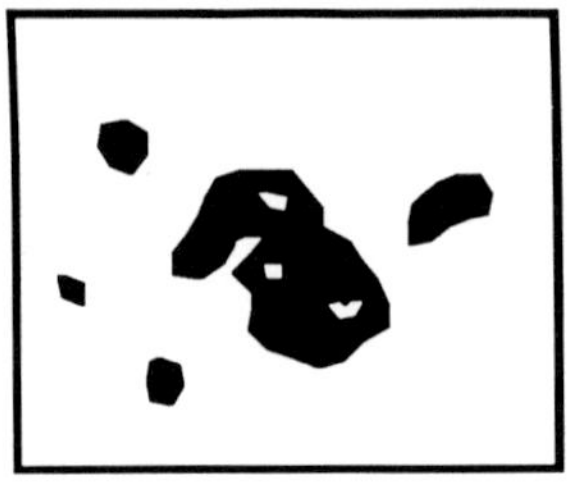

FIGURE 4. Typical appearance of thresholded binary image prior to region correction and boundary smoothing.

area are removed using a variant of the "blob coloring" algorithm;[5] subsequently, all small misclassified regions lying entirely within the object area are removed using a dual of the blob coloring algorithm. This algorithm first assigns a region number to each nonzero pixel so that all pixels belonging to a same four-connected region will have the same region number. This number-assigning procedure can be done in a single scan of the entire image, as summarized below.

A. Blob Coloring Algorithm

Given a binary (thresholded) image $\mathbf{J} = [J(i,j); 0 \leq i,j \leq (N/2) - 1]$, define a (nonbinary) array $\mathbf{R} = [R(i,j); 0 \leq i,j \leq (N/2) - 1]$ where $R(i,j)$ is the *region number* of the binary-valued pixel $J(i,j)$, according to the following pseudo-code:

```
begin
initialize R: R(i,j) = 0 for 0 ≤ i,j ≤ (N/2) − 1;
k = 1;
while (0 ≤ i ≤ (N/2) − 1 and 0 ≤ j ≤ (N/2) − 1 do
begin
  if J(i,j) = '1' then begin
    if (J(i,j − 1) = '0' and J(i − 1, j) = '0') then
    begin
        R(i,j) = k;
        k = k + 1;
    end;
    if (J(i,j − 1) = '0' and J(i − 1, j) = '1' then
        R(i,j) = R(i − 1,j);
    if (J(i,j − 1) = '1' and J(i − 1, j) = '0' then
        R(i,j) = R(i,j − 1);
    if (J(i,j − 1) = '1' and J(i − 1, j) = '1' then
    begin
        R(i,j) = R(i − 1, j);
        if (R(i,j − 1) ≠ R(i − 1, j)) then
          record R(i,j − 1) and R(i − 1, j) as equivalent region numbers;
    end
  end
end
end.
```

The output of this procedure is the array **R** containing all information regarding the connectedness of pixels having the same binary values, i.e., all pixels in **J** belonging to the same four-connected region and having the same binary value are assigned the same region

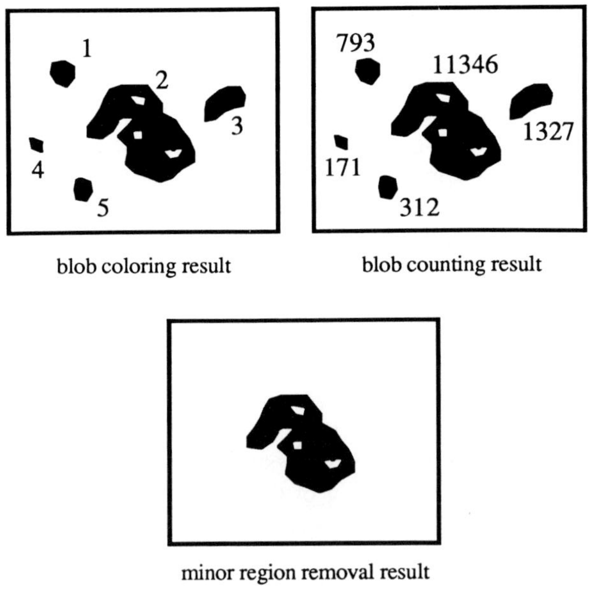

FIGURE 5. Result of blob coloring and minor region removal algorithms applied to the binary thresholded image of Figure 4.

number. Using this information the total number of regions and the area of each region can be found easily. As the implementation of the system described here is designed, for simplicity of exposition, to process images containing only a single object to be measured, only the region with the largest area (largest number of pixels) is retained. All other regions are classified as noise and are thus removed from the input binary image, according to the following simple strategy. First, the number of the region having the largest area is identified using the area information acquired in the preceding procedure. This is a trivial operation on the region number array **R**. Then the minor region deleting procedure can be done in another scan of the image as listed below.

B. Minor Region Removal Algorithm

Given a binary (thresholded) image $\mathbf{J} = [J(i,j); 0 \leq i,j \leq (N/2) - 1]$, its region-number array $\mathbf{R} = [R(i,j; 0 \leq i,j \leq (N/2) - 1]$ computed by the previous algorithm, and the number m of the largest region, then delete potentially misclassified pixels according to the following pseudocode:

```
begin
while (0 ≤ i ≤ (N/2) − 1 and 0 ≤ j ≤ (N/2) − 1) do
begin
  if J(i,j) = '1' then
    if R(i,j) ≠ m then J(i,j = '0';
end
end.
```

Figure 5 depicts the result of the blob coloring algorithm and the subsequent removal of minor regions from the binary thresholded image result in Figure 4. Of course, the above code can be modified to instead allow the largest M regions to be preserved, if $M \geq 1$ is found to be a meaningful count of the expected number of objects. However, in perhaps the majority of applications, a single object, having the largest area, will be the region desired to be extracted. For example, in all application cases of the wound area measurement system encountered thus far, currently numbering in the thousands, the above assumption that the object has the largest area has been valid.

It may be observed from Figure 5 that one application of the minor region removal algorithm does not remove those misclassified regions that lie entirely within the object of interest. However, a second application of the blob coloring/counting and minor region removal algorithms to the *logical complement* $\mathbf{J}^c$ of the binary thresholded image $\mathbf{J}$ will accomplish this in trivial manner. The complement image has elements defined according to $J^c(i,j)$ = '1' if $J(i,j)$ = '0' and $J^c(i,j)$ = '0' if $J(i,j)$ = '1'. In the complemented image, the background can be expected to be the largest "blob." Once the region counting and minor region removal has been applied to the complement image, the complement of the resulting corrected image is taken (Figure 6). The result of the two passes of these region correction algorithms is a binary image containing a single connected object or BLACK region against a WHITE background. Generally, the number of pixels contained within this corrected image will yield a fairly reasonable estimate of the true object area. However, it is possible to further improve the efficacy of the area measurement system by applying a final smoothing algorithm to the region-corrected binary image. This final step has the effect of correcting misclassifications occurring near the boundary of the object/background regions, which are not amenable to the algorithms described above. For example, the final result of Figure 6 contains a significant gulf or bay which may actually be part of the object region. Errors of this type occur frequently in applications where the boundaries of objects are diffuse, as in many microscopic images.

VII. BOUNDARY SMOOTHING BY MORPHOLOGICAL FILTERING

While the gray-level thresholding algorithms and region-correction algorithms are simple and fast, a common problem occurs when the object or the background has a varying gray level: the detected object boundary may be corroded, containing breaks, gulfs, or peninsulas not corresponding to the physically correct object boundaries. In this section, *morphological filters* are presented as a simple, elegant, and very effective solution to these problems.

Mathematical morphology, originally developed as an analytical device for the quantification of the geometric structure of objects, has been applied to image processing problems by several authors.[2,13,14] The techniques and operations of mathematical morphology have the advantages of computational simplicity and of intuitive physical meaning. Generally, *binary morphology* is a collection of operations for modifying the shapes or region coverages of objects in binary images. There is also a more general collection of operations collectively termed *gray-scale morphology,* but for our purposes the simple binary techniques suffice; these can all be defined in terms of simple binary or logical operations.

More specifically, a morphological operation is an interaction between the objects in a binary image and a *window* or *structuring element.* The window, which is usually of simple shape and local support, interacts with the image objects, transforming their shapes in a meaningful way. The type of interaction that occurs between objects and window defines the type of morphological operation occurring. The shape of the window determines the precise way in which the morphological operation modifies the shapes of objects.

A. Windows

A window $\mathbf{B}$ defines a geometric relationship between local pixels in an image; it is a device for collecting image coordinates on a local basis. Mathematically, a window $\mathbf{B}$ is a set of $2M + 1$ image coordinate shifts $B_i = (m_i,n_i)$, usually symmetrically centered around the zero-shift (0,0) (hence the odd number of elements $2M + 1$):

$$\mathbf{B} = \{B_1, \ldots, B_{2M+1}\} = \{(m_1, n_1), \ldots, (m_{2M+1}, n_{2M+1})\}$$

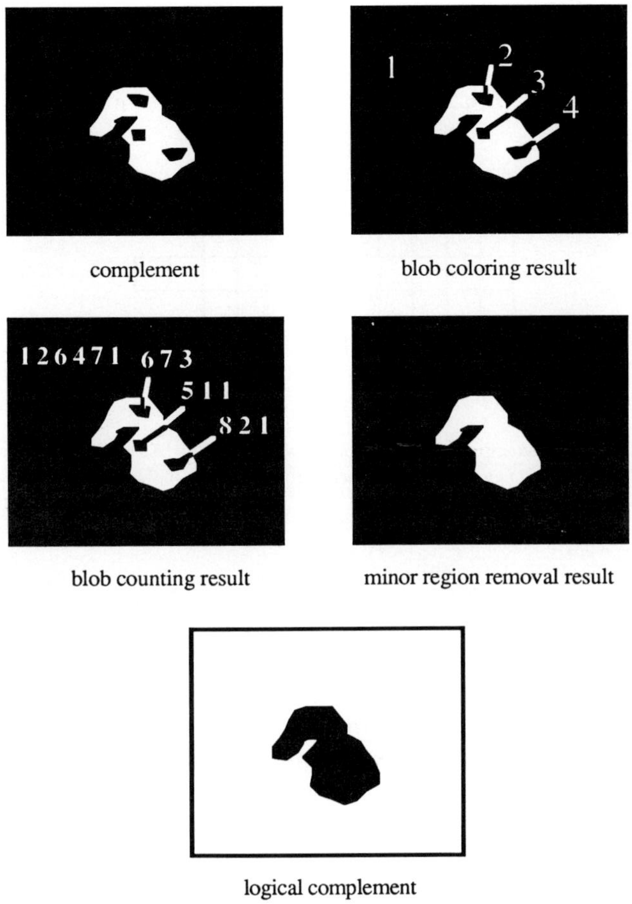

FIGURE 6. Dual of blob coloring and minor region removal algorithms applied to the partially corrected image of Figure 5.

Figure 7 depicts a few windows as geometric pixel relationships. These may also be expressed as sets of image coordinate shifts as follows:

$$\mathbf{B} = \text{SQUARE}(2M + 1) = \{(m, n): -N \leq m \leq N, \ -N \leq n \leq N\}$$

where $N = (\sqrt{2M + 1} - 1)/2$, and

$$\mathbf{B} = \text{CROSS}(2M + 1) = \{(m, 0): -M \leq m \leq M\} \cup \{(0, n): -M \leq n \leq M\}$$

B. Windowed Sets

As mentioned above, a window **B** is a device for collecting a set of image pixels according to a specified geometric relationship defined by the shape and size of **B**. This is accomplished at each image coordinate (i,j) of the binary image **J**, resulting in the indexed *windowed sets*

$$\mathbf{B} \cdot \mathbf{J}(i,j) = \{J(i - m, j - n); \quad (m, n) \in B\}$$

Thus, the windowed set $\mathbf{B} \cdot \mathbf{J}(i,j)$ has the geometric interpretation as the set of image pixels in **J** that are covered by the window **B** when it is overlaid (centered) on the image at coordinate (i,j). Thus,

$$\text{SQUARE}(2M + 1) \cdot \mathbf{J}(i,j) = \{J(i - m, j - n): -N \leq m \leq N, \ -N \leq n \leq N\}$$

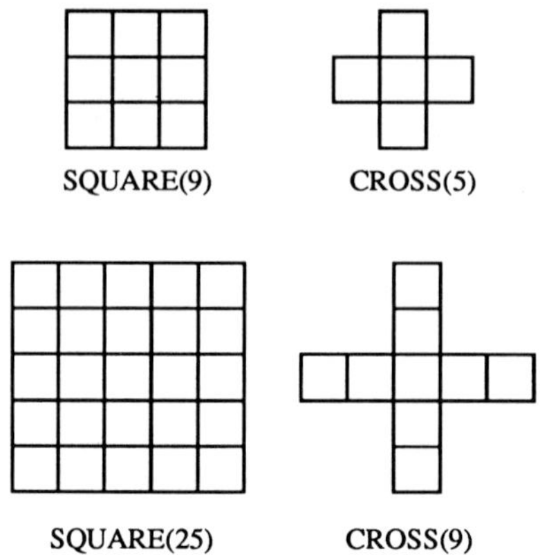

FIGURE 7. Examples of the typical windows SQUARE($2M + 1$) and CROSS($2M + 1$), where the numbers in parentheses indicate the number of coordinate shifts in the window.

where $N = (\sqrt{2M + 1} - 1)/2$ and

$$\text{CROSS}(2M + 1) \cdot \mathbf{J}(i,j) = \{J(i - m, 0): -M \leq m \leq M\} \cup \{J(0, j - n): -M \leq n \leq M\}$$

Using this notation makes it possible to define any kind of binary morphological operation on an image **J**, given a window **B**. While windows **B** other than SQUARE and CROSS can be defined, these are often useful since they are easily indexed sequentially in a computer program, and since they approximate a circular relationship between pixels. An approximately circular shape is desirable since it is usually reasonable to require the window to interact with objects in a manner that is approximately invariant to the local orientations of the object boundaries.

It should be observed that the windowed sets are not well defined at the boundaries of the image, since the coordinate shifts define pixels to be collected that lie outside of the image, i.e., outside of the prescribed range $0 \leq i,j \leq (N/2) - 1$. As with the average filtering operation defined earlier, this is easily ameliorated by replicating pixels, i.e., assigning binary values from the closest pixels to the undefined pixel elements.

C. Morphological Operations

A morphological operation F on a windowed set of image pixels is a binary or Boolean function of the pixels indexed within the window:

$$J_F(i,j) = F[\mathbf{B} \cdot \mathbf{J}(i,j)] = F\{J(i - m, j - n); (m, n) \in \mathbf{B}\}$$

If the morphological operation F is applied at every pixel coordinate $0 \leq i,j \leq (N/2) - 1$, then an output or morphologically filtered image $\mathbf{J}_F$ results:

$$\mathbf{J}_F = F[\mathbf{J}, \mathbf{B}] = [J_F(i,j); 0 \leq i,j \leq (N/2) - 1]$$

where the elements $J_F(i,j)$ are given immediately above. The resulting image $\mathbf{J}_F$ is the result of the morphological operation F applied to the image **J**, with respect to the window **B**.

While there are many other interesting morphological operations, for our purposes four basic morphological operations — *erosion, dilation, opening,* and *closing* — will suffice; interested readers may consult Serra.[2]

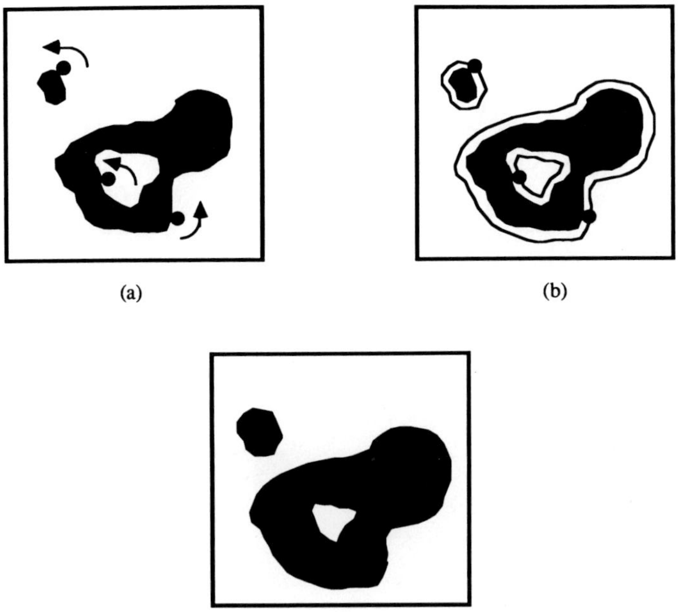

(a) (b) (c)

FIGURE 8. Qualitative description of the DILATE operation. Since the DILATE operation has a local effect only where the window **B** covers both WHITE and BLACK pixels (i.e., near or at object boundaries), is it useful to think of (a) the (approximately circular) window as rolling *outside* each boundary of BLACK objects in the image, (b) the center of **B** thus tracing a set of contours, which (c) form the boundaries of the DILATED image.

1. Dilation

Given a binary image **J** and a window or structuring element **B**, the dilation of **J** relative to **B** is defined as

$$\mathbf{J}^{\text{dilation}} = \text{DILATE}(\mathbf{J},\mathbf{B})$$

$$J^{\text{dilation}}(i,j) = \text{OR}[\mathbf{B} \cdot \mathbf{J}(i,j)]$$

where OR denotes the logical or Boolean operation

$$\text{OR}(X_1, X_2, \ldots, X_{2M+1}) = \begin{cases} \text{`0'}; & \text{if } X_1 = X_2 = \ldots = X_{2M+1} = \text{`0'} \\ \text{`1'}; & \text{otherwise} \end{cases}$$

for arbitrary binary variables $X_1, X_2, \ldots, X_{2M+1}$. Thus, at each image coordinate (i,j) the dilated image $J^{\text{dilation}}(i,j)$ will take logical value '0' if and only if the windowed set $\mathbf{B} \cdot \mathbf{J}(i,j)$ contains no logical elements '1', and logical value '1' otherwise. This operation is so termed since it tends to *increase* the sizes of object or BLACK regions (with logical value '1') in a binary image in a predictable way, as depicted in Figure 8.

2. Erosion

Given a binary image **J** and a window or structuring element **B**, the erosion of **J** relative to **B** is defined as

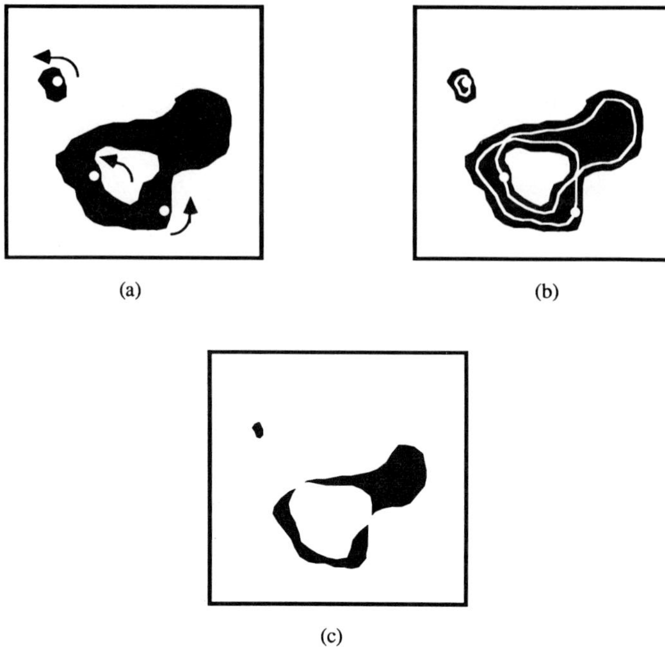

FIGURE 9. Qualitative description of the ERODE operation. Since the ERODE operation has a local effect only where the window **B** covers both WHITE and BLACK pixels (i.e., near or at object boundaries), it is useful to think of (a) the (approximately circular) window as rolling *inside* each boundary of BLACK objects in the image, (b) the center of **B** thus tracing a set of contours, which (c) form the boundaries of the ERODED image.

$$\mathbf{J}^{\text{erosion}} = \text{ERODE}(\mathbf{J},\mathbf{B})$$

$$J^{\text{erosion}}(i,j) = \text{AND}[\mathbf{B} \cdot \mathbf{J}(i,j)]$$

where AND denotes the logical or Boolean operation

$$\text{AND}(X_1, X_2, \ldots, X_{2M+1}) = \begin{cases} \text{`0'}; & \text{if } X_1 = X_2 = \ldots = X_{2M+1} = \text{`0'} \\ \text{`1'}; & \text{otherwise} \end{cases}$$

for arbitrary binary variables $X_1, X_2, \ldots, X_{2M+1}$. Thus, at each image coordinate (i,j) the eroded image $J^{\text{erosion}}(i,j)$ will take logical value '1' if and only if the windowed set $\mathbf{B} \cdot \mathbf{J}(i,j)$ contains only logical elements '1', and logical value '0' otherwise. This operation is so termed since it tends to decrease the sizes of object or BLACK regions (with logical value '1') in a binary image in a predictable way, as depicted in Figure 9.

3. Qualitative Properties of Dilation and Erosion

Generally, DILATE removes object holes of sufficiently small size (relative to the window **B**), and gulfs or bays of insufficient width, as depicted in Figure 10. Likewise, ERODE removes objects of insufficient size, and peninsulas of insufficient width, as depicted in Figure 11. From these properties, it may be deduced that the DILATION of the BLACK objects in an image is equivalent to the EROSION of the WHITE background; likewise,

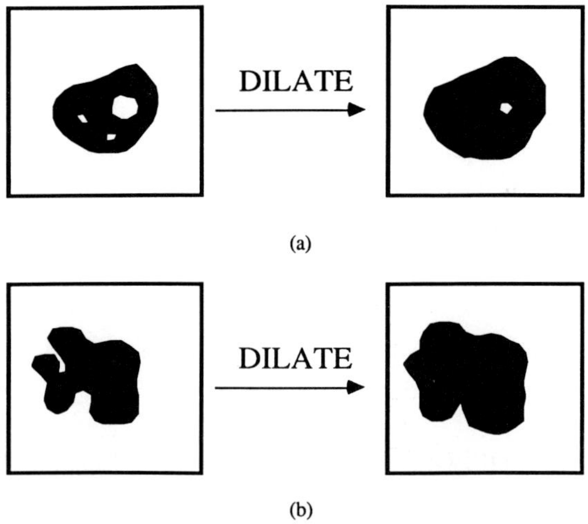

FIGURE 10. Qualitative properties of DILATE, which fill (a) object holes and (b) gulfs or bays of insufficient size (relative to the size of the window **B**).

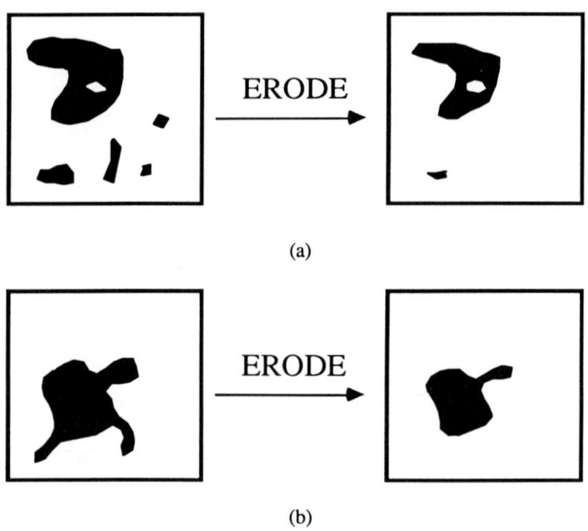

FIGURE 11. Qualitative properties of ERODE, which remove (a) extraneous objects and (b) peninsulas of insufficient size (relative to the size of the window **B**).

the EROSION of the BLACK objects in an image is equivalent to the DILATION of the WHITE background. Thus,

$$\text{DILATE}(\mathbf{J}, \mathbf{B}) = [\text{ERODE}(\mathbf{J}^c, \mathbf{B})]^c$$

and

$$\text{ERODE}(\mathbf{J}, \mathbf{B}) = [\text{DILATE}(\mathbf{J}^c, \mathbf{B})]^c$$

where 'c' denotes the image complementation. In this sense, ERODE and DILATE are actually the same operation — or dual operations with respect to image complementation. However, it is important to observe that erosion and dilation are only approximate inverses of one another; DILATING an ERODED image only rarely yields the original image. In particular, DILATE cannot recreate peninsulas eliminated by EROSION, nor can it recreate small objects eliminated by EROSION. Likewise, ERODING a DILATED image only rarely yields the original image. In particular, EROSION cannot recreate holes filled by DILATION, nor can it recreate gulfs or bays filled by DILATION. Nevertheless, certain operations defined as concatenations of EROSION and DILATION are in themselves very useful morphological operations, as described next.

4. Opening and Closing

Given a binary image **J** and a window or structuring element **B**, the opening and closing of **J** relative to **B** are respectively defined as

$$\mathbf{J}^{\text{open}} = \text{OPEN}(\mathbf{J}, \mathbf{B}) = \text{DILATE}[\text{ERODE}(\mathbf{J}, \mathbf{B}), \mathbf{B}]$$

and

$$\mathbf{J}^{\text{close}} = \text{CLOSE}(\mathbf{J}, \mathbf{B}) = \text{ERODE}[\text{DILATE}(\mathbf{J}, \mathbf{B}), \mathbf{B}]$$

Thus, the operation OPEN is the operation DILATE (relative to **B**) applied to the result of the operation ERODE (relative to **B**). Similarly, the operation CLOSE is the operation ERODE (relative to **B**) applied to the result of the operation DILATE (relative to **B**). It should be noted that ERODE and DILATE do not commute to each other, i.e., it is usually true that

$$\text{OPEN}(\mathbf{J}, \mathbf{B}) \neq \text{CLOSE}(\mathbf{J}, \mathbf{B})$$

hence the OPEN and CLOSE operations are not the same. In fact, an interesting property of OPEN and CLOSE is that, for any window **B**,

$$\text{OPEN}(\mathbf{J}, \mathbf{B}) \leq \mathbf{J} \leq \text{CLOSE}(\mathbf{J}, \mathbf{B})$$

where for any binary images $\mathbf{J}_1$ and $\mathbf{J}_2$, $\mathbf{J}_1$ '$\leq$' $\mathbf{J}_2$ denotes that $J_1(i,j)$ = '1' implies that $J_2(i,j)$ = '1' for any (i,j). Thus, the CLOSE operation must always expand object (BLACK) regions, whereas the OPEN operation must always decrease object (BLACK) regions. However, the concatenations OPEN and CLOSE generally leave the overall (gross) outlines of objects unchanged where there are no extraneous objects, holes, gulfs, or peninsulas. Thus, OPEN and CLOSE are effective object shape smoothing operations.

5. Qualitative Properties of OPEN and CLOSE

In applying morphological operators, one can choose a structuring element of any shape or size. However, most objects arising in biomedical applications have a smooth (low curvature) boundary. In our application, the wound image usually has a smooth boundary. Thus, if it is desired to smooth a boundary using the CLOSE operation, the window **B** should have a smooth boundary and should also be approximately isotropic, since it should not be biased to any particular direction. The best candidate satisfying these conditions is a circle. The effects of applying CLOSE with an approximately circular window **B** are as follows:

- No object contour will have a point with a positive curvature greater than $1/r$, where r is the radius of **B**.
- Every gap with width narrower than the diameter of **B** will be filled.
- Every hole with maximum width less than the diameter of **B** will be filled.

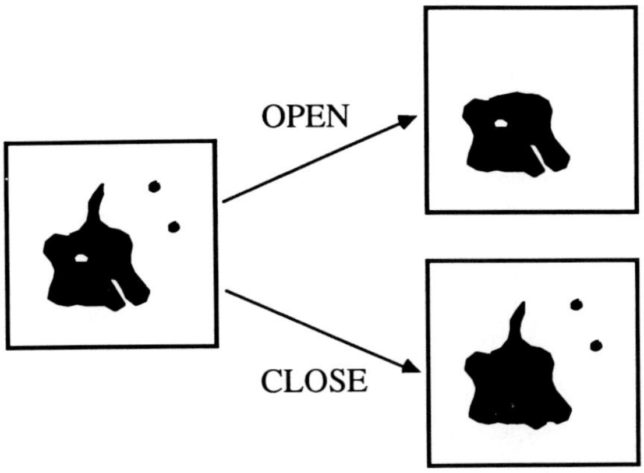

FIGURE 12. Depiction of qualitative properties of OPEN and CLOSE.

Figure 12 depicts these properties; the effects of OPEN may be deduced from the fact that OPEN is the dual of CLOSE with respect to image complementation. Note that the above properties include all effects of CLOSE in the case when the image has a single object. If the image has two or more objects, some other effects, such as merging of two separate objects, can arise. Also, note that the size of the structuring element takes an important role in the closing operation, since all of the above three properties are affected by it. In our implementation, the radius of the circle is chosen interactively by a human operator depending on the state of the thresholded image.

Finally, it is of interest to consider the implementation aspects of morphological filters. Although morphological operations can be implemented in a straightforward manner using simple logical operations defined at each image coordinate, the direct implementation of morphological operations may require excessive computation time. For example, if the circle radius (implemented as SQUARE, for example), is equal to 5, which is typical for many applications, then the area of the region of support of **B** is approximately equal to $3.14 \times 5^2 \approx 78$ pixels. Then, for an image containing $(N/2)^2$ pixels, the total number of required logical operations is approximately $78 \times (N/2)^2$ for either ERODE or DILATE. The amount of computation can be reduced considerably by using the following property: as the position of the window **B** is incremented along a horizontal line from coordinate (i,j) to coordinate $(i,j + 1)$, a large portion of the windowed set $\mathbf{B} \cdot \mathbf{J}(i,j)$ will coincide with the windowed set $\mathbf{B} \cdot \mathbf{J}(i,j + 1)$; only the elements collected on the right and left sides will be different. Hence, instead of recomputing the entire operation, the previous value computed along with the updated region may be used. This may be accomplished simply by, at each coordinate, separately storing the AND or OR of the elements contained within the overlap region $\mathbf{B} \cdot \mathbf{J}(i,j) \cup \mathbf{B} \cdot \mathbf{J}(i,j + 1)$. By using this property it is possible to reduce the amount of computation to about one fourth of a direct implementation.

VIII. EXAMPLES AND EXPERIMENTAL RESULTS

The area measurement system described so far has been applied to several types of biomedical subjects. In this section, two example applications are presented, as well as the result of an experimental performance evaluation of the system. The major problem of interest in the example application is the evaluation of surgical excision wounds created as part of a project to assess the influence of specific externally manipulated parameters on the rate of healing. Previously, the evaluation of wound healing has been performed by various manual methods such as excision and weighing of the wound tissue (which obviously

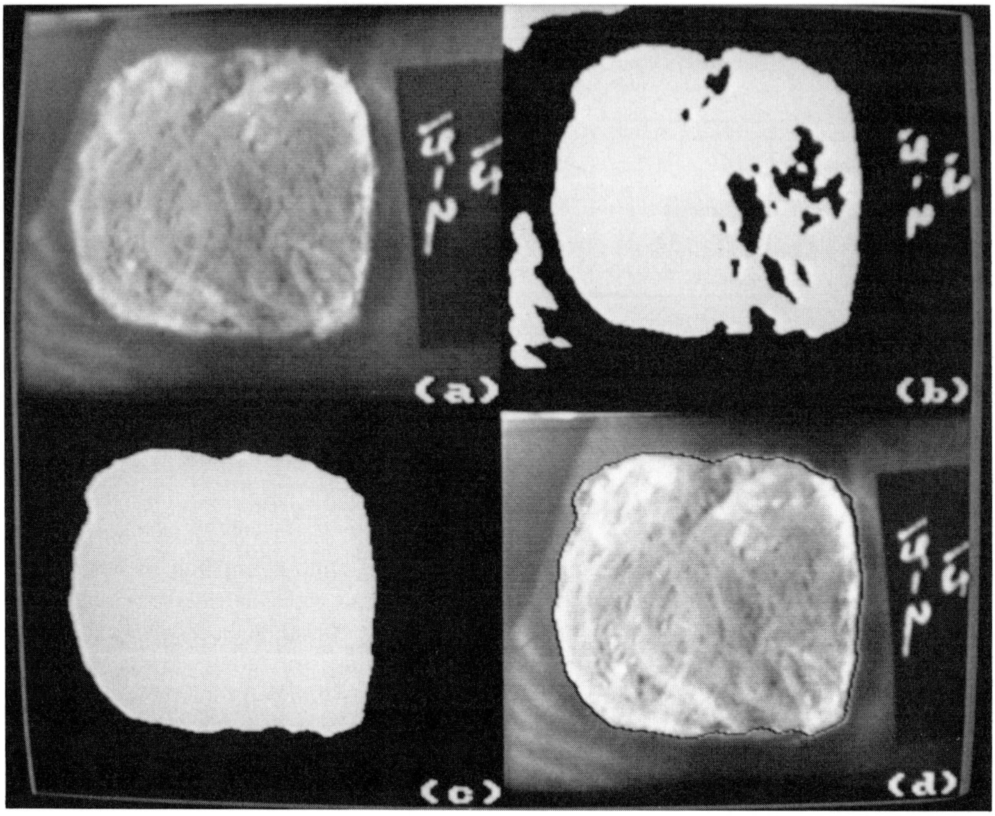

FIGURE 13. Example of wound image at each processing step. (a) Original image; (b) gray-level thresholded image; (c) region-corrected and morphologically smoothed image; (d) detected boundary superimposed on the original image. (Courtesy of Kim, N. H., Wysocki, A. B., Bovik, A. C., and Diller, K. R., *Comput. Med. Biol.*, 17, 173, 1987.)

constrains the data analysis to a single point in time since sacrifice of the animal is required), tracing the wound perimeter from the animal onto graph paper and counting included blocks, or outlining the traced wound boundary with a polar compensating radial planimeter to determine the enclosed area.

The first example depicts each processing step as executed sequentially on a wound image. Figure 13a shows an example data image. Whereas a human operator could trace the boundary of the wound without difficulty, this kind of image represents a difficult subject for a machine vision system to analyze because of its varying and overlapping gray-level distributions in both the object and the background regions. Figure 13b illustrates a binary image obtained by the gray-level thresholding process. After the region correction and morphological operations are finished, the system gives the final output as the one illustrated in Figure 13c. It is worth noting the difference between the object boundary in Figure 13b and the smoothed boundary in Figure 13c, which is noted most clearly in the lower portion of the object. The smoothed boundary matches the actual shape quite well, as can be seen in Figure 13d, where the detected boundary is superimposed on the original image.

For an objective assessment of the performance of the present system, each of the areas was measured twice for over 650 photographic images of healing wounds: once using a polar compensating radial planimeter, which represents the most sensitive of the traditional, mechanical approaches to the problem; and once using the present system. A statistical analysis of two data sets yielded an overall simple Pearson correlation of 0.9975 (ρ = 0.0001).

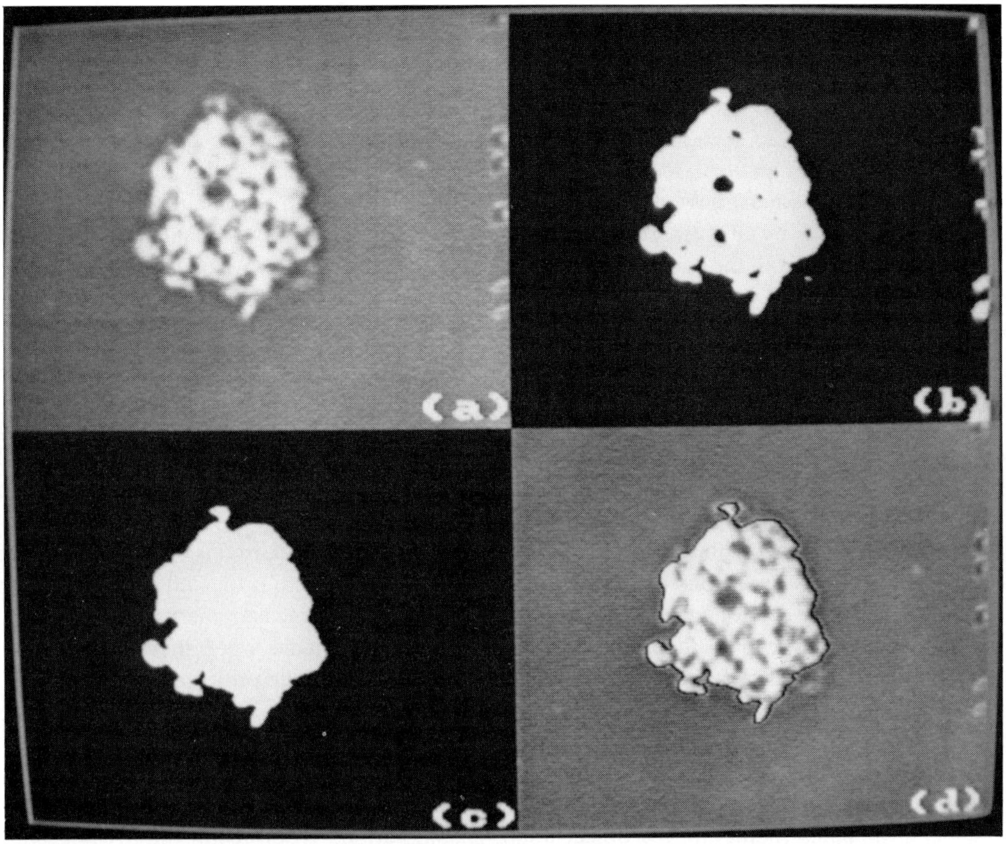

FIGURE 14. A micrograph of pancreas islet. (a) Original image; (b) gray-level thresholded image; (c) region-corrected image; (d) detected boundary superimposed on the original image. (Courtesy of Kim, N. H., Wysocki, A. B., Bovik, A. C., and Diller, K. R., *Comput. Med. Biol.*, 17, 173, 1987.)

A second example shows the application of the system to the cross-sectional area measurement of a bright-field light micrograph of a pancreas islet. The islet, shown in Figure 14a, presents a difficult image to analyze due to its complex shape, which somewhat resembles a cluster of grapes. Thus, it is not easy even to subjectively identify an outline just by casual observation. Figure 14b illustrates the thresholded binary image. Since the islet area usually has a tortuous boundary with large local variations in curvature, the morphological filters were not used for boundary smoothing; instead, only the blob coloring and subsequent region correction procedures were applied. Figure 14c shows the output of this region correction algorithm and Figure 14d the detected boundary superimposed on the original image.

IX. SPECIMEN PREPARATION AND COMPUTATION OF TRUE OBJECT AREA OR CHANGE OF AREA

In biological work a great deal of attention should be paid towards the techniques applied to prepare a specimen for quantitative morphometric area analysis measurements. The reason for this is that, while there exist quite powerful digital image processing techniques for effecting corrections to errors made in the physical image process, certain errors, or errors of large magnitude, can be very difficult to correct. Particular care must be taken to keep any fixation, dehydration, and embedding procedures as constant and minimal as possible in order to avoid considerable volume and area changes that can occur, especially due to shrinkage. A small difference in the histological technique can result in marked differences in the microscopic appearance of cells or tissue.

If the measurements are being done on tissue sections, then consideration must be given to the thickness of the sections and staining density variation in order to avoid errors caused by the "Holmes effect." An ideal specimen is an infinitely thin section; however, in practice, of course, sections must have a finite thickness. This gives rise to two main problems. First, the area occupied by a given section may be overestimated owing to contributions to the image from portions of the specimen lying outside the plane of focus. Second, the components being measured will have an imperfect contrast with the environment, particularly when small intracytoplasmic organelles are being imaged. For components such as these that do not contrast well with the surrounding medium, there will be a tendency to underestimate the area occupied on the surface of a section. Fortunately, these two effects tend to cancel each other out on many occasions, but it is essential to be aware of the problems involved and take steps to minimize and possibly correct any potential errors. A detailed account of these and associated problems is available in Aherne et al.[15] In general, sections thicker than 10 μm and measurements over a large number of the objects of interest are recommended. Conversely, if the isolated cells etc. are the objects under consideration, then the use of smears or suspensions of isolated cells in order to measure projected area avoids the problem of partly-sectioned organelles being excluded in the measurements. Particular attention must also be paid to the nature of the suspending media when measurements are being made on isolated cells. For example, the volume (and hence projected area) occupied by red blood cells suspended in plasma may be different than when suspended in a phosphate buffer.

Another point of interest for certain applications is the method of staining employed for the preparation of a cellular specimen. Staining is desirable in order to enhance the contrast between the background and the object under consideration. When using fluorescent dyes, double labeling could significantly improve the selective demonstration and the contrast between two different objects. A variety of vital and specific fluorescent and non-fluorescent reagents are available for customized use and should be carefully chosen.[16,17]

Yet another important standardization problem for obtaining quantitative morphometric data is concerned with the calibration of the video camera in terms of optical density, scaling linearity, spectral sensitivity, and the data sampling rate.[18] Smooth curve dynamic range correction with a very minimum of runout error over the full scaling range of the camera, together with a low spatial reading error, must be established. The sampling rate is one determinant of the image resolution, since the resolution cannot be more than twice the pixel spacing distance. A general purpose CCD video camera provides a relative, rather than an absolute scaling response, and neither the camera scaling action nor the control facility is sufficiently drift free to provide a completely repeatable and exacting readout. In microscopic applications using dyes, optical density gray-level intensity calibrations can be accomplished by first adjusting the transmission histogram peak of the blank digitized image, then measuring the gray-level intensity of varying concentrations of the dye in a constant geometry container (e.g., hemocytometer). The least-squares fit of the normalized intensity data can then be determined. The normalized intensity, on a scale from 0 to 1, can then be used to represent the dynamic range of the imaging system.

In the most usual applications, it is of interest to measure or compute the true cross-sectional area of an object or objects rather than just the image-plane, or "pixel," area. The goal, then, is to transform the easily determined pixel area into true area, which requires that the general relationship between true area and pixel area, under the image projection process, be determined. This can be accomplished either by measuring the appropriate physical parameters of the imaging equipment directly, or alternately, by measuring objects

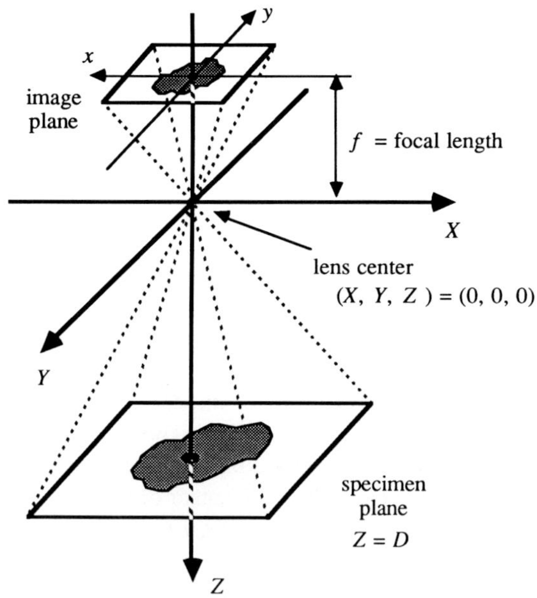

FIGURE 15. Projective imaging geometry under the pinhole lens assumption.

of precisely-known dimensions under the same imaging conditions under which the objects of interest are to be studied. In either case, it is of use to understand the assumed relationship between image coordinates and true object coordinates.

Figure 15 depicts the general projective imaging geometry under a simple "pinhole aperture" lens assumption, which will be used here. Corrections to the model under a finite-aperture or shallow depth-of-field model are beyond the current discussion as being too specific, but in any case, for specimens that are thin relative to the imaging scale, the assumption is highly accurate for most objectives of interest. In Figure 15, coordinates (X, Y, Z) denote points in true three-dimensional (3-D) space, where the origin $(X, Y, Z) = (0, 0, 0)$ is taken to be the lens center; thus the coordinate Z of an object feature is taken to be the distance of the object plane to the lens plane along the optical axis; the coordinates (X, Y) denote the orthogonal distances of the point from the optical axis. The image coordinates (x, y) denote points in the 2-D image plane, which is chosen to be parallel (congruent under projection) to the X-Y plane; the optical axis passes through the origins of both coordinate systems.

In the idealized imaging model used here, the lens is treated as a simple pinhole through which all light rays hitting the image plane pass. The image plane, which coincides with the photographic emulsion, or the CCD sensor array in a standard video camera, is a distance one focal length f from the lens. Of course, in a complex multiple-lens system such as a microscope, the focal length parameter must be replaced by an estimate of the effective focal length, which may be difficult to determine, particularly if the microscope is interfaced with a video camera through an adaptor. The specimen plane is assumed to be a distance D from the lens plane; this can be determined via mechanical measurement in most instances, although again, in a complex microscope/adaptor/camera system, determination of this quantity can be difficult. Indeed, in microscopic applications, it is usually preferred to make comparisons with an object of precisely known dimensions, as suggested below.

It is easily shown that under the simple projective geometry assumed here, a point $(X, Y, Z) = (X_0, Y_0, D)$ lying on the specimen (presumed perfectly flat) projects to the image-plane coordinates[5]

$$(x_0, y_0) = (f/D)\,(X_0, Y_0)$$

where the ratio f/D is the magnification factor, which varies with the distance D from the lens center to the specimen plane. Thus, for the distance D fixed or known, specimen coordinates are related to image coordinates by a simple scaling. It also follows that the true area A_{true} of the specimen of interest is related to the area A_{image} of its projected image by the (square of the) same scaling:

$$A_{image} = (f/D)^2 A_{true}$$

since the perspective projection is an affine transformation of the real-world coordinates. Thus, it remains to be able to measure the actual physical dimensions of points in the image plane, i.e., relate pixel areas to image plane areas. If a permanent photographic emulsion is used, this is easily accomplished by imaging the film negatives as discussed in Section IV. Otherwise, a measurement of the true pixel dimensions must be made. Commercially available CCD camera manufacturers make available the physical dimensions of the sensor arrays, from which the dimensions of individual pixels may be deduced. Other types of cameras, such as vidicons, have overall physical sensor dimensions which are also specified. Estimates of the effective pixel dimensions can then be easily determined from the specifications of the frame-grab apparatus (the sampling density).

In many applications the measurement of actual physical dimensions is not critical; it is often of interest to instead determine only the *percent change* in area, over time, of a specimen undergoing some type of regimen or evolving process such as healing. In this case the determination of the desired quantities is vastly simplified (under the pinhole assumption), since the percent change in area remains unchanged with respect to the perspective transformation, and also with respect to image-to-pixel dimension changes. Thus

$$\Delta A_{true}(\%) = \Delta A_{image}(\%) = \Delta A_{pixel}(\%)$$

where ΔA (%) represents percent change in area, and where A_{pixel} is the measured object area expressed as an integer number of pixels.

Finally, the determination of the true area of an object in a complex imaging situation, where the relative physical parameters may be difficult to measure, can be accomplished using objects of precisely known dimensions. Since the perspective projection equations of the test object and the specimen will be the same, then measurement of the test object's dimensions yields the direct relationship between the specimen's area A_{pixel} measured in pixels and the true area A_{true}. In a microscope/adaptor/camera system, calibrating the digital image in terms of pixels per unit length can be accomplished with a stage micrometer. The spatial resolution of the camera and the orthogonal equality of the image can be determined by digitizing the image of an EM grid mounted on a standard microscope slide and viewing the image. The camera is then rotated 90° on its long axis and the grid spacings are measured at right angles. When the limits of resolution are reached, the lines become deformed by a recognizable Moire pattern.

X. APPLICATION OF SHAPE MEASUREMENT TECHNIQUES TO OTHER PROBLEMS OF BIOMEDICAL INTEREST

A detailed description of the various areas in the biomedical sciences in which quantitative size and shape parameters are applied as a means of evaluating a given state, and modifications following some physical intervention or pathology, is beyond the scope of this chapter. However, a few representative areas are discussed below.

Anatomic pathology is a leading field for the application of computer-based size, shape, and pattern recognition, because the numerical parameters of the object are the prime considerations for diagnostic purposes. Fully automated systems have so far been of limited success, because identification of the cell type within the scene under analysis is still a very complicated process and is best accomplished by an interactive operator-assisted process. However, once the objects have been identified, compilation of the morphometric data and statistical evaluations can be performed automatically. Tracing the specimens of interest by any number of interactive peripherals, e.g., mouse, digitizing screens, and touch sensitive screens are by far the most popular techniques. The most common measurement is the determination of nuclear and cytoplasmic diameters.[19]

Although the effective numerical shape factors aimed at recognition of an object presented in any orientation or angle are well documented for the detection of man-made objects,[19] the application of shape factors for the interpretation of microscopic images have been of limited success. However, discrimination based on different perimeter/area relationships,[21] which determine to what extent a shape approaches or differs from a circle, has been widely used. Some investigators have used calibrated imaginary circles as graphics standards and have computed the ratio of the real perimeter measured from a cell to the standard as a measure of circularity.[19,22] Of course, this and other circularity factors that are based on perimeter/area relationships will shift whenever shape or size changes occur; they are also highly specific for a given shape. Semiautomated and automated methods using the "four-side connected region" concept for bone histiomorphometry have been developed by Beigbeder et al.[23] Their algorithm was designed to flood-fill simple geometric figures (polygons) with sharp, well-defined boundaries displayed on a video monitor. Simple gray-level thresholding has been used by Garrison et al.[24] for the determination of biomass and to compute the areas of developing colonies of chondriocytes during cartilage differentiation. It is likely that this, and other similar application domains, will benefit from the well-defined and more generic techniques developed in this chapter.

Cryomicroscopy is an area of active research interest wherein it is of interest to rigorously quantify the shape and size changes in areas of cells and cell clusters undergoing freeze/thaw processes.[9,21,25,26] Different types of cells ranging from yeast, human granulocytes,[27] and pancreas endocrine β-cells[28] to multicellular islets of Langerhans[29] have been investigated using a variety of techniques.

In the algorithm for the automated determination of the area of a cell following a defined regimen of cooling and thawing, the segmentation of the digitized images into cells and background is accomplished by a sequence of operations.[9] A modification of the Hough transform is applied as a means of evaluating all of the candidate object edges, in order to determine the one most likely of being the cell boundary.[5,25] When a series of images during a freezing sequence is analyzed, the Hough transform is continually updated so that in any frame of the image sequence, the program searches for an element having a shape identified for the cell in the immediately preceding frame. This step results in a reasonably good estimate of the cell outline in many cases. The result is a defined 2-D area, assumed to lie in a single plane. A center of gravity is identified for this area, from which the distance to all boundary pixel locations is measured. The average statistics of these distances can be used to determine cell irregularity, area, degree of circularity, etc.

XI. CONCLUSION

This chapter reviewed fundamental aspects of current microcomputer-based image area analysis techniques and their biological applications. A rather generic machine vision system which has been successfully applied to the measurement of the areas of complex, biomedically-generated shapes was demonstrated as a means for illustrating the basic techniques of object extraction and isolation, region correction and smoothing, and area measurement. The example system has been demonstrated to perform at a high level of reliability with a minimum amount of intervention from a human operator. The system described requires on-line human input appropriate to the subject of interest, only as far as for determining an appropriate gray-level threshold and, possibly, for specifying the radius of the window or structuring element used in the morphological region smoothing operations. Experimental results obtained from the processing of the wound images support the efficacy of this approach as well as its reliability.

Another merit of the techniques described and implemented for example purposes is that they can be implemented on a microcomputer that can be used as a stand-alone workstation for the image processing task. In the current example system implementation (not optimized for computational efficiency) the program requires from 4 to 8 min to process a single image. While the computation time is not particularly fast, a more efficient implementation could probably reduce the computation by at least an order of magnitude. Further considerable increases in computation speed are feasible if the example system is implemented on any of the more powerful workstations that are rapidly becoming widely available at a reasonable cost. Still, the system as described herein represents a very considerable reduction of manpower and man-hours for the simple yet important general task of measuring the areas of biological objects.

REFERENCES

1. **Baker, D. G., Hwang, H. H., and Aggarwal, J. K.,** Detection and segmentation of objects in outdoor scenes: concrete bridges, *J. Opt. Soc. Am.*, A6, 938, 1989.
2. **Serra, J.,** *Image Analysis and Mathematical Morphology,* Academic Press, New York, 1982.
3. **Bovik, A. C., Clark, M., and Geisler, W. S.,** Multichannel texture analysis using localized spatial filters, *IEEE Trans. Pattern Anal. Mach. Intelligence,* 12, 55, 1990.
4. **Rosenfeld, A. and Kak, A. C.,** *Digital Picture Processing,* Vol. 2, Academic Press, New York, 1982.
5. **Ballard, D. H. and Brown, C. M.,** *Computer Vision,* Prentice-Hall, Englewood Cliffs, NJ, 1982.
6. **Ballard, D. H. and Sklansky, J.,** A ladder-structured decision tree for recognizing tumors in chest radiographs, *IEEE Trans. Comput.*, 25, 503, 1976.
7. **Chien, Y. P. and Fu, K. S.,** A decison function method for boundary detection, *Comput. Graph. Image Process.*, 3, 125, 1974.
8. **Lester, J. M., Williams, H. A., Weintraub, B. A., and Brenner, J. F.,** Two graph searching techniques for boundary finding in white blood cell images, *Comput. Biol. Med.*, 8, 293, 1978.
9. **Dietz, T. E., Davis, L. S., Diller, K. R., and Aggarwal, J. K.,** Computer recognition and analysis of freezing cells in noisy, cluttered images, *Cryobiology,* 19, 539, 1982.
10. **Pavlidis, T.,** *Structural Pattern Recognition,* Springer-Verlag, New York, 1977.
11. **Kim, N. H., Wysocki, A. B., Bovik, A. C., and Diller, K. R.,** A microcomputer-based vision system for area measurement, *Comput. Biol. Med.*, 17, 173, 1987.
12. **Thompson, K. F. and Diller, K. R.,** Use of computer image analysis to quantify contraction of wound size in experimental burns, *J. Burn Care Rehabil.*, 2, 307, 1981.
13. **Lantuejoul, C. and Serra, J.,** M-filters, *Proc. IEEE Int. Conf. Acoust. Speech Signal Process.*, 2063, 1982.
14. **Maragos, P. A.,** Tutorial on advances in morphological image processing and analysis, *Opt. Eng.*, 26, 623, 1987.
15. **Aherne, W. A. and Dunhill, M. A.,** *Morphometry,* Edward Arnold, London, 1982.
16. **Haugland, R. P.,** *Molecular Probes: Handbook of Fluorescent Probes and Research Chemicals,* Molecular Probes, Eugene, OR, 1989.
17. **Green, F. J.,** *The Sigma-Aldrich Handbook of Stains, Dyes and Indicators,* Aldrich Chemical, Milwaukee, 1990.
18. **Inoue, S.,** *Video Microscopy,* Plenum Press, New York, 1986, 191.
19. **Marchevesky, A. M., Gil, J., and Jeanty, H.,** Computerized interactive morphometry in pathology, *Hum. Pathol.*, 18, 320, 1987.
20. **Horn, B. K. P.,** *Robot Vision,* MIT Press, Cambridge, MA, 1986.
21. **Diller, K. R. and Aggarwal, S. J.,** Computer automated cell size and shape analysis in cryomicroscopy, *J. Microsc.*, 146, 209, 1987.

22. **Steponkus, P. L.,** Role of plasma membrane in freezing injury and cold acclimation, *Annu. Rev. Plant Physiol.*, 35, 543, 1984.
23. **Beigbeder, M., Chappard, D., Alexandre, L., Vico, L., Palle, S., and Riffat, G.,** Improved algorithms for automatic bone histomorphometry on a numerised image analysis system, *J. Microsc.*, 150, 151, 1988.
24. **Garrison, J. C., Peterson, P., and Uyeki, E. M.,** Computerized-based image analysis of cartilage differentiation in embryonic limb bud micromass cultures, *J. Microsc.*, 156, 353, 1989.
25. **Dietz, T. E., Diller, K. R., and Aggarwal, J. K.,** Automated computer evaluation of time-varying cryomicroscopical images, *Cryobiology*, 21, 200, 1984.
26. **Diller, K. E. and Knox, J. M.,** Automated computer analysis of cell size changes during cryomicroscope freezing: a biased trident convolution technique, *Cryo-Letters*, 4, 77, 1983.
27. **Schwartz, G. J. and Diller, K. R.,** Osmotic response of individual cells during freezing. I. Experimental volume measurements, *Cryobiology*, 20, 61, 1983.
28. **Aggarwal, S. J., Diller, K. R., and Davidson, I.,** Microscopic response of dog pancreas b cells to freezing and osmotic stresses, *Cryo-Letters*, 7, 218, 1986.
29. **Macías-Garza, F., Diller, K. R., Bovik, A. C., Aggarwal, S. J., and Aggarwal, J. K.,** Obtaining a solid model from optical serial sections, *Pattern Recogn.*, 22, 577, 1989.

Chapter 4

DIGITAL FILTERS IN IMAGE ANALYSIS

Roberto Marangoni, Giuliano Colombetti, and Paolo Gualtieri

TABLE OF CONTENTS

I. INTRODUCTION

The final result of several scientific experiments is often represented by an image. In order to be meaningful to the human observer, this image must often be subsequently elaborated; think, for example, of a satellite, microscope, or X-ray picture.

The very fast development in the field of electronics has made available, at a relatively low cost, integrated systems which are capable of acquiring, storing, and analyzing images. These systems can be easily coupled to personal computers and thus are widely used in research laboratories.

If we restrict ourselves to the biological sciences, we find that image analysis techniques are mainly employed in the field of microscopy. Optical, electron, scanning tunneling, and atomic force microscopic images are very often presented to the reader in a more or less elaborated form. These elaborations make it possible to reduce the noise of a scene; to enhance or even reveal details that could not be seen in the original picture; to correct optical aberration; to measure quantitatively morphological parameters, such as areas, dimensions, etc.; and to determine concentrations of chemical species (e.g., Ca-sensitive dyes). It is also possible to use similar techniques to analyze sequences of images, in other words to follow the dynamical evolution of a system. It is thus possible to study motion parameters of a population of cells, measuring automatically their positions and velocities.

More recently, artificial intelligence and neural network techniques have contributed to the setup of expert systems which are capable of recognizing patterns present in a scene. Automatic pattern recognition is thus becoming possible, and we expect dramatic developments in this field in the near future.

II. IMAGE FORMATION

In most cases, a biological image is obtained by means of a microscope (see above). We will limit ourselves to the study of optical images; in fact, other systems do not form an image *sensu strictu.*

We will not consider the physical principles that underlie image formation and their mathematical description; the interested reader is referred to the literature.[4,8,9] The knowledge of this may, however, be very useful in the development of image restoring algorithms.

Independent of the system of image formation, the final result may be described by a light intensity function f(x,y) where the dominion of the (x,y) points coincides with the image area and the value of f at a certain point is proportional to the brightness of the image at that point. On the basis of this definition, the function f(x,y) represents a real image.

We will assume to be dealing with monochromatic images; this is not too restrictive an assumption. All the results that we will discuss in the following sections are, in principle, also valid for color images, even if the situation in that case may be much more complicated. The acquisition of a color image may be useful, for instance, to have a better scene-discriminating capacity. From a technical viewpoint, a color-discriminating acquisition may be obtained in two different ways: either by using a commercial device or by acquiring three different monochromatic images taken with three different filters that form a base for colors (such as red — green — blue). It is also possible to digitize a monochromatic image and subsequently artificially color it in order to enhance certain features. These images are usually known as "false color images," since the colors visualized do not correspond to the real colors. A typical example is given by colored pictures taken from a meteorological satellite, where the different colors correspond to the different temperatures recorded. We will not consider these aspects in further details; the interested reader may consult the appropriate literature.[4,7,15]

III. DIGITIZATION

Digitization is defined as the process by which the continuous function f(x,y) defined above is transformed into a discrete function l(m,n) defined over a set of discrete values, m,n. This is a very important step since the use of a digital computer requires a discrete set of data to be processed. Digitization includes both spatial and amplitude values. Digitization in space is known as sampling; in amplitude it is known as quantization.

A. Sampling and Quantization

In order to convert f(x,y) into l(m,n) the following steps have to be taken:

1. A usually square grid, formed by N · N squares, is superimposed onto the f(x,y). Each square is identified by a couple of integers m,n with $0 < m < N$ and $0 < n < N$ (see Figure 1).
2. The average brightness is calculated over each square.
3. A function g(m,n), having as value the average brightness previously calculated, is defined over each square. Its values are bound by two real numbers, g_{min} and g_{max}.
4. A set of discrete values is chosen equally spaced between 0 and an integer maximum L. These values are given by $(g(m,n) - g_{min})/(g_{max} - g_{min}) * L$ and are usually known as gray levels.
5. The discrete function l(m,n) is defined over the same dominion as g(m,n) with the values given to 4. above and is the function we are looking for.

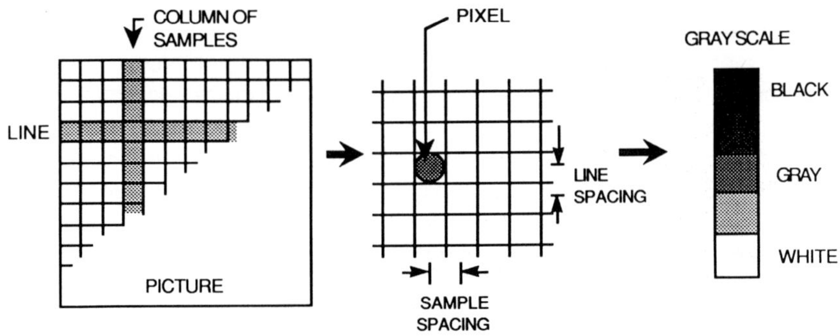

FIGURE 1. Sampling and quantization procedures (see text).

The discrete points of coordinates (m,n), elementary parts of the picture, are called pixels (from PICture Elements). The sampling and quantization described above are uniform, since a uniform square grid and equally spaced values have been used. It is also possible to use nonuniform sampling and quantization.[1,7] This is usually done for particular purposes. For instance, a fine sampling obtained by using a finely spaced grid for the central part of the image and a coarsely spaced grid elsewhere simulates the spatial sampling of the human retina better than a uniform sampling.[16,17] A nonlinear correspondence between brightness values and gray-level ranges may be useful to increase the visibility of poorly contrasted details (tapering).

B. Sampling and Quantization Techniques

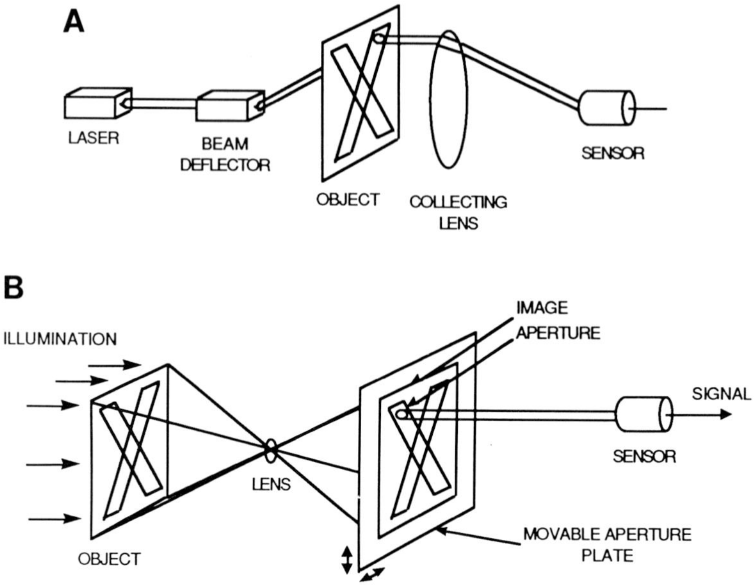

FIGURE 2. Scan-in (A) and scan-out (B) techniques (see text).

As we have seen in the previous paragraphs, a digital image is an image f(x,y) discretized both spatially and in brightness. There are two major techniques that are currently used to digitize an image; they are known respectively as scan-in and scan-out (see Figure 2).[4]

In the scan-in procedure (Figure 2A), the scene is scanned by a light beam. A light sensor measures, one pixel at a time, the amount of transmitted light that gives the brightness value at that point. In the scan-out procedure (Figure 2B), the whole scene is illuminated at the same time and the sensor reads the image one point after the other.

From a formal viewpoint, both procedures are not different, but the results obtained are different. In the scan-in procedure, the movement is mechanically driven, whereas in the scan-out, it is electronically driven. The first system has the advantage of a good signal-to-noise ratio, the second of a very high scanning speed. It is also possible to use more refined methods which utilize both the techniques briefly mentioned above.

There are many theoretical studies aimed at determining optimal procedures for sampling and quantization;[1,4,14] the average user has no need, however, to take care of such problems, since commercial scanners and TV cameras are built in such a way as to optimize image digitization.

IV. DIGITAL IMAGES

A digital image obtained according to the procedures described above can easily be represented by a square N · N matrix (a common value for N is 512), whose values are integer numbers proportional to the brightness values and ranging from 0 to an upper limit L (typically 255) in the so-called gray scale. In this representation it is possible to quantitatively analyze the image, for instance, in terms of the histogram of the gray levels or of the spatial frequencies.

A. Gray-Level Histogram

A gray-level histogram is determined by counting all the pixels of an image that have a certain gray value and by plotting on the abscissa the values of the gray levels and on the ordinate the frequencies of the different gray levels. From this histogram it is possible to obtain some information such as, the average gray level gives an idea of the overall luminosity of the image; from the shape of the histogram it is possible to have some hints for further image processing. Let us assume, for instance, we are dealing with the microscopic image of cells on a strongly contrasted background. The gray-level histogram is bimodal, the background gray levels corresponding to the greater peak and the gray levels belonging to the second, smaller peak corresponding to the cell pixels. In order to determine automatically whether some pixels belong to a cell, it is sufficient to measure their values.

B. Spatial Frequency Content

Fourier analysis, as known from the theory of signal analysis, can also be applied to the study of digitized images. According to it, every image can be decomposed into a series of sinusoidal gratings with certain amplitudes and phases, the sum of which restores the original image. The spatial frequencies contained in an image are given by the reciprocal of the periods of these sinusoidal functions. It is thus possible to build a histogram having on the abscissa the different spatial frequencies and on the ordinate the frequency with which each spatial frequency is present in the image. The importance of this transformation lies in the fact that different spatial frequencies correspond to different regions of an image; the contours are the regions with higher spatial frequencies, the homogeneous portions of an image those with the lowest frequencies. It is thus possible to develop algorithms acting selectively on determined parts of an image.

V. IMAGE PROCESSING

It is usually necessary to process a digitized image in the following cases:

1. Image restoration — An image may be affected by factors such as optical system aberrations and noise.
2. Image enhancement — It is a common wish to enhance some details present in an image.
3. Information extraction — It is very useful to be able to measure automatically and in a quantitative manner some properties of the image. For example, it may be desirable to measure the surface of cells on a microscope slide, to count automatically the number of cells present in a field, to follow moving cells and determine their motion parameters, and to measure the concentration of intracellular probes, etc.

In the following sections we will briefly discuss these topics, giving some examples for each of them. Other chapters in this book deal in detail with specific applications.

A. Digital Image Transformations

Let $l(m,n)$ be the value of a randomly chosen pixel, and H_F the gray level histogram of the original image, $l'(m,n)$ and H_G their respective values after image processing, and T a generic transformation. An image can be processed in three different ways: point transformation, local transformation, and generic transformation. In the case of point transformation, a pixel is modified taking into account only its actual value. A typical example of this is image thresholding: given a value x belonging to the gray-level interval, we have

$$l'(m, n) = \begin{cases} 0 & \text{if} \quad l(m, n) < x \\ L & \text{elsewhere} \end{cases} \tag{1}$$

It should be noted that this transformation law is concerned only with pixel $l(m,n)$, and there is no consideration given to the other pixels belonging to the image. In the case of local transformation, a pixel is transformed taking into account the values of the neighboring pixels. We can schematically represent the pixel and its neighbors, for instance, with a 3 · 3 matrix such as in Figure 3,

$$\begin{pmatrix} A & B & C \\ D & E & F \\ G & H & I \end{pmatrix}$$

FIGURE 3. Labels of a 3 · 3 neighbor of a pixel (see text).

where the pixel to be modified is the central element in E. A law such as $l'(m,n) = (A + B + C + D + E + F + G + H + I)/9$ is considered a local modification. Usually a local transformation is a weighted average of the pixels of a small neighborhood; it is mathematically known as a convolution. When the transformation law is chosen, a matrix, known as a mask, is defined; its elements are the weights of the pixels. Let $K(s,t)$ be such a matrix. It can be shown that the correlation function between neighboring pixels becomes very small beyond the fifth neighbor.[4,7,14,15] It is thus sufficient to choose a 5 · 5 or 3 · 3 masking matrix. In order to save computing time, as small a mask as possible is usually chosen. The convolution between the mask and the image is then carried out; this means that $l(m,n)$ will be replaced by:

$$l'(m, n) = l*K = \sum_i \sum_j l(i, j)\, K(i - m, j - n) \tag{2}$$

In order to visualize this operation, think of the mask as sliding over the image, superimposing each time to a 3 · 3 neighborhood of a point, and then compute the convolution products. It is desirable that the sum of the elements of the masking matrix be equal to one, in order not to introduce artificial variations of the average brightness. This procedure is usually known as image filtering: filters may be used for different purposes, such as selection of determined bands of spatial frequencies, approximation, error dispersion, evaluation, etc. Most of the problems we will consider in what follows deal with appropriate filters.

In the case of global transformation, the new pixel value depends on the values of all the pixels of the image. Such transformations are known as transforms and are widely used in signal analysis. The Fourier transform, which gives the frequency spectrum of an image, is probably the most well-known transform, but there also other transforms that are useful to extract other types of information.[7,14,15] These transforms deal with all images, and usually require heavy calculations; they have, therefore, usually been replaced by digital filters, which are much simpler and faster. The situation might change in the near future since there are commercially available systems which can perform these transforms in real time.

VI. IMAGE RESTORING

The techniques of image restoring are used to improve image quality; they can be divided into two categories:

1. One is noise cleaning. This technique aims at an improvement of the signal-to-noise ratio.
2. The other is image restoring *sensu strictu*. It is used each time an image is systematically distorted and the distortion can be mathematically modelled. This is true, for instance, for the spherical aberrations of a lens or for a motion blurred picture. In these cases it is possible to develop an algorithm that, by distorting the image in the opposite way, restores it to its original aspect.

We will limit ourselves to the technique of noise cleaning. Among the different types of noise that affect an image at the different spatial frequencies, we will consider in particular the high-frequency noise, which causes loss of sharpness and alters the image contours. We will describe some low-pass filters that can reduce or even eliminate this type of noise.

FIGURE 4. Original image (A) and superimposed Gaussian noise (B) (see text).

In the examples shown, the low-pass filters are obtained by averaging the pixel value with the values of the pixels in its 3 · 3 neighborhood. These filters are easily obtained, but they have certain drawbacks; for instance, the image contours become blurred since the high spatial frequencies are indiscriminately cut off; also, there is a broadening of the contours, so the image loses sharpness and appears a bit out of focus.

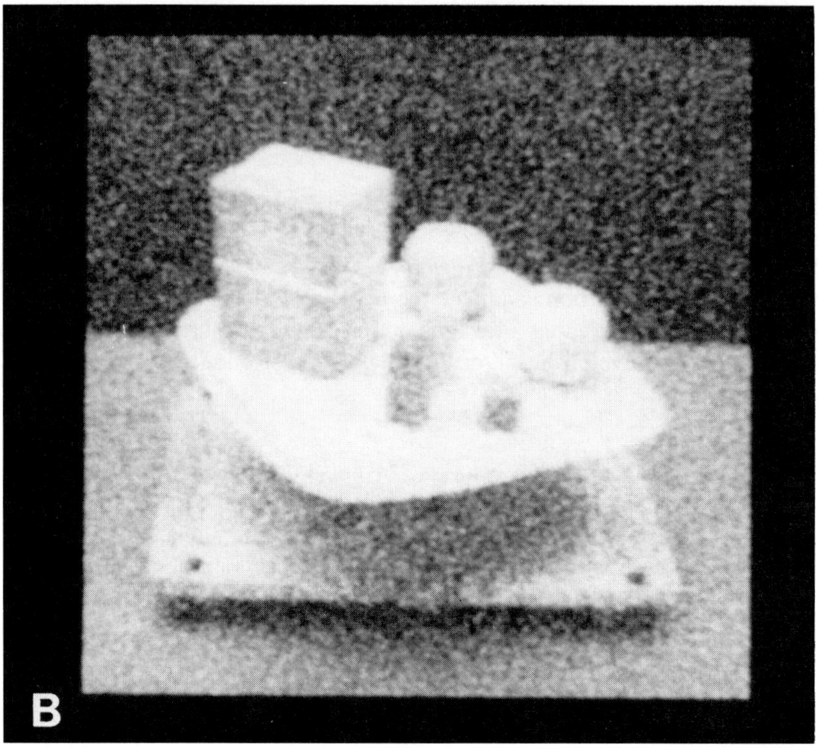

FIGURE 4B.

$$A = \frac{1}{9}\begin{pmatrix}1&1&1\\1&1&1\\1&1&1\end{pmatrix} \quad B = \frac{1}{10}\begin{pmatrix}1&1&1\\1&2&1\\1&1&1\end{pmatrix} \quad C = \frac{1}{16}\begin{pmatrix}1&2&1\\2&4&2\\1&2&1\end{pmatrix}$$

FIGURE 5. Examples of low-pass filters (see text).

Figure 4A shows an image and 4B shows the same image with superimposed Gaussian noise; Figure 5 shows three examples of low-pass filters; and Figure 6A and 6B show the results of the application of filter A on the two images. It is possible to limit the broadening effect by choosing the pixels to be filtered by means of a previous thresholding such as:

$$l''(m, n) = \begin{cases} l'(m, n) & \text{if } [8E - (A + B + C + D + F + G + H + I)] > \varepsilon \\ l(m, n) & \text{elsewhere} \end{cases} \tag{3}$$

A filter that does not bring about a broadening effect is a median filter. The median of a population of values is defined as the value that is greater than 50% of the population values and smaller than 50% of them. A median filter may, for example, be obtained by taking the values of a $3 \cdot 3$ neighborhood of a pixel, sorting them in ascending or descending order and assigning to the original pixel the fifth value of the sorted sequence. Unlike the

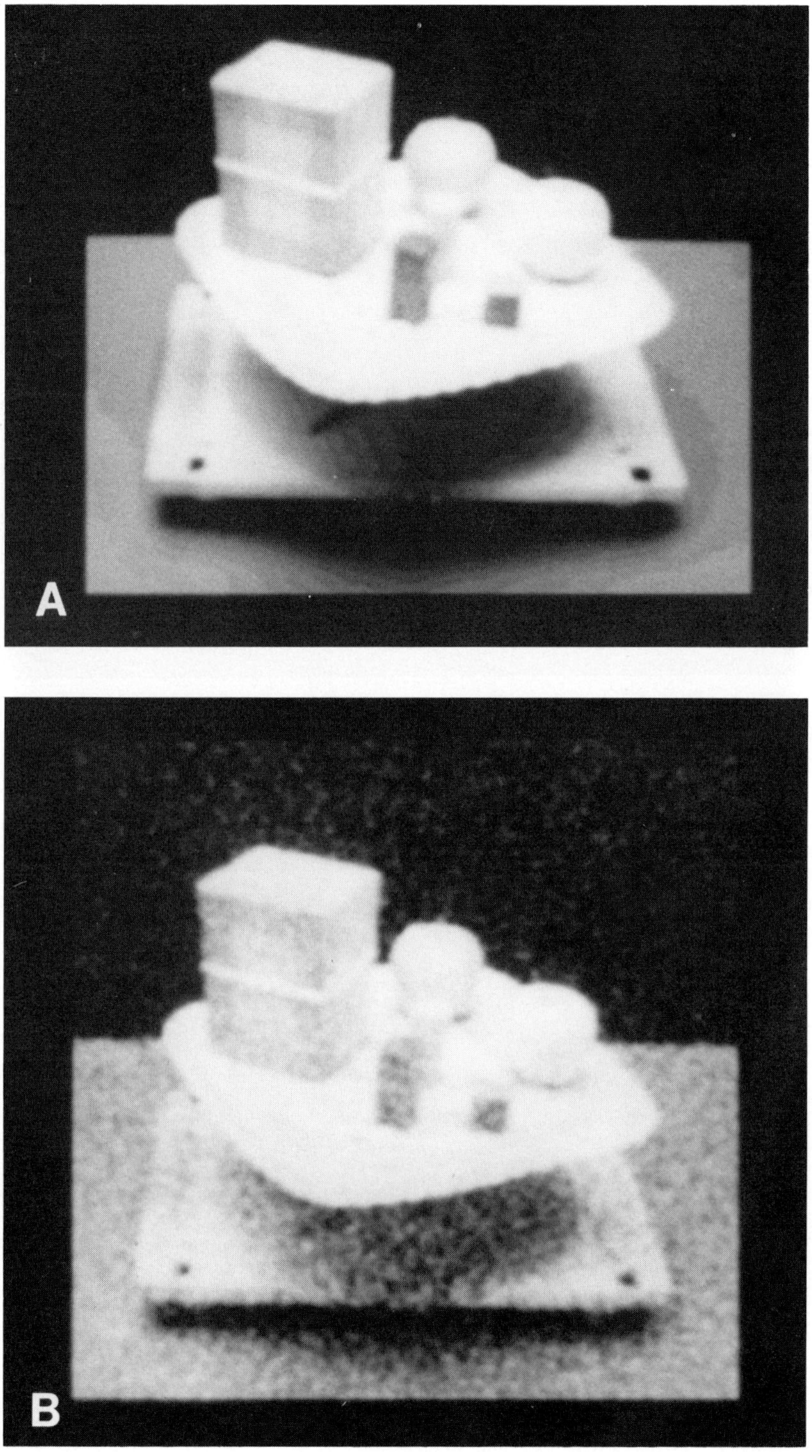

FIGURE 6. Result of convolution of images in Figure 5 with filter A in Figure 4 (see text).

filters based on average values, this filter is relatively insensitive to noise; its main disadvantage is that being a non-linear filter it requires much longer computing times than the filters considered above. For the results, see Figures 7A and 7B.

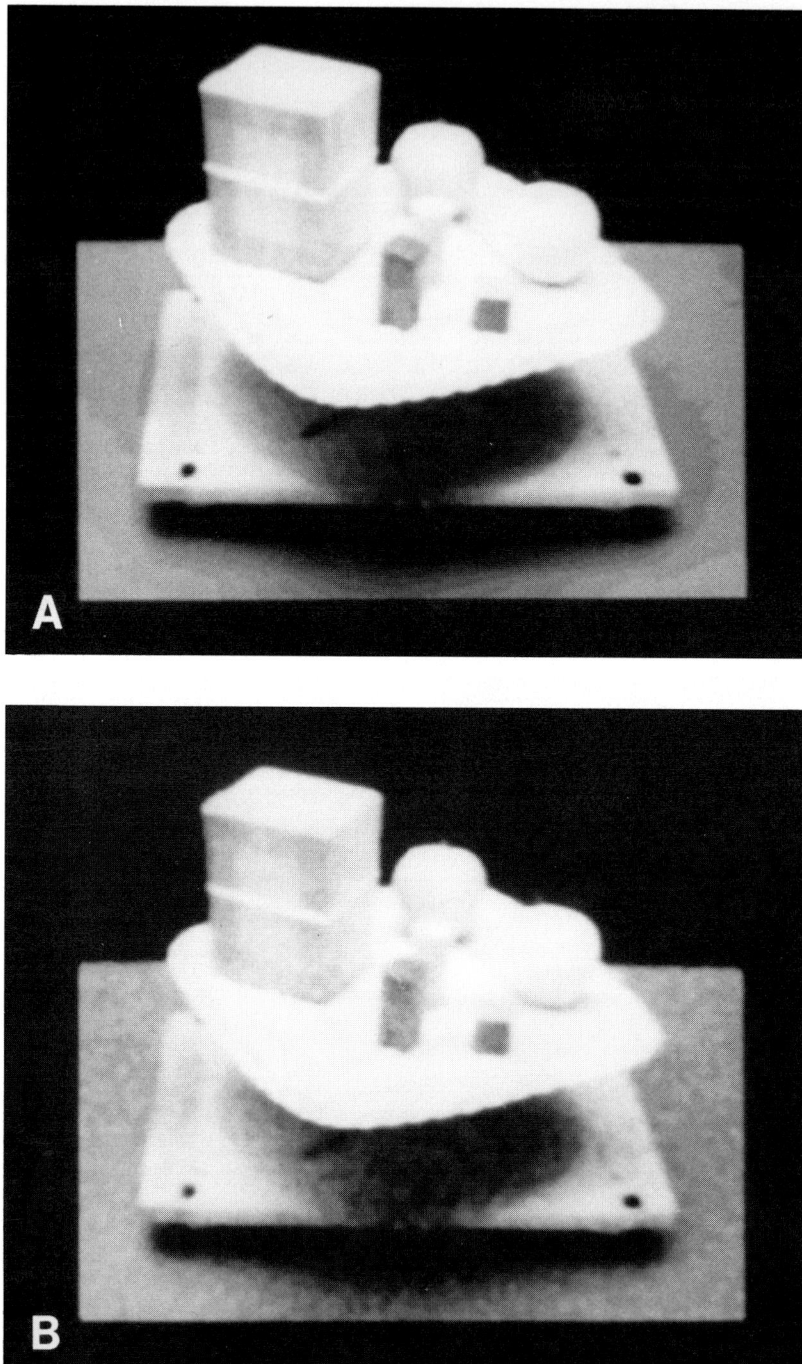

FIGURE 7. Result of the application of the medium filter over the images in Figure 4 (see text).

VII. IMAGE ENHANCEMENT

Even if an image has a good signal-to-noise ratio and is not distorted, it may still need to be elaborated in order to be suitable for the required analysis. In general, for instance, some features may not be clear enough, either because of a low contrast or a wrong gray-scale choice. We will consider some aspects of this problem.

A. Edge Crispening

In order to selectively enhance contours, it is necessary to operate in an opposite way to that described in the previous section; in other words, high-pass filters are required.[5]

$$A = \begin{pmatrix} 0 & -1 & 0 \\ -1 & 5 & -1 \\ 0 & -1 & 0 \end{pmatrix} \quad B = \begin{pmatrix} -1 & -1 & -1 \\ -1 & 9 & -1 \\ -1 & -1 & -1 \end{pmatrix} \quad C = \begin{pmatrix} 1 & -2 & 1 \\ -2 & 5 & -2 \\ 1 & -2 & 1 \end{pmatrix}$$

FIGURE 8. Examples of high-pass filters (see text).

In Figure 8 some high-pass masks are given; as can be seen from the numerical coefficients used, they increase the brightness difference between a pixel and its neighbors. Figures 9A and 9B show the results of a high-pass filter.

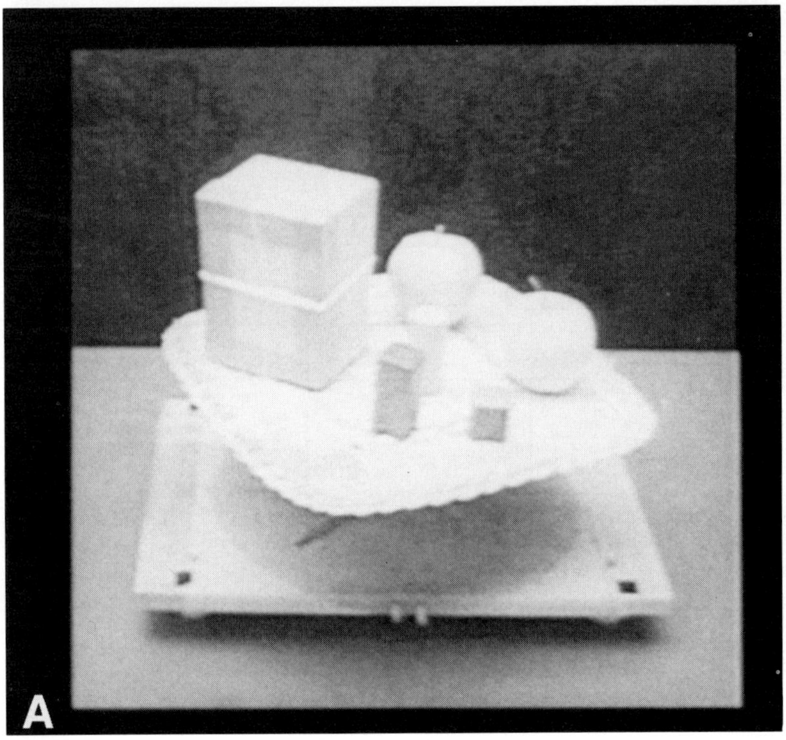

FIGURE 9. Result of the application of filter A in Figure 8 on the images in Figure 4 (see text).

B. Contrast Optimization

Whenever an image has a low contrast, it is very difficult to recognize the finest details. The techniques that deal with this kind of problem are based on suitable manipulations of the gray-level histogram.

Contrast expansion is obtained by means of a linear expansion of a certain interval (between a max and a min determined by the user) of the gray-level histogram of the original image.

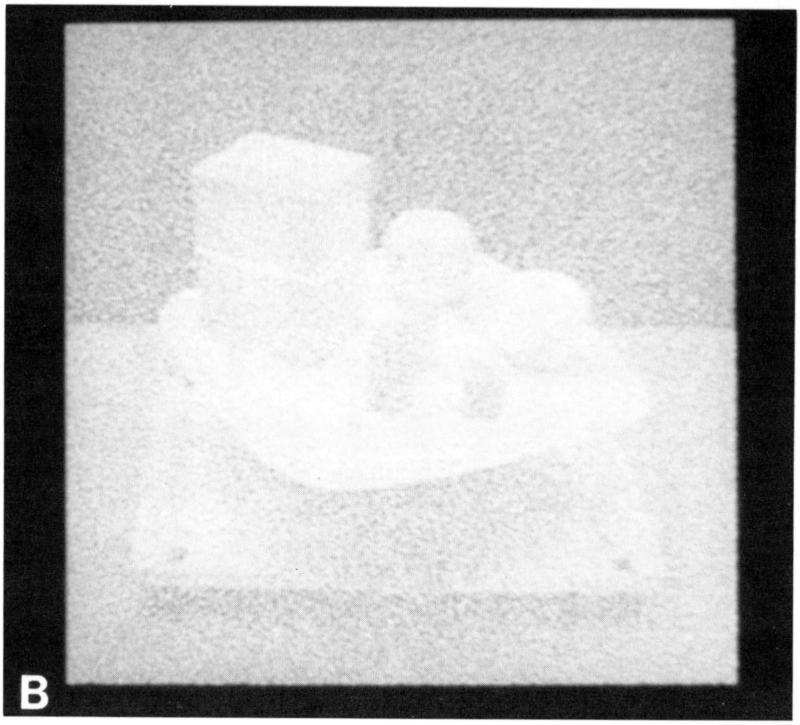

FIGURE 9B.

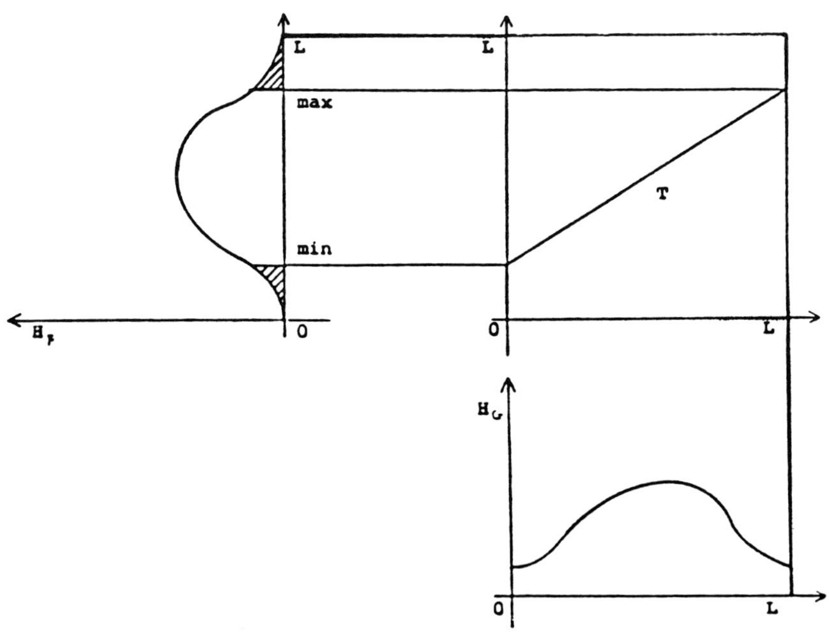

FIGURE 10. A representation of the transformation induced by contrast-expansion techniques on the gray-level histogram of an image (see text).

This is shown in Figure 10. It can be seen that this causes a loss of information in the regions of higher or lower brightness. This transformation is mathematically represented by

$$l'(m, n) = \frac{l(m, n) - min}{max - min} \cdot \mathrm{L} \tag{4}$$

Figure 11 shows the result of contrast expansion on the sample image.

FIGURE 11. Contrast expansion on the non-blurred image of Figure 4 (see text).

Another type of histogram manipulation is its equalization, so that it becomes rectangular. This transformation is more difficult to implement, but is much more useful because it maximizes image information.[7,15] Figure 12 shows the type of transformation that is brought about by the histogram equalization. Mathematically, the transformation is:

$$l'(m, n) = \frac{\mathrm{L}^2}{\mathrm{N}} \sum_{\mathrm{k}=0}^{\mathrm{a}} H_{\mathrm{F}}(k), \qquad \text{where} \quad \mathrm{a} = l(m, n) \tag{5}$$

Histogram equalization requires time-consuming calculations, but it is worthwhile. Figure 13 has been obtained by equalizing the histogram of the sample image.

VIII. FEATURE EXTRACTION

For a long time, image analysis dealt mainly with the problem of the development of edge detection algorithms. Nowadays, there is a certain number of good algorithms for this purpose, even if there is no definite conclusion on the type of algorithm or on the method to use.[2,3,10,13] The question of feature extraction has, however, gone far beyond this original purpose. The development of pattern recognition techniques, associative memories, and neural networks has made it possible to implement, even if at a still experimental level,

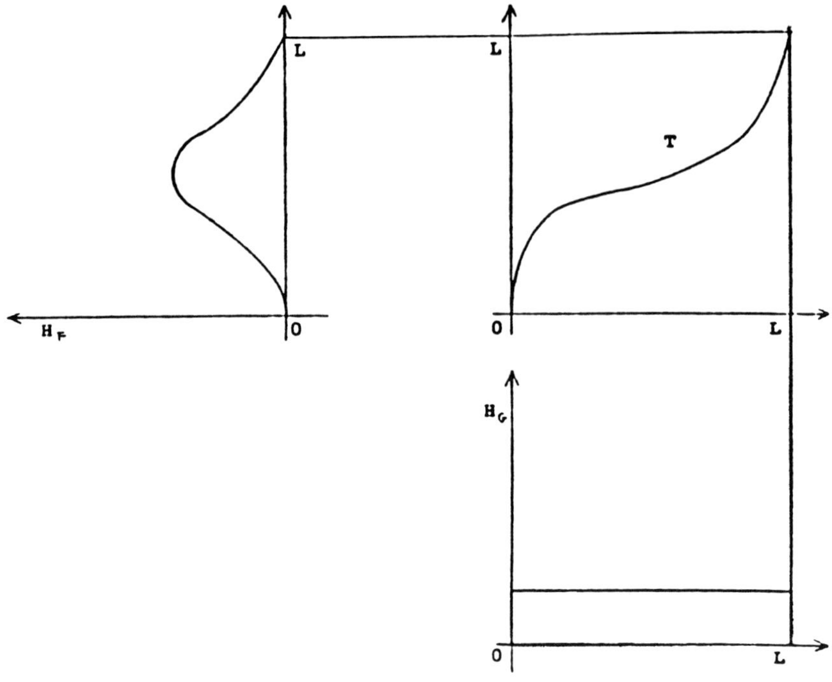

FIGURE 12. A representation of transformation induced by histogram equalization techniques on the gray-level histogram of an image (see text).

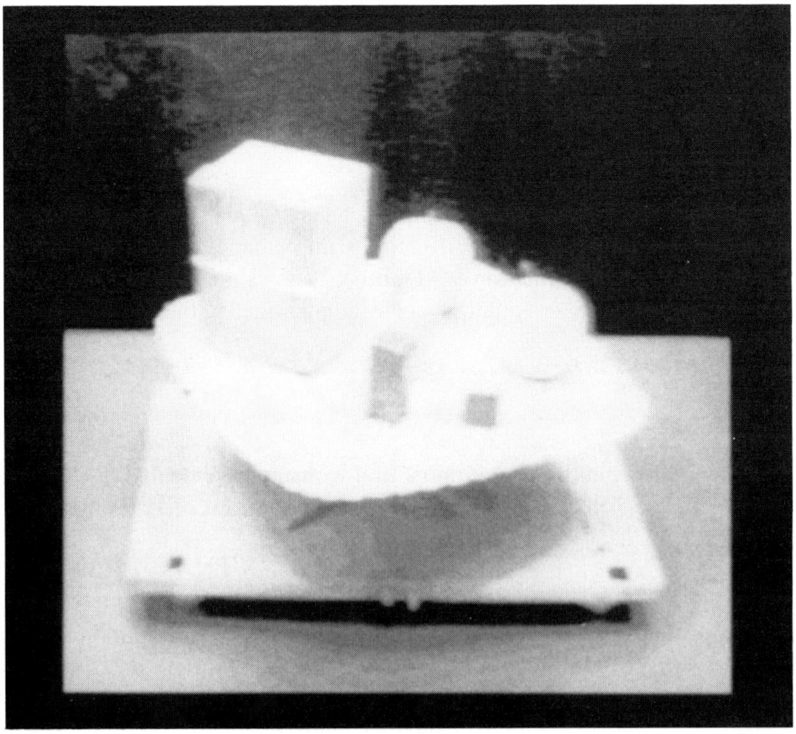

FIGURE 13. Result of histogram-equalization techniques on nonblurred image of Figure 4 (see text).

expert systems capable of recognizing several complex patterns.[6] The development of relational semantics makes it possible to name an identified object, thus allowing computer recognition.[6] We will restrict ourselves to the techniques of edge detection, referring the interested reader to the appropriate literature.[1,7]

The edge, defined as the spatial region where the object ends, is characterized by the region of strongest brightness variation. The function $f(x,y)$, the brightness of a real image, has in fact a discontinuity in the edge region. It has no meaning to look for a discontinuity of the discrete function $l(m,n)$; one should rather look for a very high incremental ratio. The first researchers in the field moved along this line of thought, using linear filters such as derivatives or laplacian.[3,11,12]

$$A = \begin{pmatrix} 0 & -1 & 0 \\ -1 & 4 & -1 \\ 0 & -1 & 0 \end{pmatrix} \quad B = \begin{pmatrix} -1 & -1 & -1 \\ -1 & 8 & -1 \\ -1 & -1 & -1 \end{pmatrix} \quad C = \begin{pmatrix} 1 & -2 & 1 \\ -2 & 4 & -2 \\ 1 & -2 & 1 \end{pmatrix}$$

FIGURE 14. Examples of edge-detecting Laplacian-like filters (see text).

The three masks presented in Figure 14 are based on the methods of finite difference enhancement or laplacian calculation.These masks are simple, but, unfortunately, are very sensitive to noise since they enhance the high spatial frequencies. The details of an image are strongly affected by such filters. One should compare Figure 15A with Figure 15B to see the result of the applications of such filters when a strong noise is present. Because of this, it is recommended to smooth the image before the utilization of these edge-detecting filters.

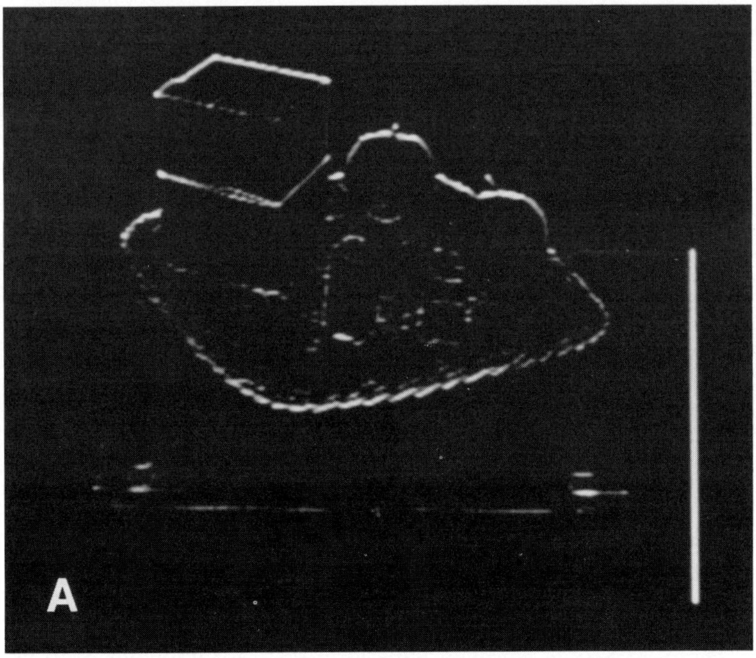

FIGURE 15. Result of the application of filter A in Figure 14 on the two images in Figure 4 (see text).

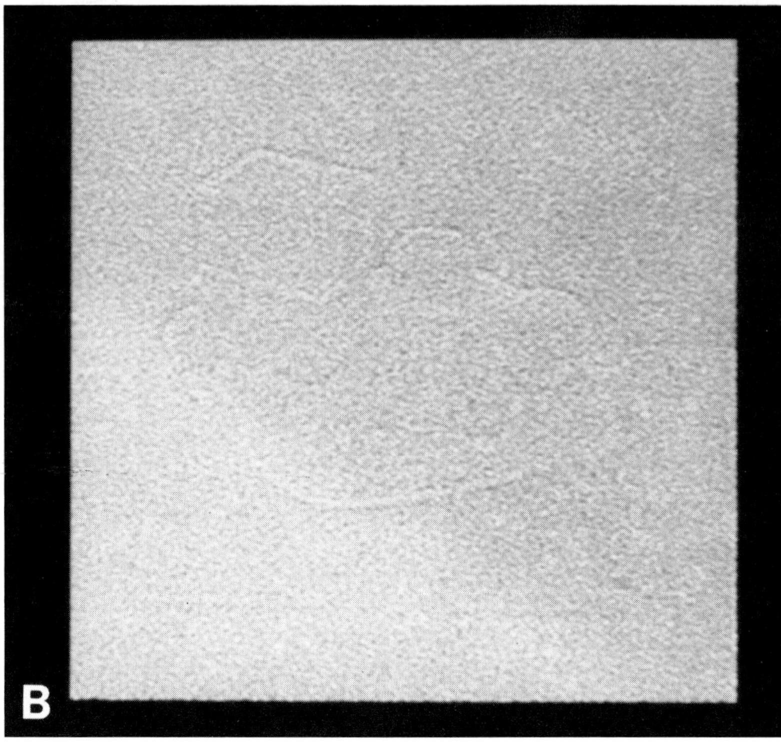

FIGURE 15B.

Nonlinear filters, which are much more robust to noise, have been developed in order to eliminate these drawbacks; they are known as the gradient and the Sobel filters. They are local filters whose mathematical expression for the gradient filter is given by

$$l'(m, n) = |(A + B + C) - (G + H + I)| + |(A + D + G) - (C + F + H)| \tag{6}$$

$$l'(m, n) = \sqrt{[(C + 2F + I) - (A + 2D + G)]^2 + [A + 2B + C) - (G + 2H + I)]^2} \tag{7}$$

for the Sobel one; the terms A, B, etc. refer to the 3 · 3 neighborhood defined in Section V. These nonlinear filters are quite time consuming, but also quite efficient. One should compare Figure 16, obtained applying these filters to sample images with Gaussian noise added, with Figure 15, where Gaussian masks have been employed.

ACKNOWLEDGMENTS

We are grateful to Scuola Normale Superiore, Pisa, for allowing us to use computing and image analysis facilities.

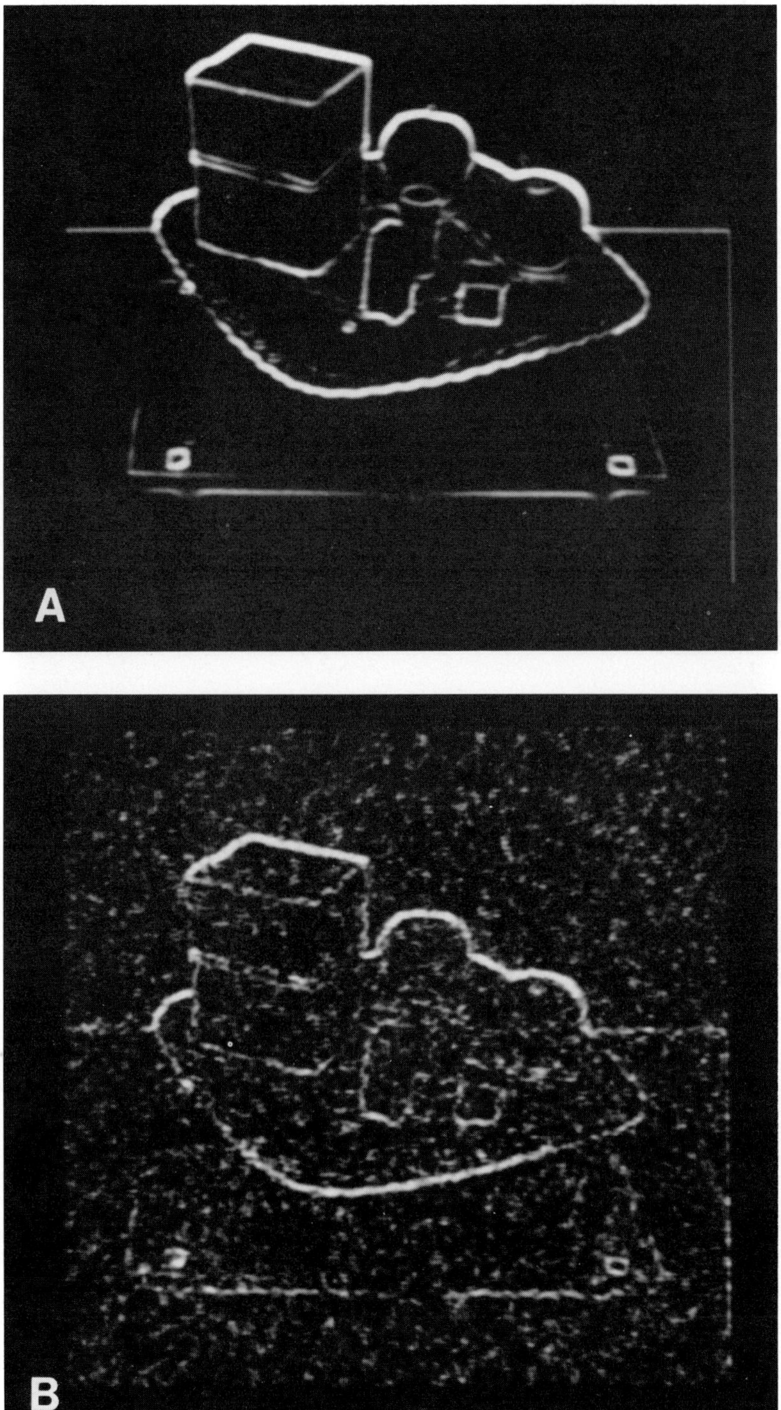

FIGURE 16. Result of the application of gradient (A, B) and Sobel (C, D) filters on the images in Figure 4 (see text).

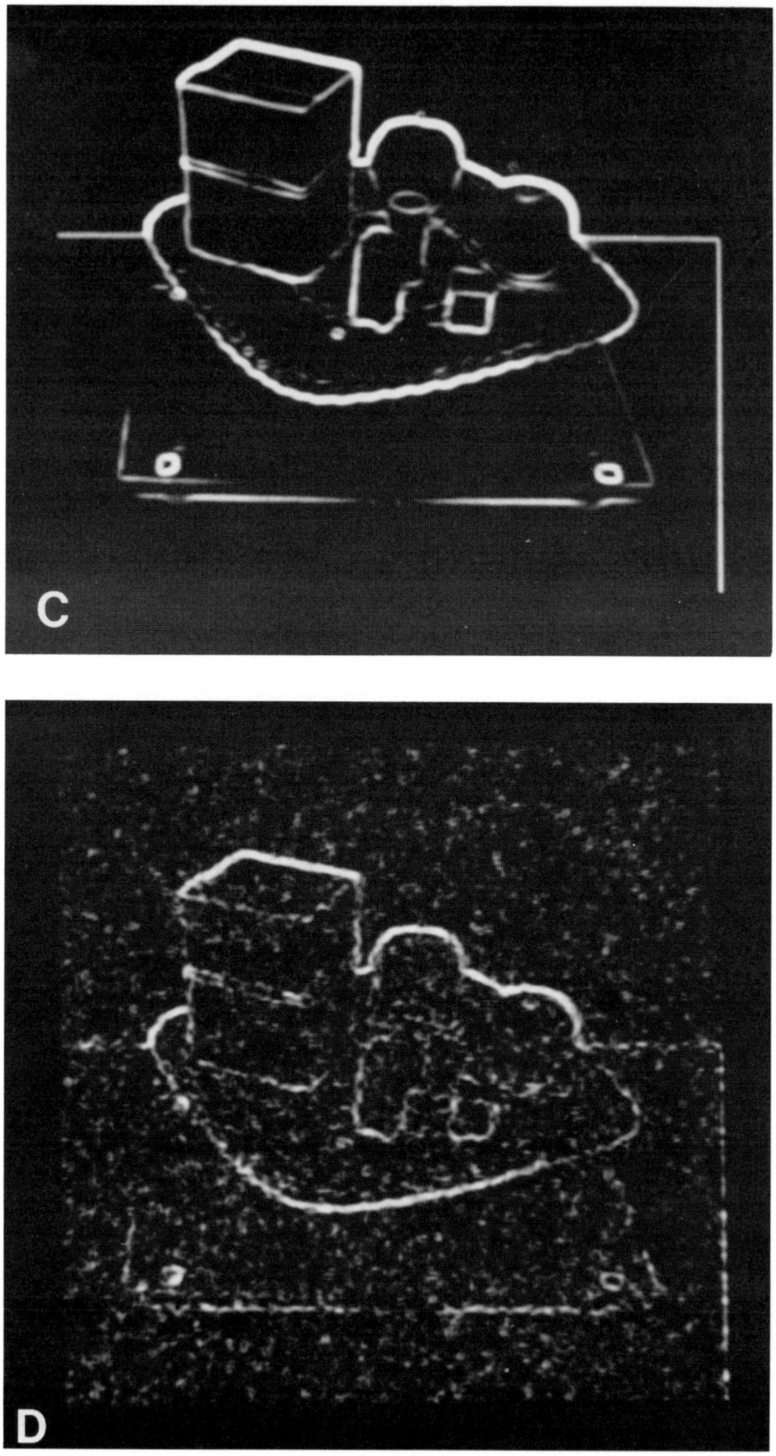

FIGURE 16C and D.

REFERENCES

1. **Ballard, D. H. and Brown, C. M.,** *Computer Vision,* Prentice-Hall, Englewood Cliffs, NJ, 1984.
2. **Burr, D. C. and Morrone, M. C.,** Feature detection in biological and artificial visual systems, in *Vision Coding and Efficiency,* Blakemore, C., Ed., Cambridge University Press, London, 1990, 56.
3. **Canny, J. F.,** *Finding Edges and Lines in Images,* MIT Press, Cambridge, 1983.
4. **Castleman, K. R.,** *Digital Image Processing,* Prentice-Hall, Englewood Cliffs, NJ., 1979.
5. **Davis, L. S.,** A survey of edge detection techniques, *Comput. Graph. Image Process.,* 4, 248, 1975.
6. **Gonzales, R. C. and Thomason, M. G.,** *Syntactic Pattern Recognition,* Addison-Wesley, Reading, MA, 1978.
7. **Gonzales, R. C. and Wintz, P.,** *Digital Image Processing,* Addison-Wesley, Reading, MA, 1987.
8. **Gualtieri, P.,** Microscopia digitale, *Le Scienze,* 261, 52, 1990.
9. **Gualtieri, P., Barsanti, L., and Coltelli, P.,** Computer processing of optical microscope image, *Micron Microsc. Acta,* 16, 154, 1985.
10. **Hildreth, E. C.,** *Edge Detection,* A.I. Memo, MIT Press, Cambridge, 1985, 858.
11. **Marr, D.,** Vision, W. H. Freeman, San Francisco, 1979.
12. **Marr, D. and Hildreth, E. C.,** Theory of edge detection, *Proc. R. Soc. London, Ser B.,* 207, 169, 1980.
13. **Morrone, M. C. and Owen, R. A.,** Feature detection from local energy, *Pattern Recognition Lett.,* 6, 303, 1987.
14. **Oppenheim, A. V. and Shafer, R. W.,** *Digital Signal Processing,* Prentice-Hall, Englewood Cliffs, NJ, 1975.
15. **Pratt, W. K.,** *Digital Image Processing,* John Wiley & Sons, New York, 1977.
16. **Sandini, G. and Tagliasco, V.,** An antropomorphic retina-like structure for scene analysis, *Comput. Graph. Image Process.,* 14, 365, 1980.
17. **Sandini, G.,** Il sistema visivo, in *Metodi di Analisi dei Sistemi Neurosensoriali,* Schmidt, R., Ed., Patron, Bologna, 1986, 34.

Chapter 5

SINGLE PHOTON IMAGING

Tsuyoshi Hayakawa

TABLE OF CONTENTS

I. INTRODUCTION

Low light level imaging has been used in life sciences under various conditions including microscopes. The imaging method has recently becomes more effective in the analysis of biological objects, as image processing and laser technologies were widely and rapidly developed.

In the past, silicon intensified target (SIT) cameras or image intensifiers (II) were mainly used for imaging at low light levels. Although there are several practical problems in sensitivity, the developments of dyes and labeling techniques with these image pickup devices enable us to visualize the behavior of biomolecules in solution. For example, the distribution of calcium ions in a single cell was detected by the combination of the newly developed dye (Fura-2) and the ratio imaging;[1] and the behavior of actin filaments or DNA fragments in solution was also visualized and analyzed.[2,3] Thus, the research using fluorescence microscopes requires cameras with higher sensitivity to analyze the biological phenomena.

On the other hand, the photon counting method using photomultiplier tubes has been widely utilized in detecting small amounts of luminescent biomolecules. However, information on the spatial distribution of those molecules is lost with this method, while two-dimensional (2-D) detection of biomolecules is required.

So far, radioisotopes have been extensively used as a high sensitive method (e.g., autoradiography or electrophoresis). Alternative and more convenient methods to avoid them would probably be in high demand. This requirement will be more effectively achieved by 2-D photon counting.

An ultrahigh sensitivity TV camera,[4] whose main capability is single photon imaging, has been developed through the above mentioned stages. In principle, the distribution of an extremely small amount of biomolecules can be measured by single photon imaging, which might be called "molecular counting imaging". Single photon imaging will be useful in solving practical problems such as photobleaching, damage by illumination, and biological interference of reagents.

II. PRINCIPLES OF SINGLE PHOTON IMAGING

A. CONCEPT OF PHOTON COUNTING IMAGING

As light intensity [photons/mm^{2}s] decreases, light appears more and more localized in discrete areas on the photodetector. The degree of this problem depends on the temporal-spatial characteristics of the detector. Photon counting imaging has three main advantages: (1) ultrahigh sensitive detection, (2) excellent quantification, and (3) temporal-spatial information of a single photon can be obtained.

Figure 1 shows that there are two models of photodetection depending on the relationship between light intensity and temporal-spatial characteristics of a detector: in Figure 1a the superimposed signal (amplitude) of many photons is detected in order to measure the light intensity, while in Figure 1b each output signal corresponds to an individual photon which is counted. Therefore, these two photodetection modes might be called "analog mode" and "counting mode", respectively. From the point of view mentioned above, photodetection can generally be achieved by counting photons if a photodetector has sufficient temporal-spatial characteristics.

The counting mode is not possible in conventional low light level image pickup devices. In other words, no signal can be detected at ultralow light levels, even when accumulated for a long time.

B. SINGLE PHOTON IMAGING DEVICES

Human eyes have a considerably high sensitivity after dark adaptation. It is estimated

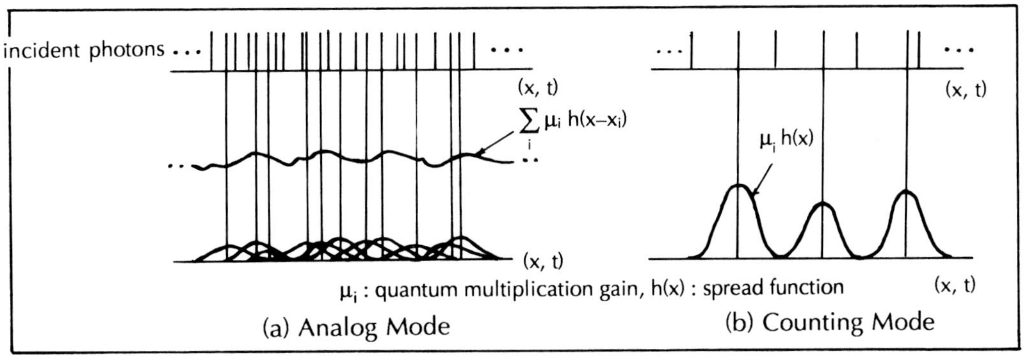

FIGURE 1. Analog mode and counting mode.

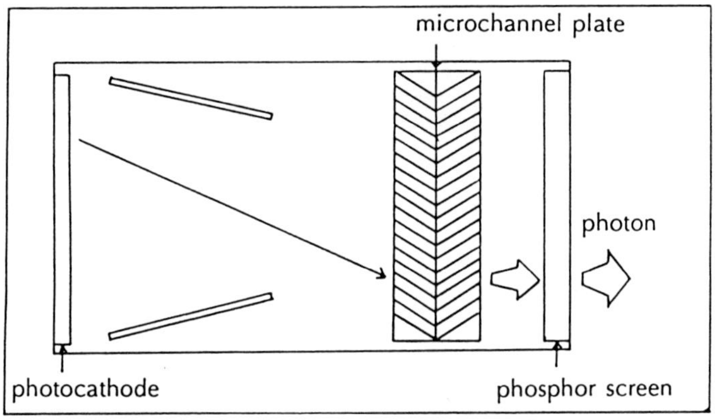

FIGURE 2. Operation of a 2-D photon counting tube. A single photon generates on an order of 10^8 photons on the phosphor screen.

that a few tens of photons can be detected by the naked eye. This could be called ultrahigh sensitivity, but quantification of light is not possible.

Single photon imaging is achieved by a specially designed intensifier tube[5] (2-D photon counting tube) with a two-stage microchannel plate (MCP)[6] (Figure 2). An MCP consists of fiber bundles. The diameter of a single channel is about 10 μm. There is an electron multiplication function in each channel (Figure 3). When an optical image focuses on the photocathode of the photon counting tube, photons are converted to photoelectrons. This photoelectron image is then focused on the MCP surface. In the two-stage MCP, the photoelectron image is amplified by a factor on the order of about 10^6, maintaining the positional information. Next the electrons strike the phosphor screen. As a result of these processes, even a single photon can generate about 10^8 photons as the output of the 2-D photon counting tube. This high density spot of a single photon can be easily detected by a successive image pickup device. Thus, single photon imaging is achieved.

Figure 4 shows several types of spectral sensitivity of the 2-D photon counting tube. Instead of using photoelectric spectral sensitivity S [A/W], it is more convenient to use quantum efficiency η [electron/photon] in the counting mode.

The relationship of S and η is

$$\eta = \frac{hc}{\lambda} \cdot \frac{S}{e} \tag{1}$$

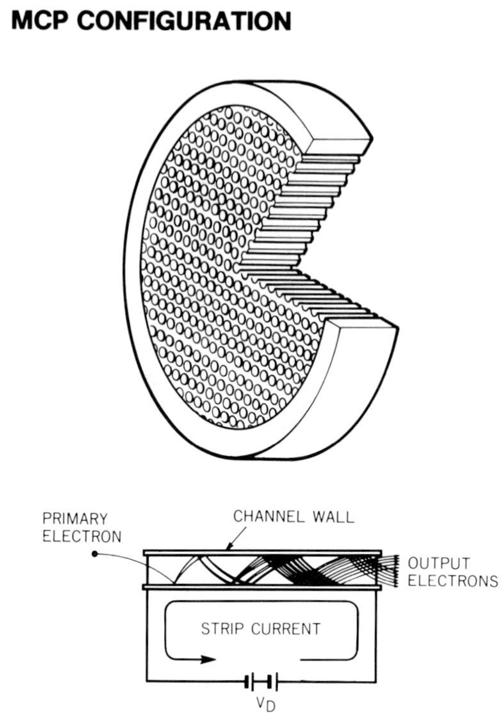

FIGURE 3. Structure of the microchannel plate and its operation.

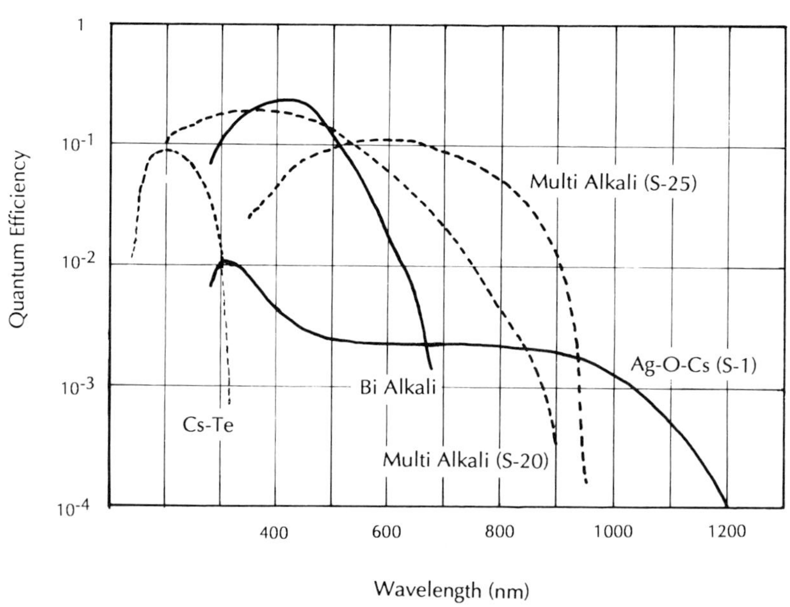

FIGURE 4. Quantum efficiency vs. wavelength (2-D photon counting tube).

$$\eta = \frac{1.24}{\lambda} \cdot S \tag{1'}$$

where h is the Planck constant 6.626×10^{-34} [s], c the velocity of light in vacuum 3×10^8 [m/s], λ the wavelength [m] in (1), [μm] in (1′), and e the electron charge 1.6×10^{-19} [coulomb].

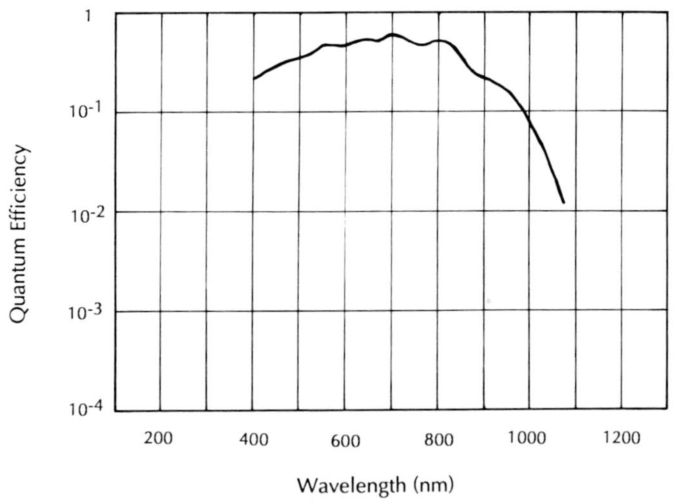

FIGURE 5. Quantum efficiency vs. wavelength (CCD).

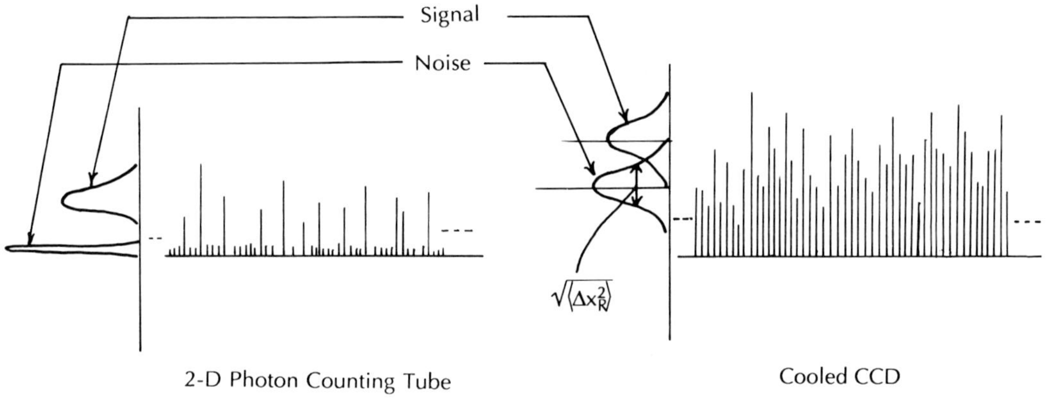

FIGURE 6. Schematic representation of photo-signal readout. The output signal of a single photon in the 2-D photon counting tube is clearly distinct. On the other hand, in a cooled CCD the possibility of photon counting imaging is strongly dependent on the readout noise $\sqrt{\langle \Delta x_R{}^2 \rangle}$.

A cooled charge coupled device (CCD) camera[7] has been developed recently for the purpose of ultra-low light level imaging. The photocathode of this device is basically the same as that of a silicon photodiode (Figure 5). In the operation of a cooled CCD camera, a signal of accumulated photoelectrons in each pixel is serially and slowly read out to form the image (ordinary readout of a frame is 1 s to 2 h). Up to now, as the readout noise is estimated to be greater than several electrons rms (root mean square value) per pixel, photon counting imaging is only partially possible.

Figure 6 shows the difference between a 2-D photon counting tube and a cooled CCD camera concerning the state of the photosignal readout.

III. PHOTON COUNTING CAMERA SYSTEM[4]

A. SYSTEM CONSTRUCTION AND CHARACTERISTICS

Figure 7 shows the schematic diagram of the photon counting camera system. The amplified image of photons on the phosphor screen of the 2-D photon counting tube is read out by a low-lag vidicon or a CCD camera and stored within an image processor. The quality

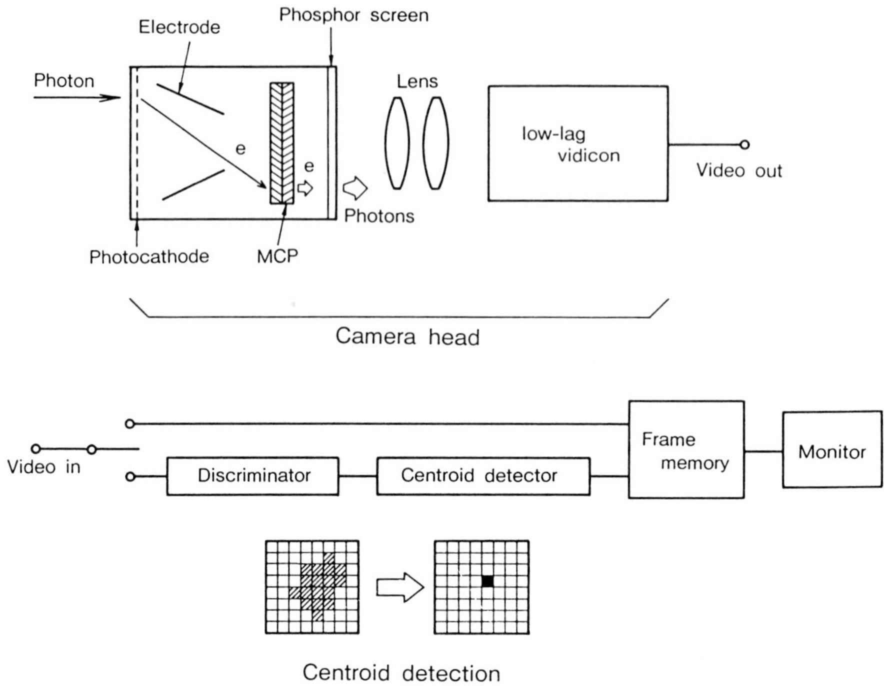

FIGURE 7. Schematic diagram of the photon counting camera.

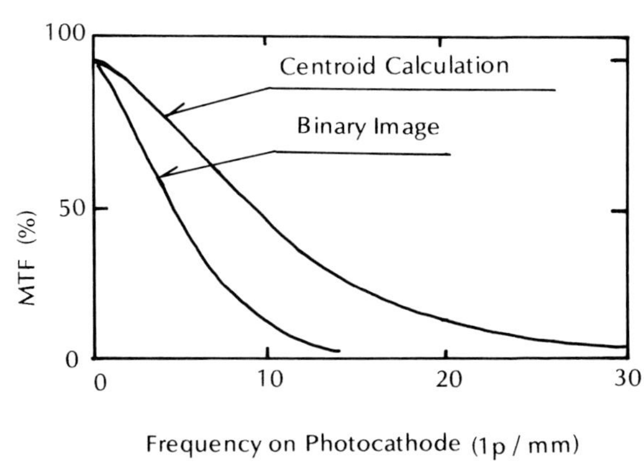

FIGURE 8. Modulation transfer function (MTF) of the photon counting camera.

of images thus obtained depends on the spatial resolution as well as on the accuracy of the number of detected photons. As the spot of a single photon on the phosphor screen occupies about several to ten pixels in the frame memory, the centroid calculation of this spot improves both the spatial resolution (Figure 8) and the accuracy of the number of detected photons. In this system, the "analog mode" is also available by operating the MCP at a relatively low gain when the light intensity is sufficient.

The image quality is also determined by the signal-to-noise ratio. The signal-to-noise ratio S/N in the counting mode is given by

$$\frac{S}{N} = \sqrt{\frac{L\eta AT}{1 + (2x_d + x_b)/L\eta}} \tag{2}$$

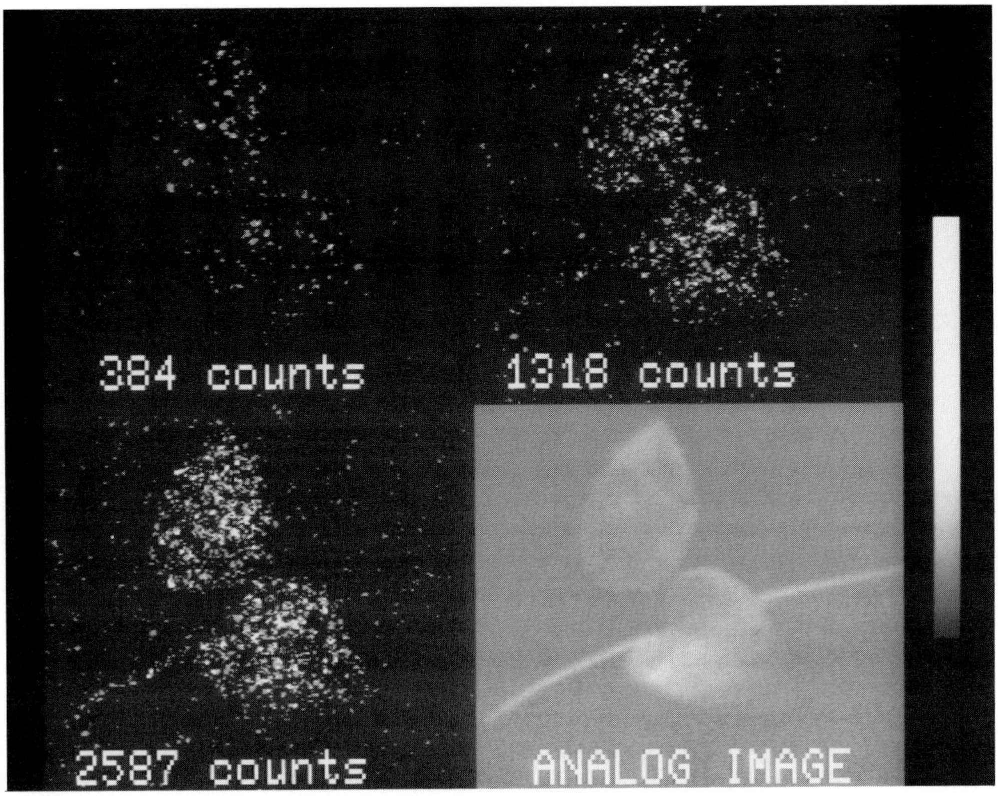

FIGURE 9. Examples of image quality vs. counts number.

where L is the light intensity [photons/mm^2 s], η the quantum efficiency of photocathode [electron/photon], A the signal detection area [mm^2], T the signal integration time [s], x_d the dark noise [counts/mm^2 s], and x_b the background noise [counts/mm^2 s].

The dark noise of the photon counting camera is less than 20 [counts/cm^2 s] (with a bialkali photocathode at 25°C, or a multialkali photocathode at −20°C). An example of the dependence of the photon count number on the image quality is given in Figure 9, showing delayed fluorescence images of leaves. The lower right image was obtained in ''analog mode'' by weak illumination.

A cooled CCD camera system has been developed recently. As described above, the photon counting mode is not available in the strict sense of the word. However, if the readout noise $\sqrt{\langle \Delta x_R{}^2 \rangle}$ in Figure 6 is less than a few electrons, the counting mode becomes available. The pixel size of a cooled CCD camera is larger than 12 μm × 12 μm, and currently the number of picture elements is 1000 × 1000. The signal-to-noise ratio S/N in one pixel is given by

$$\frac{S}{N} = \sqrt{\frac{L\eta_{CCD}T}{1 + \frac{1}{L\eta_{CCD}}\left(\frac{2\langle \Delta x_R{}^2 \rangle}{T} + 2x_d + x_b\right)}} \tag{3}$$

where L is the light intensity [photons/pixel s], η_{CCD} the quantum efficiency of the photocathode [electron/photon], T the integration time [s], $\sqrt{\langle \Delta x_R{}^2 \rangle}$ the readout noise [electrons, rms], x_d the dark noise [counts/s], and x_b the background noise [counts/s].

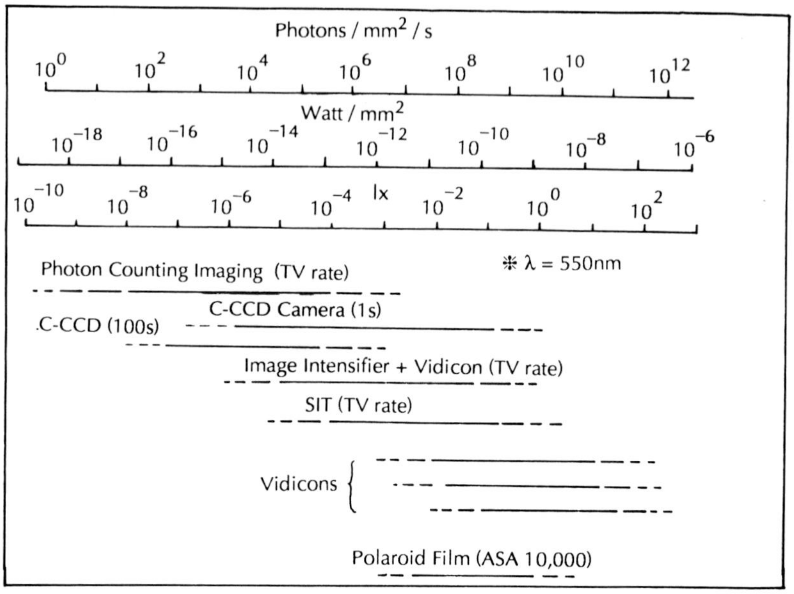

FIGURE 10. Method of image pickup vs. incident light intensity.

A typical dark count per pixel is about 0.1 [counts/s] at −30°C. The readout noise (at several seconds per frame) is at least several electrons rms at present.

A suitable imaging method can be selected corresponding to the light intensity as shown in Figure 10. The photon counting camera can detect a single photon image in real time (TV rate). A cooled CCD camera will be available in the extremely low light range (on the order of 10^2 [photons/mm^2 s]) when the frame readout time is longer than several seconds.

B. APPLICATIONS OF PHOTON COUNTING CAMERAS

When single photon imaging is used under extremely low light levels, very often the dominant noise is the background noise x_b. For example, stray light, leak of excitation light, weak fluorescence emission from optical parts, and weak luminescence caused by impurities of the samples should be reduced in order to quantify the information of light. The fundamental capability of single photon imaging in biology is to measure the distribution and function of extremely small amounts of biomolecules. Its application has just been started in the field of luminescence. There are several remarkable applications such as two-dimensional detection of the biophotons,[8,9] real time visualization of oxyradical burst activities,[10] the monitoring of gene expression[11,12] the distribution of ATP or glucose in tissues,[13] and continuous observation of the calcium ion wave moving through a fertilized egg.[14]

Figure 11a shows the result of the visualization of an oxyradical burst from a single human neutrophil stimulated by opsonized zymosan in luminol-dependent chemiluminescence; Figure 11b is a three-dimensional display of the two-dimensional distribution of the neutrophil luminescence.

The two-dimensional distribution of a permeant moving through a membrane was measured in real time using chemiluminescence.[15] Figure 12 shows the experimental system of the permeant measurement and Figure 13 shows one of the results.

An example of a simulation experiment by fluorescence on molecular number counting imaging is shown in Figure 14. Currently the limitation of molecular detection is determined not by the characteristics of the photon counting camera but by the stray light (background noise). By reducing this stray light, we can detect a single molecule with its positional information.

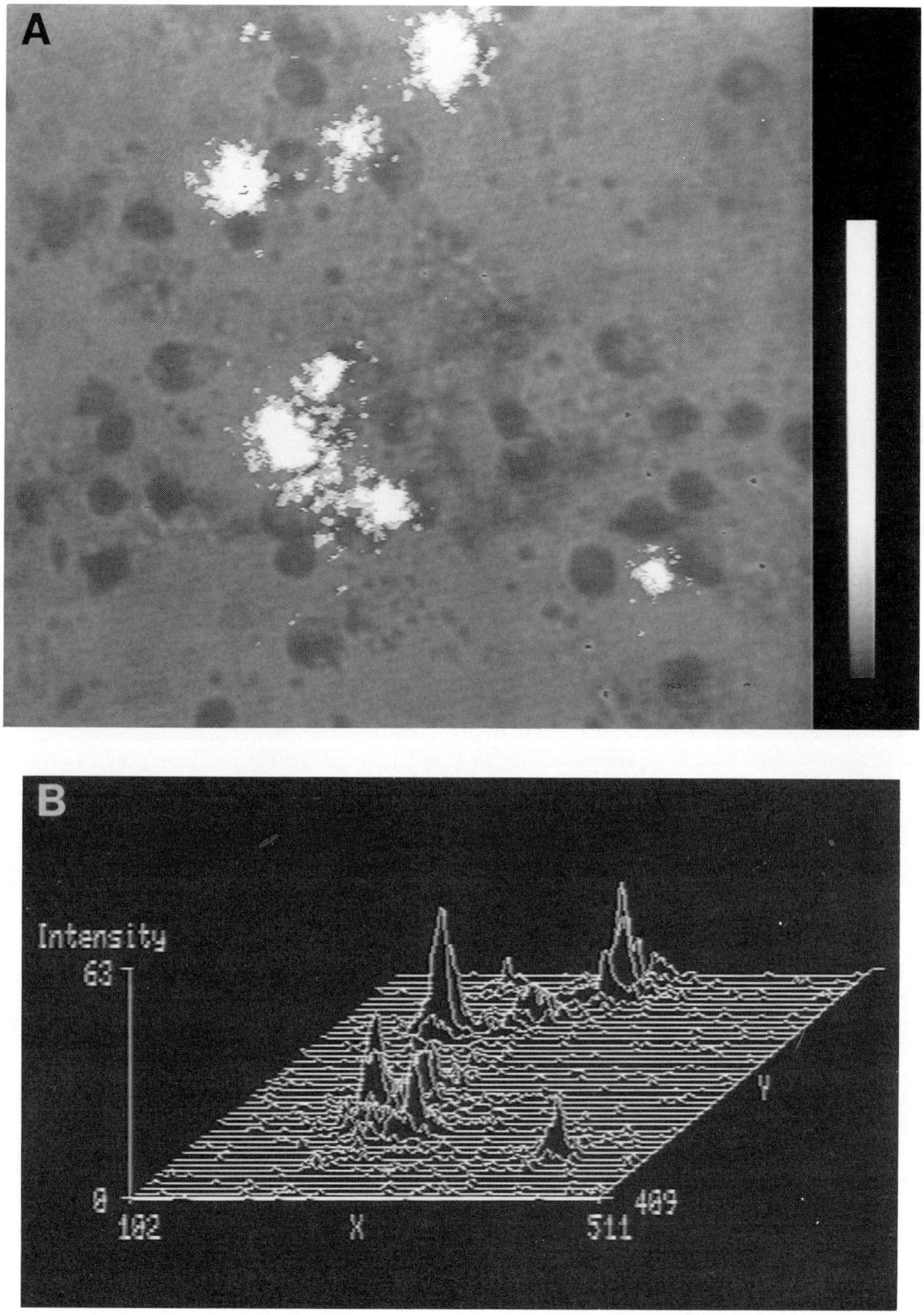

FIGURE 11. Image of neutrophil reaction (oxyradical burst) in luminol-dependent chemiluminescence. (From Suematsu, M., *Biochem. Biophys.* Res. Commun., 155, 106, 1988. With permission.)

The single photon imaging method has just been developed and started to be used in video intensified microscopy or other macroscopic imaging. It will be used more frequently in biology for the purpose of understanding biological phenomena *in vivo*. The transition from a ''single photon'' to a ''single molecule'' or from a histological image to a ''function image'' will be the most important application.

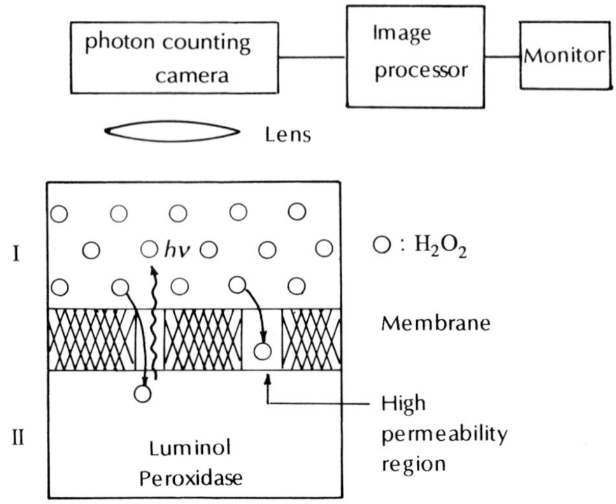

FIGURE 12. Experimental apparatus for imaging of permeants through membrane.

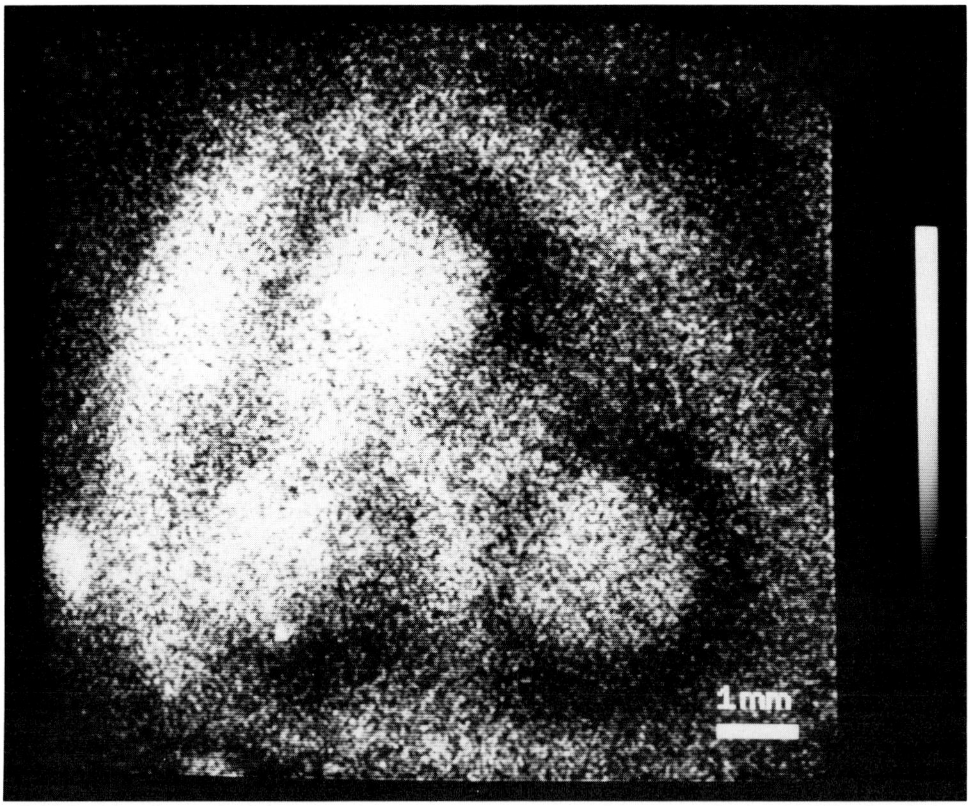

FIGURE 13. Two-dimensional distribution of permeant (hydrogen peroxide) through a polymethyl methacrylate film nonuniform in thickness. (From Hiramatsu, M., A new method to characterize film inhomogeneities: permeation of matter followed by a combination of chemiluminescence with photodetectors, *J. Polym. Sci.*, 28, 133. Copyright 1990, John Wiley & Sons. With permission.)

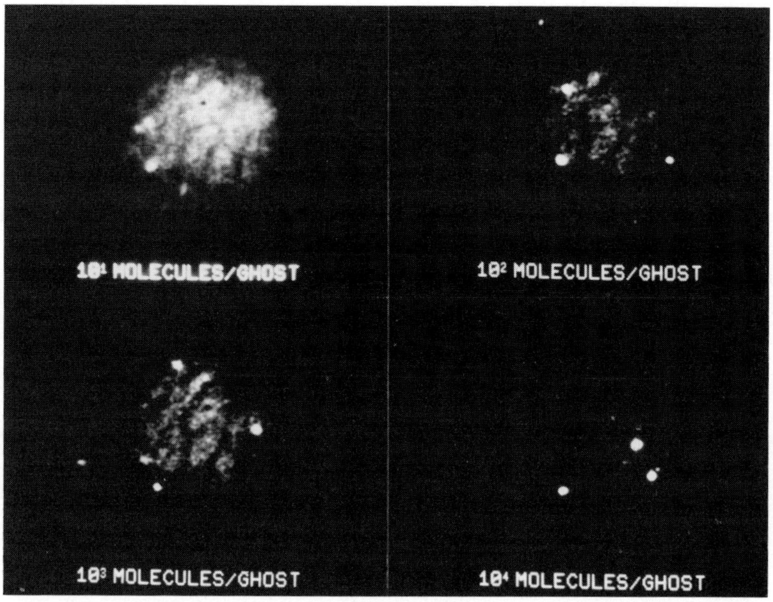

FIGURE 14. Experimental results of simulation of molecular counting imaging by the photon counting camera (FITC-Dextran in ghost of erythrocyte, objective 40 ×, excitation 100-W Hg lamp.

REFERENCES

1. **Williams, D. A., Fogarty, K. E., Tsien, R. Y., and Fay, F. S.,** Calcium gradients in single smooth muscle cells revealed by the digital imaging microscope using Fura-2, *Nature,* 318, 558, 1985.
2. **Yanagida, T., Nakase, M., Nishiyama, K., and Oosawa, F.,** Direct observation of motion of single F-actin filaments in the presence of myosin, *Nature,* 307, 58, 1984.
3. **Matsumoto, S., Morikawa, D., and Yanagida, Y.,** Light microscopic structure of DNA in solution studied by the 4′,6-diamidino-2-phenylindole staining method, *J. Mol. Biol.,* 152, 501, 1981.
4. **Hayakawa, T., Kinoshita, K., Miyaki, S., Fujiwake, H., and Ohsuka, S.,** Ultra-low-light level camera for photon counting imaging, *Photochem. Photobiol.,* 43, 95, 1986.
5. **Kinoshita, M., Kinoshita, K., Yamamoto, K., and Suzuki, Y.,** A two-dimensional photon counting tube, *Adv. Electron. Electron Phys.,* 64B, 323, 1985.
6. **Matsuura, S., Umebayashi, S., Okumura, C., and Oba, K.,** Characteristics of the newly developed MCP and its assembly, *IEEE Trans. Nucl. Sci.,* 32, 1, 1985.
7. **Chandler, C. E., Bredthauer, R. A., Janesick, J. R., Westphal, J. A., and Gunn, J. E.,** Subelectron noise charge coupled devices, *SPIE,* 1242, 238, 1990.
8. **Scott, R. Q., Usa, M., and Inaba, H.,** Ultraweak emission imagery of mitosing soybeans, *Appl. Phys.,* B48, 183, 1989.
9. **Ichimura, T., Hiramatsu, M., Hirai, N., and Hayakawa, T.,** Two-dimensional imaging of ultra-weak emission from intact soybean roots, *Photochem. Photobiol.,* 50, 283, 1989.
10. **Suematsu, M., Oshio, C., Miura, S., and Tsuchiya, M.,** Real-time visualization of oxyradical burst from single neutrophil by using ultrasensitive video intensifier microscopy, *Biochem. Biophys. Res. Commun.,* 149, 1106, 1987.
11. **Gallie, D. R., Lucas, W. J., and Walbot, V.,** Visualization mRNA expression in plant protoplasts, *Plant Cell,* 1, 301, 1989.
12. **Escher, A., O'Kane, D. J., Lee, J., and Szalay, A. A.,** Bacterial luciferase fusion protein is fully active as a monomer and highly sensitive *in vivo* to elevated temperature, *Proc. Natl. Acad. Sci. U.S.A.,* 86, 6528, 1989.

13. **Mueller-Klieser, W., Walenta, S., Paschen, W., Kallinowski, F., and Vaupel, P.,** Metabolic imaging in microregions of tumors and normal tissues with bioluminescence and photon counting, *J. Natl. Cancer Inst.*, 80, 842, 1988.
14. **Yoshimoto, Y., Iwamatsu, T., Hirano, K., and Hiramoto, Y.,** The wave pattern of free calcium release upon fertilization in medaka and sand dollar eggs, *Dev. Growth Diff.*, 28, 583, 1986.
15. **Hiramatsu, M., Muraki, H., and Ito, T.,** A new method to characterize film inhomogeneities, *J. Polym. Sci. Part C Polym. Lett.*, 28, 133, 1990.

Chapter 6

ANALYSIS OF TWO-DIMENSIONAL ELECTROPHORESIS GELS

Donat-P. Häder

TABLE OF CONTENTS

I. INTRODUCTION

A. GEL ELECTROPHORESIS

The necessity to separate proteins, DNA, RNA, and other biochemically important substances for analytical or preparative reasons has led to a number of different techniques which have been developed in the past. One separation procedure relies on the different retardation of substances by percolating them through a matrix of appropriate size beads or binding them to ligands on the matrix in a vertical column.[1,2] Recent developments in this field are HPLC[3-5] and FPLC,[6-8] which differ from the simple column chromatography by using high or medium pressure, respectively, to force the substances to be separated through comparatively small columns.

A second group of techniques, gel electrophoresis, is one of the most powerful tools in modern biochemistry to separate molecules which differ in their physical and chemical properties.[9-13] Samples obtained by solubilization from tissues, cells, or cellular organelles are separated from cellular debris and solubilized using biochemical methods, such as high salt (e.g., NaCl or urea) concentrations or detergents (e.g., Triton X 100), or physical methods, such as grinding or ultrasound treatment.

The solubilized sample is applied to a gel matrix (e.g., agarose or polyacryl amide) at one end and a DC voltage on the order of several hundred or several thousand volts is applied across the gel by platinum electrodes to produce an electric field with (ideally) parallel field lines (Figure 1). The individual proteins (or other molecules) carry an electric charge so that they are electrophoretically pulled by the external field across the gel matrix (native gel electrophoresis). Different size molecules are retarded to a different extent, so that a separation is achieved. Alternatively, the proteins are treated with sodium dodecyl sulfate (SDS), which binds to the side groups of the protein and carries a negative charge; in this case different molecules move different distances (R_f values) according to the charge they carry (SDS polyacrylamide gel electrophoresis = SDS PAGE).

Usually the protein sample is applied to a stacking gel, which serves to sharpen the bands, from where it moves to a separation gel (Figure 2), which can have a constant mesh size in the polyacrylamide or form a gradient (e.g., 8 to 20% polyacrylamide).[14,15] The externally applied electric field can be pulsed (pulsed field gel electrophoresis, PFG).[16]

A different technique is based upon the fact that native proteins carry both positive and negative charges. Depending on the pH of the separation buffer, one of the charges exceeds the other and at a specific pH both charges compensate each other, which is defined as the isoelectric point (IEP). This effect is employed by the technique of isoelectric focusing (IEF),[17-19] where a pH gradient is established in the gel before the application of the sample (Figure 3). Then the proteins are applied to the gel, and an electric field is induced so that the proteins migrate in the gel until they move to a pH, which corresponds with its isoelectric point, and stop (Figure 4). Thus, the individual proteins in the sample are separated according to their IEP. A similar approach is the isoelectric titration where the proteins are applied in a line to the center of the gel and move in the electric field toward the anode or the cathode, respectively, along the pH gradient, which has been established previously perpendicular to the electric field (Figure 5). Calibration proteins have been added before the establishment of the pH gradient to indicate which region of the gel corresponds to which pH. Immunoelectrophoresis, affinity gel electrophoresis, agarose gel electrophoresis,[20-22] chromatofocusing, and other methods are also used for separation of the individual components in the protein mixture.

B. 2-D SEPARATION

Especially for large numbers of proteins in the sample[23,24] two-dimensional (2-D) techniques are being employed for better resolution.[25-33] Separation is a two-step process where

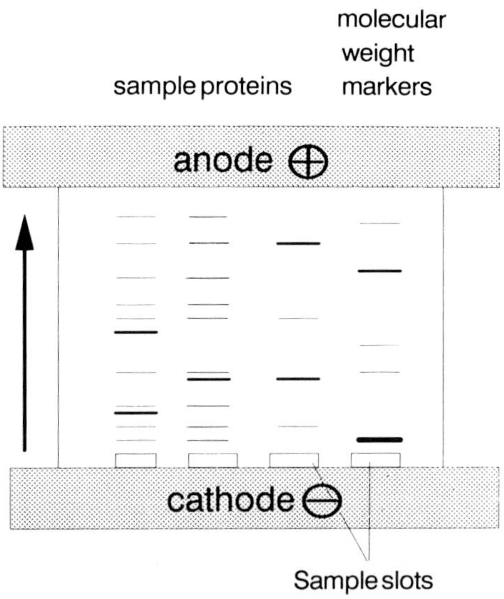

FIGURE 1. Gel electrophoresis of proteins in a gel using an external DC field. The proteins are applied in three sample slots while the fourth is reserved for marker proteins of known molecular weight.

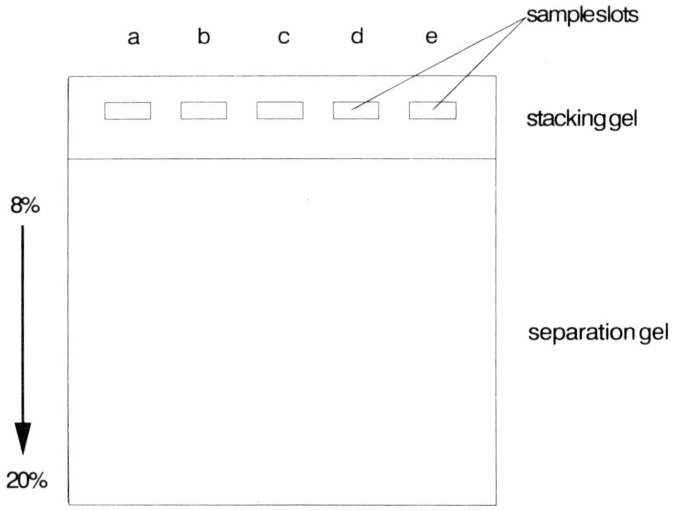

FIGURE 2. SDS PAGE of proteins. The proteins are denaturated by SDS and carry a negative charge so that they move from the sample slots toward the anode (bottom). They are first focused in the stacking gel and then separated in the separation gel, in which the pore size gradually decreases while the polyacrylamide gradient increases from 8 to 20%.

IEF or nonequilibrium pH gradient electrophoresis (NEPHGE) are used in the first direction[34-36] and usually SDS PAGE in the second direction perpendicular to the first (Figure 6).[37,38] Flatbed gel electrophoresis[39,40] can be run horizontally or vertically and has a number of advantages such as an efficient cooling, small protein loss, and high resolution during separation in thin[41,42] or ultrathin gels.[40,43,44]

C. STAINING AND VISUALIZATION

It is difficult to detect the location of individual proteins after separation in the gel even

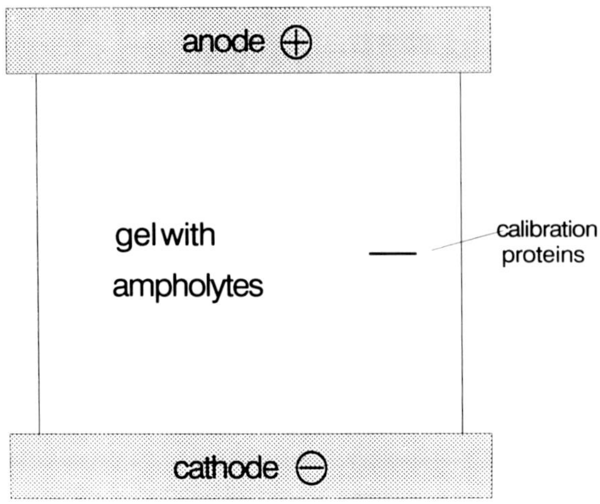

FIGURE 3. Prefocusing in an electric field establishes a pH gradient in the gel indicated by calibration proteins of known IEP.

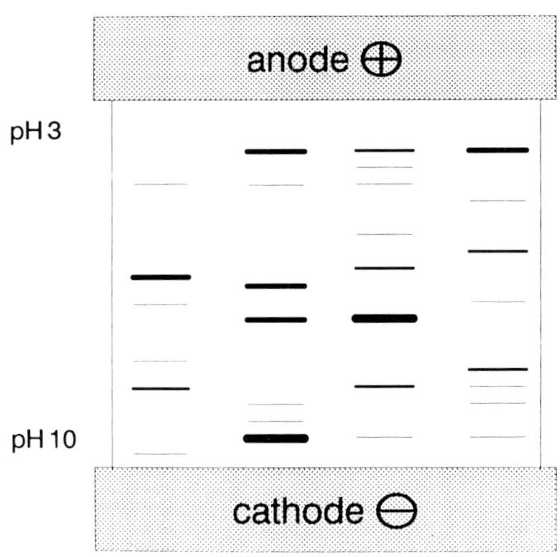

FIGURE 4. Isoelectric focusing of proteins running in an electric field. The proteins move to a position where their isoelectric point coincides with the pH in the gel.

when they carry a chromophoric group (chromoprotein). One possibility to visualize the spots is using fluorescence of the proteins or the fluorescence of an added dye.[45] The alternative is to use one of a number of different stains including Coomassie Brilliant Blue R-250,[46,47] silver,[27,48-50] violet 17, india ink,[51] or ethidium bromide (a fluorescent label for DNA).[52,53] A third possibility is to radioactively label the substances and perform an autoradiography by placing the gel after separation on a photographic film in the dark.[17,38,54-57]

The separated protein spots can be transferred (blotted) onto suitable carriers such as nitrocellulose or nylon membranes[58-62] where they are stained or made visible by immunological techniques[63-65] including the addition of colloidal gold.[66,67]

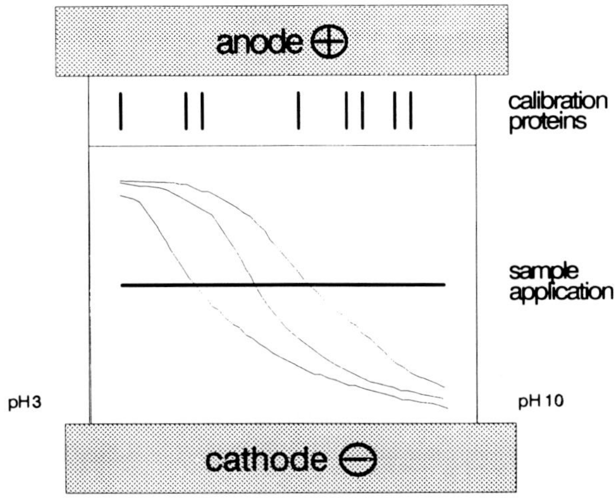

FIGURE 5. Isoelectric titration curves of proteins moving at different pH in an electric field.

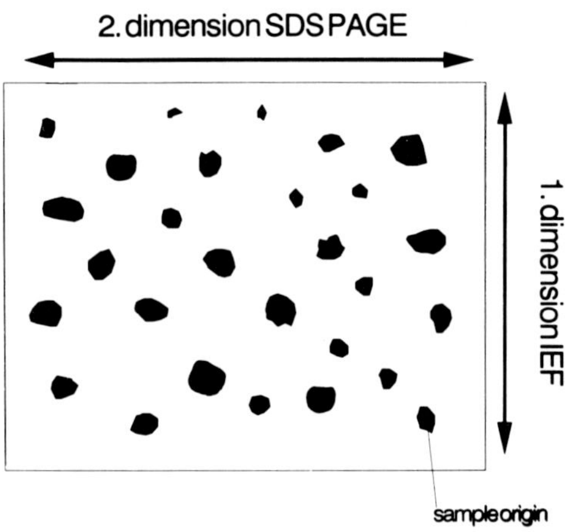

FIGURE 6. 2-D gel electrophoresis using IEF in the first dimension and SDS PAGE in the second.

II. DETECTION OF PROTEIN SPOTS BY IMAGE ANALYSIS

Due to the large amount of protein spots in a 2-D gel, automatic analysis of the samples was regarded a necessity early in the development (see review by Dunn[68]). Early attempts to achieve some kind of automation in protein spot analysis was the manual input into a computer using a digitization tablet. One-dimensional (1-D) gels were evaluated using a densitometer where the gel (or a photographic film) is placed on a stage which is moved across a light beam.[69] A photocell measures the attenuation when a stained protein spot passes the light beam. The signal is recorded on a chart strip recorder. The next step in automation is the digitization of the analog signal.

Automatic image analysis of 2-D gels is basically performed as is the electronic analysis of any other image.[70-73] These techniques are used also for cell, particle or colony counting,[74] shape and size determination,[75] and tracking of motile microorganisms.[76-79] Light and electron micrographs are processed[80-82] by extracting features of interest and determining areas of objects.[83-85] A special form of image analysis is its use for the three-dimensional (3-D) reconstruction of objects such as plant dictyosomes or animal nuclei, chromosomes, or even single macromolecules from serial cuttings.[86-90] A review on biomedical image processing and future trends can be found in Dhawan.[91]

We have developed a 2-D gel electrophoresis analysis system which on the one hand meets the existing industry standard and on the other hand makes use of the existing hardware in the medical or biological laboratory. For this purpose, a number of minimal requirements can be defined: the hardware should fit into the AT environment and the software should run on a 80 × 86 CPU and make use of a 80 × 87 coprocessor for fast and easy mathematical treatment; the software should run under common operating systems such as MS-DOS, OS/2, or UNIX; the graphic representation of the images should be based on high resolution color graphic adapters such as VGA or enhanced VGA; automatic spot detection and labeling should require minimal time (e.g., under 1 min for 500 spots). Also, the analysis of the key features such as position, area, and volume should be as fast as possible preferably by running these algorithms in the background during image loading and preediting. The results of the analyses need to be documented graphically and numerically on different periphery devices including dot matrix printers, plotters, and laser printers in camera-ready quality. Another most important requirement is the development of a user friendly surface to facilitate operation by untrained personnel to guarantee error-free and efficient utilization of the built-in features.

An addition advantage of automatic image analysis of 2-D gels is that the data are available in an electronic form after evaluation.[92-95] These data can be used to develop a database which allows comparison between the results from various laboratories provided the experimental procedures have been standardized.[96,97]

A. HARDWARE FOR DIGITIZING ELECTROPHORESIS GELS

Several hardware approaches have been developed to digitize the gel image or a high resolution photograph (or autoradiograph): A low cost approach uses a video camera or a charge coupled device (CCD) camera to record the image.[98-102] The advantage of video scanning is that the image is available in real time. However, the disadvantages are the low spatial resolution and the small range of optical densities limited to about 2.4 O.D. (optical density). Values above this point are cut off even when the automatic gain control (AGC) of the camera is disabled.[103] In some applications, the low spatial resolution can be circumvented by using a macro lens to record a small portion of the gel at a time. For low light level monitoring, special photon detection camera systems have been developed.[104]

After recording, the image is digitized (analog-to-digital [A/D] conversion) either in real time using flash converters, or it may take up to a few minutes in sequential A/D conversion depending on the available hardware. The digitizer may be a stand-alone unit or it can be a card within the host computer.[105-107] The spatial resolution of digitization is defined by the number of distinct image points (pixels) in x and y directions. The sample density is given by its darkness defined for each pixel. The gray level is expressed by a digital value which ranges between black and white in a number of distinct steps which encode the absorption in a linear or logarithmic fashion.[79-103] The gray-level resolution is limited by the number of bit planes during A/D conversion.

Higher resolution of especially 2-D gels can be achieved by using a laser scanner. The measuring beam is produced from a well-focused laser beam. The gel is scanned by mechanically advancing it in x and y directions across the beam or by moving the laser beam by use of turning mirrors. The photoreceptor can be a photomultiplier which has the capability of measuring a wide range of optical densities. In these machines the resolution is determined by the step width of the stepper motor with which the gel or the beam is moved with respect to the photodetector; resolutions of 5 μm have been achieved. One of the disadvantages of modern laser densitometers is the long time required for scanning a gel at high resolution, which can be on the order of several hours. Another suitable detector is a CCD array[53,108] with a number (e.g., 2048) of parallel channels which measure the absorbance in adjacent pixels.[109]

In our system the high resolution scanning of the gels or photographic negatives is performed with a laser scanner (Ultrascan, Pharmacia LKB, Bromma, Sweden). The gel is placed on a backlit screen[108] where the section to be analyzed is selected by motor-driven rulers. The resolution is defined for x and y directions and the scanning is performed line by line with high precision stepper motors allowing for a spatial resolution of <40 μm. The attenuation of the laser beam by the sample is determined for each pixel in the range between 0 and 4 O.D. using a linear scaling curve.

After digitization the pixel matrix is transferred to the host computer and stored in a dedicated video RAM (read and write memory), in the internal RAM of the computer or on a magnetic storage medium (hard disk drive, floppy disk, magnetic tape) for analysis and feature extraction. We used a microcomputer (IBM Personal System/2, model 60) and transferred the digitized image via the serial input channel under software control and stored it on the hard disk drive (42 MB, Seagate type ST252-1 with 28 ms access time). Depending on the scanned area and the required resolution, the size of the data files can be up to 4 MByte. The computer uses a 80286 CPU running at 10 MHz with zero wait states and a 80287 coproccessor. The terminal screen (IBM Personal System/2, color display, type 8515 S/N) serves the dual purpose to run the program package and to display the gel images in pseudocolor representation.

The image of the gel can be displayed on a suitable screen which can be the computer terminal or a separate display as a black and white binary or gray-level image. Current standards for high resolution color representation are the VGA color adapter which can display 16 colors at 640×480 pixels or 256 colors at low resolution (320×200). Graphic adapters with more memory can reach higher resolution (800×600) with 256 colors (VEGA Wonder).

Better graphical output is achieved by using pseudocolor representation in which the gray values are color coded using look-up tables (LUTs): each gray value is assigned to a specific mixture of the three basic colors (cyan, green, red).[79,103] This technique can be described by looking up the gray value of each individual pixel and assigning it a new value looked up in a previously defined look-up table. For instance, a gray value of 0 (black) is replaced by one of 255 (white), a value of 1 by 254, etc., which in essence inverses the image into its negative. LUTs are also suitable to expand of compress part of the gray scale to increase or reduce the contrast, to suppress the background, or to isolate image elements of specific gray levels. The digitized image stored in memory can be accessed by the host computer by reading and writing individual pixels.[71,110,111] Thus the image can be analyzed and also manipulated.

B. SOFTWARE APPROACHES FOR 2-D GEL ELECTROPHORESIS

Computer Aided Evaluation of 1- and 2-D electrophoresis Gels (CAEG) was developed soon after the introduction of the electrophoresis techniques themselves.[38,53,99,112,113] The early computer program systems ran on large mainframe computers.[114,115] Later packages were developed for smaller machines.[116,117] Spot detection is the first step in gel electrophoresis analysis: a threshold level is defined which separates background and spot areas by gray levels. This simple method may not work reliably, especially when the background is not uniform. In these cases a nonlinear background adjustment may be necessary before spot detection. The areas occupied by the protein spots are identified using one of several algorithms such as edge detection,[77] chain coding,[118,119] or area filling.[120] The chain coding algorithm is explained in more detail in the chapter on real time tracking of microorganisms by Häder and Vogel in this volume.

Celis and his coworkers[121] have used the PDQUEST software (Protein Databases Inc.,) running on a MASSCOMP CPV 5520 computer to analyze fluorescence images of 2-D gels of human embryonic lung MRC-5 fibroblast proteins. Editing the raw gel images with 1482 individual proteins took about 1.5 d. In the analyzed image, ellipses are drawn representing the area and position of the proteins in the original gel labeled and numbered for identification.[113,122,123] Spatial Data Analysis Laboratory at Virginia Polytechnic Institute developed a general image processing system which was used by Tyson and Haralick[124] to process 2-D gels of proteins from *Physarum polycephalum* photographed from the original gel and scanned with a laser scanner. This package is capable of removing even a nonhomogeneous background, separating merged protein spots, and estimating the volume of the individual spots. Mathematical operations such as adding or removing Gaussian components or curve-of-growth models to estimate peak volumes indicative of protein loading are used to improve the analysis.[100]

Automatic image analysis is also used to read and analyze DNA sequence gels.[53,109,125,126] Häder and Truß[127] used a UV-B emitting light source (transilluminator) to detect the fluorescence of ethidium bromide employed to stain nucleic acids. Automatic band detection algorithms have been developed to automatically read DNA restriction fragments in high resolution images obtained by 1-D scanning of agarose gels.[53,109]

C. FILE COMPRESSION AND PREEVALUATION

In order not to exceed the hard disk storage capacity in a short time, it is necessary to compress the data files since images of electrophoresis gels can occupy several megabytes (MB) in storage.[128] Compression techniques take advantage of the fact that many pixels of the same optical density may be in sequence, which in the scanned image are coded individually. During compression the number of uniform pixels and their gray values are stored. In addition, if only 16 gray values are necessary, they can be stored in nibbles (4 bit, half a byte) rather than a full byte; thus the file size can be reduced by up to 80% without losing any information as compared to the original file. Simultaneously to data compression, a smoothing is performed to remove noise (bubbles, dirt, or tears on the gel) from the original file. This technique is based on an algorithm which determines the density distribution of the whole gel (Figure 7) and then replaces individual pixels which strongly deviate from their neighbors by an average value in this range.

The pseudocolor coding of the gray values for graphic representation is scaled according to a Gaussian or exponential fitting. The Gaussian scaling curve is adapted to the density distribution in the gelscan by controlling the σ and μ parameters.

$$f(x) = \frac{1}{\delta\sqrt{2\pi}} e \left\{ -\frac{1}{2} \left(\frac{x - \tau}{\delta} \right)^2 \right\} \tag{1}$$

Exponential fitting curves are used automatically if this produces better results; for adjustment the exponent n and the multiplication factor a are varied.

$$f(x) = ax^n \tag{2}$$

Both the compression and the nonlinear scaling are written in Assembler language called from C routines so that the speed could be greatly enhanced as compared to the much slower high level languages; a 1-MB file can be preevaluated and compressed to 25 KB in about 1.5 min.

The whole program package runs under GEM (Digital Research Inc., Monterey, CA), which allows it to incorporate Assembler language routines for fast display and manipulation of large amounts of pixels. The image is displayed in a window on the screen and can be

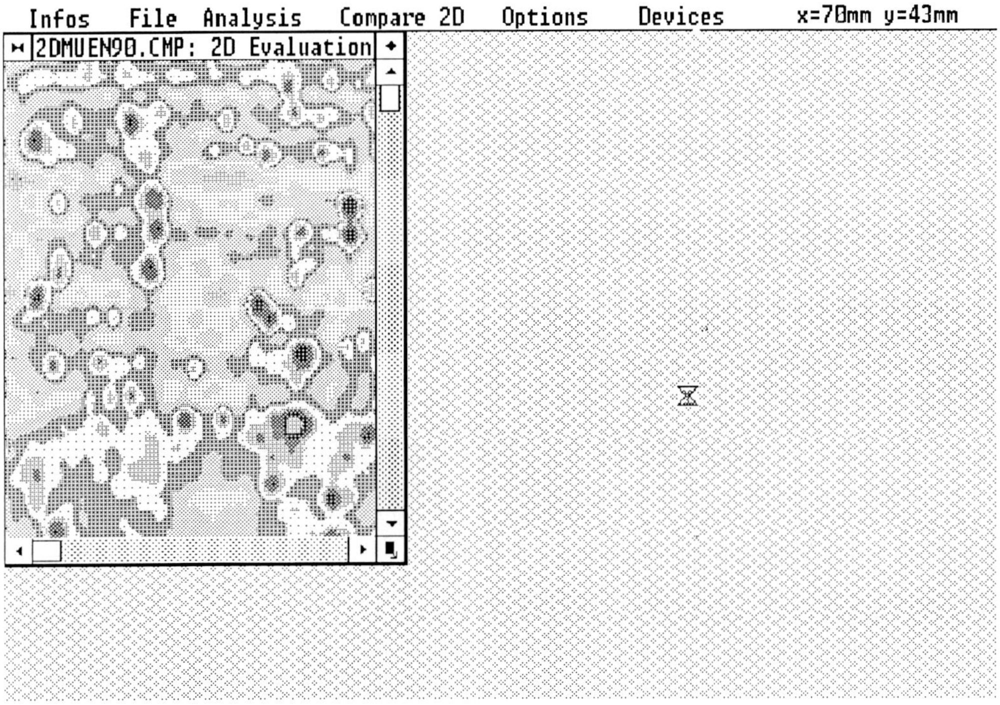

FIGURE 7. Graphical representation of a scanned photographic film showing protein spots after 2-D gel electrophoresis in 16 gray levels (shown in 16 colors on screen).

easily moved around and adjusted in size and proportions using rubber-band techniques controlled by a mouse as input device. Pull-down menus and pop-up windows are used to guide the user through the various steps of analysis (Figure 8).

Zooming is used to isolate and magnify specific sections of interest in the gel which also speeds up processing. Noninteger zooming is achieved by independent zooming in x and y directions rather than using complicated matrix interpolation techniques. Zooming (and shrinking) is based on nonlinear interpolation between existing data points first in x and subsequently in y direction. The options incorporated allow zooming without correction of the x and y ratio, zooming with respect to a chosen x or y zooming factor or zooming with fixed independent factors. During zooming, the section of interest can be displayed in a window or fill the whole screen. Multiple zooming in sequence is possible. Overlapping spots can be separated manually using a cross-hair cursor; in addition, spot separation can be performed automatically by the erosion technique in which shells are removed sequentially from the circumference of the spots until they are reliably separated. Afterwards the spots are dilated again to their full size leaving a separation line of at least one pixel width between the spots.

D. SPOT FINDING

The chain code algorithm[118,119,129,130] modified for the identification of protein spots in electrophoresis gels is used to determine the position and size of the spots. When the background is relatively homogeneous (or has been removed beforehand), spots are easily identified by scanning the image line by line until a pixel is detected which exceeds the background by a predefined threshold. The coordinates of this pixel are then stored for future calculations. The chain code algorithm is now used to determine the next pixel on the periphery of the spot. The direction vector from the first pixel P to the second outline pixel is encoded and the vector number is stored as the first link in the chain.

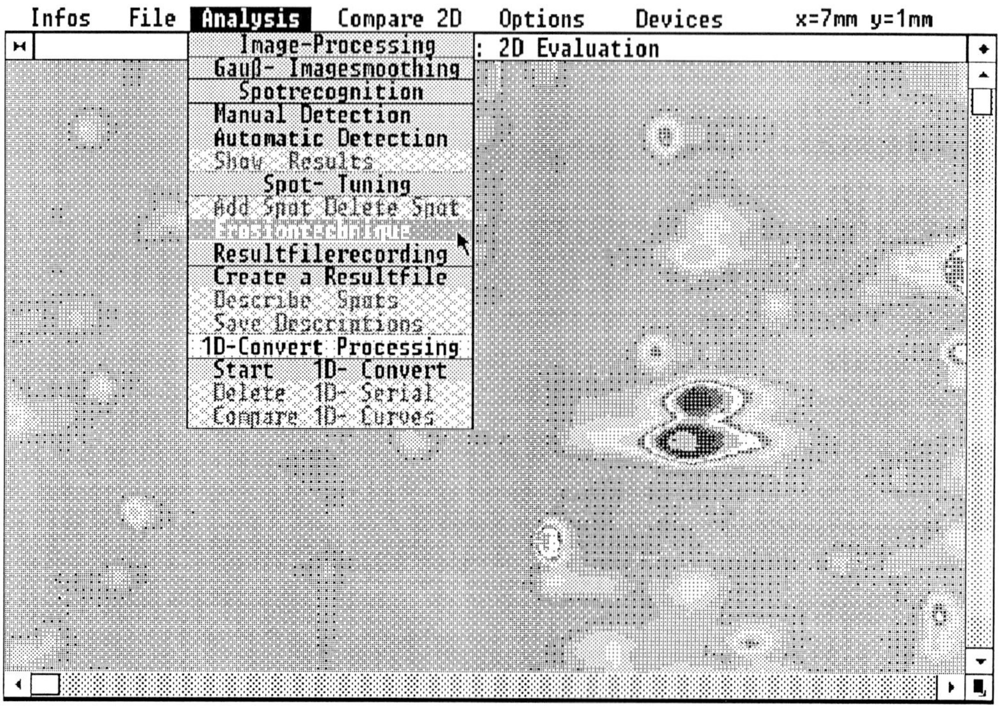

FIGURE 8. User surface with pull-down and pop-up submenus to facilitate the analysis of 2-D electrophoresis gels.

This procedure is repeated for all outline pixels until the first pixel on the periphery is found again. The stored direction vectors constitute the chain which uniquely defines the form and position of the spot. It is not necessary to analyze all eight neighbors for being possible outline candidates; rather it is sufficient to look two positions counterclockwise from the previous search direction. If this is not a spot pixel, the chain code is decremented until the next point has been identified. The diameters and the areas as well as their positions of the spots in x and y directions can be easily determined by using techniques described by Häder and Vogel elsewhere in this volume.

The chain code algorithm has the advantage of being very efficient in memory use and speeding up the calculation. After the first spot has been identified in the image, the search is continued from the first found pixel on the periphery until all spots have been found. In order not to recalculate an already identified spot, the periphery pixels are all set to an arbitrary value otherwise not used in the gray scale which also has the advantage that defined spots are visually characterized by an outline in a different color. Also, all spots are numbered sequentially on screen (Figure 9). The chain code can also directly be used to drive a plotter or printer to obtain a hard copy which faithfully reproduces the original shape of the spots.

E. VOLUME CALCULATION

In some 2-D gel calculations, the area of the spots is used to determine the amount of protein present. However, the protein loading depends on the absorption of the individual pixel. Since the absorption is an exponential function of the concentration, the Lambert-Beer law can be used to determine the amount of protein:

$$A(\lambda) = \log_{10}(I_0/I) = \epsilon(\lambda)\, Cl \qquad (3)$$

where $A(\lambda)$ is the absorbance at the wavelength λ, $\epsilon(\lambda)$ the molar extinction coefficient, I_0

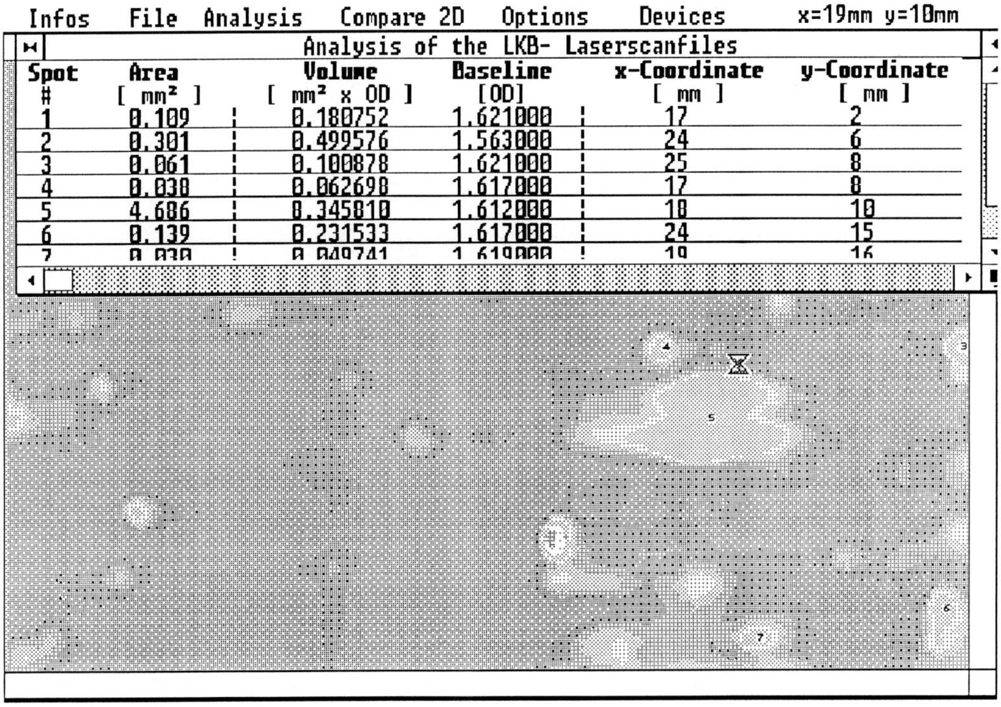

Spot #	Area [mm²]	Volume [mm² x OD]	Baseline [OD]	x-Coordinate [mm]	y-Coordinate [mm]
1	0.109	0.180752	1.621000	17	2
2	0.301	0.499576	1.563000	24	6
3	0.061	0.100878	1.621000	25	8
4	0.038	0.062698	1.617000	17	8
5	4.686	0.345810	1.612000	18	10
6	0.139	0.231533	1.617000	24	15
7	0.030	0.049741	1.619000	19	16

FIGURE 9. After spot finding, all identified spots are set to a specific color and numbered sequentially on screen. The protein concentration of the individual spots can be shown numerically in a separate window.

and I are the incident and transmitted light intensities, C is the molar concentration (mol/l) of the absorbing species, and l is the optical path length.

Since the gray values of the measured pixels are obtained by digitization, filling in new values requires some interpolation. This is done by using the Lagrange equation

$$(x\text{-}x_2)\ ((x\text{-}x_3)\ldots(x\text{-}x_n) \tag{4}$$

$$P(x) = \frac{(x - x_2)(x - x_3)\ldots(x - x_n)}{(x_1 - x_2)(x_1 - x_3)\ldots(x_1 - x_n)} y_1$$

$$+ \frac{(x - x_1)(x - x_3)\ldots(x - x_n)}{(x_2 - x_1)(x_2 - x_3)\ldots(x_2 - x_n)} y_2 + \ldots$$

$$+ \frac{(x - x_1)(x - x_2)\ldots(x - x_{n-1})}{(x_n - x_1)(x_n - x_2)\ldots(x_n - x_{n-1})} y_n$$

Interpolation is difficult when there are only a few basepoints over long intervals, because the precision decreases and the high degree interpolation polynomes tend to oscillate especially at the end of the set interval producing density artifacts.[131] For this reason the volume calculation in 2-D gels is performed by cubic spline interpolations which use a third-degree polynomial

$$f(x) = ax^3 + bx^2 + cx + d \tag{5}$$

During calculation, the algorithm determines the parameters a, b, c, and d for a curve fitted through the measured data points located at given x and y coordinates making sure that the inclination and curvature of adjacent curve segments are identical at the data points. The cubic spline interpolation is derived from the commonly used one[132] modified for volume integration. The nodes for the cubic spline interpolation are of importance since they determine the efficiency and speed of fitting. In order to determine these nodes, a preprocessing algorithm detects the optimal nodes by scanning along the density distribution before the spline interpolation takes over. The cubic spline interpolation is not constantly engaged, but is rather cancelled when the scan deviates just a little from the baseline; it is only used when a sudden change in absorption near a spot is detected. This dynamic adaptation to the conditions in the gel scan increase the calculation speed considerably. The cubic spline interpolation uses the following equation (for further details see Press et al.[132]); consider basepoints with coordinates at

$$y_i = y(x_i); \qquad \{i = 1 \ldots N\} \tag{6}$$

with the corresponding abscissa values

$$x_i < x_{i+1} \tag{7}$$

Differentiation of the special case of the Lagrange equation (3rd degree) yields the solution for the N-2 linear equations. Area and volume values are displayed in numerical form in a different window on screen and are also printed as a table (Figure 9).

F. 1-D GELS

In principle, 1-D gels can be evaluated using the same software package developed to analyze 2-D gels. Scanning a 1-D gel only along one scan line produces a gel scan which may be distorted by noise both in the bands and the gaps. Treating the image as a 2-D window and integration over the width of the bands smoothes out accidental flaws. However, the operator has to ensure that smaller bands fill the lane completely in order not to distort the analysis. This is facilitated by using the rubber-band technique (Figure 10). Further analysis of spot area and volume utilize the same algorithms described above for 2-D analysis.

G. SPOT MATCHING

In several biological or medical applications, it is useful and necessary to compare the spot patterns between two or more gels. E.g., the comparison of the protein patterns between the wild type and the mutant of an organism can identify the missing gene product. Likewise this technique is used to identify the induction of specific gene products during specific developmental stages or by external factors such as light or UV radiation.[38,122,133,134] The prerequisite for a successful matching of corresponding spots in two parallel 2-D gels is a distance matching in both directions. This can be done by rubber-banding techniques, in which one gel is stretched or compressed with respect to the other in one or two directions until a close alignment is found.[113] This process may be complicated by nonlinear distortions between the gels.

Several search strategies have been developed to match corresponding spots including hierarchical nearest neighbor analysis,[113] multidimensional classification,[38] heuristic clustering,[134] least square fitting,[133] and multivariance analysis.[122] Some software approaches need the manual input of the operator to start the search by matching "landmark" spots. Others attempt a fully automatic analysis. The main disadvantage of most programs developed so far is the enormous time requirement which can be on the order of hours or days.

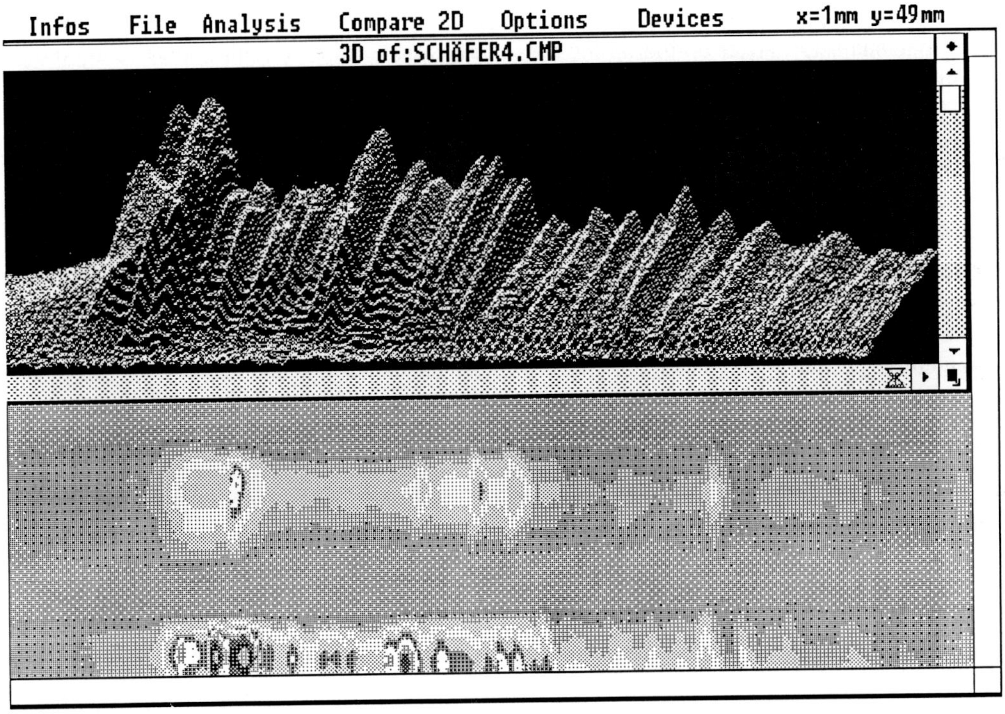

FIGURE 10. 1-D processing of the lower lane. The area of interest is selected with a rubber band and the density is shown in 3-D.

III. COMMERCIALLY AVAILABLE PROGRAM PACKAGES

Today a number of commercial 2-D gel analysis packages are marketed, e.g., HERMeS II (Biolog Diffusion, Versailles); Gelsim (Biosoft, Cambridge, UK); Visage 1-D and 2-D gel software (Bio Image, Ann Arbor, MI, U.S.A.); MCID (Imaging Research, Ontario, Canada); PhastImage (Pharmacia LKB, Uppsala, Sweden); and GS365W (Hoefer Scientific Instruments, San Francisco, CA, U.S.A.). Other programs include the Clip system,[135] the MasterScan (Scanalytic, Billerica, MA, U.S.A.), the QUEST sytem,[136] and a system to identify mutant proteins from 2-D gels after video recording and digitization of the image.[94] The performance of these packages varies widely from mere digitization to a full analysis and spot matching. One of the programs (QUEST) is designed for the analysis of radiolabeled proteins. 2-D Gaussian fitting is used for spot quantitation and separation of overlapping spots. Spot matching is not restricted to two gels, but a whole set called matchset within each gel is compared to every other gel. The authors claim that with an average of 2000 spots per gel the system matches 97% correctly and that fewer than 1% are matched inconsistently.

IV. CONCLUSIONS

Image analysis is an appropriate tool to analyze and quantify sample spots (e.g., proteins, DNA, RNA) after separation in two dimensions by gel electrophoresis. Various systems are commercially available today for different tasks and one developed in the author's lab, is described in more detail in this chapter. Different separation techniques are used in the two directions to take advantage of the specific properties of the sample. Photographic films of gels can also be analyzed. All techniques for automatic analysis of 2-D electrophoresis gels by computer-aided image analysis start out with a 2-D scan of the gel using a laser scanner or a video snapshot. After digitizing the optical densities of the individual pixels, the image is stored on disk or tape in a computer. The data can be compressed and preevaluated before the individual sample spots are identified and quantified either in the whole field or in a zoomed portion. The found spots are numbered and shown in pseudocolor representation or in 3-D graphical output. A cubic spline algorithm, which automatically switches on and off depending on the necessity, is used to interpolate between individually measured points. The program package has been developed in the computer language C, but time-consuming routines are written in Assember language in order to achieve high speed calculation and graphical output. In addition to 2-D gels, 1-D gels can also be analyzed with the same package which allows averaging over a strip of adjustable width. The results can be documented on periphery devices (e.g., laser printer) for camera ready output.

ACKNOWLEDGMENTS

The development of the software package has been supported by equipment and financial aid from Pharmacia LKB (Freiburg, FRG) and a grant from the Bundesminister für Forschung und Technologie (KBF 57). The skillful technical assistance of J. Schäfer, U., Schleier, and H. Vieten is gratefully acknowledged. Host modules were programmed by G. Vaues.

REFERENCES

1. **Söderberg, L., Müller, R.-M., and Fägerstam, L.,** Chromatofocusing of desialylated human transferrin on Mono P, *Protides Biol. Fluids,* 30, 661, 1983.
2. **Moreno-Lopez, J., Kristiansen, T., and Karsnäs, P.,** BVDV: affinity chromatography on *Crotalaria juncea* lectin, *J. Virol. Methods,* 2, 293, 1981.
3. **Tam, M., Taylor, A., and Giri, L.,** Purification of *E. coli* ribosomal (30S subunit) proteins by HPLC, paper presented at 4th Int. Symp. HPLC of Proteins, Peptides, and Polynucleotides, Baltimore, MD, 1984.
4. **Burchiel, S. W., Billman, J. R., and Alber, T. R.,** Rapid and efficient purification of mouse monoclonal antibodies from ascites fluid uing HPLC, *J. Immunol. Methods,* 69, 33, 1984.
5. **Welling, G. W., Groen, G., and Welling-Wester, S.,** Isolation of Sendai virus F protein by anion exchange high performance liquid chromatography in the presence of Triton X-100, *J. Chromatogr.,* 266, 629, 1983.
6. **Markey, F.,** Rapid purification of deoxyribonuclease I using FPLC, *FEBS Lett.,* 167, 155, 1984.
7. **Johns, E. H., Cooper, E. H., and Turner, R.,** Separation of cerebrospinal fluid by fast protein liquid chromatography (FPLC), paper presented at 3rd Int. Symp. HPLC of Proteins, Peptides and Polynucleotides, Monaco (Abstr.), 914, 1983.
8. **Karsnäs, P., Moreno-Lopez, J., and Kristiansen, T.,** BVDV: purification of surface proteins in detergent-containing buffers by FPLC, *J. Chromatogr.,* 266, 643, 1983.
9. **Lammel, B.,** A method for continuous preparative electrophoresis using a supporting medium, *Electrophoresis,* 2, 39, 1981.
10. **Rabilloud, T., Vincens, P., Hubert, M., Pennetier, J. L., and Tarroux, P.,** Changes in nuclear proteins induced by heat shock in *Drosophila* culture cells, *FEBS Lett.,* 184, 278, 1985.
11. **Rabilloud, T., Hubert, M., and Tarroux, P.,** Procedures for two-dimensional electrophoretic analysis of nuclear proteins, *J. Chromatogr.,* 351, 77, 1986.
12. **Andrews, A. T.,** *Electrophoresis: Theory, Techniques, and Biochemical and Clinical Applications,* Clarendon Press, Oxford, 1986.
13. **Chrambach, A., Dunn, M. J., and Radola, B. J.,** *Advances in Electrophoresis,* Vol. 1, VCH Publishers, Weinheim, Germany, 1989.
14. **Shapiro, A. L., Vinuela, E., and Maizel, J. V.,** Molecular weight estimation of polypeptide chains by electrophoresis in SDS-polyacrylamide gels, *Biochem. Biophys. Res. Commun.,* 28, 815, 1967.
15. **Righetti, P. G.,** On the pore size and shape of hydrophilic gels for electrophoretic analysis, in *Electrophoresis '81,* Allen, R. C. and Arnaud, P., Eds., Walter de Gruyter, Berlin, 1981, 3.
16. **Schwartz, D. C. and Cantor, C. R.,** Separation of yeast chromosome-sized DNAs by pulsed field gradient gel electrophoresis, *Cell,* 37, 67, 1984.

17. **Patel, K., Dunn, M. J., Günther, S., Postel, W., and Görg, A.,** Dual-label autoradiographic analysis of human skin fibroblast and myoblast proteins by two-dimensional polyacrylamide gel electrophoresis using immobilized pH gradient in the first dimension, *Electrophoresis,* 9, 547, 1988.
18. **Righetti, P. G., Morelli, A., Gelfi, C., and Westermeier, R.,** Direct recovery of proteins into a free-liquid phase after preparative isoelectric focusing in immobilized pH gradients, *J. Biochem. Biophys. Methods,* 13, 151, 1986.
19. **Morel, M. H. and Autran, J. C.,** Separation of durum wheat proteins by ultrathin-layer isoelectric focusing — a new tool for the characterization and quantification of low molecular weight glutenins, *Electrophoresis,* 11, 392, 1990.
20. **Serwer, P.,** Electrophoresis of duplex deoxyribonucleic acid in multiple-concentration agarose gels: fractionation of molecules with molecular weights between 2×10^6 and 110×10^6, *Biochemistry,* 19, 3001, 1980.
21. **Maniatis, T., Fritsch, E. F., and Sambrook, J.,** *Molecular Cloning, A Laboratory Manual,* Cold Spring Harbor Laboratory, Cold Spring Harbor, NY, 1982.
22. **McDonnel, M. W., Simon, M. N., and Studier, F. W.,** Analysis of restriction fragments of T7 DNA and determination of molecular weights by electrophoresis in neutral and alkaline gels, *J. Mol. Biol.,* 110, 119, 1987.
23. **Scheele, G. A.,** Two-dimensional gel analysis of soluble proteins, *J. Biol. Chem.,* 250, 5375, 1975.
24. **Roggero, P. and Pennazio, S.,** Two-dimensional polyacrylamide gel electrophoresis of extracellular soybean pathogenesis-related proteins using PhastSystem, *Electrophoresis,* 11, 86, 1990.
25. **Celis, J. E. and Bravo, R.,** *Two Dimensional Gel Electrophoresis of Proteins,* Academic Press, New York, 1984.
26. **Pahlic, M. and Tyson, J. J.,** Analysis of *Physarum* proteins throughout the cell cycle by two-dimensional PAGE, *Exp. Cell. Res.,* 161, 533, 1985.
27. **Görg, A., Postel, W., Weser, J., Günther, S., Strahler, J. R., Hanash, S. M., and Somerlot, L.,** Horizontal two-dimensional electrophoresis with immobilized pH gradients in the first dimension in the presence of nonionic detergent, *Electrophoresis,* 8, 45, 1987.
28. **Hanash, S. M., Strahler, J. R., Somerlot, L., Postel, W., and Görg, A.,** Two-dimensional electrophoresis with immobilized pH gradients in the first dimension: protein focusing as a function of time, *Electrophoresis,* 8, 229, 1987.
29. **Strahler, J. R., Hanash, S. M., Somerlot, L., Weser, J., Postel, W., and Görg, A.,** High resolution two-dimensional polyacrylamide gel electrophoresis of basic myeloid polypeptides: use of immobilized pH gradients in the first dimension, *Electrophoresis,* 8, 165, 1987.
30. **Görg, A., Postel, W., Weser, J., Günther, S., Strahler, J. R., Hanash, S. M., Somerlot, L., and Kuick, R.,** Approach to stationary two-dimensional pattern: influence of focusing time and immobiline/carrier ampholytes concentrations, *Electrophoresis,* 9, 37, 1988a.
31. **Görg, A., Postel, W., and Günther, S.,** The current state of two-dimensional electrophoresis with immobilized pH gradients, *Electrophoresis,* 9, 531, 1988b.
32. **Dunn, M. J. and Burghes, A. H. M.,** High resolution two-dimensional polyacrylamide gel electrophoresis. I. Methodological procedures, *Electrophoresis,* 4, 97, 1983a.
33. **Dunn, M. J. and Burghes, A. H. M.,** High resolution two-dimensional polyacrylamide gel electrophoresis. II. Analysis and applications, *Electrophoresis,* 4, 173, 1983b.
34. **O'Farrell, P. Z., Goodman, H. M., and O'Farrell, P. H.,** High resolution two-dimensional electrophoresis of basic as well as acidic proteins, *Cell,* 12, 1133, 1977a.
35. **O'Farrell, P. H. and O'Farrell, P. Z.,** Two-dimensional polyacrylamide gel electrophoretic fractionation, *Methods Cell Biol.,* 16, 407, 1977b.
36. **Mattila, K., Pirttilä, T., and Frey, H.,** Horizontal two-dimensional electrophoresis of cerebrospinal fluid proteins with immobilized pH gradients in the first dimension, *Electrophoresis,* 11, 91, 1990.
37. **O'Farrell, P. H.,** High resolution two-dimensional electrophoresis of proteins, *J. Biol. Chem.,* 250, 4007, 1975.
38. **Tarroux, P.,** Analysis of protein patterns during differentiation using 2-D electrophoresis and computer multidimensional classification, *Electrophoresis,* 4, 63, 1983.
39. **Dunn, M. J. and Patel, K.,** Immobilized pH gradients for the first dimension of 2-D page, *Electrophoresis '86,* Dunn, M. J., Ed., VCH Publishers, Weinheim, Germany, 1986b, 574.
40. **Görg, A., Postel, W., Günther, S., and Weser, J.,** Electrophoretic methods in horizontal systems, in *Electrophoresis '86,* Dunn, M. J., Ed., VCH Publishers, Weinheim, Germany, 1986, 435.
41. **Dunn, M. J., Burghes, A. H. M., Patel, K., Witkowski, J. A., and Dubowitz, V.,** Analysis of genetic neuromuscular diseases by 2D-page: A system for comparisons, in *Electrophoresis '84,* Neuhoff, V., Ed., VCH Publishers, Weinheim, Germany, 1984, 281.
42. **Burghes, A. H. M., Dunn, M. J., and Dubowitz, V.,** Enhancement of resolution in two-dimensional gel electrophoresis and simultaneous resolution of acidic and basic proteins, *Electrophoresis,* 4, 354, 1982.

43. **Görg, A., Postel, W., and Westermeier, R.,** Ultrathin-layer horizontal electrophoresis, isoelectric focusing, and protein mapping in polyacrylamide gels on cellophane, in *Electrophoresis '79,* Radola, B. J., Ed., Walter de Gruyter, Berlin, 1980, 67.
44. **Görg, A., Postel, W., and Westermeier, R.,** SDS electrophoresis of legume seed proteins in horizontal ultrathin-layer pore gradient gels, *Z. Lebensm. Unters.-Forsch.,* 174, 282, 1982.
45. **Jackson, P., Urwin, V. E., and Mackay, C. D.,** Rapid imaging, using a cooled charge-coupled-device, of fluorescent two-dimensional polyacrylamide gels produced by labeling proteins in the first-dimensional isoelectric focusing gel with the fluoro-phore 2-methoxy-2,4-diphenyl-3(^{3}H)furanone, *Electrophoresis,* 9, 330, 1988.
46. **Neuhoff, V., Arold, N., Taube, D., and Erhardt, W.,** Improved staining of proteins in polyacrylamide gels including isoelectric focusing gels with clear background at nanogram sensitivity using Coomassie Brilliant Blue G-250 and R-250, *Electrophoresis,* 9, 255, 1988.
47. **Neuhoff, V., Stamm, R., Pardowitz, I., Arold, N., Ehrhardt, W., and Taube, D.,** Essential problems in quantification of proteins following colloidal staining with Coomassie Brilliant Blue dyes in polyacrylamide gels, and their solution, *Electrophoresis,* 11, 101, 1990.
48. **Merril, C. R., Goldman, D., Sedman, S. A., and Ebert, M. H.,** Ultrasensitive stain for proteins in polyacrylamide gels show regional variation in cerebrospinal fluid proteins, *Science,* 211, 1437, 1981.
49. **Sammons, D. W., Adams, L. D., and Nishizawa, E. E.,** Ultrasensitive silver-based color staining of polypeptides in polyacrylamide gels, *Electrophoresis,* 2, 135, 1981.
50. **Hoffman, E. P., Brown, R. H., and Kunkel, L. M.,** Dystrophin: the protein product of the Duchenne muscular dystrophy locus, *Cell,* 51, 919, 1987.
51. **Aebersold, R. H., Pipes, G., Hood, L. H., and Kent, S. B. H.,** Amino-terminal and internal sequence determination of microgram amounts of proteins separated by isoelectric focusing in immobilized pH gradients, *Electrophoresis,* 9, 520, 1988.
52. **Laskey, R. A. and Mills, A. D.,** Quantitative film detection of ^{3}H and ^{14}C in polyacrylamide gels by fluorography, *Eur. J. Biochem.,* 56, 335, 1975.
53. **Gray, A. J., Beecher, D. E., and Olson, M. V.,** Computer-based image analysis of one-dimensional electrophoretic gels used for the separation of DNA restriction fragments, *Nucleic Acids Res.,* 12, 473, 1984.
54. **Johnston, R. F., Pickett, S. C., and Barker, D. L.,** Autoradiography using storage phosphor technology, *Electrophoresis,* 11, 355, 1990.
55. **Easty, D. J., Patel, K., Dover, R., Evans, D. J., and Dunn, M. J.,** A comparative study of protein synthesis by keratinocytes and fibroblasts in vitro using two-dimensional gel electrophoresis and dual isotope autoradiography, *Electrophoresis,* 9, 227, 1988.
56. **Bateman,J. E.,** Quantitative autoradiographic imaging using gas-counter technology, *Electrophoresis,* 11, 367, 1990.
57. **Sjöberg, S., Carlsson, P., Enerbäck, S., and Bjursell, G.,** A compact, flexible and cheap system for acquiring sequence data from autoradiograms with a digitizer and transferring it to an arbitrary host computer, *CABIOS,* 5, 41, 1989.
58. **Southern, E. M.,** Detection of specific sequences among DNA fragments separated by gel electrophoresis, *J. Mol. Biol.,* 98, 503, 1975.
59. **Beisiegel, U.,** Protein blotting, *Electrophoresis,* 7, 1, 1986.
60. **Burnette, W. N.,** "Western blotting": electrophoretic transfer of proteins from sodium dodecyl sulfate-polyacrilamide gels to unmodified nitrocellulose and radiographic detection with antibody and radioiodinated protein A, *Anal. Biochem.,* 137, 195, 1981.
61. **Jägersten, C., Edström, A., Olsson, B., and Jacobson, G.,** Blotting from PhastGel media after horizontal sodium dodecyl sulfate-polyacrylamide gel electrophoresis, *Electrophoresis,* 9, 662, 1988.
62. **Scherberich, J. E., Fischer, P., Bigalke, A., Stangl, P., Wolf, G. B., Haimerl, M., and Schoeppe, W.,** Routine diagnosis with PhastSystem compared to conventional electrophoresis: automated sodium dodecyl sulfate-polyacrylamide gel electrophoresis, silver staining and western blotting of urinary proteins, *Electrophoresis,* 10, 58, 198.
63. **Wolff, J. M., Pfeifle, J., Hollmann, M., and Anderer, A.,** Immunodetection of nitrocellulose-adhesive proteins at the nanogram level after trinitrophenyl modification, *Anal. Biochem.,* 147, 396, 1985.
64. **Heegaard, N. H. H., Hagerup, M., Thomsen, A. C., and Heegaard, P. M. H.,** Concanavalin A crossed affinity immunoelectrophoresis and image analysis for semiquantitative evaluation of microheterogeneity profiles of human serum transferrin from alcoholics and normal individuals, *Electrophoresis,* 10, 836, 1989.
65. **Macnamara, E. M. and Whicher, J. T.,** Electrophoresis and densitometry of serum and urine in the investigation and significance of monoclonal immunoglobulins, *Electrophoresis,* 11, 376, 1990.
66. **Jones, A. and Moeremans, M.,** Colloidal gold for the detection of proteins on blots and immunoblots, in *Methods in Molecular Biology,* Vol. 3, J. M. Walker, Ed., Humana Press, Clifton, NJ, 1988, 441.
67. **Surek, B. and Latzko, E.,** Visualization of antigenic proteins blotted onto nitrocullulose using the immunogold-staining (IGS)-method, *Biochem. Biophys. Commun.,* 121, 284, 1984.

68. **Dunn, M. J.,** Paper symposium — quantitative evaluation and densitometry, *Electrophoresis,* 11, 353, 1990.
69. **Pardowitz, I., Ehrhardt, W., and Neuhoff, V.,** Quantitative densitometry from the point of view of information theory, *Electrophoresis,* 11, 400, 1990.
70. **Shipton, H. W.,** The microprocessor, a new tool for the bioscience, *Annu. Rev. Biophys. Bioeng.,* 8, 269, 1979.
71. **Serra, J.,** Digitalization, *Mikroskopie Suppl.,* 37, 109, 1980.
72. **Kemnitz, H.-D. and Hougardy, H. P.,** Design of an interface for digital image storage, *Mikroskopie,* 7, 415, 1980.
73. **Skarnulis, A. J.,** A computer system for on-line image capture and analysis, *J. Microsc.,* 127, 39, 1982.
74. **Erhardt, R., Reinhardt, E. R., Schlipf, W., and Bloss, W. H.,** FAZYTAN — a system for fast automated cell segmentation, cell image analysis and feature extraction based on TV-image pickup and parallel processing, *Anal. Quant. Cytol. J.,* 2,25, 1980.
75. **Preston, K. Jr. and Dekker, A.,** Differentiation of cells in abnormal human liver tissue by computer image processing, *Anal. Quant. Cytol. J.,* 2, 1, 1980.
76. **Desai, V. and Reimer, L.,** Digital image recording and processing using an Apple II microcomputer, *Scanning,* 7, 185, 1985.
77. **Häder, D.-P. and Lebert, M.,** Real time computer-controlled tracking of motile microorganisms, *Photochem. Photobiol.,* 42, 509, 1985.
78. **Jaffe, M. J., Wakefield, A. H., Telewski, F., Gulley, E., and Biro, R.,** Computer-assisted image analysis of plant growth, thigmomorphogenesis, and gravitropism, *Plant Physiol.,* 77, 722, 1985.
79. **Bryan, S. R., Woodward, W. S., Griffis, D. P., and Linton, R. W.,** A microcomputer based digital imaging system for ion microanalysis, *J. Microsc.,* 138, 15, 1985.
80. **Boyes, E. D., Muggridge, B. J., and Goringe, M. J.,** On-line image processing in high resolution electron mircoscopy, *J. Microsc.,* 127, 321, 1982.
81. **Gravekamp, C., Koerten, H. K., Verwoerd, N. P., de Brujn, W. C., and Daems, W. T.,** Automated image analysis applied to electron micrographs, *Cell Biol. Int. Rep.,* 6, 656, 1982.
82. **Blackstad, T. W., Karagülle, T., and Ottersen, O. P.,** MORFOREL, a computer program for two-dimensional analysis of micrographs of biological specimens, with emphasis on immunogold preparations, *Comput. Biol. Med.,* 20, 15, 1990.
83. **Omasa, K. and Onoe, M.,** Measurement of stomatal aperture by digital image processing, *Plant Cell Physiol.,* 25, 1379, 1984.
84. **Kothari, M. J. and Shah, G. L.,** Epidermal structures and ontogeny of stomata in the Papilonaceae (tribe Hedysareae), *Bot. Gaz.,* 136, 372.
85. **Kincaid, D. T. and Schneider, R. B.,** Quantification of leaf shape with a microcomputer and Fourier transform, *Can. J. Bot.,* 61, 2333, 1983.
86. **Lockhausen, J., Kristen, U., Menhardt, W., and Dallas, W. J.,** 3-dimensional reconstruction of a plant dictyosome from series of ultrathin sections using computer image processing, *J. Microsc.,* 158, 197, 1990.
87. **Montag, M., Spring, H., Trendelenburg, M. F., and Kriete, A.,** Methodical aspects of 3-D reconstruction of chromatin architecture in mouse trophoblast giant nuclei, *J. Microsc.,* 158, 225, 1990.
88. **Jones, S. J., Taylor, M. L., Baarslag, M. W., Krol, J. J., Mosterd, B., Mans, A., Brakenhoff, G. J., and Nanninga, N.,** Image processing techniques for 3-D chromosome analysis, *J. Microsc.,* 158, 235, 1990.
89. **Dispro, A., Adami, M., Sartore, M., and Nicolino, C.,** IMAGO — a complete system for acquisition, processing, 2/3-dimensional and temporal display of microscopic bio-images, *Comput. Methods Progr. Biomed.,* 31, 225, 1990.
90. **Frank, J.,** Image analysis of single macromolecules, *Electron Microsc. Rev.,* 2, 53, 1989.
91. **Dhawan, A. P.,** A review on biomedical image processing and future trends, *Comput. Methods Progr. Biomed.,* 31, 141, 1990.
92. **Garrels J. I. and Franza, B. R. Jr.,** Transformation-sensitive and growth-related changes of protein synthesis in REF52 cells, *J. Biol. Chem.,* 264, 5299, 1989.
93. **Garrels, J. I. and Franza, B. R. Jr.,** The REF52 protein database, *J. Biol. Chem.,* 264, 5283, 1989.
94. **Yang, R.-C., Tsuji, A., and Suzuki, Y.,** Two-dimensional electrophoresis aided by personal computer analysis for screening of mutant proteins in inherited diseases, *Electrophoresis,* 10, 785, 1989.
95. **Neidhardt, F. C.,** Report of workshop on cellular protein databases derived from two-dimensional polyacrylamide gel electrophoresis, *Electrophoresis,* 10, 73, 1989.
96. **Neidhardt, F. C., Appleby, D. B., Sankar, P., Hutton, M. E., and Phillips, T. A.,** Genomically linked cellular protein database derived from two-dimensional polyacrylamide gel electrophoresis, *Electrophoresis,* 10, 116, 1989.
97. **Gill, P. and Werrett, D. J.,** Interpretation of DNA profiles using a computerized database, *Electrophoresis,* 11, 444, 1990.
98. **Santarén, J. F.,** Towards establishing a protein database of *Drosophila, Electrophoresis,* 11, 254, 1990.

99. **Mancini, P., Benassi, A., Valli, G., and Donato, L.,** Minimum computer system for videodensitometry and image analysis, *Med. Biol. Eng. Comput.,* 16, 542, 1978.
100. **Jansson, P. A., Grim, L. B., Elias, J. G., Bagley, E. A., and Lonberg-Holm, K. K.,** Implementation and application of a method to quantitate 2-D gel electrophoresis patterns, *Electrophoresis,* 4, 82, 1983.
101. **Freeman, S. E., Larcom, L. L., and Thompson, B. D.,** Electrophoretic separation of nucleic acids — evaluation by video and photographic densitometry, *Electrophoresis,* 11, 425, 1990.
102. **Boniszewski, Z. A. M., Comley, J. S., Hughes, B., and Read, C. A.,** The use of charge-coupled devices in the quantitative evaluation of images, on photographic film or membranes, obtained following electrophoretic separation of DNA fragments, *Electrophoresis,* 11, 432, 1990.
103. **Häder, D.-P.,** Computer-assisted image analysis in biological sciences, *Indian Acad. Sci. (Plant Sci.),* 98, 227, 1988.
104. **Wiedemann, G., Müller-Klieser, W., Walenta, S., Schleinkofer, L., and Wood, W. G.,** Low light level in vitro monitoring of cellular and antigen-antibody reactions using a photon detection camera system new perspectives for clinical diagnostics and research, *Klin. Wochenschr.,* 68, 33, 1990.
105. **Danielsson, P. E. and Kruse, B.,** PICAP II — a second generation picture processing system, *Mikroskopie Suppl.,* 37, 425, 1980.
106. **Steinbach, T., Unland, F., and Müller, K.-M.,** Kostengünstiges Microprozessorsystem zur Ergänzung eines Quantimet 720-Bildanalysegerätes, *Microsc. Acta,* 86, 139, 1982.
107. **Allen, R. D. and Allen, N. S.,** Video-enhanced microscopy with a computer frame memory, *J. Microsc.,* 129, 3, 1983.
108. **Toda, T., Fujita, T., and Ohashi, M.,** A method of microcomputer-aided two-dimensional densitometry: an apparatus equipped with a charge-coupled device camera, and an algorithm of microcomputer programming, *Electrophoresis,* 5,, 42, 1984.
109. **Elder, J. K., Green, D. K., and Southern, E. M.,** Automatic reading of DNA sequencing gel autoradiographs using a large format digital scanner, *Nucleic Acids Res.,* 14, 417, 1986.
110. **Russ, J. C. and Russ, J. C.,** Image processing in a general purpose microcomputer, *J. Microsc.,* 135, 89, 1984.
111. **Mayfield, C. I.,** A simple computer-based video image analysis system and potential applications to microbiology, *J. Microbiol. Meth.,* 3, 61, 1984.
112. **Ford-Holevinski, T. S., Agranoff, B. W., and Radin, N. S.,** An inexpensive, microcomputer-based, video densitometer for quantitating thin-layer chromatographic spots, *Anal. Biochem.,* 132, 132, 1983.
113. **Miller, M. J., Olson, A. D., and Thorgeirsson, S. S.,** Computer analysis of two-dimensional gels: automatic matching, *Electrophoresis,* 5, 297, 1984.
114. **Anderson, L.,** Identification of mitochondrial proteins and some of their precursors in two-dimensional electrophoretic maps of human cells, *Proc. Natl. Acad. Sci. U.S.A.,* 78, 2407, 1981.
115. **Celis, J. E., Ratz, G.P., and Celis, A.,** Secreted proteins from normal and SV40 transformed human MRC-5 fibroblasts: toward establishing a database of human secreted proteins, *Leukemia,* 1, 707, 1987.
116. **Klerk, H. and Jespers, A.,** Gelanal, a personal computer-program to compare protein patterns on 2-dimensional polyacrylamide gels, *Electrophoresis,* 11, 420, 1990.
117. **Czerniak, B., Herz, F., Wersto, R. P., Alster, P., Puszkin, E., Schwarz, E., and Koss, L. G.,** Quantitation of oncogene products by computer-assisted image analysis and flow cytometry, *J. Histochem. Cytochem.,* 38, 463, 1990.
118. **Freeman, H.,** Computer processing of line-drawing images, *Comput. Surv.,* 6, 57, 1974.
119. **Freeman, H.,** Analysis and manipulation of lineal map data, in *Map Data Processing,* Freeman, H. and Pieroni, G. G., Eds., Academic Press, New York, 1980, 151.
120. **Häder, D.-P.,** Automatic area calculation by microcomputer-controlled video analysis, *EDV Med. Biol.,* 18, 33, 1987.
121. **Celis, J. E., Ratz, G. P., Madsen, P., Gesser, B., Lauridsen, J. B., Brogaard-Hansen, K. P., Kwee, S., Rasmussen, H. H., Nielsen, H. V., Grüger, D., Basse, B., Leffers, H., Honore, B., Moller, O., and Celis, A.,** Computerized, comprehensive databases of cellular and secreted proteins from normal human embryonic lung MRC-5 fibroblasts: identification of transformation and/or proliferation sensitive proteins, *Electrophoresis,* 10, 76, 1989.
122. **Tarroux, P., Vicens, P., and Rabilloud, T.,** HERMeS: a second generation approach to the automatic analysis of two-dimensional electrophoresis gels. V. Data analysis, *Electrophoresis,* 8, 187, 1987.
123. **Krauss, M. R., Collins, P. J., and Blose, S. H.,** Computer-analysed 2-D electrophoresis, *Nature,* 337, 669, 1989.
124. **Tyson, J. J. and Haralick, R. H.,** Computer analysis of two-dimensional gels by a general image processing system, *Electrophoresis,* 7, 107, 1986.
125. **Mount, D. W. and Conrad, B.,** Improved programs for DNA and protein sequence analysis on the IBM personal computer and other standard computer systems, *Nucleic Acids Res.,* 14, 443, 1986.
126. **Elder, J. K.,** Maximum entropy image reconstruction of DNA sequencing gel autoradiographs, *Electrophoresis,* 11, 440, 1990.

127. **Häder, D.-P. and Truß, M.,** High resolution scanning of absorbing and fluorescent electrophoresis gels using video image analysis, *CABIOS,* 3, 339, 1987.
128. **Häder, D.-P. and Kauer, G.,** Image analysis techniques for automatic evaluation of two-dimensional electrophoresis, *Electrophoresis,* 11, 407, 1990.
129. **Danielsson, P. E.,** Encoding of binary images by raster-chain-coding of cracks, Rep. ISY-I-0497,S-581 83, Linköping, Sweden, 1982.
130. **Haberäcker, P.,** *Digitale Bildverarbeitung,* Carl Hanser Verlag, München, 1989, 1.
131. **Bronstein, I. N. and Semendjajew, K. A.,** *Taschenbuch der Mathematik,* Verlag Harri Deutsch, 1981.
132. **Press, W. H., Flannery, B. P., Teukolsky, S. A., and Vetterling, W. T.,** *Numerical Recipes,* Cambridge University Press, Cambridge, 1988.
133. **Lemkin, P. F. and Lipkin, L. E.,** 2-D electrophoresis gel data base analysis: aspects of data structures and search strategies in GELLAB, *Electrophoresis,* 4,71, 1983.
134. **Appel, R., Hochstrasser, D., Roch, C., Funk, M., Muller, A. F., and Pellegrini, C.,** Automatic classification of two-dimensional gel electrophoresis pictures by heuristic clustering analysis: a step toward machine learning, *Electrophoresis,* 9, 136, 1988.
135. **Potter, D. J.,** A review of the clip system for the quantitative analysis of 2-dimensional electrophoresis gels, *Electrophoresis,* 11, 415, 1990.
136. **Garrels, J. I.,** The QUEST system for quantitative analysis of two-dimensional gels, *J. Biol. Chem.,* 264, 5269, 1989.

Microscopy and Ultramicroscopy

Chapter 7

CONFOCAL IMAGE CYTOMETRY - QUANTITATIVE ANALYSIS OF THREE-DIMENSIONAL IMAGES OBTAINED BY CONFOCAL SCANNING MICROSCOPY

Jean Paul Rigaut, Santos Carvajal-Gonzalez, and Jany Vassy

TABLE OF CONTENTS

I. INTRODUCTION

The term "image cytometry" (ICM) defines the quantitative image analysis of cells and tissues with the help of a computerized system. Using conventional microscopy, the specimens (smears, imprints, and tissue sections) are observed in two-dimensional (2-D) projection. We propose[1-3] that the name "confocal image cytometry" (CICM) be given to the quantitative analysis of three-dimensional (3-D) images of cells and tissues, obtained by confocal microscopy.

Confocal microscopy was invented in 1955 by Marvin Minsky,[4,5] who was to become famous in another domain, artificial intelligence. Interestingly, his motivation, far from being unrelated to the latter when building his confocal microscope, arose from a superbly quixotic search for a way to study the 3-D arrangement of nerve cells in the brain! Minsky may therefore be considered as the father of confocal image cytometry, although many years would pass before this dream of obtaining 3-D microscopical images of the brain would actually become progressively true. This happened after another inventor, Mojmir Petráň, independently developed a similar type of confocal microscope,[6,7] also to observe brain cells,[8] followed by others using laser instruments.[9-13]

After a long technical evolution, confocal microscopy now represents an easy way to obtain 3-D images from thin, independent, optical "sections" deep inside unsectioned cells or tissues. Therefore, confocal imaging allows a straightforward use of the full 3-D information present in a tissue block. This offers a revolutionary approach to quantitative image analysis in cell biology. Up to recent years, the access to the third dimension, in terms of quantitative image analysis, relied mostly upon probabilistic stereological estimations from limited sets of physical sections.[14,15] Stereological methods are not devoid of severe limitations.[16] The only solutions for the 3-D imaging of thick preparations of cells or tissues were either tedious reconstructions from serial microtome-obtained sections or computationally intensive reconstructions from serial optical "sections" using deconvolution methods to eliminate the optical contributions from above and under each "section".[17]

Most work published up to now on confocal microscopy either has dealt with theoretical and instrumental aspects or has been satisfied with obtaining strikingly sharp images in 2-D confocal planes inside cells or tissues. The two major reasons for using 3-D images are reconstruction and quantitation. Although reconstruction per se does not correspond to CICM (which includes quantitation), it represents a compulsory step before CICM. Although some authors had understood early[18] that 3-D images can be produced by stacking up consecutive confocal images, relatively few results of biological applications on such images have been presented and those which include quantitation are still rare. Obviously, CICM is still in its infancy, but this is not surprising as the first confocal laser microscopes were not commercialized until 1988.

II. CONFOCAL MICROSCOPY

A. BASIC KNOWLEDGE

1. Principles of Confocal Microscopy

The reader is referred to leading articles in the field for detailed information on the fundamental principles of confocal microscopy.[19-27]

In classical optical microscopy, the depth of field is defined by the thickness of the part of the viewed specimen for which the image appears in focus. However, all the out-of-focus information participates in the imaging process and, therefore, no real resolving power exists in the direction of the optical axis.

In scanning optical microscopy (SOM), the specimen is scanned by a light beam focused to a diffraction-limited spot by a high numerical aperture (NA) objective. This scanning can be achieved by many simultaneous light beams, using a Nipkow disk (scanning reflected light microscope), or by a single laser beam (confocal scanning laser microscope). In both systems the confocal effect, i.e., the elimination of out-of-focus light contributions, is achieved by using a pinhole aperture on the imaging plane in front of the detector. With laser instruments, confocal images can be obtained only by reflectance or by fluorescence.

2. Instruments

In the first type of confocal SOM (CSOM), the scanning reflected light microscope (SRLM) invented by Minsky[4,5] and later, independently, by Petráň,[6-8] the reflected or fluorescently emitted light passes back in "tandem" (TSRLM) after beam splitting through holes in a Nipkow disk, diametrically opposed to those through which the illuminating light had passed.[28] A nontandem type of SRLM, in which both illuminating and emitted beams pass through the same Nipkow disk holes, has recently been described.[29] The first biological confocal images, obtained by TSRLM, were published in 1967 to 1968.[7,8]

The confocal scanning laser microscope (CSLM) derives from the marriage between the CSOM principle and scanning laser microscopy (SLM).[30-32] The first biological images obtained by CSLM were published in 1972 to 1973.[32,33]

Laser scanning may be brought about either by translating the specimen under a fixed beam (scanned stage SLM) or by moving the beam on a fixed specimen (scanned beam SLM). The latter system is usually achieved by rotating or vibrating mirrors, although other techniques, using acousto-optic deflectors (*vide infra*), or objective lens scanning,[34,35] are under experimentation.

Scanned beam ("off-axis") SLM, first developed in 1971,[31,36] is used in all present commercial CSLM instruments. An important advantage is that the scanning module may be easily adapted to any conventional microscope.

Scanned stage ("on-axis") SLM, first developed in 1977,[37] has the obvious advantage of minimizing optical aberrations and of offering a field of view which is less limited than with scanned beam SLM. Scanned stage SLM suffers, however, from a slow scan rate due to the speed limitation of present motorized high-precision scanning stages.[38]

A list of confocal systems which have been reported and/or commercialized up to now is shown in Table 1. The CSLM, because of its higher {x, y} and z resolutions and its ability to use much smaller light intensities, is preferable to SRLM systems for the preparation of 3-D images.

Confocal imaging with a CSLM may be achieved by reflectance or by fluorescence and only with epi-illumination, with the exception of stage-scanning instruments,[19] which are not commercially available yet.

Several commercial CSLM instruments allow simultaneous imaging at two different emission wavelengths,[39,40] a technique which produces less photobleaching than when using two consecutive imagings[41] and eliminates the risk of a misalignment produced by the change of emission filter. The emission spectra always overlap to some extent, but we have shown that the resulting "cross-talk" can be corrected.[40] Ratio imaging may be used to evaluate spectral changes induced by ion-binding[42] and fluorescence-resonance energy transfer.[24]

An additional nonconfocal image for the scanned transmitted light may be obtained by implementing a fiber-optic guide collecting the light from below the microscope condenser to a second detector. Many other imaging modes such as phase contrast, differential phase contrast, differential amplitude contrast, darkfield, polarization, optical-beam-induced contrast, and confocal interference contrast may be used.[27]

Although all commercial confocal laser microscopes may be used for CICM, the range of prices for approximately equivalent performances is stunning. Some systems (surprisingly not the most expensive) offer more complete image analysis packages. A potential buyer will be wise to purchase an instrument with which he will be able (without having to buy

TABLE 1
Confocal Microscopes

System	Commercial	Light source	UV	Scan mode	Detector	Ref.
		SRLM				
Confocal 2002	Strojimport	Lamp		b/t	tv	7
K-2	Technical Instrument Co.	Lamp		b/t	ccd	
TSM	Tracor	Lamp	+	b/t	tv	164
SSOM		Lamp		b	tv	29
—		Lamp		b	tv	165
		CSLM				
MRC-600	Bio-Rad	Laser(s) (1—2)		b	ppm	109
LSM-10	Carl Zeiss	Lasers (2)	+	b	ppm	78
2LM21	Lasertec	Lasers (2)		b	ccd	
PHOIBOS-1000	Molecular Dynamics	Laser		b	ppm	110
CLSM	Leica	Lasers (2)		b	ppm	
ODISSEY	Tracor	Laser		b	ppm	
INSIGHT	Meridian	Laser		b	tv	
—		Laser		s	ppm	31
CSLM		Laser	+	s	ppm	19
—		Laser		b + s	ppm	57
—		Lasers (2)	+	b(a)	ppm	48
SOFM		Laser		s	ppm	166
EMBL-CSSLM		Laser	+	s	ppm	18
EMBL-CBSLM		Laser		b	ppm	167
—		Laser		o	ppm	34
TNO-CLSM		Laser		b(a)	ppm	55
LSM		Laser		b	ppm	77
—		Laser		b(a)	idt	56

Note: b, beam; b(a), beam with acousto-optical deflector(s); b/t, beam/tandem (TSRLM); b + s, beam + stage; o, objective lens; tv, TV camera; ccd, CCD camera; ppm, photomultiplier; idt, image dissector tube.

another expensive machine) to enhance his images and segment, reconstruct, and measure all relevant objects.

3. Lasers for CSLM

Among the many types of continuous wave lasers,[43] it seems that, up to now, only gas lasers (argon ion, krypton ion, cadmium and helium neon) have been used for CSLM. Dye lasers have a broad tuning range. Most solid state lasers operate in the near infrared (IR). The tuneable solid state titanium sapphire, argon ion-pumped laser is replacing IR dye lasers. Solid state diode-pumped lasers, which can operate in the red at 527 nm (yttrium lithium fluoride) or at 532 nm (neodymium yttrium aluminum garnet), are an attractive alternative to helium-neon lasers. A pulsed dye laser has been used in the recently developed two-photon CSLM technique (*vide infra*).[44]

Low-powered, air-cooled argon-ion lasers are by far the most commonly used, as their two main excitation lines, at 488 and 514 nm, are very convenient for many markers and especially for the most common ones, fluorescein (with 488 nm) and rhodamine and Texas Red™ (with 514 nm). Among their many other less intense excitation wavelengths (between 458 and 529 nm), the 458-nm line has been used in CSLM for chromomycin A3[2] and for lucifer yellow.[9,11,12,45]

Krypton-ion lasers have numerous wavelengths (between 338 and 799 nm). The strongest lines are in the red range. Lines at 476,[46] 531,[47] and 647[46,48] nm have been used in CSLM.

Helium-cadmium lasers also have numerous wavelengths (between 325 and 888 nm), but only those at 325 and 442 nm are strong enough to be used in practice, with the exception of additional lines in the green and the red with "white-light" He-Cd lasers. The line at 442 nm has been used in CSLM.[42,48]

Helium-neon lasers are mostly used for red excitation. Among their numerous wavelengths (between 543 and 1523 nm), the strong 633-nm line has been used in CSLM[19] and in TSRLM,[49] and a weak line at 543 nm (wavelength close to the classic mercury lamp green line at 546 nm) has been used in CSLM.[50]

Imaging at UV wavelengths (using either a helium-cadmium or krypton-ion laser or the pumped rare-gas "excimer" laser) is to be considered with extreme care, as presently available objectives are not achromatic under UV light[51] and this may produce substantial image deformations.[52]

4. Ongoing Developments

New instruments and techniques are being developed. Most of them concentrate on increasing the image acquisition speed of CSLM systems. Cell scanning rates approaching those achieved by flow cytometry should be possible in the near future.[48] Novel scanning systems, such as fast-rotating polygons[53] or high-resolution acousto-optical deflectors are being tried.[54-56] The polygon has been used to scan in the x-direction while the specimen was translated in the y-direction,[57] but this is not easy to achieve.[48] Acousto-optical deflectors allow imaging at standard video rates.[56] They have, however, been reported to introduce small image distortions.[58] They usually require that one of the scan axes still use a conventional mirror and that the partially descanned fluorescence emission be imaged through a slit (instead of a circular pinhole) or by CCD detectors, due to the fact that the fluorescence emission cannot be sent back through the acousto-optic modulator which is wavelength specific. A new instrument which allows the use of two acousto-optical deflectors and achieves confocality by raster scanning the aperture of an image dissector tube has recently been presented.[56]

New microscope objectives which include a correction for spherochromatism in the near UV range are being developed.[48,52]

"Super resolution"[59] can be achieved by deconvoluting (*vide infra*) images obtained with an array of detectors.[60]

Preliminary results have been presented on particle trapping with a confocal microscope, the trap manipulation being done either by moving the beam or by moving in {x, y, z} a special high NA objective.[61]

A technique based on nonlinear molecular excitation by the simultaneous absorption of two photons allows an intrinsic 3-D resolution and the elimination of out-of-focus bleaching and of UV chromatic aberrations.[44] The simultaneous absorption of two photons from a stream of strongly focused femtosecond pulses of red light, generated by a colliding pulse mode-locked dye laser, stimulates fluorescence from fluorophores having single-photon absorption in UV.

Finally, confocal Raman microspectroscopy[62] should allow studies of compositional and structural information inside cells without the use of fixatives or stains.

B. CYTOLOGICAL AND HISTOLOGICAL PREPARATION METHODS

1. Reflectance

Gold immunolabeling, silver intensified[63] or not,[64] has been used for reflectance with CSLM.

We are currently quantitating the number of nucleolar organizer region-associated proteins (NORs) in normal and pathological livers. The silver-staining method for NORs[65] must be modified to include a reduction of the product to metallic silver.

2. Fluorescence

Immunofluorescence is by far the most common preparation technique used for CSLM. The quantitation of markers imaged simultaneously at different wavelengths is possible. Confocal ''sectioning'' eliminates the problems, encountered in conventional microscopy, of underestimating particle counts because of projection effects.

It is beyond our scope to cite all the fluorochromes that might be useful in confocal microscopy. The reader is referred to two major reviews[66,67] and to Table 2 for a nonexhaustive list.

Specific dyes (e.g., for DNA, cytoskeletal microfilament proteins, mitochondria) and fluorescinated monoclonal antibodies may be used. In pathology, markers for DNA,[3,68,69] oncogenes,[26] BrdU,[70] and the Ki-67 antigen[71] are especially interesting.

Specific DNA probes for *in situ* hybridization[72-74] have been used with CSLM in fluorescence. A quantitative evaluation of the hybridized sequence may be envisaged.[75] Sequences distant from less than 100 kb can be visualized separately in interphase nuclei.[76] *In situ* hybridization is increasingly used to locate specific chromosome regions and there is no doubt that multiple DNA probes will become an important way to search for chromosomal abnormalities in cancers.

Markers of DNA, which are the most routinely used fluorochromes for ICM, will certainly be in great demand as well for CICM. The most common ones are pararosanilin,[64] quinacrine,[26] propidium iodide,[70,77] acridine orange,[78,79] ethidium bromide, mithramycin,[64,78,80] and chromomycin A3.[2,3,68,69,81-83] The Hoescht 33,342 and DAPI stains require UV excitation, which presents major difficulties with CSLM instruments (*vide supra*). The same problem limits the use of Indo-1 for Ca^{2+} imaging and quantitation, which is more easily achieved with Rhod-2 or Fluo-3.[84]

The use of particular fluorochromes makes observation *in vivo* possible in some cases, for instance, in the study of living embryos,[85] molecule and drug microkinetics[86,87] and intracellular pH,[42] flavoprotein (FAD),[88] and Ca^{2+}.[84,89] Voltage-sensitive styryl dyes open the way to the imaging of the activity of single cells in a functioning brain or tissue slice preparation.[10] Experiments on living plant cells and tissues require a treatment at pH 10 to permeate the cell membrane[90] before being able to use most fluorescent dyes.[91]

For dual wavelength imaging, a convenient method consists in combining fluorescein and Texas Red™.[40] A double fluorochrome conjugate with a large Stokes shift, FR-1 (covalently-linked molecules of fluorescein and rhodamine), has recently been developed for single-wavelength double-labeling applications.[66]

Autofluorescence may also be used with suitable biological materials, for instance chloroplasts.[21,47]

Photobleaching is a major problem in conventional fluorescence imaging,[24,92] and this is also true in CSLM (*vide infra*).[2,24,67,93] When using fixed specimens, antibleaching agents may be used.[67,94] Many antioxidants, such as *p*-phenylene-diamine (PPD), diazobicyclooctane (DABCO), propyl-gallate, hydroquinone, crocetin, and etretinate can be used, but they are not devoid of problems, as the first four can have pharmacological effects, the first one is dangerous to manipulate (carcinogenic), and the last two (carotenoids) are fluorescent.

3. Penetration

Analyzing cells or tissues by CSLM requires a good penetration of the fluorochrome into the specimen. This requires a permeation treatment during the histological preparation when preparing very thick 3-D images. Saponin, for instance, can be added to the fixation solution.[2,3,68,69]

TABLE 2
Fluorochromes

Use	Fluorochrome	Excitation peak (nm)	Emission peak (nm)
Covalent labeling	Fluorescein-isothiocyanate (FITC)	490—494	520—521
	Eosin-ITC	524—525	548—550
	Erythrosin-ITC	535—540	558—560
	Tetramethylrhodamine-ITC (TRITC)	541—554	572—573
	Rhodamine X-ITC (XRITC)	578—582	601—604
	Texas Red™ sulfonyl chloride	596	615—620
	Lissamine rhodamine B sulfonyl chloride	567—570	584—590
	Cascade Blue™	375—378	423—425
		398—399	423—425
	Phycoerythrin-R	480—565	578
	Allophycocyanin	650	660
	Coumarin-phalloidin	387	470
	Bodipy™ phallicidin	505	512
	Nitrobenzoxadiazole (NBD)	478	520—550
	Dansyl hydrazine	336	531
	Indotrimethinecyanines (CY3)	530—550	575
	Indopentamethinecyanines (CY5)	630—650	670
Nucleic acid labeling	Hoechst 33342	340—343	450—483 (DNA)
	DAPI	345—350	455—470 (DNA)
	Propidium iodide	493—536	623—630 (DNA)
	Ethidium bromide	482—510	595—616 (DNA)
	ACMA	430	474 (DNA)
	Chromomycin A3	420—445	575—580 (DNA)
	Mithramycin	420—445	575—580 (DNA)
	7-Aminoactinomycin-D	523	647 (DNA)
	Acridine Orange	480—487	510—520 (DNA)
		440—470	650 (RNA)
	Thiazole Orange	453—501	480—547 (RNA)
	Pyronin Y	497	563 (RNA)
Thiol conjugates	Lucifer Yellow	426—490	525—540
	Cascade Blue™	375—378	423—425
		398—399	423—425
Membrane potential	Thiazole Orange	453	480
	$DiOC_n$ (3)	482—485	500—510
	Tetramethyl-rhodamine ethylester	549	574
Mitochondria	Rhodamine 123	505—511	534
	2-di_1ASP (DASPMI)	461	518
Lipid content	Nile Red	485	525
	NBD phosphatidyl-ethanolamine	450	530
	Diphenylhexatriene (DPH)	330, 351, 370	430
pH	BCECF	505—508	530—531 (basic pH)
	SNAFL-1™	479, 508	543 (acid pH)
		537	623 (basic pH)
	SNARF-1™	518, 548	587 (basic pH)
		574—579	636—640 (basic pH)
	DCDHB	340—360	420—440 (acid pH)
		340—360	500—580 (basic pH)
	FD	490	515
Ca^{2+}	Fura-2	335—340	512—518 (low Ca)
		360	505—510 (high Ca)
	Fura-3	368—373	462—494 (low Ca)
		341	459 (high Ca)
	Indo-1	350	482—485 (low Ca)
		330	390—410 (high Ca)

TABLE 2 (continued)
Fluorochromes

Use	Fluorochrome	Excitation peak (nm)	Emission peak (nm)	
	Quin-2	332	492	(high Ca)
		352	492	(low Ca)
	Rhod-2	550	574	(low Ca)
	Fluo-3	464	525	(low Ca)

Note: The ranges are due to determinations of wavelengths made in various solvents.[66,67]

Staining can be made on cryostat slices[40] or *en bloc*[2,3,68,69] after paraformaldehyde fixation. We have shown that there is no limitation of the penetration of chromomycin A3 up to at least 60 μm.[2] It seems difficult to obtain good results with paraffin-embedded material.

III. 3-D IMAGES

A. 3-D IMAGING

1. The Imaging Process

The axial resolution in CSLM is limited by the imaging process, i.e., the convolution of the original image with the transfer function of the instrument.[19,20,37,95-104]

The maximum theoretical z resolution (Rayleigh criterion) in CSLM systems is between 650 and 750 nm, depending upon instrumental and imaging conditions.[18,19,46,80] Experimental values are found between 700 and 750 nm in reflectance imaging[18,19,46] and between 800 and 900 nm in fluorescence imaging[21,64,79,80,103] under usual optimal conditions (planapochromat objective of 1.40 NA, immersion oil of 1.515 refraction index, excitation wavelength of 488 or 514 nm, emission wavelength above 520 nm, small detector pinhole opening). Microscope objectives of high quality and high NA (1.3 or 1.4) must be used to ensure optimal confocal "sectioning" effects.[102,105,106]

The {x,y} and z resolutions in CSLM will decrease if the detection pinhole is opened wider to increase the image intensity. The z resolution, however, is not too sensitive to a modest increase in the pinhole aperture.[64,101] The MRC-600 (Bio-Rad, U.K.) CSLM, which uses "quasi-infinity optics" by folding the optical path by means of mirrors, has an adjustable pinhole which allows the imaging of weakly fluorescent specimens; this is not the case for most other instruments.

The maximum theoretical {x, y} resolution in CSLM, increased by a factor of approximately 1.4 as compared to conventional microscopy, is between 159 and 185 nm, depending on instrumental and imaging conditions.[19,80] Experimental values are found between 190 and 300 nm.[18,19,46,79,103]

2. Construction of 3-D Images

A 3-D image is usually obtained by stacking up {x, y} images from consecutive confocal planes.[18] It is also possible with a CSLM to assemble consecutive {x, z} images.[2,18,107]

Scanning in the z direction may be achieved by a motor drive on the microscope's focus knob or by a piezoelectric translator between the microscope stage and the specimen.[101,105]

The ideal spacing (z) between confocal planes depends on the imaging conditions. According to the Nyquist theorem,[108] the {x, y} and z sampling frequencies required for an accurate digital representation should be twice the spatial resolution of the optical system. This rule, crucial when deconvolution (*vide infra*) is envisaged, leads to {x, y} and z scanning intervals of approximately 100 and 400 nm, respectively. When a deconvolution is not envisaged, larger intervals may be used, even for quantitative studies.[2,3,68,69,83,83]

The x, y calibration can, of course, be modified by the microscope objective, but also by the laser scanning step (''electronic zoom'').

The problem of determining the optimal laser power is not trivial, as all undesirable effects (*vide infra*) increase when the intensity is increased to approach the saturation of the revelant fluorochrome.[67]

We have obtained images of up to 100 μm deep at high magnification in normal and cancerous livers and esophagus.[3,69] Other workers have obtained images of up to 200 μm deep at low magnification in the brain.[11] High NA objectives have a limited working distance. Lower NA values can be used when high resolution is not essential.[81,109,110]

3. Undesirable Effects Encountered in 3-D Imaging

a. Photobleaching and Attenuation of the Fluorescence Intensity in Depth

Even in ideal conditions an attenuation of the detected fluorescence intensity with depth is observed in 3-D images.

There is no decrease of the intensity of the excitatory beam up to at least 60 μm deep, as demonstrated by the fact that the bleaching half-life does not depend upon the depth,[2] whereas it does depend of course upon the laser intensity.[2] This is at variance with an *a priori* belief that the opposite may be true.[64,81]

Therefore, the main limiting factors are the attenuation of the intensity with depth of the emission beam as it passes through the tissue and the attenuation of the intensity by photobleaching (''bleaching''). The former can be modeled by a negative exponential function.[2] Contrary to what seems to be a common belief,[111] this is not true for the latter.[2]

The chemistry of bleaching is complex.[67,92,93] Attempts at deriving analytical equations for bleaching are dependent upon limiting assumptions, for instance, bleaching from the triplet electronic energy level state only and no return from this state to the ground level singlet state.[93] Even a sum of three exponential functions, derived from theoretical considerations,[93] does not allow a good fit of experimental curves of bleaching vs. time.[2] We have obtained a very good empirical fit with a modified log-logistic function.[2,112]

The intensity of the excitatory beam at the in-focus voxel is enormously higher with CSLM than with conventional fluorescence,[81] but the total energy absorbed by the specimen may nevertheless be lower with CSLM since only a very small fraction of the field is illuminated at any given time. An important reduction of bleaching rates may be obtained by scanning fast enough to reduce the dwelling time to approximately 0.1 μs/pixel.[113] The dwelling time is 2.5 μs/pixel under our conditions for 1 scan; the use of averaging for noise reduction (10 scans per image) leads in practice to a value of 25 μs/pixel. The use of improved imaging and scanning techniques in CSLM should help in obtaining high-resolution 3-D images with shorter scanning dwell times and, therefore, less bleaching.

High excitation levels may produce images of reduced contrast.[81] This does not seem to be due to increased bleaching in excitation-saturated regions,[27] but only to saturation effects per se.[81]

A spatial heterogeneity in bleaching rate constants has been reported in cytology.[114] We did not observe this under our experimental conditions.[2]

When acquiring consecutive stacked-up confocal images, the effects of bleaching are more important in the last planes than in the first ones and it is therefore preferable, when preparing a 3-D image, to stack up confocal images from in depth to the surface. Indeed, contrary to some reports,[81,115,116] bleaching is not limited to the confocal plane being viewed. Our experimental demonstration of this fact[2] is in agreement with theoretical notions.[93]

b. Other Undesirable Effects

Photodestruction of biological structures (distinct from photobleaching) may occur, saturating with increasing laser power. It seems that the damage may be reduced by intermittent illumination, using higher intensities for shorter times.[67]

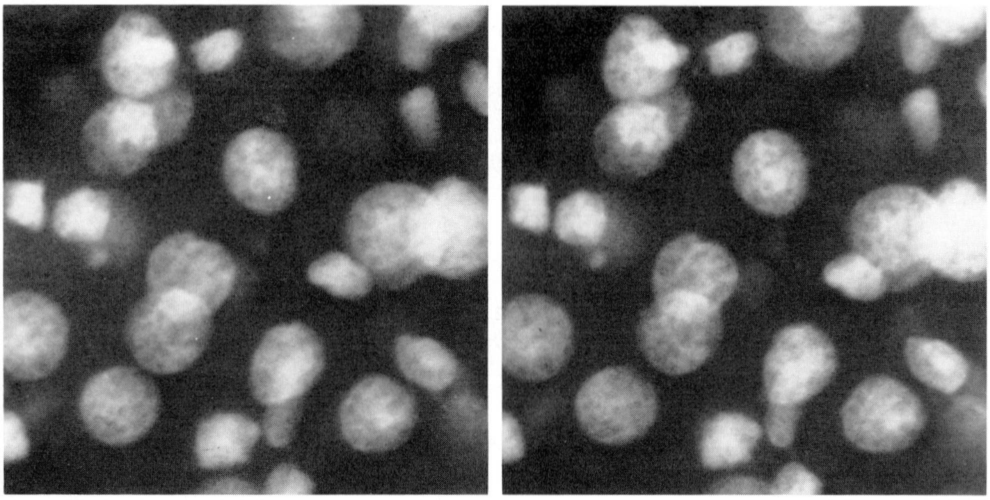

FIGURE 1. Stereopair of a 20 μm-thick region inside a 3-D image obtained with the MRC-600 (Bio-Rad, U.K.) confocal scanning laser microscope showing nuclei from a normal adult rat liver. *En bloc* staining of DNA by chromomycin A3.[2] Calibrations: {x, y} = 4 pixels/μm; z = 1 pixel/μm. (From König, D. et. al., *J. Microsc. (Oxford)*, 161, 405, 1991. With permission.)

Rayleigh scattering can produce unwanted signals, due either to excitation light passing through the dichroic mirror and the barrier filter or to imperfect monochromaticity of the excitation beam; this is often the case when using a simple interference filter to select one line from a multiline laser.[67]

Raman scattering contributes fluorescencelike signals, for instance, due to the characteristic Raman band of water (producing an emission peak at 584 nm for an excitation at 488 nm); additional wavelengths closer to the excitation may appear at high concentrations of protein or embedding media.[67]

Rayleigh and Raman scattering, being proportional to the laser power, do not saturate, as opposed to specific fluorescence.[67] Therefore, excessive laser power decreases the image contrast.

Unwanted autofluorescence may arise from endogenous fluorochromes. Flavins and flavoproteins, which absorb at 488 nm and emit in the same region as fluorescein,[67] are the main source of problems. Reduced pyridine nucleotides and lipofuscin absorb UV light and are difficult to saturate.[67]

The predominant source of noise in confocal images is photon noise.[11] A random noise component due to the use of barrier filters for dual channel imaging has also been observed.[40]

4. Visualization of 3-D Images

A 3-D image can of course be displayed as a series of consecutive {x, y} or {x, z} 2-D images, but everyone's dream is a real 3-D visualization. Real-time holograms from digital data are not quite feasible yet. Creating images with a pseudorelief effect is at present the only available solution.

The most popular way consists in preparing stereoscopic pairs ("stereopairs") by photographic means[117] or by computing orthographic projections from arbitrary viewing directions.[108] Stereomicrophotography has mainly been used with TSRLM instruments.[28,118] With CSLM systems, digital methods are particularly easy to use. The most common one consists in assigning the maximum values of the pixel intensities encountered along symmetric oblique "viewing" lines in the 3-D image to the pixels in the corresponding images for the stereopair.[80,119,120] An example is shown in Figure 1. The intensity may also be assumed to be inversely proportional to the distance from the viewer.[121] An enhancement of the stereoscopic perception can be obtained be preprocessing each confocal image by a modified Wallis transformation or a gradient operator.[122] One person out of ten is unable to perceive depth from stereopairs.[123] Stereoprojection, now used in many congresses, seems more easily accepted by most people than printed stereopairs. It requires two ordinary slide projectors with additional polarizing jelly on their optical path, a special metallized screen, and polarizing viewing glasses. The image fusion is achieved by the projectors. Stereoscopic animation may also be used.[120]

A ray-tracing algorithm[124] has been used to reconstruct a "cuberille" model.[125] A simulated fluorescence emission process has also been proposed.[126]

A 3-D effect can be obtained by a rotation, with an adequate speed, of the full 3-D image on the computer graphics screen (e.g., with the Thru-View software, Bio-Rad, U.K.).

Several types of algorithms can be used to visualize 3-D images containing segmented biological objects (*vide infra*). The need for standards has been stressed.[127] Methods producing consecutive "wires" corresponding to binarized object contours should be considered obsolete compared to state of the art algorithms including surfacing, smoothing,[128] and shading.[129,130] Three commercial software packages based on such methods, Visilog-3D (Noesis, France), Vanis (Molecular Dynamics, U.S.), and 3D-GMR (Apollo, U.S.), have been applied to CSLM, with, respectively, the Bio-Rad[83] and the Molecular Dynamics[121] commercial instruments and a noncommercial system.[126] These surfacing methods are based on smoothing and shading, following either a contour polygonation and triangulation between contour points[128] (e.g., Visilog-3D; 3D-GMR) or a voxel representation[131] (e.g., Vanis; VoxelView, Silicon Graphics, U.S.).

User-friendly procedures for 3-D data manipulation are being developed.[132,133]

5. Storage of 3-D Images

The storage requirements of 3-D images are huge. For example, the images we routinely prepare at present, which are composed of 512 × 512 × 64 voxels, require 16 Mbytes of storage space each. Standard computer hard disks will never have enough space for a complete experiment, which will usually require a large number of images (especially for quantitation, where sampling becomes crucial). The solution is to use an archive medium. The possibilities are, at present, either a large Winchester disk or removable media.[134] Among the latter, erasable magnetic cartridge tapes (40 to 200 Mbytes) are the cheapest. Nonerasable laser optical disks (WORM) have a very large capacity (2 Gbytes). Erasable laser optical disks (600 Mbytes) have been recently commercialized.

Data compression procedures, such as run-length encoding and octree representations[135] may be used. A vector representation may be used after data reduction.[45]

B. ANALYSIS OF 3-D IMAGES

1. Introduction

Image analysis includes image enhancement, segmentation of relevant objects, and their measurement. Segmentation is facilitated by image enhancement. The latter may be achieved by nonspecific means,[132] but it is always preferable to start by restoring the image, i.e., applying methods based on the knowledge of the distortions due to the imaging process.

The image processing of large 3-D images requires a more powerful workstation interfaced with the host computer of the CSLM system, which then remains free for image evaluation and acquisition. Our image analysis network, for instance, is composed of an MRC-600 (Bio-Rad, U.K.) controlled by a Nimbus (R.M., U.K.) microcomputer, interfaced by Ethernet to a Vax/VMS cluster (D.E.C., U.S.) and to three image analyzers: a Deskpro (Compaq, U.S.) Unix microcomputer with an SM board (Matrox, Canada) is dedicated to direct 3-D image analysis and reconstruction, and an IMCO-1000 and a DEC/IPS (Kontron,

Germany) are used for fast 2-D image processing (i.e., separately for each confocal plane). A user who would be less oriented toward development than we are would still be well advised to acquire a separate workstation for image processing and reconstruction, interfaced by Ethernet to the host computer of his CSLM system. The Unix system is preferable to MS/DOS because of the constraints specific to 3-D analysis and reconstruction packages.

2. Image Restoration

a. Correction for the Attenuation of Intensity in Depth

Correcting mathematically for depth- and bleaching-related attenuation of intensity is useful when preparing images for 3-D visualization and before the automated segmentation of 3-D images (*vide infra*), where the use of a constant gray-level threshold at all depths will facilitate discrimination operations. Correcting becomes compulsory before any quantitative analysis based on gray levels, such as chomatin texture analysis of nuclei and quantitation of nuclear DNA (*vide infra*).

A correction method using a proportionally increasing factor has been described,[64] but it yields imperfect results because the attenuation of fluorescence intensity in depth is not linear.[2]

We have recently proposed a more satisfactory mathematical correction method based on a formula derived from the log-logistic equation.[2,112] An example of a correction is shown in Figure 2.

b. Deconvolution

As seen above, the axial resolution in CSLM is limited by the imaging process. As the modulation transfer function (MTF) of the instrument can be evaluated,[79,104,136] deconvolution can be achieved[60] producing an important gain in resolution. The method, however, is very computation intensive.

''Super resolution''[59] can be achieved by using an array of detectors, allowing the use of the out-of-focus information in deconvolution schemes.[60]

c. Other Restoration Methods

Enhancement by geostatistical filters has been reported.[137] This allows both a certain degree of signal restoration and an interpolation of voxels onto regular lattices (*vide infra*).

As the predominant source of noise in confocal images in photon noise,[11] better signal to noise ratios are obtained by averaging several image inputs and by increasing the laser power if it is too far from saturation.

The ''cross talk'' and the noise component observed when using dual channel imaging may be eliminated.[40]

3. Image Segmentation

Automated image segmentation, although difficult in most cases in biology,[16] is facilitated in CSLM by the fact that the intensity of scattered light is decreased by several orders of magnitude, producing highly contrasted images.

Image segmentation can of course be used separately on each 2-D confocal plane of a 3-D image.[78,79,82,125] It makes more sense, however, to use direct 3-D algorithms, notably for particle deagglomeration[83] and for noise reduction.[126] It must be noted that the CSLM signal strength is dependent on the orientation of the relevant objects with respect to the confocal plane.[104]

With CSLM-obtained images, a major difficulty in 3-D processing arises from the anisotropic character of the voxels, due to the lower resolution in z than in {x, y}. Interpolation algorithms must be used to create artificial planes between the actual ones so as to equalize the {x, y} and z resolutions.

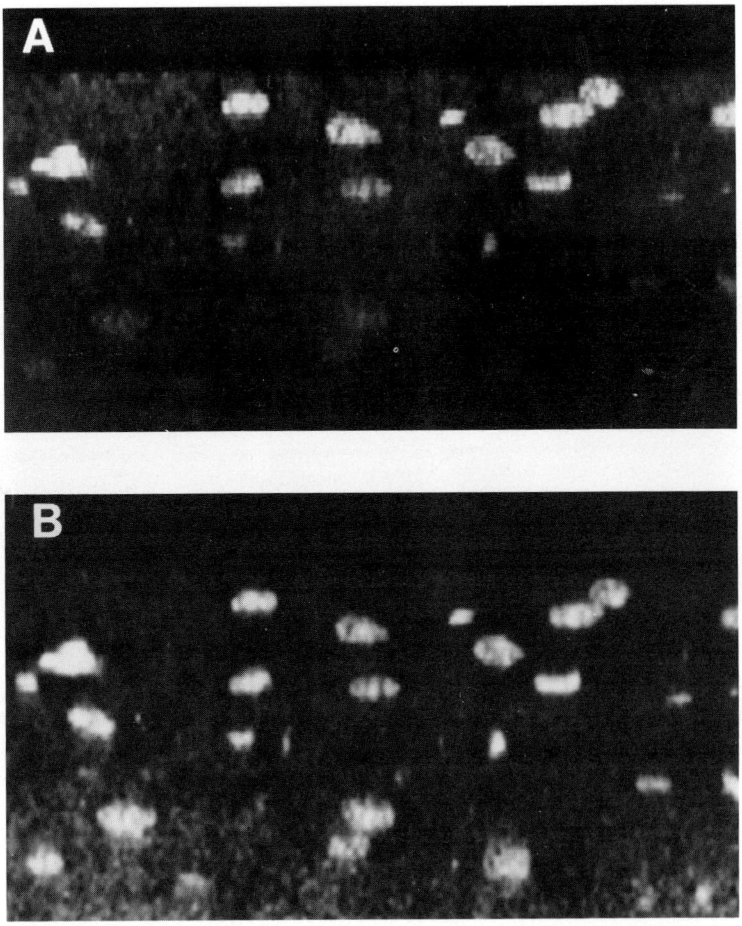

FIGURE 2. An {x, z} section through a 3-D image (50 μm-thick) of a normal adult rat liver. Same calibrations as in Figure 1. (A) The attenuation of the fluorescence intensity in depth is obvious; (B) same, after mathematical correction[2] for photobleaching and attenuation of fluorescence emission in depth.

As the cubic digital grid is ill-suited for the estimation of connectivity in space, the lattice system must be transformed into another one. The tetradecahedron (also called cuboctohedron) described by Archimedes and discussed by Kepler[138] was first used as a lattice kernel in the early 1980s[139-141] for 3-D image analysis, including skeletonization.[142]

A renewed interest in 3-D image analysis, due to confocal microscopy, has spurred the development of theoretical derivations for the optimal generalization of mathematical morphology[143,144] to 3-D.[137,145] The tetradecahedron (based on the center of the unit cube and the centers of its edges) and also the rhombododecahedron (based on the center and the vertices of the unit cube) may be used as lattice kernels and structuring elements.[143] Using a transformed lattice, the distance transformation is easily implemented with a sequential algorithm, allowing the use of watershed-derived methods[145] and of a skeletonization procedure.[126,145] Skeletonization is impossible in 3-D by the thinning method commonly used in 2-D.[145]

We have presented recently the first results of direct 3-D segmentation by gray-tone mathematical morphology on the tetradecahedral lattice (Visilog-3-D, Noesis, France);[83] rat liver cells were segmented in 3-D images obtained with an MRC-600 (Figures 3 and 4).

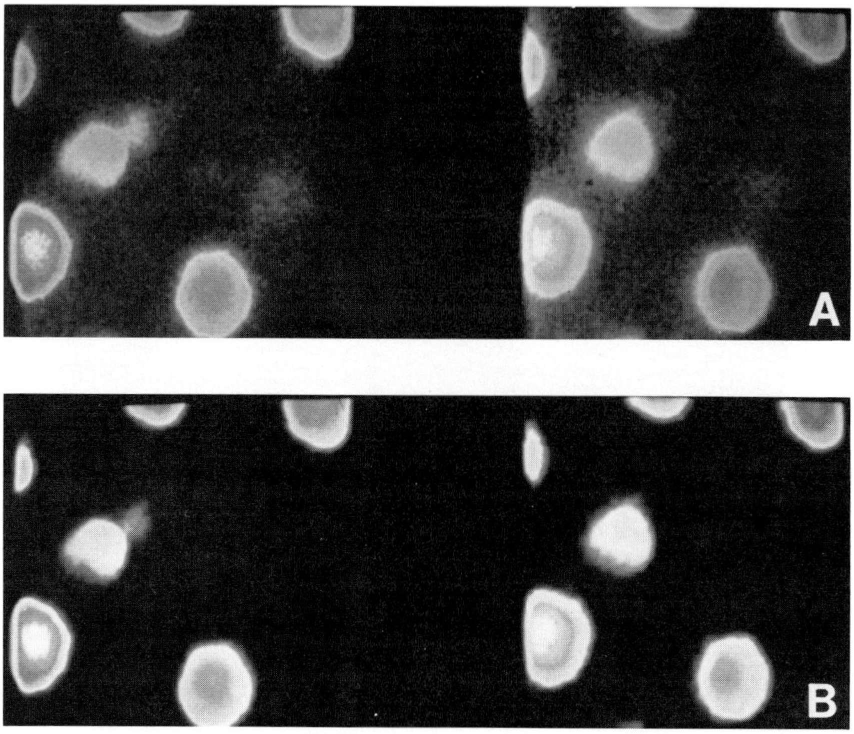

FIGURE 3. Stereopairs showing the elimination of background noise in a region (16 μm thick) inside a 3-D image of a normal adult rat liver. A 3-D mathematical morphology gray tone opening was used on a tetradecahedral lattice obtained by transformation of the original lattice.[83] Calibrations: {x, y} = 4 pixels/μm; z = 3.3 pixels/μm. (A) Before opening; (B) after opening.

Other authors have used binary mathematical morphology on a similar type of lattice to "clean" thresholded images (CSLM-obtained) of osteocyte lacunae.[146] Some earlier CSLM results had been obtained with Gauss, Laplace, Sobel, and median filters and with binary dilation and erosion, but with no modification of the original lattice; cubic filters were used, but discarding corner voxels improved the results for median filtering.[126]

4. Measurements

For any quantitation of discrete objects, a guard frame must be used[147] to avoid the frame-edge bias related to particles of different sizes.

Measurements may require some theoretical developments concerning, for instance, the estimation of surface areas, curvatures, etc., from binarized representations using either a voxel or a surface-triangulated model (*vide supra*).

The estimation of particle volume by the "Cavalieri principle"[15] is straightforward (the number of 1-voxels is an unbiased estimator). The estimation of its variance is more complex.[148]

The surface area may be estimated by the average number of chords over the major directions of the lattice with an appropriate weighting.[146]

The shell method for the correction of the frame-edge bias in the estimation of the Euler-Poincaré characteristic has been adapted to the tetradecahedral lattice.[149]

We have developed a method for the estimation of nuclear DNA amounts.[3,68,69]

Spatial statistics[150-152] and marked point process theory[153,154] can be used to quantitate histological architectural patterns.[82,83,155] Graph theory, which has been used to study tissue architecture in 2-D,[156] may also be applied to 3-D data.

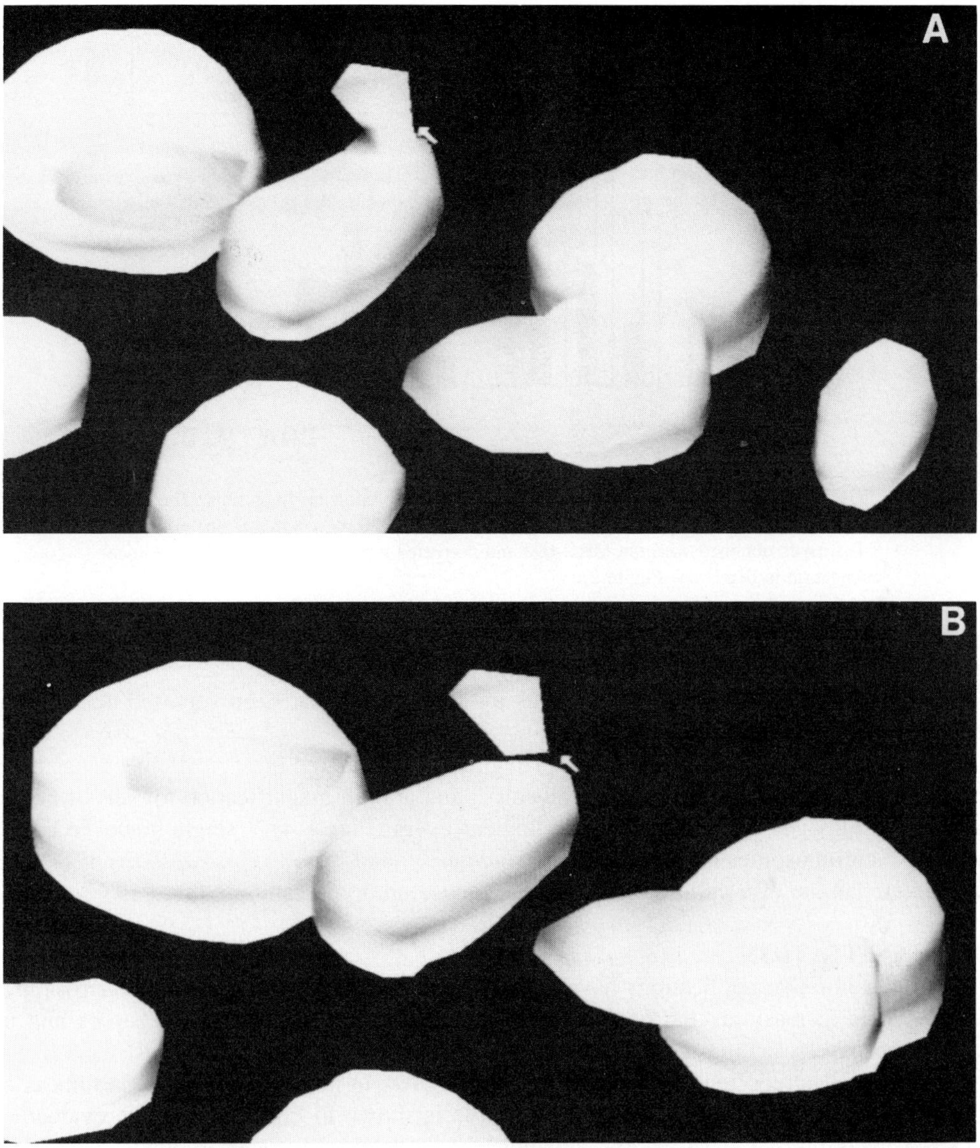

FIGURE 4. Computerized reconstructions of binarized nuclei in a region inside a 3-D image of a normal adult rat liver. (A) After segmentation by an algorithm operating in consecutive 2-D confocal planes; two nuclei (from an hepatocyte and a mesenchymal cell) are not separated (arrow); (B) after segmentation by an algorithm operating with 3-D mathematical morphology operations (Visilog-3D, Noesis, France)[83] the same nuclei are separated.

IV. APPLICATIONS

A. 3-D RECONSTRUCTION

Obtaining thick 3-D images is required for studies by CICM. Biological objects must be segmented and binarized before being measured. They may be reconstructed in 3-D surfaced and shaded mode and may then be used as masks for intensity measurements on the original 3-D gray tone image.

Constructing thick 3-D images in histology has been achieved at low magnification for buccal epithelial cells,[96] neurons,[9,11,12] and embryos,[26,27,81] and at high magnification for

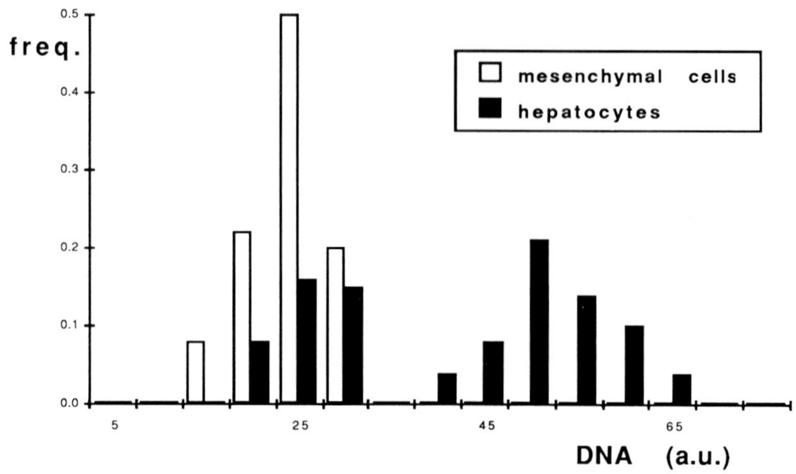

FIGURE 5. Histogram of nuclear DNA contents from adult rat liver cells. The measurements were made by CICM on 120 hepatocytic and 20 mesenchymal nuclei in four 3-D images obtained with the MRC-600 and corrected mathematically for fluorescence attenuation in depth (cf. Figure 2).

actin,[20,64] tubulin,[81] cytokeratin,[40] vimentin,[40] chloroplasts,[47] chromosomes,[20,81,132] chromatin,[20,80] and sets of nuclei from normal and pathological tissues.[2,3,68,69,82,83]

Reconstruction into a cuberille model by ray tracing has been reported for murine trophoblast giant-cell chromatin.[125]

Surfaced and shaded reconstruction of segmented, binarized biological objects has been reported at low magnification for lung alveoli[121] and at high magnification for sets of nuclei in thick 3-D images of normal and pathological livers (Figure 4),[83] single nuclei in breast tumors,[157] chromosomes from *Chironomus thummi*[77,79] and from *Crepis capillaris*,[132,158] and the nuclear lamina of a cell derived from the mouse embryocarcinoma line P19.[126]

B. QUANTITATION

Few quantitative applications have been published up to now in confocal microscopy, whereas it is increasingly hard to keep track of the number of theoretical papers and of application papers dealing with purely qualitative aspects.

Our team has reported quantitative studies on DNA in normal tissues and in tumors of the liver and of the esophagus.[3,68,69] An example is shown in Figure 5. For the evaluation of DNA histograms, CICM has the advantage, as compared to other methods, that DNA profiles can be obtained from different regions inside a tissue block. We have shown by CICM, for instance, that the hepatocytic tetraploid/diploid DNA ratio is higher in pericentrolobular zones than in periportal zones of the liver.[3] Another advantage of CICM is the possibility of analyzing different cellular types separately. This eliminates the contamination of DNA profiles by irrelevant cells[3,69] and allows an estimation of the diploid reference value from stromal cells, or from epithelial cells in a normal zone neighboring a tumor, or from mesenchymal cells in the liver (Figure 5).[3,69]

A quantitative 3-D evaluation of architectural patterns observed in normal and pathological tissues is possible by using spatial statistics[150-152] and stochastic modeling by unmarked and marked point processes[153,154] applied to the 3-D distributions of particle centroids. We have studied the architecture of the liver using the centroids of hepatocytic and mesenchymal nuclei.[68,82,83,153,154] Some examples of our results are shown in Figures 6, 7, and 8. The numerical density[147] and the spatial arrangement[155] of osteocyte lacunae have also been studied. Various quantities have been estimated including numerical density,[147] K and L

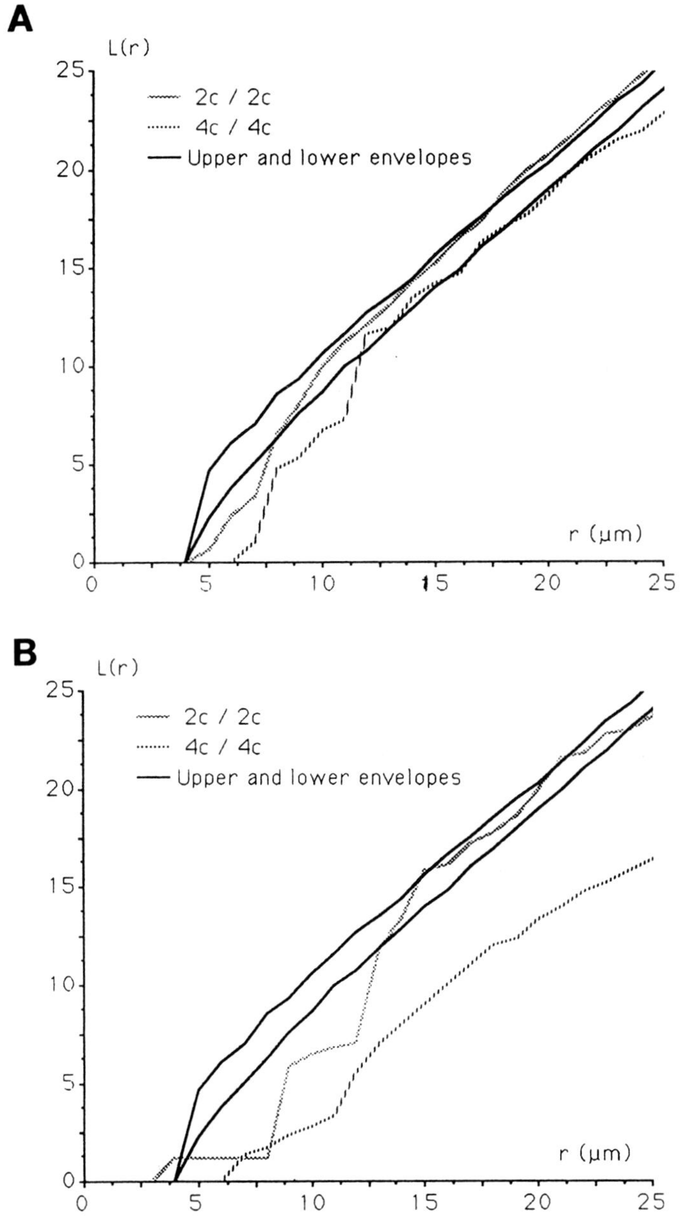

FIGURE 6. Evaluation of the histological architecture of the liver in 3-D space by using marked point process theory[154] on the coordinates of nuclear centroids in 3-D images. The L-function, a second-order quantity which evaluates the spatial arrangement of nuclei, is shown here using data from adult rat hepatocytes in 3-D images. The curves do not pass by the origin because of the "hard-core" effect (as nuclei cannot interpenetrate). The envelopes were obtained by 500 simulations, using a hard-core model. Separate curves are shown for couples of diploid (2c) and tetraploid (4c) nuclei. (A) Newborn: the curves show that the nuclei are randomly dispersed; (B) normal adult: the curve for the 4c/4c marked process, falling far below the lower envelope, demonstrates that there is a regularity in the spatial arrangement of tetraploid hepatocytes in the adult; (C) hyperplastic nodule observed during experimental carcinogenesis in an adult; as the normal architecture of the adult is disappearing, the curves resemble those of the newborn.

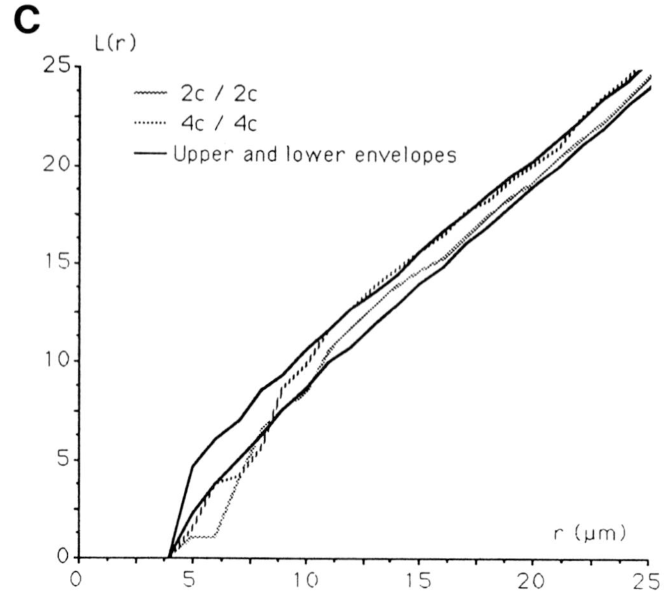

FIGURE 6C.

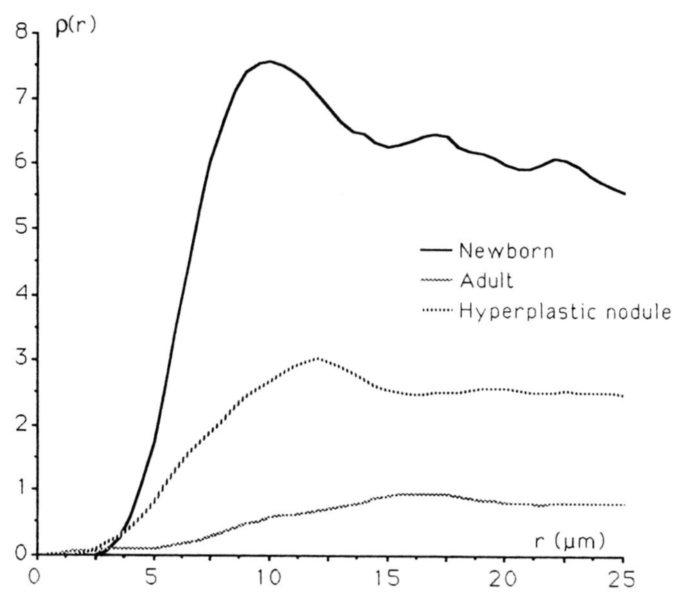

FIGURE 7. A second example of the use of point process theory.[154] Plot of the second-order product density [ρ(r)] versus the distance (r) between nuclear centroids. As in Figure 6, the result for the hyperplastic nodule is closer to the result of the newborn than to that of the adult (more regularly arranged nuclei).

functions (Figure 6),[68,82,83,154,155] pair correlation function (Figure 7), product density function, distribution of nearest neighbor distances, spheric contact distribution, mark covariance function,[82,83,154] orientation distribution function, and Schmidt projection (Figure 8).[153]

An analysis of nuclear anaphases from *Crepis capillaris* showed several nonrandom arrangements of chromosomal material inside the nuclei.[64]

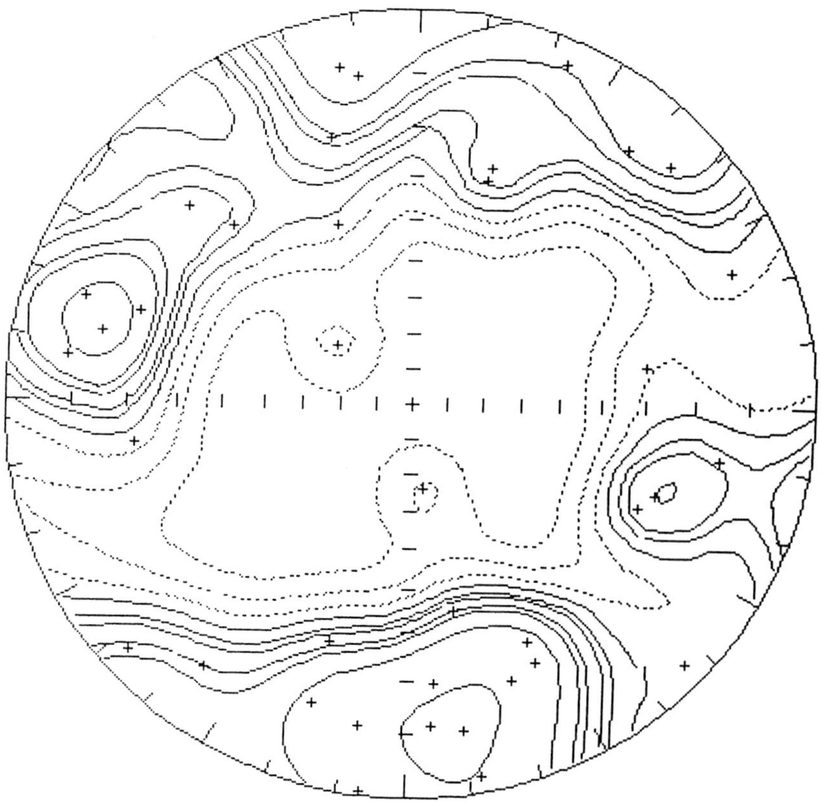

FIGURE 8. A third example of the use of point process theory.[154] Projection, according to Schmidt,[152,153] of the polar coordinates of nuclear centroids from normal adult rat hepatocytes in a 3-D image. On such a plot, if the nuclei were randomly arranged, the contours would all be concentric, whereas on the present plot there are some clusters in certain directions.

In situ hybridization was used to construct mappings of centrometric and telomeric regions in the interphase nucleus,[73,74,159] of human genomic fragments,[76,160] and of cellular RNA.[73] A statistical approach showed that chromosome 1 centromeres were not positioned randomly inside interphase blood cell nuclei.[74]

Transmitter-identified axon terminals were quantitated automatically.[13] The number of microtubules was evaluated in NGF-activated PC12 cells.[161] The levels of nuclear SV40 large-T antigen in cultures of immortalized mouse skin fibroblasts were evaluated in correlation with the increase of the cell density.[162] The phagocytosis of fluorescent microspheres ingested by murine macrophages was studied quantitatively.[163]

Intracellular Ca^{2+} amounts[84] and pH[42] have been evaluated *in vivo* by CSLM. An attempt to use voltage-sensitive dyes to quantitate voltage-dependent changes in neuronal tissues was unsuccessful (although the dye could be seen) because of a low signal to noise ratio.[10]

REFERENCES

1. **Rigaut, J. P.,** What's new from the field: a new technology is born — confocal image cytometry, *Anal. Cell. Pathol.*, 3, in press.
2. **Rigaut, J. P. and Vassy, J.,** High resolution three-dimensional images from confocal scanning laser microscopy. Quantitative study and mathematical correction of the effects due to bleaching and fluorescence attenuation in depth, *Anal. Quant. Cytol. Histol.*, in press.

3. **Rigaut, J. P., Vassy, J., Herlin, P., Duigou, F., Masson, E., Briane, D., Foucrier, J., Carvajal-Gonzalez, S., Downs, A. M., and Mandard, A. M.,** Three-dimensional DNA image cytometry by confocal scanning laser microscopy in thick tissue blocks, *Cytometry,* 12, in press.
4. **Minsky, M.,** Microscopy apparatus, U.S. Patent 3,013,467, 1961.
5. **Minsky, M.,** Memoir on inventing the confocal scanning microscope, *Scanning,* 10, 128, 1988.
6. **Petráň, M. and Hadravsky, M.,** Czechoslovakian Patents 128,936 and 128,937, 1968.
7. **Petráň, M., Hadravsky, M., Egger, M. D., and Galambos, R.,** Tandem-scanning reflected light microscope, *J. Opt. Soc. Am.,* 58, 661, 1968.
8. **Egger, M. D. and Petráň, M.,** New reflected light microscope for viewing unstained brain and ganglian cells, *Science,* 157, 305, 1987.
9. **Brodin, L., Ericsson, M., Mossberg, K., Hökfelt, T., Ohta, Y., and Grillner, S.,** Three-dimensional reconstruction of transmitter-identified central neurons by *"en bloc"* immunofluorescence histochemistry and confocal scanning microscopy, *Exp. Brain Res.,* 73, 441, 1988.
10. **Fine, A., Amos, W. B., Durbin, R. M., and McNaughton, P. A.,** Confocal microscopy: applications in neurobiology, *T. I. N. S.,* 11, 346, 1988.
11. **Carlsson, K., Wallén, P., and Brodin, L.,** Three-dimensional imaging of neurons by confocal fluorescence microscopy, *J. Microsc.,* 155, 15, 1989.
12. **Wallén, P., Brodin, L., Carlsson, K., Liljeborg, A., Mossberg, K., Ericsson, M., Grillner, S., Hökfelt, T., and Ohta, Y.,** Confocal laser scanning microscopy utilized for 3-D imaging of fluorescent neurons in the central nervous system, *Eur. J. Cell Biol.,* 48 (Suppl 25), 43, 1989.
13. **Mossberg, K. and Ulfhake, B.,** Automated analysis of axon terminals from 3D-data recorded with a confocal scanning laser microscope, *Trans. R. Microsc. Soc.,* 1, 413, 1990.
14. **Weibel, E. R.,** *Stereological Methods,* Academic Press, London, 1979.
15. **Gundersen, H. J. G.,** Stereology of arbitrary particles. A review of unbiased number and size estimators and the presentation of some new ones, in memory of William R. Thomson, *J. Microsc.,* 143, 3, 1986.
16. **Rigaut, J. P.,** Image analysis in histology — hope, disillusion, and hope again, *Acta Stereol.,* 8(1), 3, 1989.
17. **Agard, D. A. and Sedat, J. W.,** Three-dimensional architecture of a polytene nucleus, *Nature,* 302, 676, 1983.
18. **Wijnaendts van Resandt, R. W., Marsman, H. J. B., Kaplan, R., Davoust, J., Stelzer, E. H. K., and Stricker, R.,** Optical fluorescence microscopy in three dimensions: microtomoscopy, *J. Microsc.,* 138, 29, 1985.
19. **Brakenhoff, G. J., Blom, P., and Barends, P.,** Confocal scanning light microscopy with high aperture immersion lenses, *J. Microsc.,* 117, 219, 1979.
20. **Brakenhoff, G. J., van der Voort, H. T. M., van Spronsen, E. A., and Nanninga, N.,** Three-dimensional imaging in fluorescence by confocal scanning microscopy, *J. Microsc.,* 153, 151, 1989.
21. **Brakenhoff, G. J., van Spronsen, E. A., van der Voort, H. T. M., and Nanninga, N.,** Three-dimensional confocal fluorescence microscopy, in *Methods in Cell Biology, Vol. 30, Fluorescence Microscopy of Living Cells in Culture — Part B, Quantitative Fluorescence Microscopy — Imaging and Spectroscopy,* Taylor, D. L. and Wang, Y. L., Eds., Academic Press, San Diego, 1989, 379.
22. **Wilson, T. and Sheppard, C. J. R.,** *Theory and Practice of Scanning Optical Microscopy,* Academic Press, New York, 1984.
23. **Wilke, V.,** Optical scanning microscopy: the laser scan microscope, *Scanning,* 7, 88, 1985.
24. **Jovin, T. M. and Arndt-Jovin, D. J.,** Luminescence digital imaging microscopy, *Annu. Rev. Biophys. Chem.,* 18, 271, 1989.
25. **Pawley, J., Ed.,** *The Handbook of Biological Confocal Microscopy,* I. M. R. Press, Madison, 1989.
26. **Shuman, H., Murray, J. M., and DiLullo, C.,** Confocal microscopy: an overview, *Biotechniques,* 7, 154, 1989.
27. **Shotton, D. M.,** Confocal scanning optical microscopy and its applications for biological specimens, *J. Cell Sci.,* 94, 175, 1980.
28. **Boyde, A.,** Stereoscopic images in confocal (tandem scanning) microscopy, *Science,* 230, 1270, 1985.
29. **Xiao, G. Q. and Kino, G. S.,** A real-time confocal scanning optical microscope, *Proc. Soc. Photo-Opt. Instrum. Eng.,* 809, 107, 1987.
30. **Davidovits, P. and Egger, M. D.,** Scanning laser microscope, *Nature,* 223, 831, 1969.
31. **Davidovits, P. and Egger, M. D.,** Scanning laser microscope for biological investigations, *Appl. Opt.,* 10, 1615, 1971.
32. **Slomba, A. F., Wasserman, D. E., Kaufman, G. I., and Nester, J. F.,** A laser flying spot scanner for use in automated fluorescence antibody instrumentation, *J. Assoc. Adv. Med. Instrum.,* 6, 230, 1972.
33. **Davidovits, P. and Egger, M. D.,** Photomicrography of corneal endothelial cells *in vivo, Nature,* 244, 366, 1973.

34. **Hamilton, D. K. and Wilson, T.,** Scanning optical microscopy by objective lens scanning, *J. Phys. E* 19, 52, 1986.
35. **Reimer, L., Egelkamp, S., and Verst, M.,** Lock-in technique for depth-profiling and magneto-optical Kerr effect imaging in scanning optical microscopy, *Scanning,* 9, 17, 1987.
36. **Phelan, R. J. and DeMeo, N. L.,** Rapid scanning microscope for light probing and infrared mapping, *Appl. Opt.,* 10, 858, 1971.
37. **Sheppard, C. J. R. and Choudhury, A.,** Image formation in the scanning microscope, *Opt. Acta,* 24, 1051, 1977.
38. **Marsman, A. J. B., Stricker, R., Wijnaendts van Resandt, R. W., and Brakenhoff, G. J.,** Mechanical scan system for biological applications, *Rev. Sci. Instrum.,* 54, 1047, 1983.
39. **Brelje, T. C., Scharp, D. W., and Sorenson, R. L.,** Three-dimensional imaging of intact isolated islets of Langerhans with confocal microscopy, *Diabetes,* 38, 808, 1989.
40. **Vassy, J., Rigaut, J. P., Hill, A. M., and Foucrier, J.,** Analysis by confocal scanning laser microscopy imaging of the spatial distribution of intermediate filaments in foetal and adult rat liver cells, *J. Microsc.,* 157, 91, 1990.
41. **Mossberg, K. and Ericsson, M.,** Detection of doubly stained fluorescent specimens using confocal microscopy, *J. Microsc.,* 158, 215, 1990.
42. **Weinlich, M.,** Confocal laser scanning microscopy opens a new dimension in *in vivo* fluorescence microscopy, communication presented at the Bio-Rad Confocal Laser Scanning Microscope User's Meeting, Lelystad, The Netherlands, 1990.
43. **Gratton, E. and vandeVen, M. J.,** Laser sources for confocal microscopy, in *The Handbook of Biological Confocal Microscopy,* Pawley, J., Ed., I. M. R. Press, Madison, 1989, 47.
44. **Denk, W., Strickler, J. H., and Webb, W. W.,** Two-photon laser scanning fluorescence microscopy, *Science,* 248, 73, 1990.
45. **Carlsson, K. and Liljeborg, A.,** A confocal laser microscope scanner for the digital recording of optical serial sections, *J. Microsc.,* 153, 171, 1989.
46. **Brakenhoff, G. J., van der Voort, H. T. M., van Spronsen, E. A., and Nanninga, N.,** 3-Dimensional imaging of biological structures by high resolution confocal scanning laser microscopy, *Scanning Microsc.,* 2, 33, 1988.
47. **van Spronsen, E. A., Sarafis, V., Brakenhoff, G. J., van der Voort, H. T. M, and Nanninga, N.,** Three-dimensional structure of living chloroplasts as visualized by confocal scanning laser microscopy, *Protoplasma,* 148, 8, 1989.
48. **Shack, R. V., Bartels, P. H., Buchroeder, R. A., Shoemaker, R. L., Hillman, D. W., and Vukobratovich, D.,** Design for a fast fluorescence scanning microscope, *Anal. Quant. Cytol. Histol.,* 9, 509, 1987.
49. **Bianco, P. and Boyde, A.,** Alkaline phosphatase cytochemistry in confocal scanning light microscopy for imaging the bone marrow stroma, *Basic Appl. Histochem.,* 33, 17, 1989.
50. **Kapitza, H. G.,** Advanced fluorescence methods in confocal laser scan microscopy, *R. Microsc. Soc. Proc.,* 25, S55, 1990.
51. **Keller, H. E.,** Objective lenses for confocal microscopy, in *The Handbook of Biological Confocal Microscopy,* Pawley, J., Ed., I. M. R. Press, Madison, 1989, 69.
52. **Amos, W. B.,** Personal communication, 1990.
53. **Shack, R. V., Bell, B., Hillman, D., Landesman, A., Shoemaker, R. L., Vukobratovich, D., and Bartels, P. H.,** Ultrafast laser scanner microscope. First performance tests, in *Proc. Int. Workshop Physics, and Engineering in Medical Imaging,* Nalcioglu, O., Ed., Pacific Grove, 1982, 49.
54. **Suzuki, T. and Horikawa, Y.,** Development of a real-time scanning laser microscope for biological use, *Appl. Opt.,* 25, 4115, 1986.
55. **Draaijer, A., and Houpt, P. M.,** A standard video-rate confocal laser-scanning reflection and fluorescence microscope, *Scanning,* 10, 139, 1988.
56. **Goldstein, S., Hubin, Tj., Rosenthal, S., and Washburn, C.,** A confocal video-rate laser-beam scanning reflected-light microscope with no moving parts, *J. Microsc.,* 157, 29, 1990.
57. **Bartels, P. H., Buchroeder, R. A., Hillman, D. W., Jonas, J., Kessler, D., Shoemaker, R. L., Shack, R. V., Towner, D., and Vukobratovich, D.,** Ultrafast laser scanner microscope. Design and construction, *Anal. Quant. Cytol.,* 3, 55, 1981.
58. **Horikawa, Y., Yamamoto, M., and Dosaka, S.,** Laser scanning microscope: differential phase images, *J. Microsc.,* 148, 1, 1987.
59. **Bertero, M., Brianzi, P., and Pike, E. R.,** Super-resolution in confocal scanning microscopy, *Inverse Probl.,* 3, 195, 1987.
60. **Bertero, M., Boccacci, P., Brakenhoff, G. J., Malfanti, F., and van der Voort, H. T. M.,** Three-dimensional image restoration and super-resolution in fluorescence confocal microscopy, *J. Microsc.,* 157, 3, 1990.

61. **Visscher, K. and Brakenhoff, G. J.,** Single beam optical trapping integrated in a confocal microscope for biological applications, *Cytometry,* Suppl. 4, 17, 1990.
62. **Puppels, G. J., de Mul, F. F. M., Otto, C., Greve, J., Robert-Nicoud, M., Arndt-Jovin, D. J., and Jovin, T. M.,** Studying single cells and chromosomes by confocal Raman microspectroscopy, *Nature,* 347, 301, 1990.
63. **van den Pol, A. N.,** Neuronal imaging with colloidal gold, *J. Microsc.,* 155, 27, 1989.
64. **Brakenhoff, G. J., van der Voort, H. T. M., Baarslag, M. W., Mans, B., Oud, J. L., Zwart, R., and van Driel, R.,** Visualization and analysis techniques for three dimensional information acquired by confocal microscopy, *Scanning Microsc.,* 2, 1831, 1988.
65. **Thiébaut, F., Rigaut, J. P., and Reith, A.,** Improvement in the specificity of the silver staining technique for AgNOR-associated acidic proteins in paraffin sections, *Stain Technol.,* 59, 181, 1984.
66. **Haugland, R. P.,** *Molecular Probes: Handbook of Fluorescent Probes and Research Chemicals,* Molecular Probes Inc., Eugene, 1989.
67. **Tsien, R. Y. and Waggoner, A.,** Fluorophores for confocal microscopy: photophysics and photochemistry, in *The Handbook of Biological Confocal Microscopy,* Pawley, J., Ed., I. M. R. Press, Madison, 1989, 153.
68. **Rigaut, J. P., Vassy, J., Carvajal, S., Downs, A. M., Nguyen, Q., El Kebir, F. Z., and Foucrier, J.,** Three-dimensional histo-morphometry by laser scanning confocal microscopy, *Cytometry,* Suppl. 2, 86, 1988.
69. **Rigaut, J. P., Vassy, J., Briane, D., and Foucrier, J.,** DNA analysis in thick tissue blocks by laser scanning confocal microscopy (LSCM), *Cytometry,* Suppl. 4, 43, 1990.
70. **Arndt-Jovin, D. J., Robert-Nicoud, M., and Jovin, T. M.,** Probing DNA structure and function with a multi-wavelength fluorescence confocal laser microscope, *J. Microsc.,* 157, 61, 1990.
71. **Verheijen, R., Kuijpers, H. J. H., van Driel, R., Beck, J. L. M., van Dierendonck, J. H., Brakenhoff, G. J., and Ramaekers, F. C. S.,** Ki-67 detects a nuclear matrix-associated proliferation-related antigen. II. Localization in mitotic cells and association with chromosomes, *J. Cell Sci.,* 92, 531, 1989.
72. **Harders, J., Lukacs, N., Robert-Nicoud, M., Jovin, T. M., and Riesner, D.,** Imaging of viroids in nuclei from tomato leaf tissue by *in situ* hybridization and confocal laser scanning microscopy, *EMBO J.,* 8, 3941, 1989.
73. **van Dekken, H., van Spronsen, E., Bauman, J., Jonker, R., and Visser, J.,** Three dimensional analysis of intranuclear DNA and cellular RNA distribution by confocal microscopy after fluorescent *in situ* hybridization, *Cytometry,* Suppl. 2, 14, 1988.
74. **van Dekken, H., van Rotterdam, A., Jonker, R., van der Voort, H. T. M., Brakenhoff, G. J., and Bauman, J. G. J.,** Confocal microscopy as a tool for the study of the intranuclear topography of chromosomes, *J. Microsc.,* 158, 207, 1990.
75. **du Manoir, S. and Brugal, G.,** Is fluorescent DNA-DNA *in situ* hybridization a quantitative approach?, *Anal. Cell. Pathol.,* 1, 325, 1989.
76. **Lawrence, J. B.,** An *in situ* hybridization approach to interphase gene mapping and chromatin organization, *Cytometry,* Suppl. 4, 13, 1990.
77. **Takamatsu, T. and Fujita, S.,** Microscopic tomography by laser scanning microscopy and its three-dimensional reconstruction, *J. Microsc.,* 149, 167, 1988.
78. **Robert-Nicoud, M., Arndt-Jovin, D. J., Schormann, T., and Jovin, T. M.,** 3-D imaging of cells and tissues using confocal laser scanning microscopy and digital processing, *Eur. J. Cell Biol.,* 48 (Suppl. 25), 49, 1989.
79. **Schormann, T. and Jovin, T. M.,** Optical sectioning with a fluorescence confocal SLM: procedures for determination of the 2-D digital modulation transfer function and for 3-D reconstruction by tesselation, *J. Microsc.,* 158, 153, 1990.
80. **Brakenhoff, G. J., van der Voort, H. T. M., van Spronsen, E. A., Linnemans, W. A. M., and Nanninga, N.,** Three-dimensional chromatin distribution in neuroblastoma nuclei shown by confocal scanning laser microscopy, *Nature,* 317, 748, 1985.
81. **White, J. G., Amos, W. B., and Fordham, M.,** An evaluation of confocal versus conventional imaging of biological structures by fluorescence light microscopy, *J. Cell Biol.,* 105, 41, 1987.
82. **Carvajal-Gonzalez, S., König, D., Downs, A. M., Nguyen, Q., Vassy, J., and Rigaut, J. P.,** Analysis of histological architecture by point process modelling and spatial statistics applied to three-dimensional-images from laser scanning confocal microscopy, *Acta Stereol.,* 8 (2), 407, 1989.
83. **Carvajal-Gonzalez, S., Rigaut, J. P., Vassy, J., and König, D.,** Three-dimensional architecture analysis in cellular pathology by confocal microscopy and point process modelling, *Trans. Microsc. Soc.,* 1, 301, 1990.
84. **McNaughton, A.,** Imaging of intracellular calcium using the confocal microscope, communication presented at the Bio-Rad Confocal Laser Scanning Microscope User's Meeting, Lelystad, The Netherlands, 1990.
85. **Hyman, A. A. and White, J. G.,** Determination of cell division axes in the early embryogenesis of *Caenorhabditis elegans, J. Cell Biol.,* 105, 2123, 1987.

86. **Adler, J. and Cheema, M.**, Drug microkinetics using confocal (laser scanning) microscopy, *Eur. Microsc. Anal.*, 5, 15, 1990.
87. **Entwistle, A., Hoffmann, H. H., Noble, M., and Stroobant, P.**, Measurement of the diffusion time of small molecules into cell aggregates using confocal microscopy, *Trans. R. Microsc. Soc.*, 1, 361, 1990.
88. **Weinlich, M. and Acker, H.**, Flavoprotein-fluorescence imaging for metabolic studies in multicellular spheroids by means of confocal scanning laser microscopy, *J. Microsc.*, 160, RP1, 1990.
89. **Hernandez-Cruz, A., Sala, F., and Adams, P. R.**, Subcellular dynamics of $[Ca^{2+}]_i$ monitored with laser scanned confocal microscopy in a single voltage-clamped vertebrate neuron, *Biophys. J.*, 55, 216a, 1989.
90. **van der Valk, H. C. P. M., Blaas, J., van Eck, J. W., and Verhoeven, H. A.**, Vital DNA-staining of agarose-embedded protoplasts and cell suspensions on *Nicotiana plumbaginifolia, Plant Cell Rep.*, 7, 489, 1988.
91. **Verhoeven, H. A.**, CLSM of intact and living cells, communication presented at the Bio-Rad Confocal Laser Scanning Microscope User's Meeting, Lelystad, The Netherlands, 1990.
92. **Linden, S. M. and Neckers, D. C.**, Bleaching studies of rose bengal onium salts, *J. Am. Chem. Soc.*, 110, 1257, 1988.
93. **Wells, K. S., Sandison, D. R., Strickler, J., and Webb, W. W.**, Quantitative fluorescence imaging with laser scanning confocal microscopy, in *The Handbook of Biological Confocal Microscopy*, Pawley, J., Ed., I. M. R. Press, Madison, 1989, 23.
94. **Johnson, G. D., Davidson, R. S., McNamee, K. C., Russell, G., Goodwin, D., and Holborow, E. J.**, Fading of immuno-fluorescence during microscopy: a study of the phenomenon and its remedy, *J. Immunol. Methods*, 55, 231, 1982.
95. **Sheppard, C. J. R. and Wilson, T.**, Depth of field in the scanning microscope, *Opt. Lett.*, 3, 115, 1978.
96. **Sheppard, C. J. R.**, Confocal microscopy of thick structures, *Eur. J. Cell Biol.*, 48 (Suppl. 25), 33, 1989.
97. **Sheppard, C. J. R.**, Axial resolution of confocal fluorescence microscopy, *J. Microsc.*, 154, 237, 1989.
98. **Sheppard, C. J. R. and Cogswell, C. J.**, Three-dimensional image formation in confocal microscopy, *J. Microsc.*, 159, 179, 1990.
99. **Wilson, T.**, Three-dimensional imaging in confocal systems, *J. Microsc.*, 153, 161, 1989.
100. **Wilson, T.**, Optical sectioning in confocal fluorescent microscopes, *J. Microsc.*, 154, 143, 1989.
101. **Wilson, T. and Carlini, A. R.**, Three-dimensional imaging in confocal imaging systems with finite sized detectors, *J. Microsc.*, 149, 51, 1988.
102. **Wilson, T. and Carlini, A. R.**, The effect of aberrations on the axial response of confocal systems, *J. Microsc.*, 154, 243, 1989.
103. **van der Voort, H. T. M., Brakenhoff, G. J., and Janssen, G. C. A. M.**, Determination of the 3-dimensional optical properties of a confocal scanning laser microscope, *Optik*, 78, 48, 1988.
104. **van der Voort, H. T. M. and Brakenhoff, G. J.**, 3-D image formation in high-aperture fluorescence confocal microscopy: a numerical analysis, *J. Microsc.*, 158, 43, 1990.
105. **Wilson, T. and Hamilton, D. K.**, Dynamic focusing in the confocal scanning microscope, *J. Microsc.*, 128, 139, 1982.
106. **Cogswell, C. J., Sheppard, C. J. R., Moss, M. C., and Howard, C. V.**, A method for evaluating microscope objectives to optimize performance of confocal systems, *J. Microsc.*, 158, 177, 1990.
107. **Stelzer, E. H. K. and Wijnaendts van Resandt, R. W.**, Applications of fluorescence microscopy in three dimensions: microtomoscopy, *Proc. Soc. Photo-Opt. Instrum. Eng.*, 602, 63, 1986.
108. **Castleman, K.**, *Digital Image Processing*, Prentice Hall, Englewood Cliffs, NJ, 1979, 368.
109. **White, J. G. and Amos, W. B.**, Confocal microscopy comes of age, *Nature*, 105, 328, 183, 1987.
110. **Carlsson, K. and Åslund, N.**, Confocal imaging for 3-D digital microscopy, *Appl. Opt.*, 26, 3232, 1987.
111. **Koppel, D. E., Carlson, C., and Smilowitz, H.**, Analysis of heterogeneous fluorescence photobleaching by video kinetics imaging: the method of cumulants, *J. Microsc.*, 155, 199, 1989.
112. **Rigaut, J. P. and Vassy, J.**, Mathematical models for the enhancement of three-dimensional images from laser scanning confocal fluorescent microscopy, *Anal. Quant. Cytol. Histol.*, 12, 210, 1990.
113. **Webb, J. P., McColgin, W. C., Peterson, O. G., Stockman, D. L., and Eberly, J. H.**, Intersystem crossing rate and triplet state lifetime for a lasing dye, *J. Chem. Phys.*, 53, 4227, 1970.
114. **Benson, D. M., Bryan, J., Plant, A. L., Gotto, A. M., Jr., and Smith, L. C.**, Digital imaging fluorescence microscopy: spatial heterogeneity of photobleaching rate constants in individual cells, *J. Cell Biol.*, 100, 1309, 1985.
115. **Amos, W. B.**, Results obtained with a sensitive confocal scanning system designed for epifluorescence, *Cell Motil. Cytoskel.*, 10, 54, 1988.
116. **van Oostveldt, P. and Bauwens, S.**, Quantitative fluorescence in confocal microscopy. The effect of the detector pinhole aperture on the re-absorption and inner filter phenomena, *J. Microsc.*, 158, 121, 1990.
117. **Böhme, N.**, Neue Methoden der Mikro-Stereoskopie und ihre mikrophotographische Auswertung, *Photo-Techn. Wirtsch.*, 6, 218, 1953.

118. **Boyde, A.,** Direct recording of stereoscopic pairs obtained directly from disk-scanning confocal light microscopes, in *The Handbook of Biological Confocal Microscopy,* Pawley, J., Ed., I. M. R. Press, Madison, 1989, 147.
119. **Cox, I. J. and Sheppard, C. J. R.,** Digital image processing of confocal images, *Image Vision Comput.,* 1, 52, 1983.
120. **Carlsson, K., Danielsson, P. E., Lenz, R., Liljeborg, A., Majlöf, L., and Åslund, N.,** Three-dimensional microscopy using a confocal laser scanning microscope, *Opt. Lett.,* 10, 53, 1985.
121. **Forsgren, P. O.,** Visualization and coding in three-dimensional image processing, *J. Microsc.,* 159, 195, 1990.
122. **Schormann, T., Robert-Nicoud, M., and Jovin, T. M.,** Improved stereovisualization method for confocal laser scanning microscopy, *Eur. J. Cell Biol.,* 48 (Suppl. 25), 53, 1989.
123. **Richards, W.,** Stereopsis and stereoblindness, *Exp. Brain Res.,* 10, 380, 1970.
124. **Herman, G. T., Reynolds, R. A., and Udupa, J. K.,** Computer techniques for the representation of three-dimensional data on a two-dimensional display, *Proc. Soc. Photo-Opt. Instrum. Eng.,* 367, 3, 1982.
125. **Montag, M., Spring, H., Trendelenburg, M. F., and Kriete, A.,** Methodical aspects of 3-D reconstruction of chromatin architecture in mouse trophoblast giant nuclei, *J. Microsc.,* 158, 225, 1990.
126. **van der Voort, H. T. M., Brakenhoff, G. J., and Baarslag, M. W.,** Three-dimensional visualization methods for confocal microscopy, *J. Microsc.,* 153, 123, 1989.
127. **Huijmans, D. P.,** Integration of software computer-aided three-dimensional reconstruction from parallel serial sections; a plea for standards, *Eur. J. Cell Biol.,* 48 (Suppl. 25), 57, 1989.
128. **Keppel, E.,** Approximating complex surfaces by triangulation of contour lines, *IBM. J. Res. Dev.,* 19, 2, 1975.
129. **Gouraud, H.,** Continuous shading of curved surfaces, *IEEE Comput.,* 20, 623, 1971.
130. **Phong, B. T.,** Illumination for computer generated pictures, *Commun. ACM,* 18, 311, 1975.
131. **Goldwasser, S. M. and Reynolds, R. A.,** Real-time display and manipulation of 3-D medical objects: the voxel processor architecture, *Comput. Vision Graph. Image Proc.,* 39, 1, 1987.
132. **Houtsmuller, A. B., Oud, J. L., van der Voort, H. T. M., Baarslag, M. W., Krol, J. J., Mosterd, B., Mans, A., Brakenhoff, G. J., and Nanninga, N.,** Image processing techniques for 3-D chromosome analysis, *J. Microsc.,* 158, 235, 1990.
133. **Leeman, T., Baggiolini, V., Walt, H., Emmerich, P., and Anliker, M.,** Computer aided 3-D localization of chromosome 1 within nuclei of human sperm cells, *Trans. R. Microsc. Soc.,* 1, 409, 1990.
134. **Carrington, W. A., Fogarty, K. E., Lifschitz, L., and Fay, F. S.,** Three-dimensional imaging of confocal and wide-field microscopes, in *The Handbook of Biological Confocal Microscopy,* Pawley, J., Ed., I. M. R. Press, Madison, 1989, 137.
135. **Smith, J., Jongenelen, J., Lamers, W. H., Los, J. A., and Strackee, J.,** Octree representation for 3D image analysis, *Eur. J. Cell Biol.,* 48 (Suppl. 25), 97, 1989.
136. **Shaw, P. J. and Rawlins, D. J.,** Measurements of the point spread function and its use in deconvolution of confocal microscope images, *Trans. R. Microsc. Soc.,* 1, 227, 1990.
137. **Conan, V., Howard, C. V., Jeulin, D., Renard, D., and Cummins, P.,** Improvements of 3D confocal microscope images by geostatistical filters, *Trans. R. Microsc. Soc.,* 1, 281, 1990.
138. **Kepler, J.,** *Omnia Opera,* Vol. 5, *Harmonices Mundi,* 1619.
139. **Preston, K., Jr.,** The crossing number of a three-dimensional dodecamino, *J. Combin. Info. Sys. Sci.,* 5, 281, 1980.
140. **Preston, K., Jr. and Duff, M. J. B.,** *Modern Cellular Automata — Theory and Applications,* Plenum Press, New York, 1984, 49.
141. **Sternberg, S.,** Cellular computers and biomedical image processing, in *Lecture Notes in Medical Informatics,* Vol. 17, *Biomedical Images and Computers,* Sklansky, J. and Bisconte, J. C., Eds., Springer-Verlag, Berlin, 1980, 274.
142. **Hafford, K. J. and Preston, K., Jr.,** Three-dimensional skeletonization of elongated solids, *Comput. Vision Graph. Image Proc.,* 27, 78, 1984.
143. **Serra, J.,** *Image Analysis and Mathematical Morphology,* Academic Press, London, 1982.
144. **Serra, J., Ed.,** *Image Analysis and Mathematical Morphology,* Vol. 2, *Theoretical Advances,* Academic Press, London, 1988.
145. **Meyer, F.,** 3D mathematical morphology, *Trans. R. Microsc. Soc.,* 1, 275, 1990.
146. **Gesbert, S., Howard, V., Jeulin, D., and Meyer, F.,** The use of basic morphological operations for 3D biological image analysis, *Trans. R. Microsc. Soc.,* 1, 293, 1990.
147. **Howard, V., Reid, S., Baddeley, A., and Boyde, A.,** Unbiased estimation of particle density in the tandem scanning reflected light microscope, *J. Microsc.,* 138, 203, 1985.
148. **Mattfeldt, T.,** Volume estimation of biological objects by systematic sections, *J. Math. Biol.,* 25, 685, 1987.
149. **Bhanu Prasad, P., Jernot, J. P., Lantuéjoul, C., and Chermant, J. L.,** Use of the shell correction method for quantification of three-dimensional digitized images, *Trans. R. Microsc. Soc.,* 1, 297, 1990.

150. **Ripley, B. D.,** *Spatial Statistics,* John Wiley & Sons, New York, 1981.
151. **Diggle, P. J.,** *Statistical Analysis of Spatial Point Patterns,* Academic Press, London, 1983.
152. **Fisher, N. I., Lewis, T., and Embleton, B. J. J.,** *Statistical Analysis of Spherical Data,* Cambridge University Press, Cambridge, 1987.
153. **König, D., Blackett, N., Clem, C. J., Downs, A. M., and Rigaut, J. P.,** Orientation distribution for particle aggregates in 3-D space based on point processes and laser scanning confocal microscopy, *Acta Stereol.,* 8 (2), 213, 1989.
154. **König, D., Carvajal-Gonzalez, S., Downs, A. M., Vassy, J., and Rigaut, J. P.,** Modelling and analysis of 3-D arrangements of particles by point processes with examples of application to biological data obtained by confocal scanning laser microscopy, *J. Microsc.,* 161, 405, 1991.
155. **Baddeley, A. J., Howard, C. V., Boyde, A., and Reid, S.,** Three-dimensional analysis of the spatial distribution of particles using the tandem-scanning reflected light microscope, *Acta Stereol.,* 6 (2), 87, 1987.
156. **Kayser, K., Stute, H., Bubenzer, J., and Paul, J.,** Combined morphometrical and syntactic structure analysis as tools for histomorphological insight into human lung carcinoma growth, *Anal. Cell. Pathol.,* 2, 167, 1990.
157. **Theeuwes, J. W., Smith, G., and Baak, J. P. A.,** Reconstruction and visualization of 3D nuclear images from breast tumors with confocal laserscan microscope (CLSM), communication presented at the 6th Int. Symp. Diagn. Quant. Pathol., Basel, 1990.
158. **Oud, J. L., Mans, A., Brakenhoff, G. J., van der Voort, H. T. M., van Spronsen, E. A., and Nanninga, N.,** Three-dimensional arrangement of *Crepis capillaris* in mitotic prophase and anaphase as studied by confocal scanning laser microscopy, *J. Cell Sci.,* 92, 329, 1989.
159. **Bartholdi, M. F.,** Higher order structure in the cell nucleus, *Cytometry,* Suppl. 4, 43, 1990.
160. **Lichter, P., Boyle, A., Ferguson, M., Ballard, G, and Ward, D. C.,** High resolution mapping of DNA probes and detection of chromosome aberrations by non-isotopic *in situ* hybridization, *Cytometry,* Suppl. 4, 13, 1990.
161. **Stevens, J. K., Trogadis, J., Mills, L., and Leatio, C.,** Three-dimensional volume investigation of serial confocal data sets, *Trans. R. Microsc. Soc.,* 1, 375, 1990.
162. **Entwistle, A. and Noble, M.,** The semi-quantitative measurement of nuclear SV-40 large-T antigen employing confocal microscopy, *Trans. R. Microsc. Soc.,* 1, 405, 1990.
163. **Hook, G. R. and Odeyale, C. O.,** Confocal scanning fluorescence microscopy: a new method for phagocytosis research, *J. Leukocyte Biol.,* 45, 277, 1989.
164. **McCarthy, J. J., Aeschbach, J. F., and Shattuck, J. A.,** A new tandem scanning reflected light microscope, *Acta Stereol.,* 6 (3), 639, 1987.
165. **Cheng, P. C., Chen, V. H. K., Kim, H. G., and Pearson, R. E.,** An epi-fluorescent spinning-disk confocal microscope, *Proc. 47th. Annu. Meet. E.M.S.A.,* 1989.
166. **Cox, I. J., Sheppard, C. J. R., and Wilson, T.,** Super-resolution by confocal fluorescent microscopy, *Optik,* 60, 391, 1982.
167. **Stelzer, E. H. K.,** Considerations on the intermediate optical system in confocal microscopes, in *The Handbook of Biological Confocal Microscopy,* Pawley, J., Ed., I. M. R. Press, Madison, 1989, 83.

Chapter 8

VIDEO-MICROSCOPIC TECHNIQUES TO STUDY THE LIVING CYTOPLASM

Dieter G. Weiss and Günther Galfe

TABLE OF CONTENTS

I. INTRODUCTION

The cytoplasm of eukaryotic cells is highly structured. It consists of the fluid phase (cytosol) and the cytoskeleton made up from several types of proteinaceous filaments. Embedded in this so-called cytomatrix, which in animal cells usually has a high viscosity,[1] are the various kinds of membrane-bounded organelles ranging from 30-nm vesicles to the mitochondria which are up to several micrometers long. Eukaryotic cells need ATP-consuming mechanisms to move actively and redistribute the organelles. The cytoskeletal filaments serve as tracks, as part of the force generating mechanism, and they organize the distribution of organelles to their places of destination.[2,3]

Today all cytoplasmic objects are structurally well characterized in the fixed and dehydrated state by electron microscopy. The study of most of their details in the living cell was, however, impossible for almost all but the largest organelles due to the limited resolution of light microscopy. When using visible light, this limit is around 200 nm, thus making the observation of most organelles such as secretory vesicles, synaptic vesicles, peroxisomes, and small lysosomes, and of essentially all cytoskeletal filaments such as actin filaments (6-nm diameter), intermediate filaments (10 nm), and microtubules (25 nm) practically impossible. Considerable progress in the study of cytoplasmic structures and motility had, therefore, to await new techniques capable of visualizing these structures, identifying their chemical nature, and quantitatively analyzing their motile behavior.

The power of light microscopy was restricted until very recently by the limitations imposed by the laws of physics and the properties of the human eye. It is now not only possible to create electronic images at much lower light levels than required for human vision, but also to digitize and process such microscope images in "real time", i.e., at video rates and at high spatial resolution. This led to dramatic improvements in resolution, brightness, and contrast of microscopic images. This so-called "electronic revolution in light microscopy"[4-7] now allows the additional application of new optical techniques and the use of living cells in many cases where previously only electron microscopy of dehydrated material could be used. The basic principles of video microscopy and the type of equipment required, as well as an overview of some applications, will be discussed here.

Since quantitative data for different morphological and biochemical parameters can be calculated from the live images in real time and encoded in the form of grey-shaded or pseudo-color images, they can also be continuously recorded and yield video films of the intracellular events. Furthermore, video microscopy allows multiparametric studies to be performed especially with cultured cells which yield a wealth of information that was inaccessible in the past. A synopsis of some of the present video-microscopic techniques is given in Table 1. It is to be expected that video microscopy will rapidly find increasing application in modern biology.

Considerable progress has also been made in increasing the resolution in the vertical or z-axis (optical sectioning). While classical differential interference contrast (DIC) microscopy allowed already resolutions down to 300 nm, this can be increased by video-enhancement (VEC-DIC) to 100 nm or better.[8-10] Also, series of digitized images can be subjected to advanced three-dimensional (3-D) reconstruction algorithms that process full grey-tone images.[10] In the case of fluorescence microscopy, the depth resolution is relatively poor (several micrometers). This can be improved to a resolution of about 0.7 μm by confocal light microscopy.[11,12] Although in confocal microscopy, image processing routines similar to those in video microscopy are used, this technique is due to advancements in microscopic design and therefore is not generally understood as part of the family of video microscopy techniques; it will therefore not be covered in this chapter.

TABLE 1
Parameters Which Can be Measured by Video Microscopy in Living Cells or Tissue Slices

I. Morphology-based parameters
 1. Visualization of cells and cellular processes in thick tissue slices (IR VEC-DIC, anaxial illumination)[21,56]
 2. Number, position, size, and shape parameters of free cells (e.g., video microinterferometry)[117]
 3. Number, position, total area, size, and shape of organelles (organelle-specific vital dyes):[14,21] nucleus,[14,118] mitochondria,[71,72] endoplasmic reticulum,[30,33,70] Golgi apparatus,[66] lysosomes,[73] endosomes[14]
 4. Motion analysis of organelles and microtubules:[29,93,94] velocity, fluctuation of velocity, straightness of path, rhythmicity, directionality, pauses, contribution of Brownian motion
 5. Cytoplasmic viscosity (Brownian motion) (requires microinjection)[102]
 6. Elastic properties of organelles, microtubules, membranes, and cells[104,105]
 7. Local motile activity in cells (root-mean-square average of differences of pixel brightness values)[119]
 8. Determination of dry matter in free cells (video microinterferometry):[117] total dry matter, dry matter distribution, pattern of flow of dry matter
 9. Measurement of forces in the microscopic and molecular domain (''magnetodrome'', microscopy in controlled magnetic fields)[120]

II. Biochemical parameters
 1. Thickness profile of cells (optical pathlength)[107,109]
 2. Distribution and amount of antigens (requires microinjection or fixation):
  Fluorescence-labeled antibodies[13]
  Gold-labeled antibodies (also for receptor labeling and receptor or antigen transport studies)[39,69]
 3. Concentration of substances (monochromatic illumination):[13]
  Concentration of free intracellular Ca^{2+}, Mg^{2+}, Zn^{2+}, K^+, Na^+, Cl^- [13,43,113,114]
  Intracellular H^+-concentration (pH-value)[113]
  Endogenous fluorescent or absorbing compounds
 4. Membrane potential (plasma membrane and endomembranes)[14,121]
 5. Metabolites can be measured by coupled bioluminescence assays; ''metabolic imaging'' (requires frozen sections or microinjection):[61,62] lactate, glucose, ATP
 6. Transfer of fluorescence-labeled compounds between neighboring cells[116]
 7. Measurement of the activation of translation of genes associated with a given promoter (by promoter-controlled expression of the luciferase gene)[111,112]

Note: The references are given as examples only.

II. THE VIDEO-MICROSCOPIC TECHNIQUES

A. VIDEO-INTENSIFIED MICROSCOPY (VIM)

Video-microscopic techniques aimed at the visualization and quantitation of weak monochromatic images obtained by fluorescence microscopy are generally called video-intensified microscopy (VIM). Such microscope images are two-dimensional (2-D) arrays of fluorescence measurements and contain information on the amounts and distribution of intracellular metabolites, dyes, antigens, or exogenously added agents. Typical parameters for VIM measurements include Ca^{2+}-concentration, pH value, metabolites, and membrane potential as determined with the use of specifically developed fluorescence indicator dyes, and the distribution of fluorescence-labeled antibodies. Specific, nontoxic fluorescent dyes are also available to verify the biochemical identity of the cellular structures seen by conventional or video-microscopic techniques.[13-14] With the use of so-called photon-counting cameras (see Chapter 5 of this volume), it is possible to image extremely weak fluorescence and luminescence signals at illumination conditions up to six orders of magnitude below the threshold of human vision or photographic films.[15]

The technical specifications of low light level cameras and further information on the procedures and equipment needed have been described in much detail elsewhere.[8,15-23] An overview of a suitable video microscopy setup is shown in Figure 1.

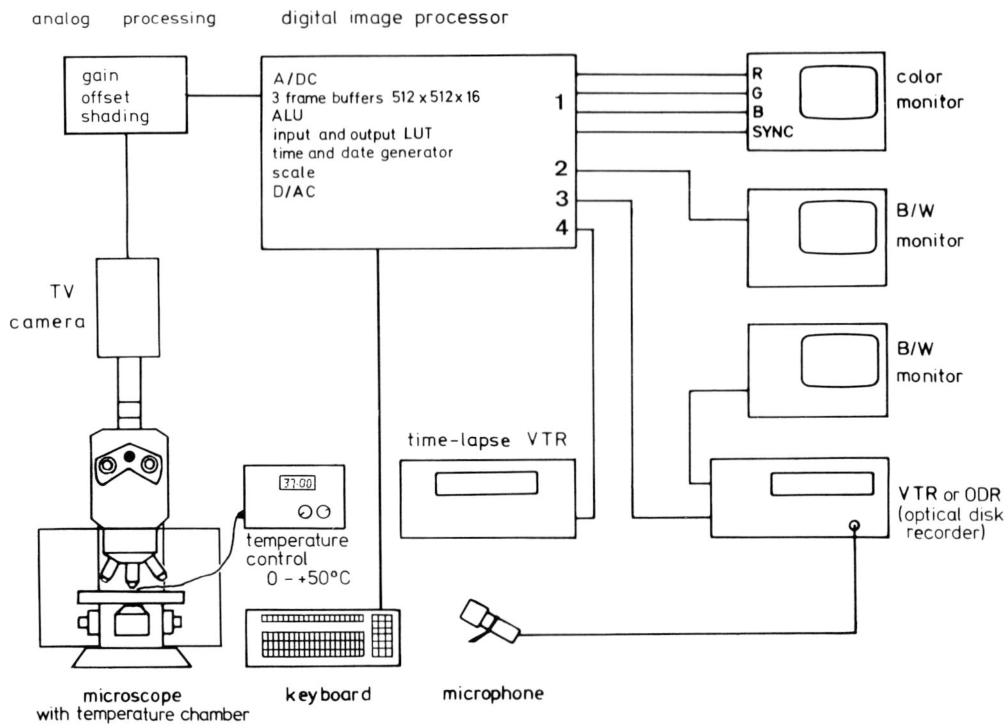

FIGURE 1. Video microscopy setup showing the two-stage image processor and the periphery required for video microscopy. Color monitor, raw image monitor and time lapse recorder are not essential, but are very useful. VTR, video tape recorder; ODR, optical disk recorder; 1, pseudocolor processed image; 2, raw image for focusing; 3,4, processed image B/W; A/DC, analog to digital converter; ALU, arithmetic logic unit; LUT, look-up table, D/AC, digital to analog converter. (Modified from Weiss, D. G., Maile, W., and Wick, R. A., in *Light Microscopy in Biology. A Practical Approach,* Lacey, A. J., Ed., IRL Press, Oxford, 1989, 221.)

B. VIDEO-ENHANCED CONTRAST (VEC) MICROSCOPY

A similarly dramatic improvement of image quality is obtained with video-enhanced contrast (VEC) microscopy. By increasing contrast and magnification, it is possible to extend the limits of both true resolution and visualization and to visualize and analyze in the living state positions and movements of biological objects smaller than one tenth of the limit of resolution of conventional light microscopy. All membrane-bounded organelles and part of the cytoskeleton can be imaged, and their motion and assembly can be quantitatively described by a number of parameters such as velocity, straightness of path, reversal of direction, and others.[24] Thus, video microscopy becomes a powerful new tool to study all forms of cellular and cytoplasmic motility.

VEC microscopy,[8,17,21] and especially the type introduced by R. D. Allen (AVEC microscopy),[25-27] should be applied when finest details are to be visualized with bright-field, differential interference contrast (DIC) or with polarization optics. The required procedures are described by Allen et al. and Weiss et al.,[7,21,27-29] and a schematic representation of the equipment needed is shown in Figure 1. As a result, resolution of objects of about 150 nm and visualization of even smaller biological objects such as endoplasmic reticulum[30-34] (Figure 2), vesicles[35,36] (Figure 3a and c), actin bundles[32,33,37] (Figure 3a), and microtubules[7,35] (Figure 3e to h), down to 20 nm is achieved with visible light and in situations where the limit of resolution of conventional microscopy is about 250 nm.[9,21,38] Individual colloidal gold particles, which are used as tags for antibody molecules, can be visualized in the 5- to 10-nm range,[39] in theory even down to 1 to 3 nm.[38,40]

The practical resolution is increased in VEC microscopy by a factor of almost two, provided that optimal optics are used. This is partly because it is possible to use the highest working numerical aperture since the resulting excessive image brightness due to stray light can be suppressed electronically by an offset voltage. Partly this improvement is also due to the fact that for a video camera, Rayleigh's criterion of the limit of resolution is replaced by the more advantageous Sparrow criterion. According to Rayleigh's criterion, two objects are resolved if the depression between them (''trough'') in their summed intensity distribution is at least 15%. Using the Sparrow criterion, the objects can be closer because the intensity distribution between them may almost approach a horizontal line at the limit of resolution, i.e., the trough may be infinitely shallow.[8,9,21]

Since the subresolution objects themselves are not imaged, but their much larger diffraction patterns are, one has to remember that very small objects appear inflated by diffraction up to the size of the resolution limit. However, given a separation of more than 200 nm from neighboring structures, such objects can be clearly seen and their position and movements can be determined to nanometer accuracy.[7,35,41]

The cells depicted in Figures 2 and 3 were observed with high-grade research microscopes with an internally corrected 100× planapochromatic oil immersion objective (NA 1.30), with a matching condenser, and with a mercury arc lamp. Additional magnification changers have to be used to reach the necessary magnification on the TV monitor of up to 10,000× (at a screen width of 25 cm). Real-time analog contrast enhancement and digital image processing with the Hamamatsu C 1966 Photonic Microscope System (Hamamatsu Photonics, Herrsching, Germany; and Hamamatsu Photonics, Bridgewater, NJ, U.S.) which is used in our laboratory include the following steps:

1. AVEC-DIC analog enhancement of the full aperture image (at an instrumental compensator setting of one nineth of a wavelength in the AVEC-DIC case) with gain and offset adjusted from about one third to one half of their respective ranges
2. Digital subtraction of mottle, i.e., fixed pattern noise arising from uneven illumination, optical imperfections, or dust particles in the light path (Figure 2a and b)
3. Manipulation of the grey scale by stretching of the pixel brightness histogram
4. In some cases, reduction of pixel noise by a real-time digital moving or rolling average operation (recursive filtering) over two frames[21]

C. EQUIPMENT

Except for the cameras, similar electronic equipment is required for the two basic types of video microscopy. However, the microscope also requires modifications to allow the use of the new techniques to their full extent.[8,21] Especially when working at the extremely high magnifications of VEC microscopy, a massive, stable microscope stand, a very bright illumination system, and optical magnification of 2 or 4× in addition to a 100× oil immersion objective (NA 1.25 or better) are recommended.[8,21]

For VEC microscopy, one needs a geometrically distortion-free video camera (Chalnicon, Newvicon, or Pasecon) or a high precision CCD (charge-coupled device) camera of highest spatial resolution. The cameras need to have externally adjustable offset (black level) and gain to allow for analog contrast enhancement.[8,21,25,26,42] Automatic gain control (AGC), if present, must be disabled. For VIM, a low light level camera is required such as a one- or two-stage silicon intensifier target (SIT or ISIT) camera or a microchannel plate-intensifier device[15,18,21] with a light sensitivity at the limit of human vision or below. When extremely weak signals, i.e., luminescence or autofluorescence, need to be imaged, photon-counting cameras will be required[15,19] (see Chapter 5 of this volume).

For most applications it is important that the image processing equipment contains a unit for analog contrast enhancement of the analog camera signal (Figure 1). Using this one

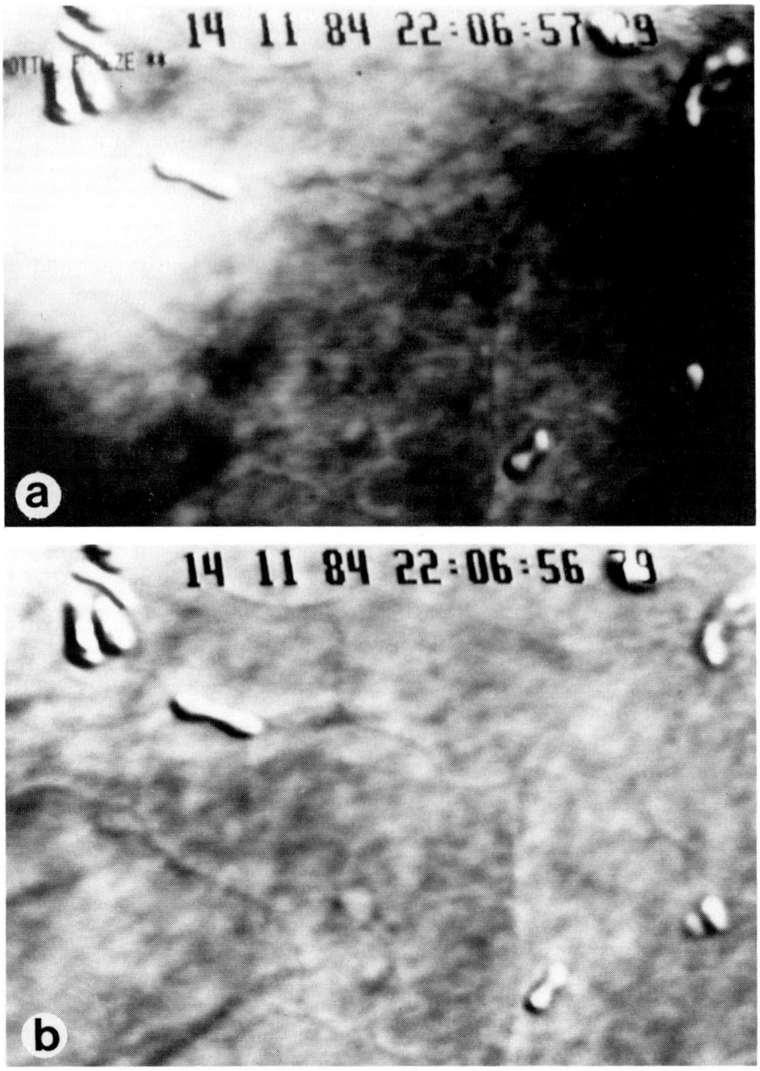

FIGURE 2. Visualization of subresolution structures by image processing. (a) Analog contrast enhancement of the microscopic image brings about more image detail, but also enhances unevenness of illumination and sometimes a disturbing mottle pattern. (b) A "cleaned" image is obtained after subtraction of the background pattern. This is achieved by storing and subtracting an out-of-focus frame and then applying digital contrast enhancement with the image processor. (c) Sometimes static objects such as cell-wall fibers become so prominent after contrast enhancement that small details of interest in the cytoplasm cannot be differentiated when they are in the same focal plane. However, the image in (a) can be stored in the computer frame memory while still in focus and then subtracted from the incoming video frames so that a picture containing only the moving cytoplasmic elements is derived (differential image). The thin, tubular endoplasmic reticulum can be observed in this way to form a motile polygonal network that would have remained invisible otherwise.[32] Bar, 5 μm.

yields a high contrast image even for very low contrast or low light objects. The image is then digitized (usually 512 × 512 picture elements (pixels) with 256 possible grey-levels each) and subjected to arithmetic operations in the frame memory of a digital real-time processor (Figure 1). For VEC microscopy one needs frame averaging to remove pixel noise,

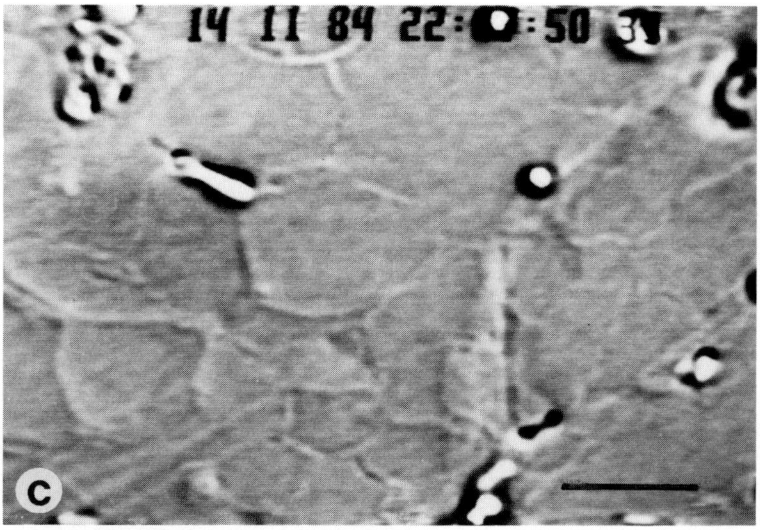

FIGURE 2c.

continuous background subtraction to remove unwanted fixed pattern noise (mottle),[27] or, if required, nonmoving image details[32] (Figure 2a and b). Digital contrast enhancement (stretching of the grey-level histogram) is another essential function usually required after image subtraction (Figure 2b and c). For the dual wavelength fluorescence ratio imaging techniques (e.g., Ca^{2+} imaging with FURA-2[43,44]), the image processor needs to have the additional capability of fast division of full video frames.

When the images have been created in real time by video-microscopic image processing and have been recorded, they may be subjected to classical image analysis in order to obtain information on morphometric parameters. It should be noted that this is usually done after thresholding, i.e., with binary images, *not* in real time, and that it would require conventional image analysis systems.[45-48] Video microscopy as discussed here involves the processing of full grey-level images at video frequency (40 ms), so that we need real-time image processors.[21,22,27] More details on the equipment required and the detailed steps of the recommended procedures are published elsewhere.[8,21,23]

III. CELL-BIOLOGICAL APPLICATIONS OF VIDEO MICROSCOPY

A. GENERAL COMMENTS

It is evident from the above description that the video microscopic techniques yield most impressive image improvements in situations where the specimen has either virtually no contrast or emits almost negligible amounts of light. Video enhancement, therefore, is best used with unstained material such as living cells[2,32,34,49-52] or extruded cytoplasmic droplets.[36,37,53,54]

With VEC-DIC microscopy, the image quality is highest in the top cell layer of any tissue samples, and out-of-focus material leads to a deterioration of image quality only to some extent, provided that birefringent or depolarizing material such as large numbers of small particles, beads, or birefingent inclusions are absent from the light path. However, the VEC techniques allow improved optical penetration of tissue slices or small organisms, especially when anaxial illumination[55] or very long wavelengths of light such as near infrared (IR) is used[56] (Figure 4). The presence of stains or very high contrast inclusions does not

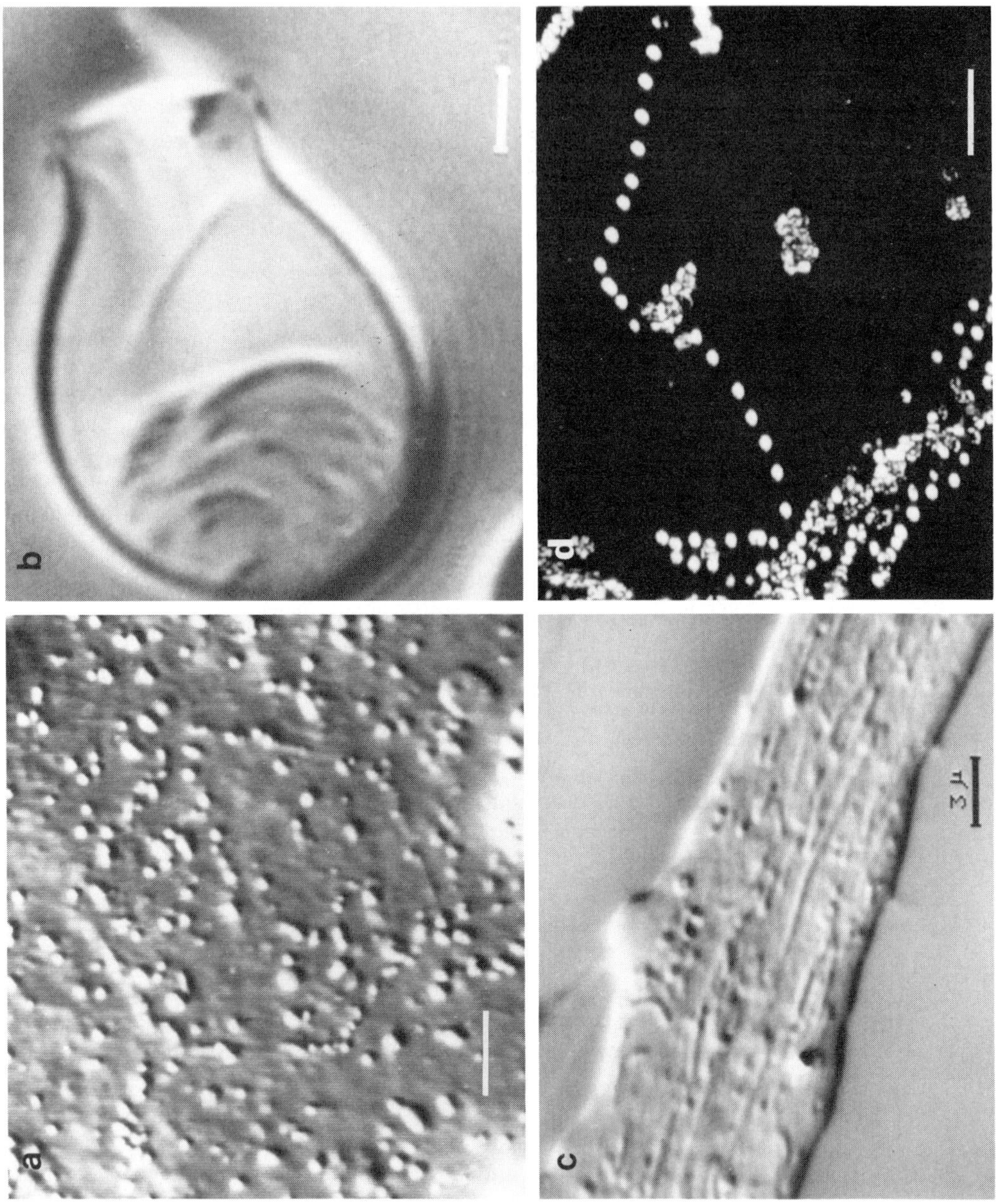
a
b
c
d
3 μ

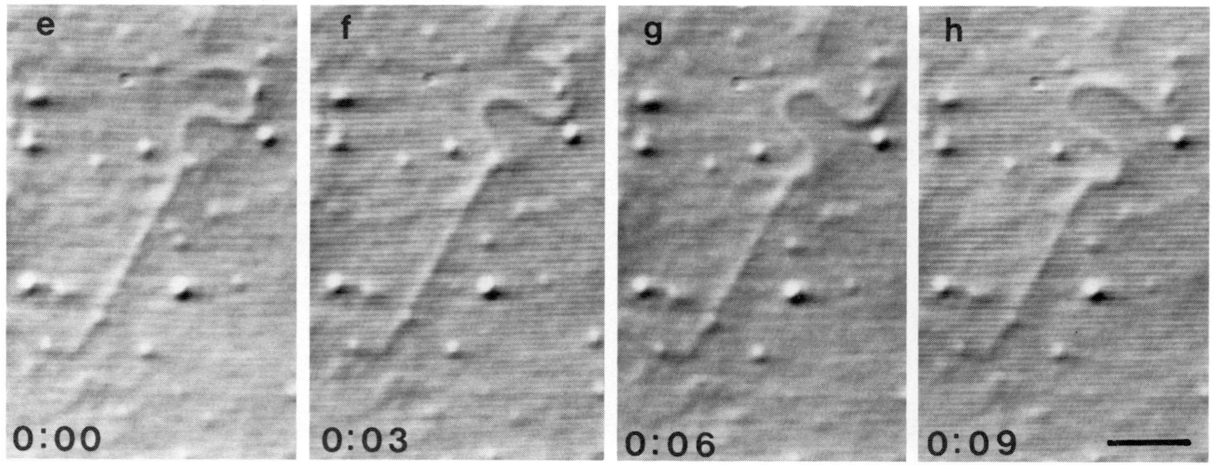

FIGURE 3. Examples of applications of VEC microscopy. (a) A rotating hexagon consisting of a microfilament bundle forms in a drop of extruded cytoplasm of *Nitella*. Bar, 5 μ. (b) Isolated nematocyte of *Hydra* which can be stimulated to explosively eject its tubular contents. Bar, 2 μm. (Courtesy of T. Holstein and W. Maile.) (c) Neurite of a N18 neuroblastoma cell in culture. Most of the organelles move either anterogradely or retrogradely with velocities of 0.5 to 2 μm/s. The smallest objects are synaptic or similarly sized membrane precursor vesicles. (From Weiss, D. G., in *Recent Advances in Neurocytochemical Methods*, Calas, A., Ed., Springer-Verlag, Berlin, 1991. With permission.) (d) Organelle motion made visible by adding differential images to a frame memory ("trace" function). Organelles are spherosomes moving along actin filaments in an onion cell. Regions with multiple moving objects can be seen at the left and right margins. One organelle traversing the center region is captured every second. Two particles in Brownian motion (center) show a clearly different pattern of motion. Interval between grabbed frames is 1 s. Bar, 5 μm. (a to d) Polyvar microscope, Reichert (Vienna, Austria), 100× Planapo, NA 1.3, 2 to 4× additional optical magnification; processor, Hamamatsu C1966 Photonic Microscope System with Chalnicon camera; (e to h) native microtubule (diameter 25 nm) from extruded squid giant axon undergoing fishtailing motion;[35,97] Zeiss Axiomat, 100× Planapo, NA 1.30, oil, 4× additional optical magnification, 50-W mercury lamp; processor, Hamamatsu C1966 with Chalnicon camera, time in seconds; bar, 1.5 μm.

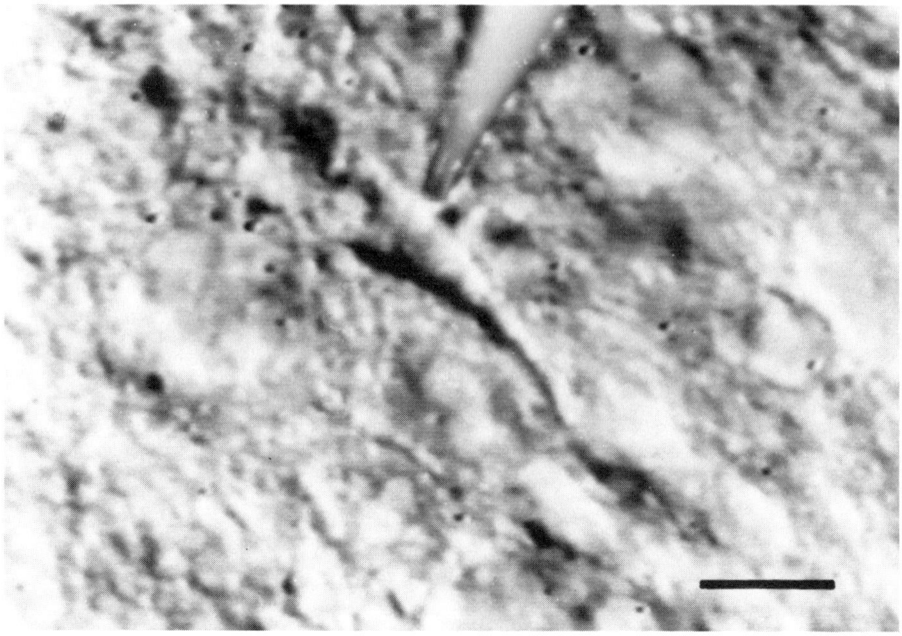

FIGURE 4. Visualization of unstained live neurons in a 300 μm thick section of brain tissue by infrared VEC-DIC microscopy. In infrared light, tissue slices are much less opaque than with shorter wavelength light and axons or dendrites of large neurons can be seen. This allows precise positioning of micropipettes for patch clamp recording or, as in this case, for iontophoretic application of pharmacological agents, antibodies or dyes. Live pyramidal cell from rat neocortex lamina II at a depth of 50 μm. Wavelength 800 nm, Zeiss Axiovert, 20× Plan Neofluar NA 0.5, 2.5× Optovar; Hamamatsu C2400 Newvicon camera with analog contrast enhancement only. Bar, 20 μm. (From Dodt, H. U. and Zieglgänsberger, W., *Brain Res.*, 537, 333, 1990. With permission.)

allow the application of contrast enhancement to a great extent because the image often gets distorted before the small details become visible.

Images can now be produced by VIM at lower illumination intensity and with lower concentrations of potentially cytotoxic dyes so that this technique has its greatest promise in vital microscopy where the living cells need to be protected from harmful radiation or chemicals.[33,43,44,57-59] For VIM the specimen needs to be thin and as close to 2-D as possible. Otherwise out-of-focus fluorescence would make the use of confocal microscopes[11,12] necessary.

Taken together, these techniques are best suited for applications involving living cells or extruded cytoplasm, and they are helpful for a better understanding of the cells' physiology,[43,44,60] biochemistry,[61,62] molecular architecture,[33,58,60,65] or molecular dynamics.[58,59,63,64]

B. VISUALIZING INTRACELLULAR FINE STRUCTURES

The systematic study of the different cell types in the living state by VEC microscopy has only started, but many new insights have already changed our view of the cytoplasm.[7,65] It is most advantageous that corresponding pairs of videomicrographs can be obtained with one of the images at high resolution (VEC) and one obtained by VIM after the addition of vital dyes which are organelle-specific markers[14,30,60,66] or of fluorescence-labeled antibodies which report the location and amount of all kinds of antigens. These compounds can be applied after the observations of live cells are completed and the cells fixed. Studies on live cells are, however, equally possible if cells are microinjected or permeabilized.[58,59,67,68] The use of 1- to 5-nm colloidal gold-coupled antibodies which are visualized by video imaging yields nonbleaching images of higher spatial resolution than fluorescence immunocytochemistry or VEC-DIC microscopy.[39,69] These techniques allow the biochemical or molecular identification of the objects revealed by the high resolution VEC techniques.

Fluorescence video micrographs are also well suited for a quantitative evaluation of fine-morphological parameters. Although little work has been done, typical morphometric parameters can be derived if a suitable software package is applied to images stained with organelle-specific vital dyes or antibodies for endoplasmic reticulum network,[30,70] for the filamentous arrays of cytoskeletal elements and mitochondria (see Figure 6b),[71,72] or for spherical organelles such as lysosomes or Golgi apparatus.[66,73]

C. ANALYSIS OF INTRACELLULAR MOTION

Considerable progress in the field of cytoplasmic motility and fine structure had to await the advent of video microscopy and especially of Allen video-enhanced contrast-differential interference contrast (AVEC-DIC) microscopy,[25,26,65] which allows us to see and to study all membrane-bounded organelles and the microtubules, together with their motions and interactions. The new technique was quickly used to study one of the most prominent systems of intracellular organelle motion, namely the transport in nerve cell processes (axoplasmic transport). From these and related studies[35,36,49,54,74,75] it became clear that organelle motion in higher animal cells proceeds along microtubules.

Many more breakthroughs based on these techniques followed, i.e., the finding that microtubules move actively,[35,53] that organelles move bidirectionally on a single microtubule,[35,49,76,77] and that the force-generating enzymes or molecular motors of microtubule-based movements could be isolated.[78-82] All biochemical studies had to employ video microscopy-based motion analysis as the only means to detect and assay the force-generating proteins.

In plant cells the role of the filamentous tracks is played by actin filament bundles as can be demonstrated by the video techniques.[2,32-34,37,83] Similar studies were performed on other forms of cytoplasmic transport and cellular motility improving our understanding of muscle contraction,[84,85] flagellar motion,[87] pigment granule motion,[75] cytoplasmic transport in protists,[50,77] to name only a few examples.

In nerve cells it is now possible to study axonal transport in olfactory axons that are so thin that they cannot be resolved by conventional light microscopy[88] or in larger myelinated and unmyelinated axons.[24,29,35,51,89] Changes in transport to and from regenerating axons and growth cones were reported[90,91] as well as the specific staining and morphological monitoring of neuromuscular endplates over months with nontoxic fluorescent dyes.[57,92]

1. Visualization of Moving Objects

The motile organelles, mostly below resolution in size, were most conveniently made visible by Allen video-enhanced contrast differential interference contrast (AVEC-DIC) microscopy as described above.[7,25,26,51,65]

In order to perform quantitative motion analyses of these organelles, several laboratories have developed powerful soft- and hardware to track moving objects and to subject the resulting time series of positional data to detailed analyses.[29,93,94] In the following paragraphs our methods and some of the results obtained are summarized.[29]

2. Generation of Trajectories from Video Sequences

The trajectories of individual particles are obtained in two ways. Particles of suitable size and good contrast are tracked automatically by an X-Y tracker (X-Y Tracker C1055, Hamamatsu Photonics, Herrsching, Germany). This device detects objects whose brightness, i.e., the analog voltage from the camera, is above an adjustable threshold, and which are

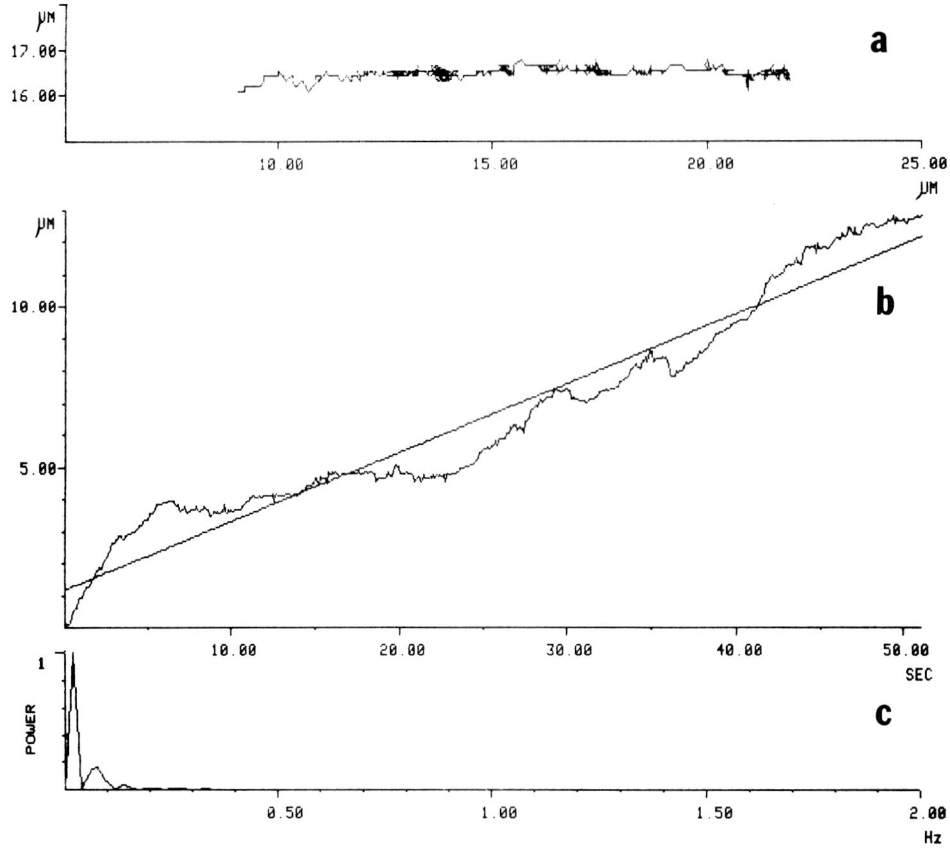

FIGURE 5. Organelle motion analysis. Series of graphs demonstrating steps of the motion analysis of a lysosome-like organelle moving along a myelinated human neuron (sural nerve biopsy material). (a) Path of the organelle moving toward the right hand side on the video screen. (b) Positions of the organelle in the main direction of movement plotted versus time. The slope along the curve indicates the instantaneous velocity. The regression line represents the average velocity which is subtracted (trend removal) prior to the FFT analysis. (c) Power spectrum of positional deviations from the hypothetical constant movement at average speed (regression line in b). The absence of a frequency signal, except for very low frequencies, indicates that the velocity fluctuations do not reflect an oscillatory mechanism but are at most frequencies mainly random. (d) Filtered instantaneous velocity is plotted versus time. (e) Filtered instantaneous velocity is plotted versus position in the main direction of movement.[29] Most of the parameters characterizing this position movement can also be averaged for a group of particles. (From Weiss, D. G., Langford, G. L., Seitz-Tutter, D., Gulden, J., and Keller, F., in *Structure and Functions of the Cytoskeleton, Colloque INSERM,* Vol. 171, Rousset, B. A. F., Ed., John Libbey Eurotext, Paris, 1988, 363. With permission.)

located in a relatively small, adjustable region of interest defined by an overlay frame on the monitor image. The frame with its center cross hair follows automatically if the object moves, and the positions of the cross hair are continuously output to the computer. Small particles with low contrast are tracked by a joy stick in connection with the X-Y tracker in its manual mode. The time series of x-y coordinates (Figure 5a) are transferred via a digital interface to an IBM compatible microcomputer. In automatic mode the coordinates are updated at video frequency (1/50 or 1/60 s depending on the video standard used); in manual mode usually the coordinates of every 10th full frame are determined. The x- and y values for the coordinates range from 0 to 511 and are measured with an accuracy which is determined empirically to be ± 2 pixels or video lines in automatic mode and ± 1 in manual mode.

Unfortunately, trajectories of particles have to be determined with this method one by

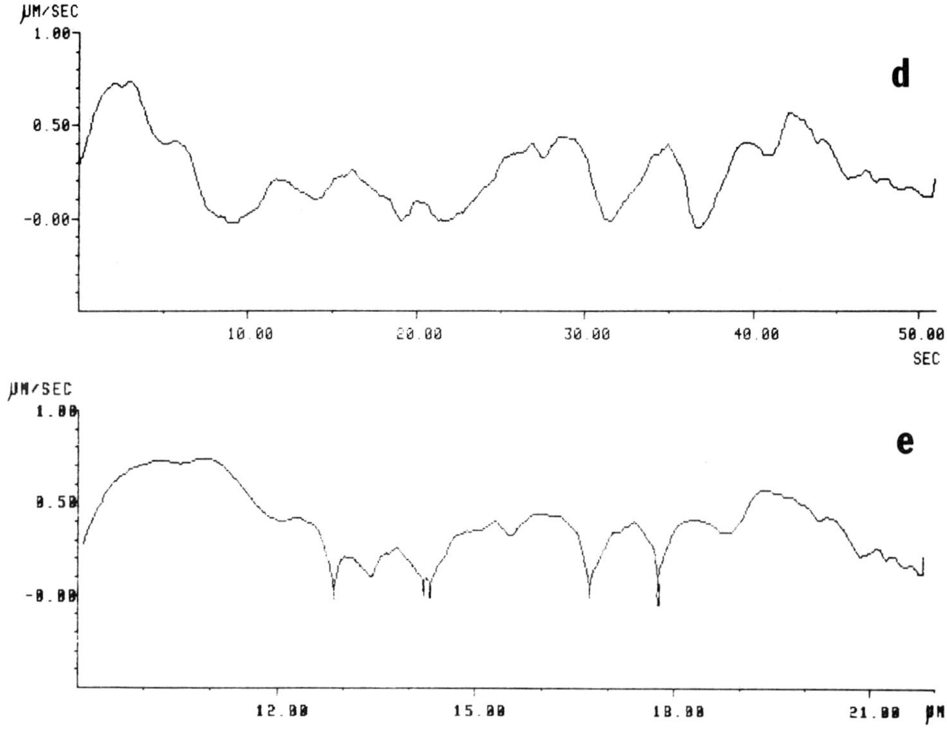

FIGURE 5d and e.

one. To overcome this restriction, we have developed a system which is capable of multiple object tracking. It is based on a digital processing system (MaxVideo, Datacube, Inc., Peabody MA, U.S.) and an area parameter accelerator board (APA512-MX, Vision Systems, Adelaide, Australia). The latter describes each object in a binary image in the form of a set of up to 20 basic descriptive parameters (e.g., number of pixels, perimeter, minimum and maximum x- and y-coordinates, etc.). Connected regions of either black or white pixels are recognized as objects. A set of parameters for up to 255 objects (object pattern) is compiled in a list for each video frame. If the particle pattern has to be generated at video frequency, that is from live video or recorder input, the system can detect and store the information on about 20 particles. If more particles occur in the predefined region of interest, frames must be skipped in order to read out the data. From a sequence of particle pattern lists, the multiple trajectories can be derived by a nearest-neighbor analysis.

3. Motion Analysis

In order to learn more about the type of motion, we analyze the time series of the individual organelle coordinates. The analysis is performed interactively either with a general purpose motion analysis software package based on the procedure developed by Koles et al.[95,96] in our laboratory,[24,29,97] or with the commercial statistics and signal analysis program package SPSS/PC + (SPSS GmbH Software, Munich, Germany).

In many cell types, the moving object and the underlying track (filament) cannot be seen simultaneously. Hence we have to estimate the track by fitting a curve to the observed x,y-positions of the object. Visualization of microtubules revealed that the majority of them form straight tracks and that their diameter is small compared to the size of the organelles being transported.[97,98] This holds true for microtubules in axons, as well as for free microtubules. Additionally, we exclude from the analysis trajectories which grossly deviate from straight tracks. Estimating the track is hence reduced to fitting a straight line to the observed positional data of the particle. One has, however, to be aware that the commonly used regression model to fit a curve is not appropriate in this case because it assumes that the y-values are measured with random errors, while the x-values are without errors. This estimated regression curve minimizes the sum of the squares of the distances of the measured y-value and the y-value of the curve. However, in our case, both the x- and the y-values are subject to measurement errors. Therefore, the sum of squares of the perpendicular distances of the x,y-positions to the line (track) has to be minimized.

The slope a and offset b of the fitted line $g(x) = \hat{y} = ax + b$ are given by equations

$$a^2S_{xy} + a(S_x^2 - S_y^2) - S_{xy} = 0 \tag{1}$$

and

$$\bar{y} = a\bar{x} + b \tag{2}$$

where S_x^2, S_y^2 and S_{xy} are the variances and the covariance, and $\bar{y}$ and $\bar{x}$ are the means of the x- and y-values. For a more detailed explanation and the discussion of confidence intervals for the estimated parameters refer to Wong.[99]

Solving the above equations we obtain an estimate for the track. For any observed position of the particle, the point of intersection of this line with the one perpendicular to it yields its estimated position on the track. If we transform the coordinate system so that the origin is transplanted to the estimated starting position of the particle on the track and the x-axis is rotated onto the estimated line (track), we obtain the desired representation for further analysis. The resulting equations for the translocation and rotation were given by Weiss et al.[29] The motion of the particle is then described by two time series, one representing its position on the track and the other its perpendicular deviation.

The most often used parameter to describe the motion of an organelle in more detail is the average velocity in the major direction of movement. This can be calculated in different ways, e.g., as the slope of the line fitted to the positional data over time, as the average of frame to frame velocities, or as the ratio of distance and time for the tracked path. The latter method is used in our work.

Since we study the molecular mechanism of the intracellular transport, the analysis of time- and track-dependent variations of the velocity in a single trajectory are in the focus of our interest. In contrast to the average velocity, this is often called "instantaneous" velocity. The methods to obtain the estimate of an instantaneous velocity vary,[95,100] so that we describe our approach in more detail.

Some difficulties will emerge when the velocity is estimated from the positional data since the coordinates are likely to be affected with measurement errors. The velocity of a particle is obtained by differentiation of the time series of x position. First-order differencing according to

$$v(t) = (x(t) - x(t - 1))/\Delta t \tag{3}$$

where Δt is the time between successive x positions, which for good reasons is considered to be a rather gross approximation to differentiation. We decided, however, to use this method for two reasons:

1. Due to our sampling frequencies (10 or 25 Hz) the maximum resolvable frequencies of the positional variation of a particle are either 5 or 12.5 Hz. In nearly all observed spectra, more than 99% of the power of the signal is well below 1 Hz. Obviously the

signal is "oversampled" and we therefore obtain, with respect to the signal, a "small" Δt and hence a better approximation to differentiation. Noise components at higher frequencies, induced by measurement errors, would, however, still degrade the estimate severely because high frequency components are "amplified" by differentiation.

2. Therefore, before differencing the time series to Equation 3, we remove critical noise components at higher frequencies in the following way. First, we transform the trend-removed time series (Figure 5) of the x-positions into the frequency domain, using the Cooley-Tukey FFT algorithm[101] (Figure 5c). Second, a cutoff frequency for low-pass filtering, usually in the range of 0.3 to 0.7 Hz, is determined from this spectrum. Third, the trend-removed time series is low-pass filtered using a low-pass filter with the respective cut-off frequency. Fourth, the trend is re-added to the filtered time series. After differencing, according to Equation 3, we finally obtain the particle's instantaneous velocity, which is plotted vs. time (Figure 5d) and vs. the main direction of movement (Figure 5e). With the methods described above, we obtain the desired variables to describe quantitatively and further analyze the displacement of organelles or microtubule ends.[24,29,63,97]

Quantitative characterization of the motile events in a cell is a good means to describe the cell's physiological situation because organelle motion is a vital and very basic property of eukaryotic cells. Once this is accomplished the motile situation can also be used as a very sensitive and multifaceted indicator to characterize cell populations in cytopharmacological and cytotoxicological studies.[102,103] Furthermore, motion analysis of intracellular objects was used successfully to gain more insight into the molecular mechanisms of force generation,[15,41,97] of actin and microtubule mechanics,[104,105] or of the assembly and disassembly dynamics of cytoskeletal filaments.[63,64,67]

D. MEASURING BIOCHEMICAL PARAMETERS IN THE LIVING CELL

Microscopes are often considered to be devices that "merely" make pictures of objects. This is certainly not true for video microscopes which generate images in which the grey or pseudocolor shades encode quantitative physical or chemical parameters of the specimen, such as motion velocity, concentrations, viscosity, birefringence, etc. (see Table 1). The video microscope can, therefore, assume features of a spectrophotometer, spectrofluorimeter, photon multiplier, etc. Usually its dynamic range is more limited than that of the other measurement devices mentioned, but this is generally made good by the advantage of yielding information not only in a punctual way (cuvette), but rather in two dimensions (x and y), three dimensions (x,y,z, in through focus series), or even in four dimensions (x,y,z,t, if video films or spatiotemporal processes are recorded).

Since the transfer function of video cameras used for quantitative microscopy is linear instead of logarithmic as in the case of the human eye and the photographic film,[8] the grey-level information in video images can be used directly to gain information on the relative amounts of fluorescent substances. Absolute values can be obtained from a series of calibration solutions which are imaged at the same instrumental setting as the unknown samples or by making cells permeable for external media containing a known concentration of dye or ligand.[106]

In addition to the intensity values, we need to know the optical pathlength, i.e., the thickness of the cell at each pixel, if information on intracellular concentrations is desired. Such optical pathlength images are obtained when cells are loaded with a dye of homogeneous and exclusively intracellular distribution, such as fluorescein diacetate or labeled dextran, and imaged.[107-109] In this case, the image brightness codes for cell thickness in a quantitative way. Dividing regular fluorescence intensity or absorption images obtained with monochromatic light by these optical pathlength images yields "concentration images".

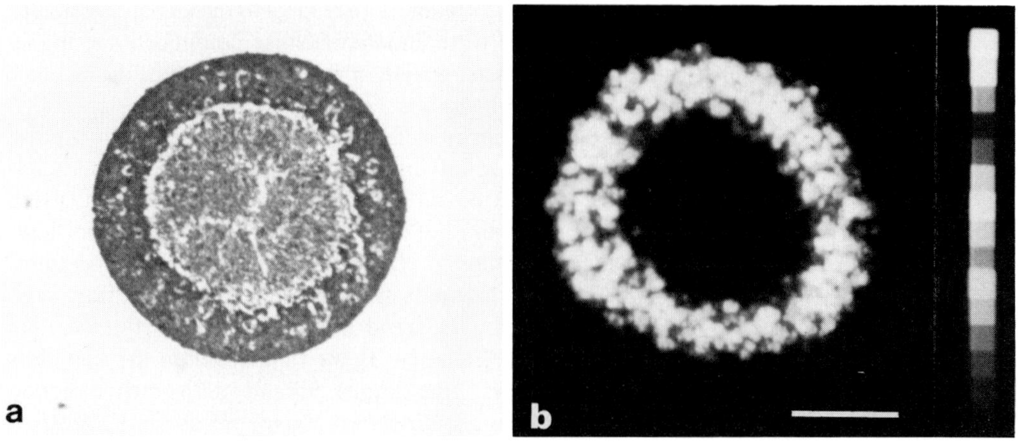

FIGURE 6. Application of the quantitative bioluminescence technique of regional ATP-measurement by single photon imaging to a central cryostat section through a multicellular EMT6 tumor cell spheroid.[61] (a) Unstained cryostat section; (b) pseudocolor-coded bioluminescence intensity distribution representing the local distribution of ATP concentrations. The color code (right) spans the concentration range from 0.1 to 1.2 mM ATP. Bar 500 μm. (Courtesy of W. Müller-Klieser.)

Based on these techniques, improved assays for regional and temporal enzyme activity could be developed which consume or synthesize absorbing or fluorescent compounds.[110] Direct enzymatic substrate measurements have already been introduced for lactate, glucose, and ATP in unfixed tissue sections. The cells are permeabilized (e.g., frozen and thawed) and exposed to or microinjected with the coupled enzyme system containing luciferase and suitable bioluminescent substrates and then imaged by photon counting[61,62] (Figure 6). An elegant technique was introduced by Szalay and collaborators to visualize gene expression in plant cells.[111,112] In rhizobium-infected soybean root nodules or in cross sections of tobacco plants, the control of gene expression by specific promoters could be directly visualized by the bacterial enzyme luciferase introduced into the plant genome as reporter gene. Light emission, i.e., activation of the promoter and expression of its associated genes, was quantitated throughout the plant, in the cambium cell layer or in defined subcellular regions (Figure 7).

At present the most rapidly expanding application of video microscopy in cell biology is the use of fluorescent chelators to measure concentrations of ions and their transient changes. These specially designed dyes cross the plasma membrane in the form of their more lipophilic acetoxymethyl esters and are trapped in the cell after intracellular cleavage of the ester bond. In the case of the Ca^{2+} chelator FURA-2 and some related ion indicators, the binding of the ion causes a concentration-dependent spectral shift of the chelator dye fluorescence. By dividing the images obtained at two different wavelengths and with the use of a calibration curve (ratio values of a standard series plotted vs. concentration), one obtains an image containing the concentration information (ratio imaging).[106,113] This procedure is independent of the optical pathlength so that no knowledge of the cell thickness is required. Besides the concentration of Ca^{2+}, H^+ (pH-value), Mg^{2+}, Cl^-, K^+, and Na^+ concentrations can also be determined with suitable dyes.[14,60,108,113-115] The image processors can be programmed to control the required filter changer and to display the ion concentration in the form of color-coded images. The temporal resolution of this technique can be less than 1 s (Figure 8).

Similar to the above-mentioned extraction of coordinates from a series of video frames, one can also extract intensity values of individual image points or image regions from such video series, so that concentration changes in individual cells or in subcellular regions can

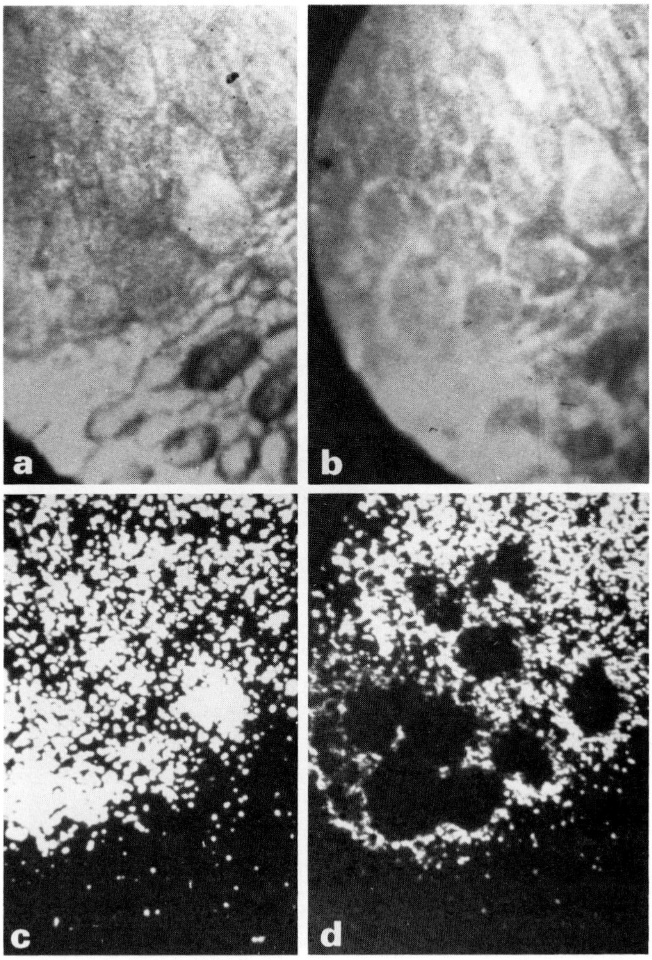

FIGURE 7. Visualization of the bioluminescence signal from a *Rhizobium*-infected soybean plant cell by photon counting imaging. The soybean root nodule cells are filled with transformed *Rhizobium* bacteroids containing a PL promoter-lux AB gene fusion. (a and b) Transmitted light images of sections through soybean stem nodule tissue. Each section has a slightly different focal plane. Thick sections through soybean root nodules show the arrangement of the cells in the nodule tissue. The bioluminescent images (c and d) specify the exact location within the root nodule of individual plant cells containing bacteroids expressing luciferase activity. (c) Reverse contrast of (d). (From Langridge, W. H. R., Escher, A., Baga, M., O'Kane, D., Wampler, J., Koncz, C., Schell, J., and Szalay, A. A., *SPIE,* 1161, 216, 1989. With permission.)

be followed over time (temporal analysis).[115,116] This makes studies of intracellular transfer, intracellular exchange, and metabolism of compounds possible.

IV. CONCLUSION

Video microscopy allows us to go beyond the classical limits of light microscopy. It enables us to see smaller objects than before, to work at extremely low light intensities, and to generate contrast where none could be generated by conventional techniques. Most of the new techniques and their applications yield best image improvement with unstained

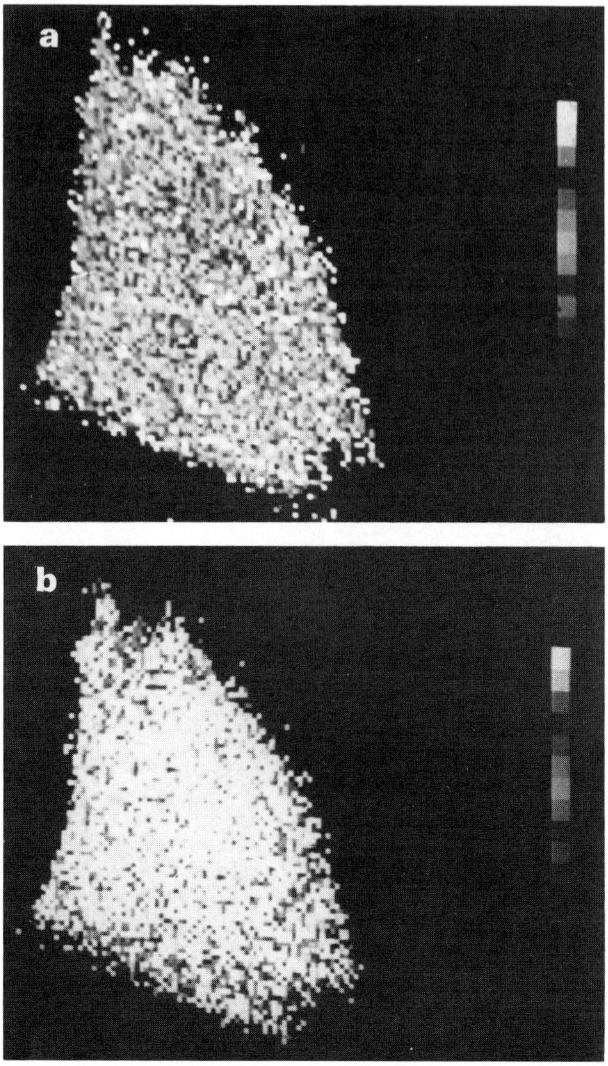

FIGURE 8. Measurement of changes in Ca^{2+} concentration over time by the ratio imaging technique using FURA-2. The specimen is excited alternatingly at 340 and 360 nm and the resulting alternating fluorescence images emitted at 500 ± 10 nm are divided. The ratio images (a to d) contain the information on Ca^{2+} concentration coded as intensity. The ratio values of 0 to 3.00 which cover the range of biological Ca^{2+} concentrations (about 0.05 to 1 μM) are first assigned to the grey levels 0 to 256 and subsequently color coded. An increase in image brightness thus means a rise in free Ca^{2+} concentration (here from level 12 to 16 in the color-code bar). The average intensity in the frame placed over the cell center (d) is measured and plotted versus time (8 s, abscissa). In this example, 100 frames were processed at reduced spatial resolution (128 × 128 pixels) at a sampling frequency of 12.5 Hz of which nos. 1 and 66 are shown. (c) Single wavelength fluorescence image at 360 nm excitation. Human fibroblast loaded with the Ca^{2+}-dependent fluorochrome FURA-2, Microscope: Polyvar, Reichert/Cambridge Instruments equipped with quartz collector, 200 W mercury arc, and 100× planfluorite oil objective; processor: ARGUS 100 of Hamamatsu; cell width: 34 μm. (From Nyman, G., Maile, W., and Weiss, D. G., unpublished.)

specimens so that live cells and their cytoplasm become amenable to study. As often happens in science, new technologies bring new insight so that already a large body of knowledge on cell structure and function could be accumulated that profoundly changed our static, electron microscopy-based understanding of the cytoskeleton and of cellular structures to a

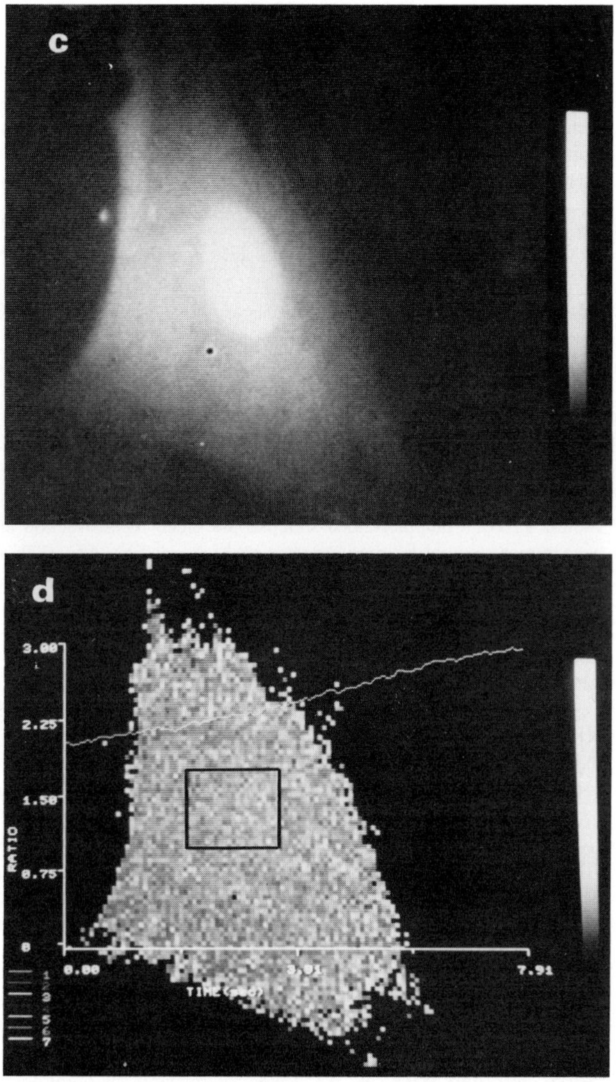

FIGURE 8c and d.

more lively, highly dynamic view. Video microscopists are presently extending the microscope's imaging power well into the molecular domain, making us eyewitnesses also of molecular dynamics, transport, and metabolism in the living cytoplasm.

ACKNOWLEDGMENTS

The video-microscopic research in the author's laboratory was supported by Deutsche Forschungsgemeinschaft, Bayer; Staatsministerien für Landesentwicklung und Umweltschutz and für Wissenschaft und Kunst. The valuable advice and help of Willi Maile and Karin Weißbacher is gratefully acknowledged. We thank H. U. Dodt, T. Holstein, W. Mueller-Klieser, and A. A. Szalay for their permission to reproduce pictures taken from their work.

REFERENCES

1. **Gross, G. W. and Weiss, D. G.,** Theoretical considerations on rapid transport in low viscosity axonal regions, in *Axoplasmic Transport,* Weiss, D. G., Ed., Springer-Verlag, Berlin, 1982, 330.
2. **Schliwa, M.,** Mechanisms of intracellular organelle transport, in *Cell and Muscle Motility,* Vol. 5, Shay, J. W., Ed., Plenum Press, New York, 1984, 1 and 403.
3. **Weiss, D. G.,** Dynamics and cooperativity in the organization of cytoplasmic structures and flows, in *Complex Systems — Operational Approaches, in Neurobiology, Physics, and Computers,* Haken, H., Ed., Springer-Verlag, Berlin, 1985, 179.
4. **Shotton, D.,** The current renaissance in light microscopy. I. Dynamic studies of living cells by video enhanced contrast microscopy, *Proc. R. Microsc. Soc.,* 22, 37, 1987.
5. **Amos, L.,** Movements made visible by microchip technology (News and Views), *Nature,* 330, 211, 1987.
6. **Webb, W. W.,** Light microscopy — a modern renaissance, *Ann. N.Y. Acad. Sci.,* 483, 387, 1986.
7. **Weiss, D. G.,** Visualization of the living cytoskeleton by video-enhanced microscopy and digital image processing, *J. Cell Sci.,* Suppl. 5, 1, 1986.
8. **Inoué, S.,** *Video Microscopy,* Plenum Press, New York, 1986.
9. **Inoué, S.,** Whither video microscopy — towards 4-D imaging at the highest resolution of the light microscope, in *Optical Microscopy for Biology,* Herman, B. and Jacobson, K., Ed., Wiley-Liss, New York, 1990, 497.
10. **Inoué, S. and Inoué, T. D.,** Computer-aided stereoscopic video reconstruction and serial display from high-resolution light-microscope optical sections, *Ann. N.Y. Acad. Sci.,* 483, 392, 1987.
11. **Shotton, D.,** Confocal scanning optical microscopy and its applications for biological specimens, *J. Cell Sci.,* 94, 175, 1989.
12. **Wilson, T.,** Trends in confocal microscopy, *Trends Neurosci.,* 12, 486, 1989.
13. **Taylor, D. L. and Wang, Y.-L.,** *Methods in Cell Biology,* Vol. 29, *Fluorescence Microscopy of Living Cells in Culture,* Part A, Academic Press, San Diego, 1989.
14. **Haugland, R. P.,** *Handbook of Fluorescent Probes and Research Chemicals,* 2nd ed., Molecular Probes, Inc., Eugene, OR, 1989.
15. **Wick, R. A.,** Quantum-limited imaging using microchannel plate technology, *Appl. Opt.,* 26, 3210, 1987.
16. **Marcus, D. A.,** High-performance optical filters for fluorescence analysis, *Cell Motil. Cytoskel.,* 10, 62, 1988.
17. **Shotton, D. M.,** Review: video-enhanced light microscopy and its applications in cell biology, *J. Cell Sci.,* 89, 129, 1988.
18. **Aikens, R. S., Agard, D. A., and Sedat, J. W.,** Solid-state imagers for microscopy, in *Methods in Cell Biology,* Vol. 29, Taylor, D. L. and Wang, Y. L., Eds., Academic Press, San Diego, 1989, 291.
19. **Spring, K. R. and Lowy, R. J.,** Characteristics of low light level television cameras, in *Methods in Cell Biology,* Vol. 29, Taylor, D. L. and Wang, Y. L., Eds., Academic Press, San Diego, 1989, 269.
20. **Wampler, J. E. and Kutz, K.,** Quantitative fluorescence microscopy using photomultiplier tubes and imaging detectors, in *Methods in Cell Biology,* Vol. 29, Taylor, D. L. and Wang, Y. L., Eds., Academic Press, San Diego, 1989, 239.
21. **Weiss, D. G., Maile, W., and Wick, R. A.,** Video microscopy, in *Light Microscopy in Biology. A Practical Approach,* Lacey, A. J., Ed., IRL Press, Oxford, 1989, 221.
22. **Herman, B. and Jacobson, K.,** *Optical Microscopy for Biology,* Wiley-Liss, New York, 1990.
23. **Spring, K. R.,** Quantitative imaging at low light levels — differential interference contrast and fluorescence microscopy without significant light loss, in *Optical Microscopy for Biology,* Herman, B. and Jacobson, K., Eds., Wiley-Liss, New York, 1990, 513.
24. **Weiss, D. G., Keller, F., Gulden, J., and Maile, W.,** Towards a new classification of intracellular particle movements based on quantitative analyses, *Cell Motil. Cytoskel.,* 6, 128, 1986.
25. **Allen, R. D., Travis, J. L., Allen, N. S., and Yilmaz, H.,** Video-enhanced contrast polarization (AVEC-POL) microscopy: a new method applied to the detection of birefringence in the motile reticulopodial network of *Allogromia laticollaris, Cell Motil.,* 1, 275, 1981.
26. **Allen, R. D., Allen, N. S., and Travis, J. L.,** Video-enhanced contrast, differential interference contrast (AVEC-DIC) microscopy: a new method capable of analyzing microtubule-related motility in the reticulopodial network of *Allogromia laticollaris, Cell Motil.,* 1, 291, 1981.
27. **Allen, R. D. and Allen, N. S.,** Video-enhanced microscopy with a computer frame memory, *J. Microsc.,* 129, 3, 1983.
28. **Weiss, D. G., Maile, W., and Shotton, D. M.,** Principles and practice of video-enhanced contrast microscopy, in *Electronic Light Microscopy: The Principles and Practice of Intensified Fluorescence, Video-Enhanced Contrast and Confocal Scanning Optical Microscopy,* Shotton, D. M., Ed., Wiley-Liss, New York, in press.

29. **Weiss, D. G., Galfe, G., Gulden, J., Seitz-Tutter, D., Langford, G. M., Struppler, A., and Weindl, A.,** Motion analysis of intracellular objects: trajectories with and without visible tracks, in *Biological Motion. Lecture Notes in Biomathematics,* Vol. 89, Alt, W. and Hoffmann, G., Eds., Springer-Verlag, Berlin, 1990, 95.
30. **Lee, C. and Chen, L. B.,** Behavior of endoplasmic reticulum in living cells, *Cell,* 54, 37, 1988.
31. **Lichtscheidl, I. K. and Url, W. G.,** Video-enhanced phase contrast microscopy: a tool for the study of endoplasmic reticulum in *Allium cepa* inner epidermal cells, *Inst. Phys. Conf. Ser.,* 89, 675, 1989.
32. **Lichtscheidl, I. K. and Weiss, D. G.,** Visualization of submicroscopic structures in the cytoplasm of *Allium cepa* inner epidermal cells by video-enhanced contrast light microscopy, *Eur. J. Cell Biol.,* 46, 376, 1988.
33. **Quader, H., Hoffmann, A., and Schnepf, E.,** Shape and movement of the endoplasmic reticulum in onion bulb epidermis cells: possible involvement of actin, *Eur. J. Cell Biol.,* 44, 17, 1987.
34. **Allen, N. S. and Brown, D. T.** Dynamics of the endoplasmic reticulum in living onion epidermal cells in relation to microtubules, microfilaments, and intracellular particle movement, *Cell Motil. Cytoskel.,* 10, 153, 1988.
35. **Allen, R. D., Weiss, D. G., Hayden, J. H., Brown, D. T., Fujiwake, H., and Simpson, M.,** Gliding movement of and bidirectional transport along native microtubules from squid axoplasm: evidence for an active role of microtubules in cytoplasmic transport, *J. Cell Biol.,* 100, 1736, 1985.
36. **Vale, R. D., Schnapp, B. J., Reese, T. S., and Sheetz, M. P.,** Organelle, bead and microtubule translocations promoted by soluble factors from the squid giant axon, *Cell,* 40, 559, 1985.
37. **Kachar, B. and Reese, T. S.,** The mechanism of cytoplasmic streaming in Characean algal cells: sliding of endoplasmic reticulum along actin filaments, *J. Cell Biol.,* 106, 1545, 1988.
38. **Mizushima, Y.,** Detectivity limit of very small objects by video-enhanced microscopy, *Appl. Opt.,* 27, 2587, 1988.
39. **De Brabander, M., Nuydens, R., Geerts, H., Nuyens, R., Leunissen, J., and Jacobson, K.,** Using nanovid microscopy to analyse the movement of cell membrane components in living cells, in *Optical Microscopy for Biology,* Herman, B. and Jacobson, K., Eds., Wiley-Liss, New York, 1990, 345.
40. **Inoué, S.,** Imaging of unresolved objects, superresolution, and precision of distance measurement with video microscopy, in *Methods in Cell Biology,* Vol. 30, Taylor, D. L. and Wang, Y. L., Eds., Academic Press, San Diego, 1989, 85.
41. **Sheetz, M. P., Turney, S, Qian, H., and Elson, E. L.,** Nanometre-level analysis demonstrates that lipid flow does not drive membrane glycoprotein movements, *Nature,* 340, 284, 1989.
42. **Inoué, S.,** Video image processing greatly enhances contrast, quality, and speed in polarization-based microscopy, *J. Cell Biol.,* 89, 346, 1981.
43. **Tsien, R. Y.,** Fluorescent probes of cell signaling, *Annu. Rev. Neurosci.,* 12, 227, 1989.
44. **Tsien, R. Y.,** Fluorescent indicators of ion concentrations, in *Methods in Cell Biology,* Vol. 30, *Fluorescence Microscopy of Living Cells in Culture,* Part B: *Quantitative Fluorescence Microscopy — Imaging and Spectroscopy,* Taylor, D. L. and Wang, Y. L., Eds., Academic Press, San Diego, 1989, 127.
45. **Anon.,** *Image Analysis. Principles and Practice,* Joyce Loebl Ltd./IRL Press, London, 1985.
46. **Häder, D.-P.,** Computer-assisted image analysis in biological sciences, *Proc. Indian Acad. Sci. (Plant Sci.),* 98, 227, 1988.
47. **Swedenburg, C. E. and Conklin, J. J.,** *Imaging Techniques in Biology and Medicine,* Academic Press, San Diego, 1988.
48. **Bradbury, S.,** Micrometry and image analysis, in *Light Microscopy in Biology: A Practical Approach,* Lacey, A. J., Ed., IRL Press, Oxford, 1989, 200.
49. **Hayden, J. H. and Allen, R. D.,** Detection of single microtubules in living cells: particle transport can occur in both directions along the same microtubule, *J. Cell Biol.,* 99, 1785, 1984.
50. **Travis, J. L. and Bowser, S. S.,** Optical approaches to the study of foraminiferan motility, *Cell Motil. Cytoskel.,* 10, 126, 1988.
51. **Weiss, D. G., Meyer, M., and Langford, G. M.,** Studying axoplasmic transport by video microscopy and using the squid giant axon as a model system, in *Squid as Experimental Animals,* Gilbert, D. L., Adelman, W. J., Jr., and Arnold, J. M., Eds., Plenum Press, New York, 1990, 303.
52. **Fisher, G. W., Conrad, P. A., DeBiasio, R. L., and Taylor, D. L.,** Centripetal transport of cytoplasm, actin, and the cell surface in lamellipodia of fibroblasts, *Cell Motil. Cytoskel.,* 11, 235, 1988.
53. **Allen, R. D. and Weiss, D. G.,** An experimental analysis of the mechanisms of fast axonal transport in the squid giant axon, in *Cell Motility: Mechanism and Regulation,* Vol. 10, (Yamada Conference, September 1984, Nagoya), Ishikawa, H., Hatano, S., and Sato, H., Eds., University of Tokyo Press, Tokyo, 1985, 327.
54. **Brady, S. T., Lasek, R. J., and Allen, R. D.,** Fast axonal transport in extruded axoplasm from squid giant axon, *Science,* 218, 1129, 1982.
55. **Kachar, B.,** Asymmetric illumination contrast: a method of image formation for video light microscopy, *Science,* 277, 766, 1985.

56. **Dodt, H. U. and Zieglgänsberger, W.,** Visualizing unstained neurons in living brain slices by infrared DIC-video microscopy, *Brain Res.*, 537, 333, 1990.
57. **Purves, D. and Voyvodic, T.,** Imaging mammalian nerve cells and their connections over time in living animals, *Trends Neurosci.*, 10, 398, 1987.
58. **Sanger, J. M., Mittal, B., Pochapin, M. B., and Sanger, J. W.,** Stress fiber and cleavage furrow formation in living cells microinjected with fluorescently labeled alpha-actinin, *Cell. Motil. Cytoskel.*, 7, 209, 1987.
59. **Wang, Y.-L.,** Fluorescent analog cytochemistry: tracing functional protein components in living cells, in *Methods in Cell Biology*, Vol. 29, Taylor, D. L. and Wang, Y. L., Eds., Academic Press, San Diego, 1989, 1.
60. **DeBiasio, R., Bright, G. R., Ernst, L. A., Waggoner, A. S., and Taylor, D. L.,** Five-parameter fluorescence imaging: wound healing of living Swiss 3T3 cells, *J. Cell Biol.*, 105, 1613, 1987.
61. **Mueller-Klieser, W., Schaefer, C., Walenta, S., Rofstad, E. K., Fenton, B. M., and Sutherland, R. M.,** Assessment of tumor energy and oxygenation status by bioluminescence, NMR spectroscopy, and cytospectrophotometry, *Cancer Res.*, 50, 1681, 1990.
62. **Walenta, S., Dötsch, J., and Mueller-Klieser, W.,** ATP concentrations in multicellular tumor spheroids assessed by single photon imaging and quantitative bioluminescence, *Eur. J. Cell Biol.*, 52, 389, 1990.
63. **Seitz-Tutter, D., Langford, G. M., and Weiss, D. G.,** Dynamic instability of native microtubules from squid axons is rare and independent of gliding and vesicle transport, *Exp. Cell Res.*, 178, 504, 1988.
64. **Cassimeris, L., Pryer, N. K., and Salmon, E. D.,** Real-time observations of microtubule dynamic instability in living cells, *J. Cell Biol.*, 107, 2223, 1988.
65. **Allen, R. D.,** New observations on cell architecture and dynamics by video-enhanced contrast optical microscopy, *Annu. Rev. Biophys. Biophys. Chem.*, 14, 265, 1985.
66. **Pagano, R. E., Sepanski, M. A., and Martin, O. C.,** Molecular trapping of a fluorescent ceramide analogue at the golgi apparatus of fixed cells, interaction with endogenous lipids provides a *trans*-golgi marker for both light and electron microscopy, *J. Cell Biol.*, 109, 2067, 1989.
67. **Sammak, P. J. and Borisy, G. G.,** Direct observation of microtubule dynamics in living cells, *Nature*, 332, 724, 1988.
68. **Hayden, J. H., Allen, R. D., and Goldman, R. D.,** Cytoplasmic transport in keratocytes: direct visualization of particle translocation along microtubules, *Cell Motil.*, 3, 1, 1983.
69. **Bajer, A. S., Sato, H., and Mole-Bajer, J.,** Video microscopy of colloidal gold particles and immunogold labeled microtubules in improved rectified DIC and epi-illumination, *Cell Struct. Function*, 11, 317, 1986.
70. **Terasaki, M., Song, J., Wong, J. R., Weiss, M. J., and Chen, L. B.,** Localization of endoplasmic reticulum in living and glutaraldehyde-fixed cells with fluorescent dyes, *Cell*, 38, 101, 1984.
71. **Johnson, L. V., Walsh, M. L., and Chen, L. B.,** Localization of mitochondria in living cells with rhodamine 123, *Proc. Natl. Acad. Sci. U.S.A.*, 77, 990, 1980.
72. **Bereiter-Hahn, J., Seipel, K.-H., Vöth, M., and Ploem, J. S.,** Fluorimetry of mitochondria in cells vitally stained with DASPMI or rhodamine 6GO, *Cell Biochem. Function*, 1, 147, 1983.
73. **Kreis, T. E.,** Role of microtubules in the organization of the Golgi apparatus, *Cell Motil. Cytoskel.*, 15, 67, 1990.
74. **Allen, R. D., Metuzals, J., Tasaki, I., Brady, S. T., and Gilbert, S. P.,** Fast axonal transport in squid giant axon, *Science*, 218, 1127, 1982.
75. **McNiven, M. A. and Porter, K. R.,** Microtubule polarity confers direction to pigment transport in chromatophores, *J. Cell Biol.*, 103, 1547, 1986.
76. **Schnapp, B. J., Vale, R. D., Sheetz, M. P., and Reese, T. S.,** Single microtubules from squid axoplasm support bidirectional movements of organelles, *Cell*, 40, 455, 1985.
77. **Koonce, M. P. and Schliwa, M.,** Bidirectional organelle transport can occur in cell processes that contain single microtubules, *J. Cell Biol.*, 100, 322, 1985.
78. **Brady, S. T.,** A novel brain ATPase with properties expected for the fast axonal transport motor, *Nature (London)*, 317, 73, 1985.
79. **Scholey, J. M., Porter, M. E., Grissom, P. M., and McIntosh, J. R.,** Identification of kinesin in sea urchin eggs and evidence for its localization in the mitotic spindle, *Nature (London)*, 318, 483, 1985.
80. **Vale, R. D., Reese, T. S., and Sheetz, M. P.,** Identification of a novel, force-generating protein, kinesin, involved in microtubule-based motility, *Cell*, 42, 39, 1985.
81. **Paschal, B. M., Shpetner, H. S., and Vallee, R. B.,** MAP 1C is a mircrotubule-activated ATPase which translocates microtubules in vitro and has dynein-like properties, *J. Cell Biol.*, 105, 1273, 1987.
82. **Warner, F. D. and McIntosh, J. R.,** *Cell Movement*, Vol. 2, *Kinesin, Dynein, and Microtubule Dynamics*, Alan R. Liss, New York, 1989.
83. **Sheetz, M. P. and Spudich, J. A.,** Movement of myosin-coated fluorescent beads on actin cables *in vitro*, *Nature*, 303, 31, 1983.

84. **Sellers, J. R. and Kachar, B.,** Polarity and velocity of sliding filaments: control of direction by actin and of speed by myosin, *Science,* 249, 406, 1990.
85. **Kron, S. J. and Spudich, J. A.,** Fluorescent actin filaments move on myosin fixed to a glass surface, *Proc. Natl. Acad. Sci. U.S.A.,* 83, 6272, 1986.
86. **Toyoshima, Y. Y., Toyoshima, C., and Spudich, J. A.,** Bidirectional movement of actin filaments along tracks of myosin heads, *Nature,* 341, 154, 1989.
87. **Vale, R. D. and Toyoshima, Y. Y.,** Rotation and translocation of microtubules *in vitro* induced by dyneins from *Tetrahymena* cilia, *Cell,* 52, 459, 1988.
88. **Weiss, D. G. and Buchner, K.,** Axoplasmic transport in olfactory receptor neurons, in *Molecular Neurobiology of the Olfactory System,* Margolis, F. L. and Getchell, T. V., Eds., Plenum Press, New York, 1988, 217.
89. **Llinás, R., Sugimori, M., Lin, J.-W., Leopold, P. L., and Brady, S. T.,** ATP-dependent directional movement of rat synaptic vesicles injected into the presynaptic terminal of squid giant synapse, *Proc. Natl. Acad. Sci. U.S.A.,* 86, 5656, 1989.
90. **Kater, S. B., Mattson, M. P., Cohan, C., and Connor, J.,** Calcium regulation of the neuronal growth cone, *Trends Neurosci.,* 11, 315, 1988.
91. **Goldberg, D. J. and Burmeister, D. W.,** Looking into growth cones, *Trends Neurosci.,* 12, 503, 1989.
92. **Herrera, A. A. and Banner, L. R.,** The use and effects of vital fluorescent dyes: observation of motor nerve terminals and satellite cells in living frog muscles, *J. Neurocytol.,* 19, 67, 1990.
93. **Cohn, S. A., Ingold, A. L., and Scholey, J. M.,** Correlation between the ATPase and microtubule translocating activities of sea urchin kinesin, *Nature (London),* 328, 160, 1987.
94. **Gelles, J., Schnapp, B. J., and Sheetz, M. P.,** Tracking kinesin-driven movements with nanometre-scale precision, *Nature (London),* 331, 450, 1988.
95. **Koles, Z. J., McLeod, K. D., and Smith, R. S.,** The determination of the instantaneous velocity of axonally transported organelles from filmed records of their motion, *Can. J. Physiol. Pharmacol.,* 60, 670, 1982.
96. **Koles, Z. J., McLeod, K. D., and Smith, R. S.,** A study of the motion of organelles which undergo retrograde and anterograde rapid axonal transport in *Xenopus, J. Physiol. (London),* 328, 469, 1982.
97. **Weiss, D. G., Langford, G. M., Seitz-Tutter, D., and Keller, F.,** Dynamic instability and motile events of native microtubules from squid axoplasm, *Cell Motil. Cytoskel.,* 10, 285, 1988.
98. **Langford, G. M., Allen, R. D., and Weiss, D. G.,** Substructure of sidearms on squid axoplasmic vesicles and microtubules visualized by negative contrast electron microscopy, *Cell Motil. Cytoskel.,* 7, 20, 1987.
99. **Wong, M. Y.,** Likelihood estimation of a simple linear regression model when both variables have error, *Biometrika,* 76, 141, 1989.
100. **Breuer, A. C., Lynn, M. P., Atkinson, M. B., Chou, S. M., Wibourn, A. J., Marks, K. E., Culver, J. E., and Fleegler, E. J.,** Fast axonal transport in amyotrophic lateral sclerosis, an intra-axonal organelle traffic analysis, *Neurology,* 37, 738, 1987.
101. **Cooley, J. W. and Tukey, J. W.,** An algorithm for the machine calculation of complex Fourier series, *Math. Comput.,* 19, 297, 1965.
102. **Weiss, D. G.,** Videomicroscopic measurements in living cells: dynamic determination of multiple end points for *in vitro* toxicology, *Mol. Toxicol.,* 1, 465, 1987.
103. **Maile, W., Lindl, T., and Weiss, D. G.,** New methods for cytotoxicity testing: quantitative video microscopy of intracellular motion and mitochondria-specific fluorescence, *Mol. Toxicol.,* 1, 427, 1987.
104. **Kishino, A. and Yanagida, T.,** Force measurements by micromanipulation of a single actin filament by glass needles, *Nature,* 334, 74, 1988.
105. **Kamimura, S. and Takahashi, K.,** Direct measurement of the force of microtubule sliding in flagella, *Nature (London),* 293, 566, 1985.
106. **Grynkiewicz, G., Poenie, M., and Tsien, R. Y.,** A new generation of Ca^{2+} indicators with greatly improved fluorescence properties, *J. Cell Biol. Chem.,* 260, 3440, 1985.
107. **Möncke, E., Lucas, C., and Weiss, D. G.,** Distribution, diffusion and transport of fluorescent dyes in cultured cells, *Eur. J. Cell Biol.,* 36 (Suppl. 7), 44, 1985.
108. **Giuliano, K. A., Nederlof, M. A., DeBiasio, R., Lanni, F., Waggoner, A. S., and Taylor, D. L.,** Multi-mode light microscopy, in *Optical Microscopy for Biology,* Herman, B. and Jacobson, K., Eds., Wiley-Liss, New York, 1990, 543.
109. **Luby-Phelps, K., Taylor, D. L., and Lanni, F.,** Probing the structures of cytoplasm, *J. Cell Biol.,* 102, 2015, 1986.
110. **Van Noorden, C. J. F.,** Enzyme reaction rate studies in tissue sections, *Proc. R. Microsc. Soc.,* 23, 93, 1988.
111. **Langridge, W. H. R., Escher, A., Baga, M., O'Kane, D., Wampler, J., Koncz, C., Schell, J., and Szalay, A. A.,** Use of low light image microscopy to monitor genetically engineered bacterial luciferase gene expression in living cells and gene activation throughout the development of a transgenic organism in *New methods in microscopy and low light imaging,* Wampler, J. E., Ed., *SPIE,* 1161, 216, 1989.

112. **Langridge, W. H. R., Fitzgerald, K. J., Koncz, C., Schell, J., and Szalay, A. A.,** Dual promoter of *Agrobacterium tumefaciens* mannopine synthase genes is regulated by plant growth hormones, *Proc. Natl. Acad. Sci. U.S.A.*, 86, 3219, 1989.
113. **Bright, G. R., Fisher, G. W., Rogowska, J., and Taylor, D. L.,** Fluorescence ratio imaging microscopy, in *Methods in Cell Biology, Vol. 30, Microscopy of Living Cells in Culture,* Part B: *Quantitative Fluorescence Microscopy — Imaging and Spectroscopy,* Taylor, D. L. and Wang, Y. L., Eds., Academic Press, San Diego, 1989, 157.
114. **Baju, B., Murphy, E., Levy, L. A., Hall, R. D., and London, R. E.,** A fluorescent indicator for measuring cytosolic free magnesium, *Am. J. Physiol.*, 256, C540, 1989.
115. **Prpic, V., Cowlen, M. S., and Adams, D. O.,** Application of digital imaging microscopy to studies of ion fluxes in murine peritoneal macrophages, in *Optical Microscopy for Biology,* Herman, B. and Jacobson, K., Eds., Wiley-Liss, New York, 1990, 337.
116. **Wade, M. H., Trosko, J. E., and Schindler, M.,** A fluorescence photobleaching assay of gap junction-mediated communication between human cells, *Science,* 232, 525, 1986.
117. **Brown, A. F. and Dunn, G. A.,** Microinterferometry of the movement of dry matter in fibroblasts, *J. Cell Sci.*, 93, 56, 1989.
118. **Montag, M., Wild, A., Spring, H., and Trendelenburg, M. F.,** Visualization of chromatin arrangement in giant nuclei of mouse trophoblast by videomicroscopy and laser scanning microscopy, in *Nuclear Structure and Function,* Harris J. R. and Zbarsky, J. B., Eds., Plenum Press, New York, 1990, 272.
119. **Forscher, P., Kaczmarek, L. K., Buchanan, J., and Smith, S. J.,** Cyclic AMP induces changes in distribution and transport of organelles within growth cones of *Aplysia* bag cell neurons, *J. Neurosci.*, 7, 3600, 1987.
120. **Petermann, H., Weiss, D. G., Bachmann, L., and Petersen, N.,** Motile behavior and measurement of the magnetic moment of magnetotactic bacteria in rotating magnetic fields, in *Biological Motion, Lecture Notes in Biomathematics,* Vol. 89, Alt, W. and Hoffmann, G., Eds., Springer-Verlag, Berlin, 1990, 387.
121. **Gross, D. and Loew, L. M.,** Fluorescent indicators of membrane potential: microspectrofluorometry and imaging, in *Methods in Cell Biology,* Vol. 30, Taylor, D. L., and Wang, Y. L., Eds., Academic Press, San Diego, 1989, 193.
122. **Weiss, D. G., Langford, G. L., Seitz-Tutter, D., Gulden, J., and Keller, F.,** Motion analysis of organelle movements and microtubule dynamics, in *Structure and Functions of the Cytoskeleton, Colloque INSERM* Vol. 171, Rousset, B. A. F., Ed., John Libbey Eurotext, Paris, 1988, 363.

Chapter 9

IMAGE PROCESSING AND ANALYSIS RELATING TO STRUCTURE AND FUNCTION IN EUKARYOTIC CILIA AND FLAGELLA

Michael E. J. Holwill and Peter Satir

TABLE OF CONTENTS

I. INTRODUCTION

The movement of eukaryotic cilia and flagella is based on a sliding microtubule mechanism[1] in which the dynein arms on one peripheral doublet of the axoneme exert a tipward force on the neighboring doublet.[2,3] (For convenience, and because it is impossible to differentiate between cilia and flagella on the basis of either their structure or patterns of beating, we shall refer to the two types of organelle collectively as cilia.) The shearing forces produced by the dynein-tubulin interaction are converted into bending moments by interdoublet linkages which oppose free sliding, and the coordinated activity of the dynein-microtubule interaction results in the bending patterns seen on the intact organelles. While there is considerable evidence to support the sliding microtubule model, the mechanism by which bends are initiated and propagated has yet to be described in detail, although many of the physical constraints required can be specified.

The relationships between structure and function in cilia are investigated through images obtained by light and electron microscopy, and a range of techniques of image processing and analysis has been used to interpret the data. The analytical procedures concerned with wave shape determination and the interpretation of electron micrographs rely on digitization and some processing (to reduce noise, for example) of the image prior to analysis, while computer modeling has been used effectively to produce images for comparison with raw data obtained by both light and electron microscopy. In this chapter we shall review these procedures and comment on their general effectiveness in relation to the study of cilia.

II. LIGHT MICROSCOPE IMAGES

The beat patterns of cilia observed through the light microscope and recorded photographically or using video techniques reflect the mechanochemical events which are responsible for the internal forces which deform the organelle. A knowledge of the precise shape adopted by a cilium, together with its beat frequency, can therefore provide information about the macromolecular mechanisms responsible for motility. The beat frequency is a parameter that can be measured directly using a stroboscope,[4] photodetector methods,[5] or laser doppler techniques,[6] while microtubule sliding rates in a reactivated axoneme have been determined by observing the movement of gold beads which stick to the doublets.[7] The determination of wave shape requires sophisticated interpretive techniques, usually involving the use of a computer. To apply the techniques, it is necessary that the spatial coordinates of the flagellar image are supplied to the computer. This can be achieved by tracking the image manually, e.g., using a bit pad; this is a time-consuming process, which can be prone to operator error and bias, but one which has been used to good effect.[8]

To gain the advantage that the coordinates are acquired objectively, with no bias being introduced by an observer, various techniques for the partially automated collection of data have been reported. Since flagellar images are usually recorded by dark-field microscopy, so that an image appears as a bright line on a dark background, simple thresholding procedures can be employed to detect the position of the line. However, the images are often imperfect, with variations in width and intensity along their lengths, and spurious scattering material may be present in the neighborhood of the flagella. Further, the region in which the flagellum is attached to the cell is usually obscured by spurious intensities resulting from the high scattering power of the cell body. These factors render it difficult to employ a totally automatic system for digitization, and all the techniques described so far require the intervention of a manual operator at some stage.

In an electro-optical curve follower[9,10] designed specifically to record coordinates of flagellar waveforms, a four-quadrant photodiode was used to monitor the intensity of light from an image on a back-projection screen. When the detector was positioned manually at

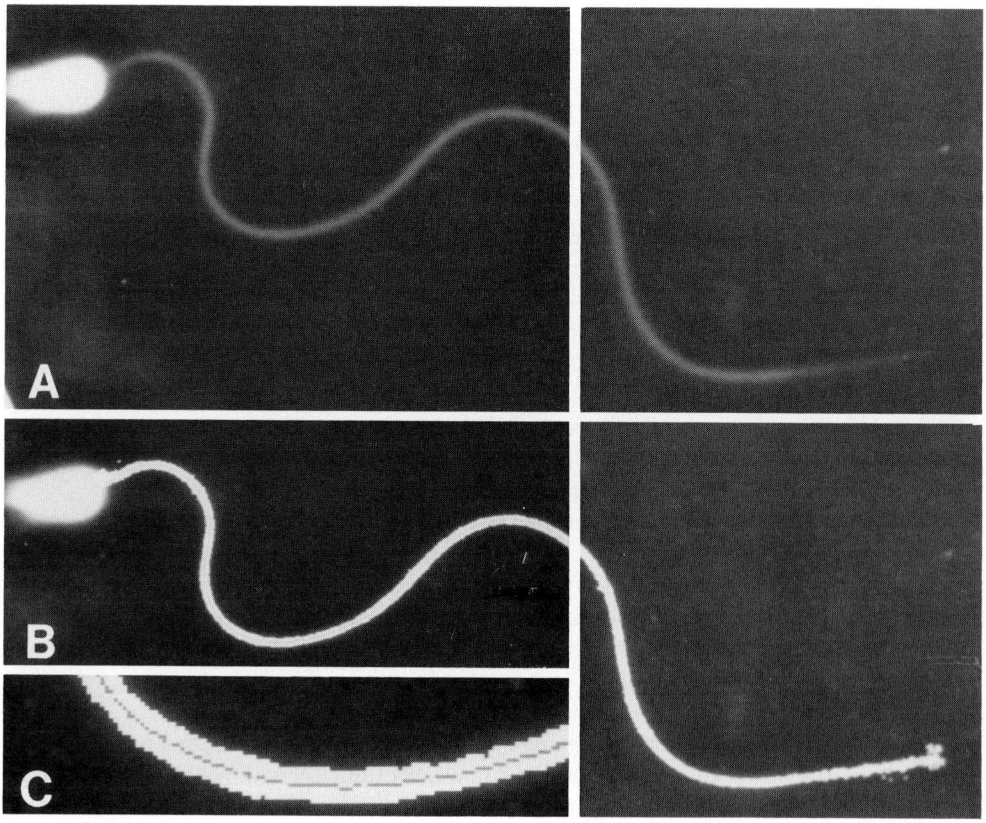

FIGURE 1. (A) Dark-field optical micrograph of a spermatozoon. (B and C) Reconstruction of sperm image following digital processing.[18]

one end of the flagellar image, the intensities of light falling on the quadrants provided feedback to stepping motors in such a way that the detector followed the image. Cartesian coordinates of the flagellum were recorded as voltages from linear potentiometers at the edges of the screen. While the technique was found to be as accurate as manual tracking methods, and has been used successfully,[11] the development of image digitization techniques has led to alternative procedures which allow coordinate recording to be achieved more rapidly than, and as accurately as, the mechanical tracking system.

Digitized images of cilia are derived from 35- or 16-mm photographs,[12,13] using an appropriate camera, or from high-speed video recordings, using an appropriate computer interface and software.[14] In extracting information from a digitized image, advantage can be taken of the fact that the representation of the cilium should be a narrow bright line on a dark background (Figure 1A). In the technique developed by Baba and Mogami,[12] an arc-shaped strip approximately normal to the flagellum was generated using a computer program, the pixels within the strip were examined, and those with intensities greater than a set threshold were identified. The distribution of average intensities across the strip when it straddles a line image has a maximum in the central region and drops to zero on either side; such an intensity profile serves to identify the cilium image and was used as a comparator in the computer program. The starting position of the strip on the cilium image was chosen by an operator, and the coordinates of the center of the cilium were calculated from the average position of the significant pixels. By moving the strip along the cilium under computer control, the coordinates of the ciliary center line could be established. If the intensity distribution differed from that expected, the operator was alerted so that appropriate action could be taken. Figures 1B and C show the results of this procedure applied to the cilium image in Figure 1A following initial digitization.

In an alternative approach that provides data in a form that requires no further processing for certain analyses, Brokaw[13] has fitted the digitized image of a sperm with a model consisting of a stylized head joined to a series of connected straight-line segments 0.5 or 1 μm long representing the sperm tail. In fitting the model to the image, the optimum position of the head is first established by monitoring the distribution of intensities in the image, and subsequently, using a similar intensity criterion, the angle between adjacent segments is altered until an optimum fit is found. Coordinate data are therefore recorded as distance (s) of a segment along the image together with the accumulated angle (ϕ) of that segment relative to the basal segment.

The digitized image is not a perfect reproduction of the original because of noise introduced by the recording techniques used. It is therefore usual to perform some averaging procedure by repeatedly tracking an individual cilium, sometimes with associated rotation of the digitizing camera to compensate for nonuniformities in its detector.

Having obtained the coordinate data, several analyses have been directed toward establishing the precise form of the bend shape on a cilium, since this has implications for the internal mechanochemistry. If it has not already been done as part of the digitization process, the coordinates are converted into (ϕ,s) format, where ϕ is the tangent angle referred to an arbitrary axis at a distance s along the cilium from some arbitrary origin. The axis often chosen is the tangent to the cilium at its base, while the origin is taken as either the tip or the base of the organelle.

The nature of the (ϕ,s) curve has been investigated graphically in terms of the variation of curvature ($= d\phi/ds$) with s and interpreted in terms of the internal mechanism of motility. In earlier studies, visual curve-fitting procedures led to ciliary wave shapes being described as arc-line,[15] meander-like,[16] and sine-generated;[17] each shape has a different implication for the internal mechanism responsible for motility. An arc-line shape, with its transition from the circular to the linear configuration, or vice versa, may reflect abrupt conformational changes at the macromolecular level within the axoneme. The meander is the minimum-energy shape adopted by a bent elastic rod freely hinged at both ends; while such constraints may not be truly representative of ciliary dynamics, if the cilium is found to adopt the shape of a meander, it may imply that the elasticity of the axoneme plays a major role in wave formation or propagation. The simple relation between ϕ and s ($\phi_\propto \sin s$) which characterizes the sine-generated wave may, if this is the shape of a cilium, have implications for the molecular oscillator involved in bend generation. Additional information is often available from the analysis of bend forms. For example, Baba and co-workers[12,18] report abrupt changes in the curvature of bends, which they take to indicate that the bending process is quantized. The visual curve-fitting approach can be somewhat subjective. Since the distinctions between forms are relatively subtle, it is desirable to have an objective analytical procedure to investigate wave shapes.

One such objective approach is to consider the coefficients of the Fourier series which represents the (ϕ,s) curve.[19] A given curve has a unique set of values for the coefficients so that ciliary bend shapes can be described unambiguously and compared with known theoretical wave shapes (Table 1). In *Crithidia,* for example, this analysis has shown that waves on *in vivo* cells fit the arc-line form,[11] while those on demembranated flagella change to the meander.[20] This implies that the internal elastic and external viscous constraints are different for the two types of preparation. The Fourier method tends to smooth out irregularities in the (ϕ,s) curves, which can be useful where the irregularities originate from noise (see below).

TABLE 1
Relative Values of Fourier Coefficients for Different Wave Shapes

Curve	b_1	b_2	b_3	b_4
Sinusoid	100	13.5	3.9	1.0
Arc-line	100	−10.41	3.26	−1.31
Meander	100	0.3	0	0
Sine-generated	100	0	0	0

III. ELECTRON MICROSCOPE IMAGES

Electron microscopy adds increased resolution to the analysis of axonemal form and construction, but relies on fixed or frozen rather than real-time motion. The general construction of the axoneme as revealed by electron microscopy is well understood, but certain details of the fine structure are more controversial, and there is a range of situations where a single interpretation has yet to emerge. As with the case of the light microscope images, efforts have been made to apply objective techniques to the interpretation of electron micrographs. In early studies,[21] optical diffraction from the electron micrographs was used to obtain a diffraction pattern from which periodicities were deduced. If the diffraction pattern, recorded on film, is masked to retain only those intensities due to the repeating structure, i.e., intensities originating from ''noise'' in the micrograph are obscured, reconstruction of the repeating structure is possible by placing the masked pattern in a suitable optical system.[22,23] The modern equivalent of this procedure is to filter the Fourier transform of those areas of a digitized micrograph which contain repeating structures and to compute the inverse Fourier transform, thereby creating a ''reconstructed'' image of the average structure.[24]

An application of the Fourier transform approach to the study of outer dynein arm structure in an avian sperm flagellum is shown in Figure 2.[25] Figure 2A is the digitized image of part of a row of outer dynein arms, all at the same orientation, of the sperm flagellum of the domestic cock (*Gallus domesticus*); digitization in this case was achieved using a microdensitometer with a sampling interval of 1.5 nm (half the estimated resolution of the image). The Fourier transform (Figure 2B) was then computed and the intensity maxima identified. The filtered transform (Figure 2C), which has an improved signal-to-noise ratio compared to the original, was obtained by retaining only the data points lying on lines associated with the maxima. Finally, Figure 2D shows the inverse Fourier transform, calculated from Figure 2C, and represents the ''average'' outer dynein arm structure for this sperm tail. On the basis of these and similarly treated images obtained from dynein arms at two other orientations, Burgess et al.[25] have constructed a model for the outer dynein arm from cock sperm which has many features in common with the dynein arm structure reported for sperm of the sea urchin *Strongylocentrotus*.[26]

The objective of the Fourier transform approach is to provide a reconstructed image that contains less noise, i.e., spurious information, than the original. Noise can be derived from a number of different sources, such as specimen distortion or excess metal deposited locally during the preparative process. The image reconstructed by the Fourier transform method is necessarily periodic, irrespective of the nature of the original image, and each repeating element will be represented in precisely the same way by an average of the corresponding elements in the original image. While this may be valuable if the axoneme has been prepared so that periodic structures, such as the spokes, can be expected to have identical configurations for long distances, it can be a disadvantage in preparations, such as an active axoneme, where the periodic structures may adopt different conformations. This is especially true for active dynein arms, which cannot be captured by this method. In such a case, averaging

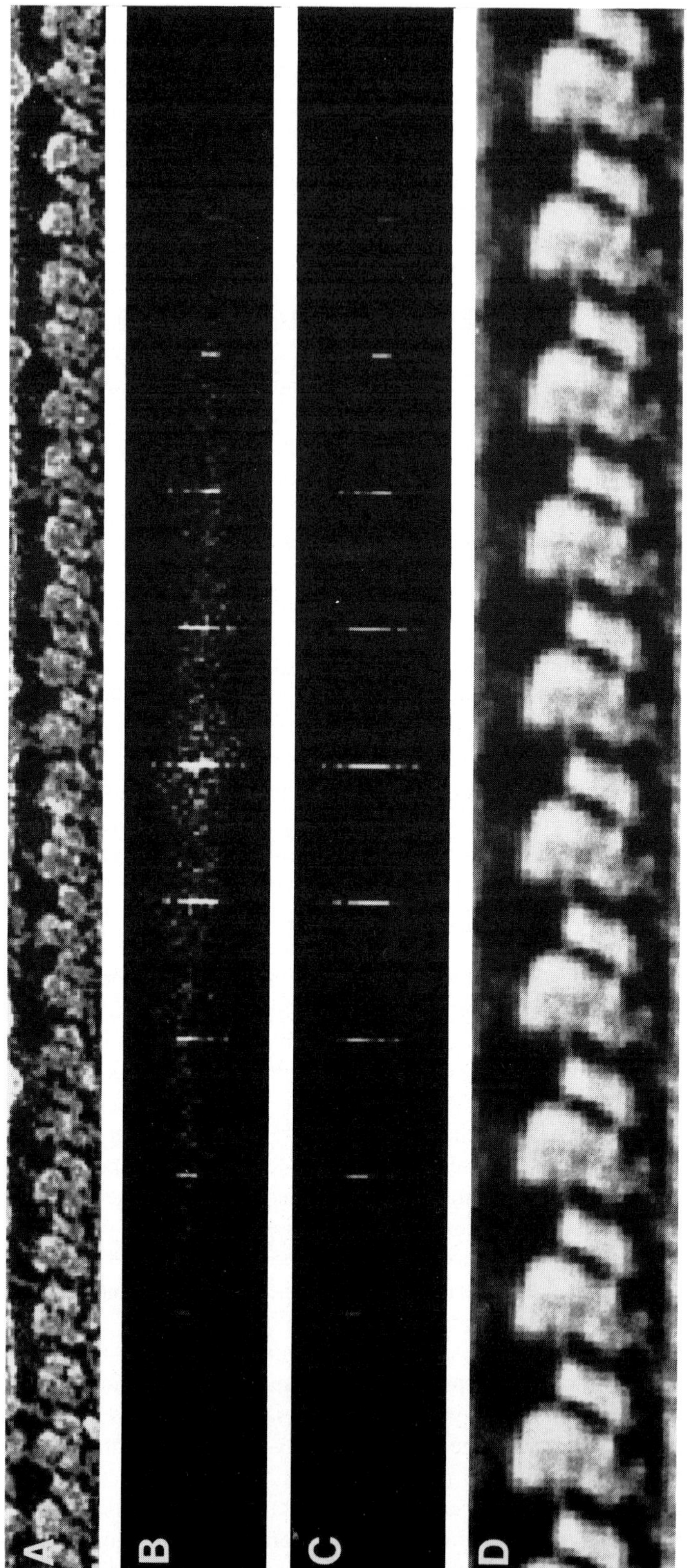

FIGURE 2. Application of Fourier transform filtering to the study of dynein arm structure.[25] (A) Digitized micrograph of a row of outer dynein arms from the cock *Gallus domesticus*; (B) Fourier transform of (A); (C) filtered Fourier transform; (D) image reconstructed from (C).

procedures will eliminate any differences, and it is necessary to use a technique, such as the computer modeling to be discussed below, which allows the conformation of individual arms to be studied.

Selection of a suitable portion of an axonemal image for study is to some extent subjective, since it is made on the basis that it contains periodicities and is straight, and may therefore not be representative of the whole axoneme. In applying the technique, components of the Fourier transform are deliberately removed so that the reconstruction may contain less information than the original; the information discarded is usually regarded as spurious, but it is possible that data relating to actual structures are eliminated, and the method should be applied with care.

The information about structures containing rotational and translational periodicities can be enhanced by using reinforcement techniques, whereby the superposition of images subjected to rotation or translation by the repeating angle or distance leads to increased prominence of the periodic structure. Superposition can be achieved by the physical translation or rotation of a negative at the appropriate level in an optical system, such as a photographic enlarger.[23] For cilia,[27] the relevant operations would be performed on the image of an individual axoneme selected for examination, so giving information about the average periodicities and periodic structures in a single organelle. The equivalent operations can be effected by computer on digitized images. In a development of the method, Afzelius et al.[28] have used computer-aided analysis to superimpose correspondingly oriented images of different doublets, revealing a nonuniform arrangement of protofilaments which may reflect attachment sites for the spokes and other doublet structures. As with the technique based on Fourier transforms, the reinforcement approach may obscure real structures in its averaging process. Additionally, the incremental translation or rotation for reinforcement is selected subjectively, and is necessarily imposed upon the final image, so that structures with a different repeat, or no periodicity at all, may be imaged at the imposed periodicity. Caution must accordingly be exercised in the interpretation of these images.

Another approach to the interpretation of electron micrographs involves the construction of three-dimensional (3-D) models of the structure. Models made from clay or other substances have been used for this purpose,[29] but have the disadvantage that they cannot easily be amended. With the development of computer graphics has come the capability to construct, using appropriate software, complex 3-D models to be viewed on a visual display monitor; the images so produced are two-dimensional (2-D), but the 3-D character is revealed by presenting different orientations of the structure on the monitor screen. This type of modeling allows for easy comparison of the proposed configuration with the electron micrograph, as both are 2-D projections of a 3-D structure.

A computer-modeling software package called SURREAL (SUrface REndering ALgorithms) designed specifically to investigate the structure of ciliary axonemes has been written by Dr. A. N. Glazzard (now of 5-D Solutions, London). SURREAL has been used to generate images of a model axoneme at 4-nm resolution with the aim of reconciling different interpretations of electron micrographs.[30] The model (Figure 3) that has been produced is consistent with available evidence obtained from electron microscopy and optical diffraction of electron micrographs. Nevertheless, there are still structures, such as the interdoublet links, where more than one interpretation of the raw data are possible. Computer modeling has the advantage of flexibility, so that different interpretations can be modeled and viewed from different angles in efforts to identify the correct model. This type of modeling can also be used to investigate dynamic aspects of axonemal behavior; a preliminary study of a possible dynein arm cycle has been proposed[30] (Figure 4), and, by incorporating elastic and other mechanical parameters into the model, it should be possible to specify a complete working model of the ciliary axoneme. While computer modeling provides a convenient structural representation of the axoneme, it cannot replace experimental studies

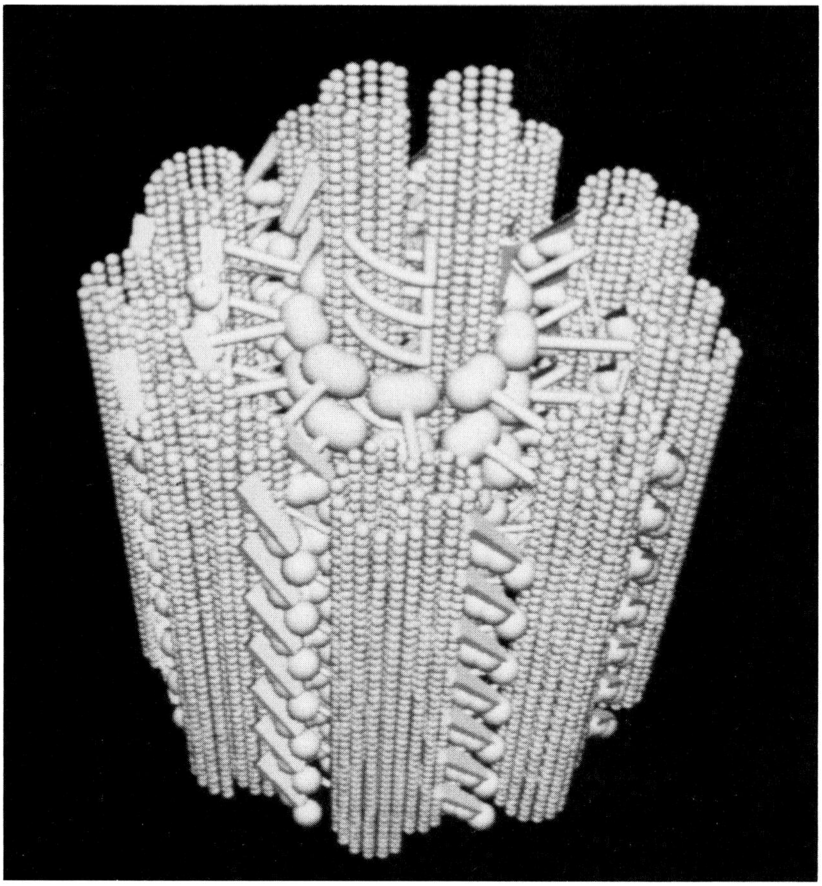

FIGURE 3. Perspective view of computer-generated model of the ciliary axoneme at 4-nm resolution. The α and β-tubulin are represented as spheres 4 nm in diameter.[30]

using electron microscopy and other techniques designed to reveal the axonemal structure. To generate the computer model, a precise knowledge of the relative positions, sizes, and shapes of all the individual structures is required, so that this approach reveals the areas in which the data are deficient and suggests appropriate experiments to rectify the deficiencies.

IV. FUTURE PERSPECTIVES

The investigation of ciliary movement involves a wide range of imaging procedures, some well established, others, such as X-ray microscopy, yet to be fully applied to the problem. The established techniques of image analysis, together with some procedures designed specifically for the study of the axoneme, are yielding important information about the mechanism of motility. Further developments in image processing will be utilized in the solution of the many unresolved problems associated with the mechanochemical events which are responsible for the cyclic motion of the cilium.

The application of modern image analysis procedures relies on the availability of a digitized image, i.e., one that can be stored numerically in a computer; such images are recorded most conveniently by video techniques. In many studies of cilia, the full potential of video procedures has yet to be realized. As we have discussed, digitization for the study of beat forms has usually been achieved by processing a photographic image. Although the cilium image is known to be a simple curve, and as such can be recorded simply despite

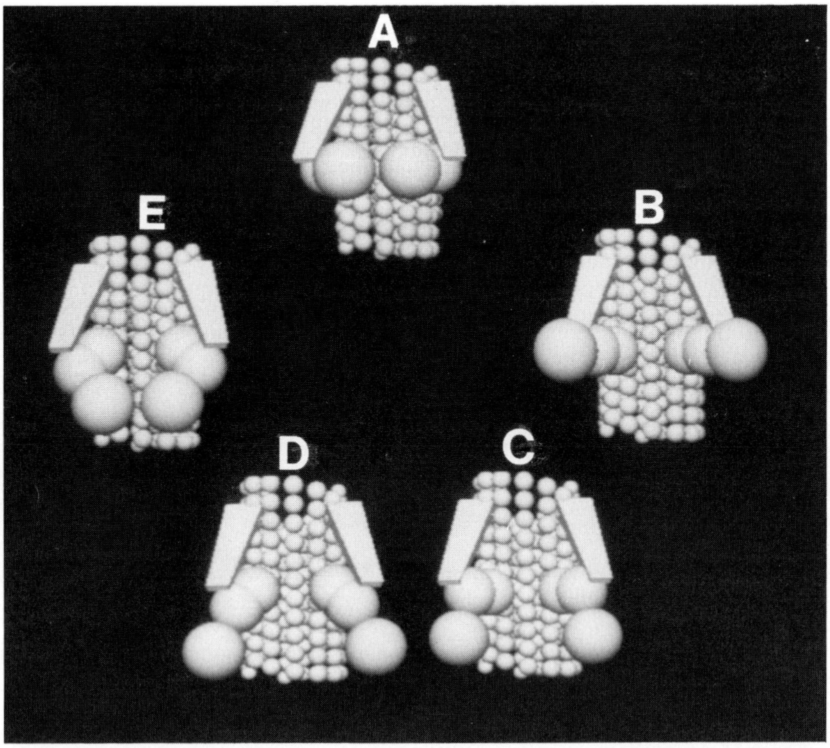

FIGURE 4. Computer simulation of a possible cycle for the dynein arm, viewed from the neighboring doublet, which is not shown. For convenience, the two arms are shown to execute the same cycle, with one the mirror image of the other (there is no experimental evidence to support such synchronization). In (A), the heads of the arms are attached to the neighboring doublet in the rigor configuration. The arms detach from the B-tubule and the heads swing away from each other (B) and then downward (C). A tilting of the body of the arm (D) is followed by a movement of the heads toward each other and their attachment to the neighboring doublet. The heads remain attached as the conformational change from (E) to (A) occurs, so that relative sliding of the doublets is induced. The cycle is based on the different dynein arm configurations revealed by electron microscopy, but is not the only one possible.[30]

some discontinuities, a procedure involving the processing of an intermediate image has the inherent disadvantage that additional noise and distortion can be introduced.

In light microscopy, photographic rather than video recording is used because standard video framing rates of 25 or 30 Hz do not allow adequate time resolution of the ciliary beat (which is usually about 20 Hz, but can be considerably higher). High-speed video techniques are available and have been used to record images of cilia at framing rates of 200 per second;[20,31] higher speeds can be achieved by modifying the line scanning control of the camera, but this degrades the image resolution. It would be highly desirable to couple high-speed video with high resolution in the detailed study of beat form. In addition, limited use has been made of image processing capabilities, such as contrast enhancement and background subtraction, since the recordings are made in a nonstandard format. Technological developments in SIT and CCD video cameras, with appropriate modifications to allow higher than standard recording rates to be achieved, and the appropriate adaptation of image-processing software will allow high-quality time-resolved video recordings of cilia to be made routinely.

Perhaps video imaging is less important for electron microscopy at the moment, but image-enhancement techniques and analyses such as that of Afzelius et al.[28] have great potential for the provision of new structural details and need to be explored further. With its higher resolution, the electron microscope also has the potential to provide improved wave-shape information compared with that available from optical microscopy. Surprisingly, this potential has been little exploited. It depends on accurate preservation of waveform during preparation for electron microscopy, but there are good indications that this can be achieved. Studies of whole-mount specimens, either in scanning or transmission electron microscopy may give the detail of overall form required, as well as indications of internal structural changes.

Prospects for time-resolved imaging at high magnification using either X-ray microscopy or time-resolved X-ray diffraction lie in the future, but have high potential for providing information on questions of structural change during bend generation and propagation that are virtually unanswerable at present.

REFERENCES

1. **Satir, P.,** Switching mechanisms in the control of ciliary motility, *Mod. Cell Biol.,* 4, 1, 1985.
2. **Sale, W. S. and Satir, P.,** Direction of active sliding of microtubules in *Tetrahymena* cilia, *Proc. Natl. Acad. Sci. U.S.A.*, 74, 2045, 1977.
3. **Woolley, D. M. and Brammall, A.,** Direction of sliding and relative sliding velocities within tripsinized sperm axonemes of *Gallus domesticus, J. Cell Sci.*, 88, 361, 1987.
4. **Brokaw, C. J.,** Movement of the flagella of *Polytoma uvella, J. Exp. Biol.*, 40, 149, 1963.
5. **Sanderson, M. J. and Dirksen, E. R.,** A versatile and quantitative computer-assisted photoelectronic technique used for the analysis of ciliary beat cycles, *Cell Motility,* 5,, 267, 1985.
6. **Lee, W. I. and Verdugo, P.,** Laser light-scattering spectroscopy. A new application in the study of ciliary activity, *Biophys. J.*, 16, 1115, 1976.
7. **Brokaw, C. J.,** Direct measurements of sliding between outer doublet microtubules in swimming sperm flagella, *Science,* 243, 1593,1989.
8. **Brokaw, C. J.,** Automated methods for estimation of sperm flagellar bending parameters, *Cell Motility,* 4, 417, 1984.
9. **Silvester, N. R. and Johnston, D.,** An electo-optical curve follower with analogue control, *J. Phys. E.*, 9, 990, 1976.
10. **Johnston, D. N. and Silvester, N. R.,** A digitally controlled curve follower, *J. Phys. E.*, 12, 235, 1979.
11. **Johnston, D. N., Silvester, N. R., and Holwill, M. E. J.,** An analysis of the shape and propagation of waves on the flagellum of *Crithidia oncopelti, J. Exp. Biol.*, 80, 299, 1979.

12 **Baba, S. A. and Mogami, Y.,** An approach to digital image analysis of bending shapes of eukaryotic flagella and cilia, *Cell Motil.*, 5, 475, 1985.

13. **Brokaw, C. J.,** Computerized analysis of flagellar motility by digitization and fitting of film images with straight segments of equal length, *Cell Motil. Cytoskel.*, 17, 20, 1990.
14. **Glazzard, A. N., Hirons, M. R., Mellor, J. S., and Holwill, M. E. J.,** The computer assisted analysis of television images as applied to the study of cell motility, *J. Submicrosc. Cytol.*, 15, 305, 1983.
15. **Brokaw, C. J.,** Non-sinusoidal bending waves of sperm flagella, *J. Exp. Biol.*, 43, 155, 1965.
16. **Rikmenspoel, R.,** Contractile mechanisms in flagella, *Biophys. J.*, 11, 446, 1971.
17. **Hiramoto, Y. and Baba, S.,** A quantitative analysis of flagella movement in echinoderma spermatozoa, *J. Exp. Biol.*, 76, 85, 1978.
18. **Baba, S. A., Mogami, Y., and Nonaka, K.,** Discrete nature of flagellar bending detected by digital image analysis, in *Biological Motion (Lecture Notes in Biomathematics Vol. 89),* Alt, M. and Hoffmann, G., Eds., Springer-Verlag, Berlin, 1990.
19. **Silvester, N. R. and Holwill, M. E. J.,** An analysis of hypothetical flagellar waveforms, *J. Theor. Biol.*, 35, 505, 1972.
20. **Marchese-Ragona, S. P., Glazzard, A. N., and Holwill, M. E. J.,** Motile characteristics of 9 + 2 and 9 + 1 flagellar axonemes of *Crithidia oncopelti, J. Exp. Biol.*, 145, 199, 198.
21. **Amos, L. A., Linck, R. W., and Klug, A.,** Molecular structure of flagellar microtubules, in *Cell Motility,* Vol. 3, Goldman, R., Pollard, T., and Rosenbaum, J., Eds., Cold Spring Harbor Laboratory, Cold Spring Harbor, NY, 1976, 847.
22. **Klug, A. and DeRosier, D. J.,** Optical filtering of electron micrographs: reconstruction of one-sided images, *Nature,* 212, 29, 1966.

23. **Beeston, B. E. P., Horne, R. W., and Markham, R.,** Electron diffraction and optical diffraction techniques, in *Practical Methods in Electron Microscopy,* Vol. 1 (Part 2), Glauert, A. M., Ed., North Holland, Amsterdam, 1972.
24. **Misel, D. L.,** Image analysis, enhancement, and interpretation, in *Practical Methods in Electron Microscopy,* Vol. 7, Glauert, A. M., Ed., North Holland, Amsterdam, 1978.
25. **Burgess, S. A., Dover, S. D., and Woolley, D. M.,** Architecture of the outer arm dynein ATPase in an avian sperm flagellum, with further evidence for the B-link, *J. Cell Sci.,* 98, 17, 1991.
26. **Sale, W. S., Goodenough, U., and Heuser, J. E.,** The substructure of isolated and *in situ* outer dynein arms of sea urchin sperm flagella, *J. Cell Biol.,* 101, 1400, 1985.
27. **Avolio, J., Glazzard, A. N., Holwill, M. E. J., and Satir, P.,** Structures attached to doublet microtubules of cilia: computer modeling of thin-section and negative-stain stereo images, *Proc. Natl. Acad. Sci. U.S.A.,* 83, 4804, 1986.
28. **Afzelius, B. A., Bellon, P. L., and Lanzavecchia, S.,** Microtubules and their protofilaments in the flagellum of an insect spermatozoon, *J. Cell Sci.,* 95, 207, 1990.
29. **Avolio, J., Lebdusca, S., and Satir, P.,** Dynein arm substructure and the orientation of arm-microtubule attachments, *J. Mol. Biol.,* 173, 389, 1984.
30. **Sugrue, P., Avolio, J., Satir, P., and Holwill, M. E. J.,** Computer modeling of *Tetrahymena* axonemes at macromolecular resolution: interpretation of electron micrographs, *J. Cell Sci.,* 98, 5, 1991.
31. **Cosson, J., Cachon, M., Cachon, J., and Cosson, M.-P.,** Swimming behavior of the unicellular biflagellate *Oxyrrhis marina: in vivo* and *in vitro* movement of the two flagella, *Biol. Cell,* 63, 117, 1988.

Chapter 10

IMAGE ANALYSIS OF STOMATAL MOVEMENTS AND GAS EXCHANGE

Kenji Omasa and Judith G. Croxdale

TABLE OF CONTENTS

I. ANALYSIS OF THE THERMAL IMAGE

To understand how leaves function in nature, information on stomatal movement, transpiration, and absorption of gasses must be acquired without destroying the plant or changing its environment. Leaf temperature can be used to provide such physiological information because it relates to water balance and heat and mass transfer of the leaves.[1-3] By interfacing image analysis with a sensitive, noninvasive technique for measuring leaf temperature, it is possible to determine rates of transpiration and gas absorption and stomatal and boundary layer resistances at any location on the leaf. Knowing the spatial distribution of these parameters provides a powerful means of integrating functional and structural aspects of leaves and of following their responses to changes in the environment.

Thermal image instrumentation, which measures the electromagnetic waves of thermal bands, is a method for obtaining spatial leaf temperatures.[4] This method has been used to examine changes in leaf temperature distributions in response to environmental stimuli in the laboratory.[5-7] It is also used for measuring leaf temperatures in street trees and crops growing in fields.[8,9]

We have developed an image instrumentation system consisting of a thermal camera of the optical-mechanical scanning type, interfaced to a computer,[5] and used it to evaluate the spatial distributions of physiological information such as stomatal resistance, transpiration rate, and gas absorption rate from temperature distributions of attached leaves.[10-12] In this section, an image processing technique for evaluating such physiological information is introduced.

A. MODEL FOR EVALUATING PHYSIOLOGICAL INFORMATION FROM LEAF TEMPERATURE

1. Heat and Mass Transfer Model at the Local Leaf Site

Water evaporates from mesophyll and epidermal cell walls and diffuses into the atmosphere through the substomatal cavity, stomatal throat, and leaf boundary layer (Figure 1). Flux in transpiration at local sites on a leaf is fundamentally determined by the water vapor density and the diffusion resistance between the free air and the gas-liquid interface of each stoma. If the leaf temperature, stomatal conditions, and boundary layer are assumed to be equal on both sides of a leaf, the transpiration rate, W_x (g cm^{-2} s^{-1}), at the local site is

$$W_x = 2\{X_{sx}(T_{lx}) - \Phi X_s(T_a)\}/(r_{wax} + r_{wsx}) \tag{1}$$

Where T_l is leaf temperature (°C), T_a is air temperature (°C), X_s (T) is the saturated water vapor density at T °C (g cm^{-3}), Φ is relative humidity, r_{wa} is the boundary layer resistance to water vapor diffusion (s cm^{-1}), and r_{ws} is stomatal resistance to water vapor diffusion (s cm^{-1}). The subscript x denotes the values at the local site x on the leaf.

On the other hand, air pollutants and CO_2 essential for photosynthesis enter the leaf in a direction opposite to that of water vapor. To model the transfer of air pollutants and CO_2, it is necessary to consider their reactions and transfer in the tissues. These are represented by the gas concentration of the gas-liquid interface in the stomata. Therefore, the gas absorption rate, Q_x (g cm^{-2} s^{-1}), and the total of gas absorption, Q_{intx} (g cm^{-2}), at the local site are given by

$$Q_x = 2(P_a - P_{lx})/(r_{gax} + r_{gsx}) \tag{2}$$

and

$$Q_{intx} = \int_0^T Q_x \, dt \tag{3}$$

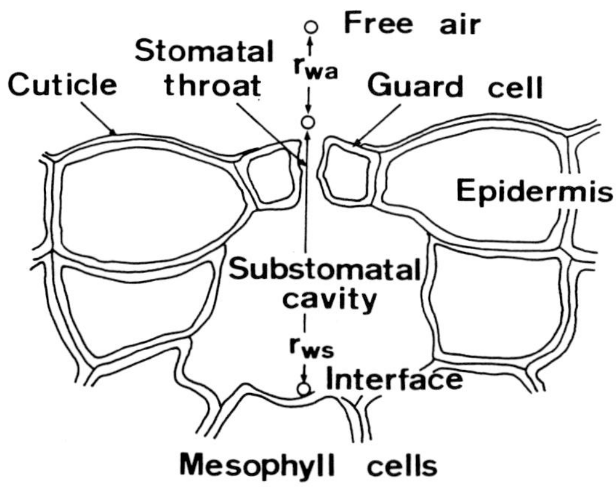

FIGURE 1. Schematic cross-section of a stoma in a leaf. r_{wa} is the boundary layer resistance and r_{ws} is the stomatal resistance to water vapor diffusion. (From Omasa, K., Abo, F., Aiga, I., and Hashimoto, Y., *Trans. Soc. Instrum. Control Eng.*, 17, 657, 1981. With permission.)

where:

$$r_{gax} = (D_w/D_g)^{2/3}\ r_{wax} \tag{4}$$

and

$$r_{gsx} = (D_w/D_g)\ r_{wsx} \tag{5}$$

since P_l is the gas concentration of the gas-liquid interface in the stomata (g cm^{-3}), P_a is the gas concentration of air (g cm^{-3}), r_{ga} is the boundary layer resistance to gas diffusion (s cm^{-1}), r_{gs} is the stomatal resistance to gas diffusion (s cm^{-1}), D_w is the air-water vapor diffusion coefficient (= 0.249 cm^2 s^{-1} at 25°C), D_g is the air-gas diffusion coefficient ($\simeq$ [molecular weight of water vapor/molecular weight of gas]$^{1/2}$ $\times$ D_w), T is exposure time (s), and t is time (s). The gas concentration of the gas-liquid interface of major air pollutants such as SO_2, NO_2, O_3, and PAN can be assumed to nearly equal zero μl l^{-1} because the metabolism and transfer in the plant's tissues are very rapid.[13-15] However, the CO_2 concentration concerned with photosynthesis and respiration varies by species and growth conditions.[3]

2. Heat Balance at the Local Leaf Site

Heat balance at a local site on the leaf is

$$c_l w_l dT_{lx}/dt = E_x + S_x - LW_x + M_x + G_x \tag{6}$$

where E is net radiation (cal cm^{-2} s^{-1}), S is sensible heat transfer by convection (cal cm^{-2} s^{-1}), M is reaction heat (cal cm^{-2} s^{-1}), G is heat conduction from other sites (cal cm^{-2} s^{-1}), c_l is specific heat (cal g^{-1} °C^{-1}), w_l is weight (g cm^{-2}), and L is latent heat of evaporation (= 583.3 cal g^{-1} at 25°C).

Net radiation, E_x in Equation 6 is

$$E_x = \alpha_p E_{sx} + \epsilon_l\{E_{wx} - 2\sigma(273.15 + T_{lx})^4\} \tag{7}$$

and sensible heat transfer, S_x, is

$$S_x = 2\rho c_p(T_a - T_{lx})/r_{kax} \tag{8}$$

where E_s is short-wavelength radiation from the environment (cal cm^{-2} s^{-1} at wavelengths $\lesssim 3$ μm), E_w is long-wavelength radiation from the environment (cal cm^{-2} s^{-1} at wavelengths $\gtrsim 3$ μm), r_{ka} is the boundary layer resistance to heat transfer (s cm^{-1}), α_p is the absorption coefficient of the short-wavelength radiation of the leaf, ϵ_l is the emissivity of the long-wavelength radiation of the leaf, σ is the Stefan-Boltzmann constant ($= 1.354 \times 10^{-12}$ cal cm^{-2} s^{-1} K^{-4}) and ρ c_o is the volumetric heat capacity of the air ($= 0.285 \times 10^{-3}$ cal cm^{-3} °C^{-1} at 25°C).

The boundary layer resistance for water vapor diffusion relative to that for heat transfer, r_{wax}/r_{kax}, is given by

$$r_{wax}/r_{kax} = (\kappa/D_w)^{2/3} \tag{9}$$

where κ is the thermal diffusion coefficient of air ($= 0.222$ cm^2 s^{-1} at 25°C).

Reaction heat, M_x, by photosynthesis, respiration, etc., is usually within 2 to 3% of short-wavelength radiation. For example, in thin sunflower leaves with a 0.2°C min^{-1} change in leaf temperature, the accumulated heat, c_l w_l dT_{lx}/dt, is about 2 to 3% of the latent heat by transpiration or net radiation. Heat conduction from other sites, G_x, is about 5% when heat conduction in a 1°C cm^{-1} heat gradient occurs in a site 0.5 cm in diameter. Usually, $c_l w^l T^l_x/dt$ and G_x at the local site of the leaf exposed to air pollutants are within the above-mentioned values. Therefore, Equation 6 can be simplified as follows:

$$E_x + S_x - LW_x = 0 \tag{10}$$

Substituting Equations 7 and 8 in Equation 10 gives the following equation for the transpiration rate at the local site:

$$W_x = \{\alpha_p E_{sx} + \epsilon_l[E_{wx} - 2\sigma(273.15 + T_{lx})^4] + 2\rho c_p(T_a - T_{lx})/r_{kax}\}/L \tag{11}$$

When environmental conditions such as air temperature, humidity, radiation, and air current are kept constant, only leaf temperature, T_{lx}, is left as a variable in the right side of Equation 11. Therefore, the transpiration rate can be evaluated by measuring the leaf temperature. Furthermore, stomatal resistance to water vapor diffusion, r_{wsx}, which is an indicator of the degree of stomatal opening, is evaluated from Equations 1 and 9.

$$r_{wsx} = 2\{X_{sx}(T_{lx}) - \Phi X_s(T_a)\}/W_x - (\kappa/D_w)^{2/3} r_{kax} \tag{12}$$

In addition, since the stomatal resistance to gas diffusion, r_{gsx}, and the boundary layer resistance to gas diffusion, r_{gax}, can be determined from the relationship between Equations 4, 5, and 9, the gas absorption rate, Q_x, at the local site is evaluated by substituting these resistances in Equation 2.

B. INSTRUMENTATION SYSTEM

1. Thermal Image Instrumentation System

Thermal infrared (IR) radiation (approximately 10 μm) from the environment is almost all absorbed by the plant. However, since thermal IR radiation is emitted from the plant according to Plank's law of radiation, the plant temperature can be obtained by measuring the radiation in this band.

For a perfectly diffuse and opaque plant surface, the spectral energy of IR radiation from the surface at temperature T, $R(\lambda,T,T_s)$, is given by:

$$R(\lambda, T, T_s) = \epsilon(\lambda, T)W(\lambda, T) + [(1 - \epsilon(\lambda, T)]E(\lambda, T_s) \tag{13}$$

where λ is the wavelength, $\epsilon(\lambda,T)$ is the spectral emissivity of the plant surface, $W(\lambda,T)$ is the spectral radiant energy of a black body at temperature T, and $E(\lambda,T_s)$ is the spectral radiant energy from the environment at temperature T_s to the plant surface.

When $R(\lambda,T,T_s)$ is measured by an IR detector with spectral sensitivity in the band from λ_i to λ_2, the output voltage $V_T(T,T_s)$ of the detector is

$$V_T(T, T_s) = \int_{\lambda_1}^{\lambda_2} f(\lambda)R(\lambda, T, T_s)d\lambda$$

$$\simeq \bar{\epsilon}(T)V_w(T) + [1 - \bar{\epsilon}(T)]V_E(T_s) \tag{14}$$

where

$$\bar{\epsilon}(T) = \left[\int_{\lambda_1}^{\lambda_2} \epsilon(\lambda, T)f(\lambda)W(\lambda, T)d\lambda\right] \Big/ \left[\int_{\lambda_1}^{\lambda_2} f(\lambda)W(\lambda, T)d\lambda\right]$$

$$V_w(T) = \int_{\lambda_1}^{\lambda_2} f(\lambda)W(\lambda, T)d\lambda$$

and

$$V_E(T_s) = \int_{\lambda_1}^{\lambda_2} f(\lambda)E(\lambda, T_s)d\lambda$$

since $f(\lambda)$ is the function describing the radiation-electricity conversion of the detector, its amplification, and the transmission and reflection of the air, lens, filter, etc.; $\bar{\epsilon}(T)$ is the average emissivity in the band from λ_1 to λ_2; $V_w(T)$ is the output voltage in the measurement of $W(\lambda, T)$; and $V_E(T_s)$ is the output voltage in the measurement of $E(\lambda, T_s)$.

When $\bar{\epsilon}(T)$ and $V_E(T_s)$ are given, $V_w(T)$ can be obtained from $V_T(T \cdot T_s)$:

$$V_w(T) = [V_T(T, T_s) - V_E(T_s)]/\bar{\epsilon}(T) + V_E(T_s) \tag{15}$$

Furthermore, the plant temperature T is evaluated from $V_w(T)$ using the characteristic relation between T and $V_w(T)$.

The emissivity $\bar{\epsilon}(T)$ of the plant in the spectral sensitivity range (8 to 13 μm) of the thermal camera is 0.95 to 0.99.[5,16,17] The influence of radiation from the environment is corrected by adjusting $V_E(T_s)$ in Equation 15. The influence of the change in $f(\lambda)$ is also corrected by continuously monitoring a standard black-body source within the camera. As a result, it is possible to measure the plant temperature with an accuracy of 0.1°C.

A thermal camera of the optical-mechanical scanning type with an HgCdTe detector (8 to 13 μm, cooled by liquid nitrogen) is now on the market. For plant measurements, the scanning type using mirror vibration has better temperature sensitivity and resolution than that using mirror rotation. Since the detector is cooled by liquid nitrogen (77 K), it is necessary to supplement it every several hours. At present, the detector with an electronic cooler (200 K) and a new image sensor of the electrical scanning type cannot be used to accurately measure plant temperature because of insufficient sensitivity and resolution.

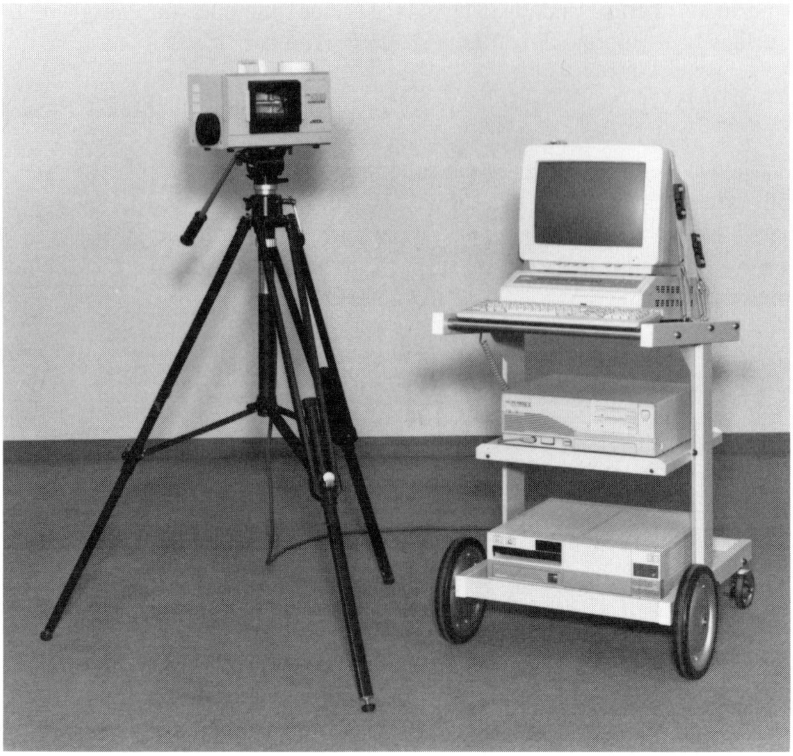

FIGURE 2. Thermal image instrumentation system consisting of the thermal camera (left) and the image processor (right).

Figure 2 shows a thermal image instrumentation system. The thermal camera (JEOL JTG-4200) is an optical-mechanical scanning type with an HgCdTe detector (8 to 13 μm, cooled by liquid nitrogen). The detected signals from the thermal camera are converted into 16-bit digital signals (512 H $\times$ 240 V) and analyzed by an image processor. The temperature-resolving power, the uniformity of the image, and the drift are within 0.05, 0.1, and 0.5°C/4 h, respectively. The image constructed during standard measurement consists of 300 lines, and the resolution is about 0.5 mm. Adding a lens in front of the detector makes it possible to improve the image resolution to 60 μm.

2. Environment Control System

An apparatus for holding an intact or model leaf horizontally and for maintaining a uniform air current on the surface was set in a growth chamber (Figure 3). The intact or model leaf was attached to a thin plastic sheet (20 $\times$ 20 cm^2) cut in the shape of the leaf and held in place with stretched fishing line (Figure 3). Air temperature, humidity, and air pollutant concentration in the chamber were controlled to within 0.1°C, 1% RH, and 0.03 $\mu l\ l^{-1}$, respectively. Metal halide lamps with heat-absorbing filters were used as a light source. The scattering filter was placed between the light source and the holding apparatus to prevent the leaf from receiving short-wavelength radiation directly from the light source. The wall in the chamber was covered with black wool paper and victoria lawn. Consequently, the spatial distribution of the short-wavelength radiation on a leaf surface in the apparatus was within 2% of the mean value. Furthermore, the temperature distribution of the model leaf (black wool paper) was kept within 0.2°C, except at the leaf edges, by controlling the

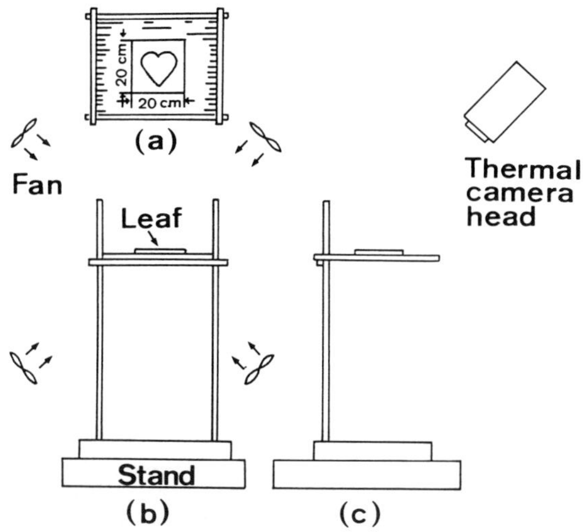

FIGURE 3. Schematic diagram of an apparatus for supporting a leaf. The apparatus is set in a growth chamber and the environmental conditions on the leaf were constant. (a) Overhead view of leaf on holding plate; (b) frontal view of holding plate attached to support stand and position of fans, (c) side view of apparatus showing position of thermal camera during measurement. (From Omasa, K., Abo, F., Aiga, I., and Hashimoto, Y., *Trans. Soc. Instrum. Control: Eng.*, 17, 657, 1981. With permission.)

air currents. This result means that local boundary layer resistance is constant throughout the leaf. Boundary layer resistances on both sides of the leaf were kept the same by adjusting the air current.

3. Other Instruments

Temperature was calibrated with a standard glass thermometer (accuracy 0.03°C) and thermocouples of 0.1 mm diameter. Humidity was measured with a dew-point instrument (EG and G, Model 660, accuracy 0.2°C), short-wavelength radiation was measured with radiometers (Eko Model MS-42 and LI-COR Model LI-185), and weight was measured with electric balances (Mettler, Model PK 16 and PL3000, precision 0.1 g).

C. EVALUATION OF STOMATAL RESISTANCE, TRANSPIRATION, AND GAS ABSORPTION

1. Parameters of the Models

Attached leaves of sunflower grown for 3 to 6 weeks were used for the experiments. The absorption coefficient of the short-wavelength radiation of the attached leaves, α_p, was 0.68 ± 0.03 and that of the model leaves (black wool paper) was 0.98 ± 0.02, measured with the Model MS-42 radiometer (effective wavelengths of about 0.4 to 3 μm). Their long-wavelength emmisivities, ϵ_l, measured with the thermal camera, were 0.98 ± 0.02. Although the spectral sensitivity of the thermal camera extends from 8 to 13 μm, the obtained emissivity is regarded as the mean value in the long-wavelength range because the spectral emissivity of the leaf is approximately constant in the range above 3 μm.

The long wavelength radiation from the environment, E_{wx}, and the boundary layer resistance to heat transfer, r_{kax}, were evaluated by the following procedures. Substituting Equations 7 and 8 in Equation 10 gives a heat balance equation:

$$\alpha_p E_{sx} + \epsilon_l E_{wx} - \epsilon_l \sigma (273.15 + T_{lx})^4 + 2\rho c_p (T_a - T_{lx})/r_{kax} - LW_x = 0 \quad (16)$$

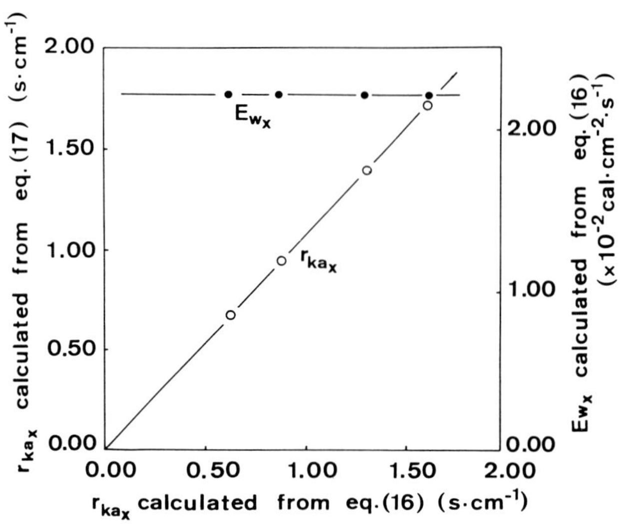

FIGURE 4. Accuracies in evaluating boundary layer resistance on the leaf to heat transfer (r_{kax}) and long wavelength radiation from the environment (E_{wx}). Environmental conditions: T_a = 25.0°C; ϕ = 59 to 62% RH; E_{sx} = 2.91 to 3.0 × 10^{-3} cal cm^{-2} s^{-1}. (From Omasa, K., Abo, F., Aiga, I., and Hashimoto, Y., *Trans. Soc. Instrum. Control Eng.*, 17, 657, 1981. With permission.)

Since E_{sx}, E_{wx}, W_x, T_{lx}, and r_{kax} are constant over the entire model leaf surface on the abovementioned apparatus, Equation 16 may be regarded as a heat-balance equation of one leaf. Therefore, E_{wx} and r_{kax} were determined by substituting the leaf temperature, transpiration rate, and short-wavelength radiation obtained from experiments on a dry or wet model leaf in Equation 16 and by solving the simultaneous equations. In order to examine the accuracy of this method, r_{kax} evaluated from Equation 16 was compared with that from the following equation obtained by substituting Equation 9 in Equation 1, where r_{wsx} was assumed to be 0 s cm^{-1}:

$$r_{kax} = 2\{X_{sx}(T_{lx}) - \Phi X_s(T_a)\}/\{W_x(\kappa/D_w)^{2/3}\} \tag{17}$$

where the T_{lx} and W_x of a wet model leaf were substituted. This result is shown in Figure 4. Although r_{kax} obtained by Equation 16 tended to be smaller than that by Equation 17, the difference between both values was within 0.1 s cm^{-1}. Furthermore, the difference in E_{wx} evaluated by substituting r_{kax} in Equation 16 was within 1 × 10^{-4} cal cm^{-2} s^{-1}.

2. Accuracy of the Evaluation

In order to examine the accuracy of the transpiration rate and stomatal resistance to water vapor diffusion of sunflower leaves, values obtained by this method were compared with those obtained by the weighing method (Figure 5). Figure 5A shows the relationship between the transpiration rate, W_w, obtained by the weighing method and W_x calculated from Equation 11. ''I'' denotes the distribution of W_x and ° denotes the mean value of the whole leaf surface in this figure. Being the mean value of one leaf, W_w was compared with the mean value of W_x. When the transpiration rate was below 0.1 × 10^{-5} g cm^{-2} s^{-1}, the leaf wilted and the leaf color was slightly altered. The difference between W_w and the mean value of W_x was within 0.02 × 10^{-5} g cm^{-2} s^{-1} except in the above case. Figure 5B shows the relationship between stomatal resistance to water vapor diffusion, r_{wsx}, obtained from W_w and that obtained from W_x. When r_{wsx} was below 10 s cm^{-1}, the difference between

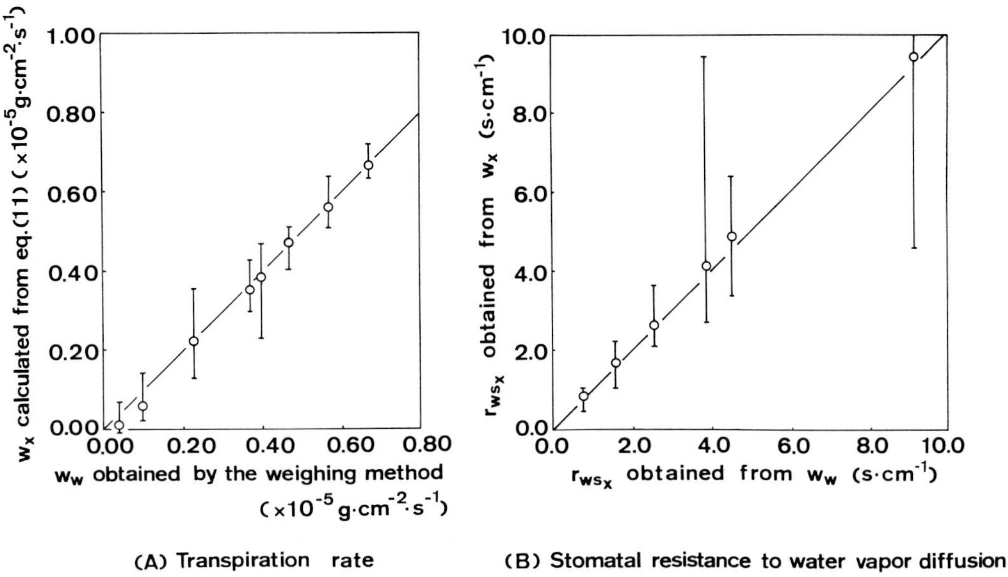

FIGURE 5. Accuracies in evaluating transpiration rate (W_x) and stomatal resistance to water vapor diffusion (r_{wsx}). (A) Transpiration rate; (B) stomatal resistance to water vapor diffusion. Environmental conditions: T_a = 24.9 to 25.4°C; ϕ = 58 to 60% RH; r_{kax} = 1.64 s cm^{-1}; E_{sx} = 3.01 × 10^{-3} cal cm^{-2} s^{-1}; E_{wx} = 22.21 × 10^{-3} cal cm^{-2} s^{-1}. (From Omasa, K., Abo, F., Aiga, I., and Hashimoto, Y., *Trans. Soc. Instrum. Control Eng.*, 17, 657, 1981. With permission.)

r_{wsx} from W_w and that from the mean value of W_x was within 0.3 s cm^{-1}. Although the accuracy of the evaluation of transpiration rate and stomatal resistance depends on the environmental conditions, we can expect sufficient accuracy under typical environmental conditions, given this result. Furthermore, the accuracy of the transpiration rate may be applied correspondingly to that of the absorption rate of air pollutants such as SO_2, NO_2, O_3, and PAN from an analogy between the models for evaluating the transpiration rate and the absorption rate of these air pollutants.

3. Responses to Air Pollutants

Figure 6 shows an example of the changes in leaf temperature distribution during exposure to about 2 μl l^{-1} SO_2. The leaf temperature ranged from 22.8 to 24.0°C before the start of SO_2 exposure, began to increase 5 min after the beginning, and reached 23.1 to 25.3°C after 60 min of exposure. From a series of images, there was an observable tendency for the leaf temperature to be high and to increase especially in the vicinity of veins and leaf edges. Water soaking began to appear slightly at the sites of the lowest leaf temperature after 60 min of exposure. However, visible leaf injury such as wilting and fading of pigments did not appear during SO_2 exposure. Figure 7 shows changes in the distributions of transpiration rate, stomatal resistance to water vapor diffusion, and SO_2 absorption rate on the leaf evaluated from the leaf temperatures, according to Equations 2, 11, and 12. Stomatal resistance, which is an indicator of stomatal opening, increased with exposure time, while the transpiration rate and SO_2 absorption rate decreased. However, the behavior varied randomly at different sites on the leaf. After entering the leaf, SO_2 caused physiological effects such as a decreased rate of photosynthesis, fading of chlorophyll and other pigments, and destruction of cells. Figure 8 shows the distribution of SO_2 absorption at the end of a 60-min exposure and the resulting visible injury photographed through an interference filter (central wavelength 0.67 μm and half-band width 0.01 μm) 1 d later. Visible injury consisted of necrosis and chlorosis. Note that healthy parts of the leaf in the photographs appear dark

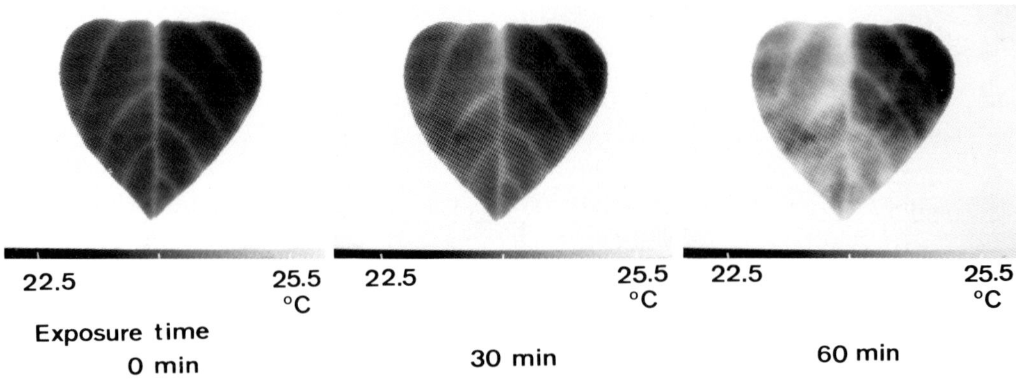

FIGURE 6. An example of changes in the temperature distribution of an attached sunflower leaf during exposure to 2 $\mu l\ l^{-1}$ SO_2. Environmental conditions: T_a = 25.0°C; Φ = 62% RH; r_{kax} = 1.5 s cm^{-1}; E_{sx} = 2.37 × 10^{-3}, cal cm^{-2} s^{-1}; E_{wx} = 2.23 × 10^{-2} cal cm^{-2} s^{-1}. (From Omasa, K., Hashimoto, Y., and Aiga, I., *Environ. Control Biol.*, 19, 59, 1981; 19, 85, 1981. With permission.)

because of their large absorption of light by leaf pigments, while areas with visible injury appear white because of small absorption of light and a high degree of reflection. There was a tendency for the visible injury to occur at sites where the total SO_2 absorption exceeded a threshold value of about 0.85 × 10^{-5} g cm^{-2}. Next, we investigated the relationships between total absorption of NO_2 or O_3 during the 60-min exposure and the visible injury photographed through the filter 1 d later (Figure 9). In the case of NO_2, results were similar to those of SO_2. Visible injury occurred at sites where the total NO_2 absorption was over a threshold value of about 1.2 × 10^{-5} g cm^{-2}. In the case of O_3, however, the degree of visible injury at local sites on a leaf was not related to total O_3 absorption. This result indicates that the degree of visible injury among local sites on the leaf is dependent not only on factors related to O_3 absorption, but also on other physiological factors among these sites.

II. ANALYSIS OF THE MICROSCOPE IMAGE

The stomatal aperture is generally measured with a ruler from a photograph taken using a scanning electron microscope or a light microscope. This method is subject to two errors: changes in stomatal aperture induced in preparing tissue for microscopic examination and the subjective determination of the edge of the stomatal pore. Therefore, an objective technique, digital image processing, was developed to evaluate images.

When the brightness of the stomatal pore differs remarkably from that of the guard cells and the epidermal cell, the pore can be easily extracted from the surrounding cells of the original image by simple thresholding, which is a digital image processing technique.[18] However, since many stomatal images, especially those observed with a light microscope, do not satisfy the above condition and the edge of the stomatal pore is not clear, simple thresholding cannot be used for pore extraction. Also, prior to the development of this method, there had been no report on the objective evaluation of the width, length, and area of a stomatal pore, which are used as common indicators of stomatal aperture. Therefore, we developed a digital image processing technique for measuring them exactly, even with images of poor quality,[19] and applied this technique to evaluate the responses of intact stomata observed by a remote-controlled light microscope system.[20-23]

A. REMOTE-CONTROL LIGHT MICROSCOPE SYSTEM

1. Outline of the System

Observation with an ordinary light microscope under a plant's growing conditions poses

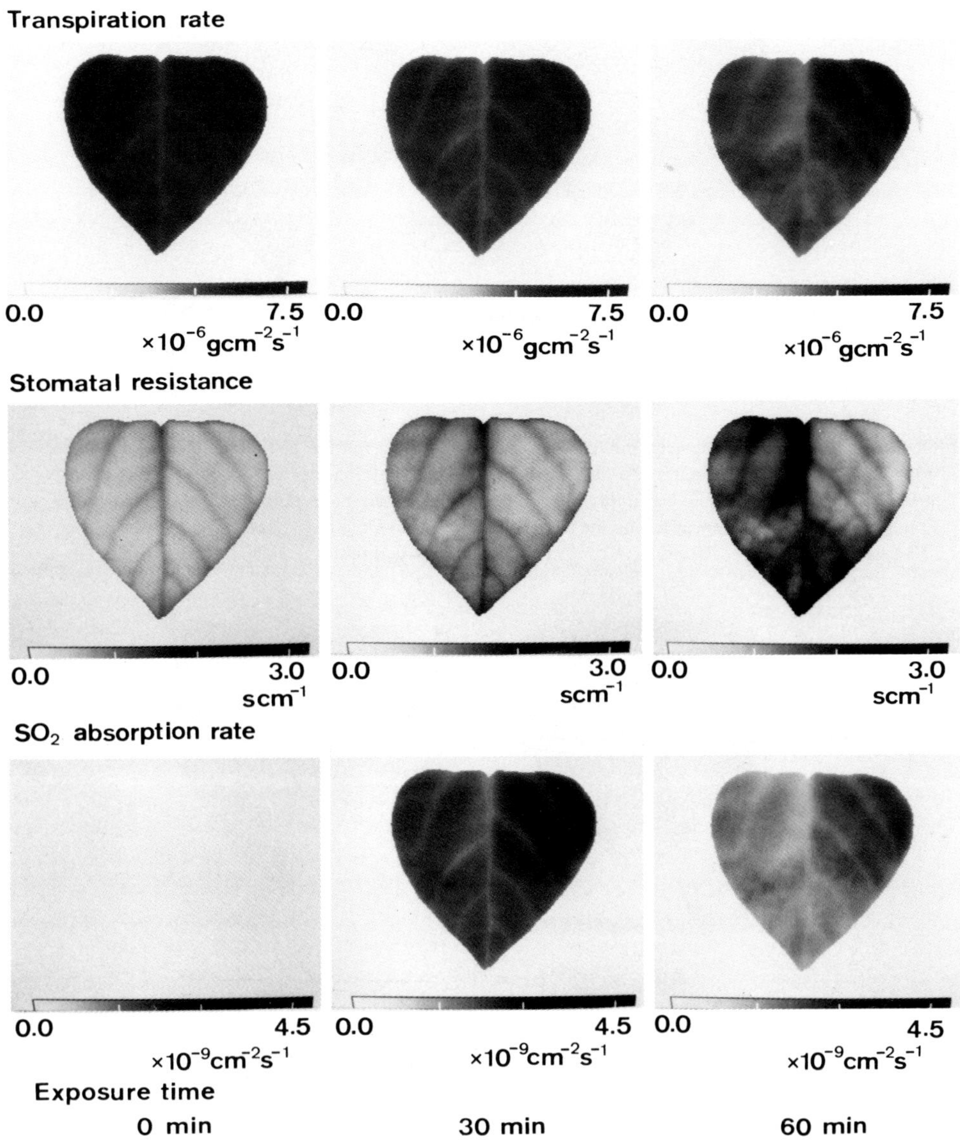

FIGURE 7. Changes in spatial distributions of transpiration rate (W_x), stomatal resistance to water vapor diffusion (r_{wsx}), and SO_2 absorption rate (Q_x) evaluated from the leaf temperature distributions in Figure 6. (From Omasa, K., Hashimoto, Y., and Aiga, I., *Environ. Control Biol.*, 19, 59, 1981; 19, 85, 1981. With permission.)

some problems.[24,25] First, visual observation under weak light is very difficult. Second, the environment of the lower side of the leaf cannot be controlled, because the leaf is directly held on the microscope stage, and the environment is also affected by human manipulation of the microscope. Third, the working distance, i.e., the distance between the leaf and the objective during the observation, is very small at high magnification, thus subjecting the leaf to the danger of sticking to the objective during focusing and destroying the environment between the leaf and the objective. Recently, Omasa et al.[20] and Kappen et al.[26] have solved the above problems of the ordinary light microscope.

Figure 10 shows our remote-controlled light microscope system.[20,22] This system consists of a light microscope (Bausch & Lomb, MicroZoom) with a long working distance (about 13 mm) at high magnification (a 50× objective, 1.5 and 2× amplifiers, and a TV adapter

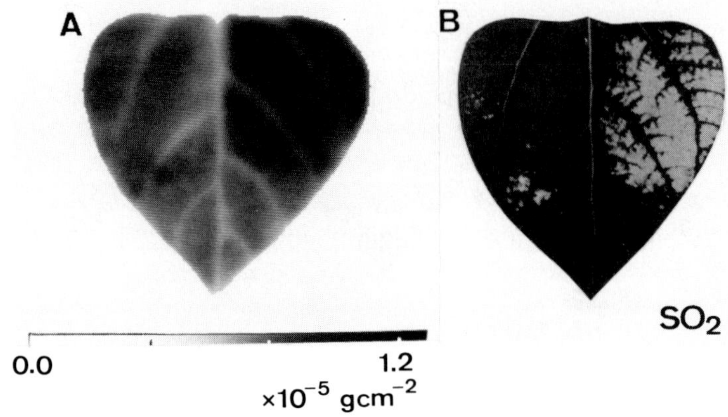

FIGURE 8. Relationship between the total SO_2 absorption (Q_{intx}) at the end of the 60-min exposure in Figure 6 and visible leaf injury 1 d later. (A) Total SO_2 absorption evaluated from a series of leaf temperature images; (B) visible leaf injury photographed through an interference filter of 0.67 μm. The visible leaf injury did not appear during the SO_2 exposure. (From Omasa, K., Hashimoto, Y, and Aiga, I., *Environ. Control Biol.*, 19, 59, 1981; 19, 85, 1981. With permission.)

FIGURE 9. Relationship between the total absorption (Q_{intx}) of NO_2 or O_3 during the 60-min exposure and visible leaf injury 1 d later. (A) total absorption of NO_2 or O_3; (B) visible leaf injury. Visible injury did not appear during the O_3 exposure, but was present shortly after the 50-min NO_2 exposure. The leaf began to wilt and develop chlorotic areas. Environmental conditions: Pa (NO_2) = about 7 μl l^{-1}; Pa (O_3) = about 1.2 μl l^{-1}. Other conditions were the same as those in Figure 6. (From Omasa, K., Hashimoto, Y., and Aiga, I., *Environ. Control Biol.*, 19, 59, 1981; 19, 85, 1981. With permission.)

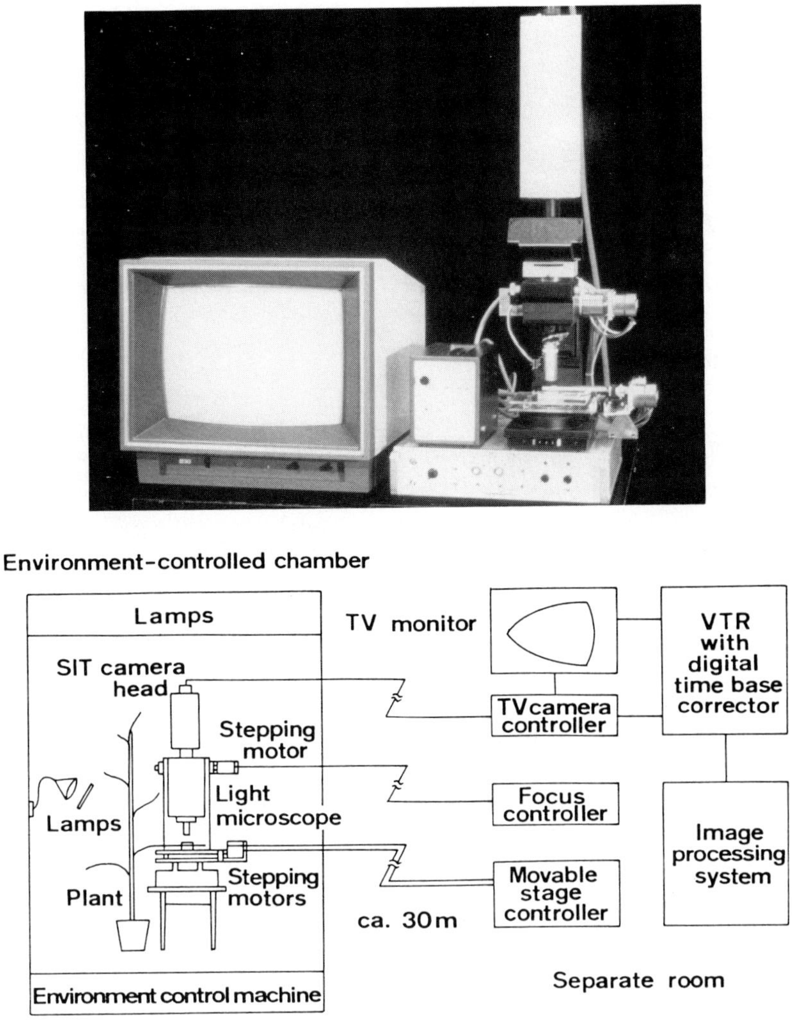

FIGURE 10. Remote-control light microscope system.

lens; about 1600-fold magnification on a TV monitor), SIT (silicon intensifier target) video camera (Hamamatsu TV, Model C1000-12) with a high sensitivity S20-type spectral response, image resolution of over 600 TV lines, distortion within 2%, and shading within 20% as a detector of the microscope image, a monochromatic TV monitor (Chuomusen, Model MD2002A) with an image resolution of about 1000 TV lines and distortion within 3%, and remote controllers (in a separate room) for adjusting camera sensitivity, microscope focus, and movement of the microscope stage. The microscope images are projected on the TV monitor in a separate room and recorded photographically on black and white (B&W) film or by a VTR (SONY, BVU820: horizontal resolution, 340 TV lines in B&W mode; S/N, about 50 dB) with a digital time base corrector (SONY, BVT800) and a time code generator/reader. The stomatal aperture is evaluated by an image-processing system with array processors for floating-point operations.

2. Performance of the System

The observation was usually carried out with light from the environment, but a halogen lamp is sometimes used as a supplementary light source for observation with transmitted

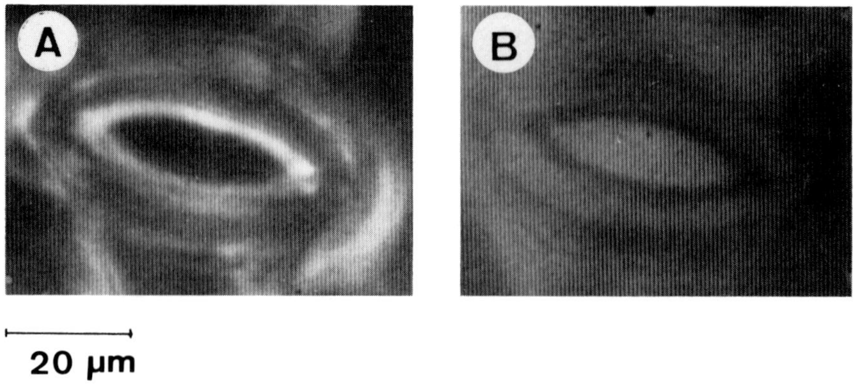

FIGURE 11. Photomicrographs of an intact sunflower stoma observed with reflected (A) or transmitted light (B) using the light microscope system. (From Omasa, K., Hashimoto, Y., and Aiga, I., *Plant Cell Physiol.*, 24, 281, 1983. With permission.)

light (Figure 11). The stomatal image was clear at high magnification. The stoma was observed with reflected light and then rapidly observed with transmitted light, and the stomatal aperture was found to be the same for both. Although this system could provide stomatal images from a mixture of reflected and transmitted light, the clearest images were obtained with either reflected or transmitted light alone. The stomata could be observed with either reflected or transmitted light above about 0.1 mW cm^{-2} (environmental illumination about 2 klx [≃0.5 mW cm^{-2}] with reflected light and about 0.5 klx [≃0.5 mW cm^{-2}] with transmitted light). If the observation is done in low light (0.1 mW cm^{-2}), with the naked eye looking through the eyepiece instead of the SIT camera, the eye must be adapted to the dark room. The attachment of an image intensifier to the SIT camera produces a further increase in sensitivity, although the image quality becomes poor. The resolution of the microscope images was within 1 μm. However, it was improved within 0.3 μm standard error by a digital image-processing technique, even when the microscope image was of poor quality.

B. EVALUATION OF STOMATAL APERTURE

1. Principle and Procedure of Image Processing

Figure 12 shows a flow diagram of digital image processing for evaluating pore width, length, and area as indicators of stomatal aperture. The original digital image obtained from the TV monitor could be improved using the image processing system. Simetimes the brightness of the stomatal pore in the image did not differ sufficiently from that of the guard cells and the surrounding epidermal cells to allow thresholding. Filtering by discrete Fourier transformation and unsharp masking were used to remove the TV scanning lines and to enhance the image as the first step of image processing.

Given f (x, y) as the gray level at the pixel (x, y = 0, 1, 2, . . . , N − 1), in the original digital image, the two-dimensional discrete Fourier transform F (u − N/2, v − N/2) of f (x, y), where the origin after transformation was moved to the center of the frequency square, was given by

$$F(u - N/2, v - N/2) = \frac{1}{N} \sum_{x=0}^{N-1} \sum_{y=0}^{N-1} (-1)^{x+y} f(x, y) \exp[-j2\pi(ux + vy)/N] \qquad (18)$$

for u, v = 0, 1, 2, . . . , N − 1.

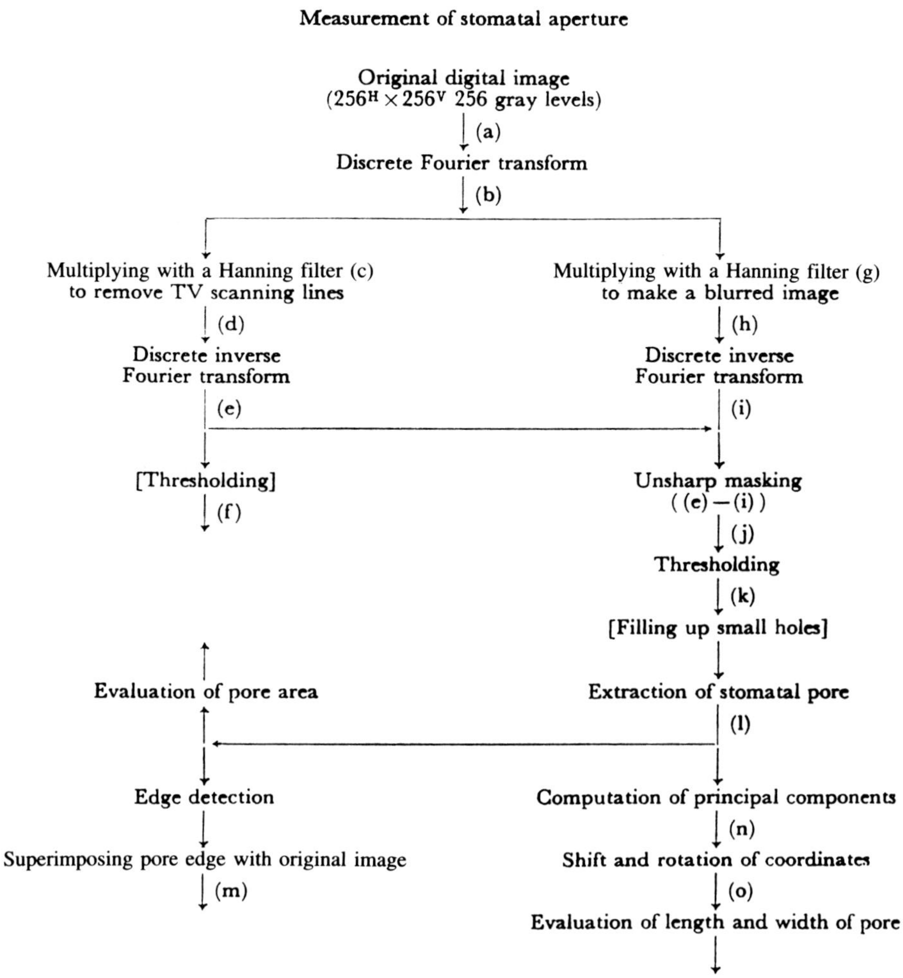

FIGURE 12. Flow diagram of digital image processing for evaluating width, length, and area of a stomatal pore. (From Omasa, K. and Onoe, M., *Plant Cell Physiol.*, 25, 1379, 1984. With permission.)

After multiplying F (u − N/2, v − N/2) by the Hanning filter H (u − N/2, v − N/2, f_c) to remove the TV scanning lines or to obtain a blurred image for the unsharp masking, each inverse Fourier transform g (x′, y′, f_c) could be expressed by:

$$g(x', y', f_c) = \frac{1}{N}(-1)^{x+y} \sum_{u=0}^{N-1} \sum_{v=0}^{N-1} (-1)^{u+v} H(u - N/2, v - N/2, f_c)$$

$$\text{x } F(u - N/2, v - N/2) \exp[j2\pi(ux + vy)/N] \qquad (19)$$

for

$$x' = x - N/2 \quad \text{and} \quad y' = y - N/2$$

where:

$$H(u - N/2, v - N/2, f_c) = \begin{cases} 0 & \text{if} \quad f_{uv} \geq f_c \\ \frac{1}{2}[1 + \cos(\pi f_{uv}/f_c)] & \text{if} \quad f_{uv} < f_c \end{cases}$$

and

$$f_{uv} = [(u - N/2)^2 + (v - N/2)^2]^{1/2}$$

The parameter f_c of the Hanning filter was selected according to need. The unsharp masking, in which the blurred image g (x', y', f_{c2}) (i in Figure 12) was subtracted from image g (x', y', f_{c1}) (e in Figure 12) after removal of the TV scanning lines, had the effect of increasing the average ramp steepness and the contrast just at the pore edge. The unsharp masked image g_u (x', y') was given by:

$$g_u(x', y') = g(x', y', f_{c1}) - g(x', y', f_{c2}) \quad (20)$$

The pore region of a given stoma could be extracted from the neighboring cells by thresholding g_u (x', y'). The result could be expressed as the two-valued image g_T (x', y').

$$g_T(x, y) = \begin{cases} 1 & \text{if} \quad g_u(x', y') \geqslant T \quad \text{at appointed stoma} \\ 0 & \text{otherwise} \end{cases} \quad (21)$$

where T is a threshold. The threshold was determined to be about zero in the gray levels of the unsharp masked image and to give the maximum area when the pore region was extracted elliptically. When there were small holes in the threshold region resulting from poor quality of the original image (although this was unusual), the holes were completely filled by thresholding.

The area A_r of the stomatal pore was obtained by counting pixels in the extracted region.

$$A_r = k^2 A \quad (22)$$

where

$$A = \sum_{x' = -(N/2)}^{(N/2)-1} \sum_{y' = -(N/2)}^{(N/2)-1} g_T(x, y)$$

and k is a coefficient for correcting the magnification.

The pore width l_a and the pore length l_b were evaluated from the frequency distributions to the two rectangular principal components of the extracted region (Figure 13). The principal components X, Y were defined by:

$$X = (x' - xo)\cos\theta + (y' - yo)\sin\theta \quad (23)$$

and

$$Y = (x' - xo)\sin\theta + (y' - yo)\cos\theta \quad (24)$$

where

$$xo = \frac{1}{A} \sum_{x' = -(N/2)}^{(N/2)-1} \sum_{y' = -(N/2)}^{(N/2)-1} x' g_T(x', y')$$

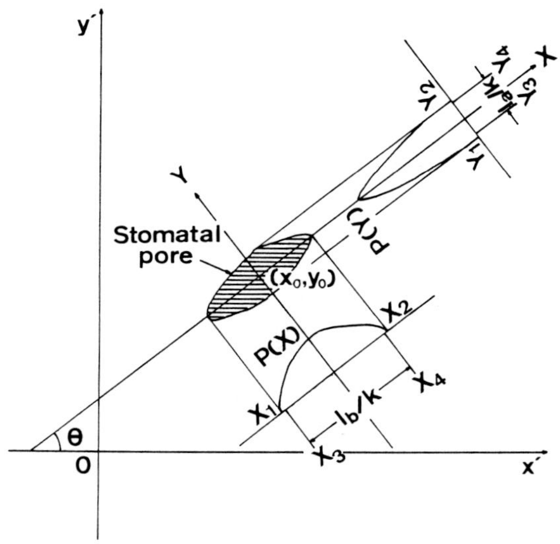

FIGURE 13. Diagram for evaluating width and length of a stomatal pore. (From Omasa, K. and Onoe, M., *Plant Cell Physiol.*, 25, 1379, 1984. With permission.)

$$yo = \frac{1}{A} \sum_{x'=-(N/2)}^{(N/2)-1} \sum_{y'=-(N/2)}^{(N/2)-1} y' g_T(x', y')$$

$$\theta = \frac{1}{2} \tan^{-1}[2S_{xy}/(S_{xx} - S_{yy})]$$

$$S_{xx} = \sum_{x'=-(N/2)}^{(N/2)-1} \sum_{y'=-(N/2)}^{(N/2)-1} (x' - xo)^2 g_T(x', y')$$

$$S_{yy} = \sum_{x'=-(N/2)}^{(N/2)-1} \sum_{y'=-(N/2)}^{(N/2)-1} (y' - yo)^2 g_T(x', y')$$

and

$$S_{xy} = \sum_{x'=-(N/2)}^{(N/2)-1} \sum_{y'=-(N/2)}^{(N/2)-1} (x' - xo)(y' - yo) g_T(x', y')$$

P(X) and P(Y), the frequency distributions to the principal components X, Y, were evaluated by computer after the shift and rotation of coordinates from x′ and y′ to X and Y, and then l_a and l_b were determined from

$$l_a = k(Y_4 - Y_3) \tag{25}$$

$$l_b = k(X_4 - X_3) \tag{26}$$

where X_1 to X_4 and Y_1 to Y_4 were defined by:

$$P(X_1) = P(X_2) = 0 \qquad X_2 > X_1$$

$$P(Y_1) = P(Y_2) = 0 \qquad Y_2 > Y_1$$

$$\sum_{X=X_1}^{X_3} P(X) = \sum_{X=X_4}^{X_2} P(X) = 0.005A \qquad X_4 > X_3$$

and

$$\sum_{Y=Y_1}^{Y_3} P(Y) = \sum_{Y=Y_4}^{Y_2} P(Y) = 0.005A \qquad Y_4 > Y_3$$

The degree of stomatal opening k_l was expressed by the ratio l_a/l_{bmax}, where l_{bmax} was the length l_b of the stomatal pore at its maximal value. The edge of a stomatal pore could be detected from $g_T(x', y')$ by a simple difference operator.

2. Accuracy of the Evaluation

Figure 14 shows the sequence of digital image processing for constructing the stomatal aperture according to the flow diagram in Figure 12. The original digital image (a) contained TV scanning lines of vertical B&W stripes. Being clear and regular, the lines appeared as specific frequency components with sharp peaks at $u \simeq \pm 68$ in the Fourier spectrum (b) [F (u − N/2, v − N/2] obtained from the magnitude of each of the transform terms. These components differed remarkably from low-frequency components associated with stomatal form. Therefore, the TV scanning lines were easily removed by a Hanning filter of f_c = 60 (c), which was a low-pass filter with the restraint function of frequency extension and leakages. The Fourier spectrum after the filtering and its inverse transform image are shown in (d) and (e), respectively. Since the brightness of the stomatal pore did not differ from that of the guard cells or the epidermal cells, the pore region could not be extracted by simple thresholding (f). Thus, unsharp masking was used for image sharpening. The blurred image (i) for the unsharp masking was obtained by the inverse transform of a Fourier spectrum (h) after multiplication with a Hanning filter of f_c = 10 (g). The unsharp masked image (j) increased in contrast at the pore edge and in brightness between the pore and its neighboring cells. Therefore, the pore region of the stoma could be easily extracted from the neighboring cells after thresholding: (k) was a two-valued image after the thresholding and (l) was a pore region extracted from the image (k). From the image (m) obtained by aligning the pore edge with the original image, we confirmed that the pore region had been exactly extracted. The pore area was computed from the extracted image (l). The pore width and the pore length were evaluated after the shift and rotation of coordinates to the principal components of the pore region; (n) expressed principal components and (o) the shift and rotation of the coordinates.

The quality of the original digital image affects the accuracy of the stomatal aperture evaluation and depends upon the display image itself, the grain size and development of the film, and other factors. Therefore, the accuracy was first examined using ten original images of the same stomatal aperture, but differing in quality, and also using continuous images of different quality during exposure to about 1 $\mu l\ l^{-1}$ O_3. Table 1 shows the standard errors in the evaluation of pore area, pore length, and pore width of the control images. The errors of the area, length, and width were 10 μm^2, 0.3 μm, and 0.3 μm, respectively. Figure 15A shows the changes in the pore area A_r and the degree k_l of opening of a stoma during the O_3 exposure. The slow increase of both A_r and k_l began shortly before the exposure and continued until 10 min (b) after its start. Thereafter, both began to decrease rapidly and reached a constant level at about 25 min (e). Figure 15B shows the original digital images superimposed with the pore edge at time points (a) to (f) in Figure 15A. They confirmed

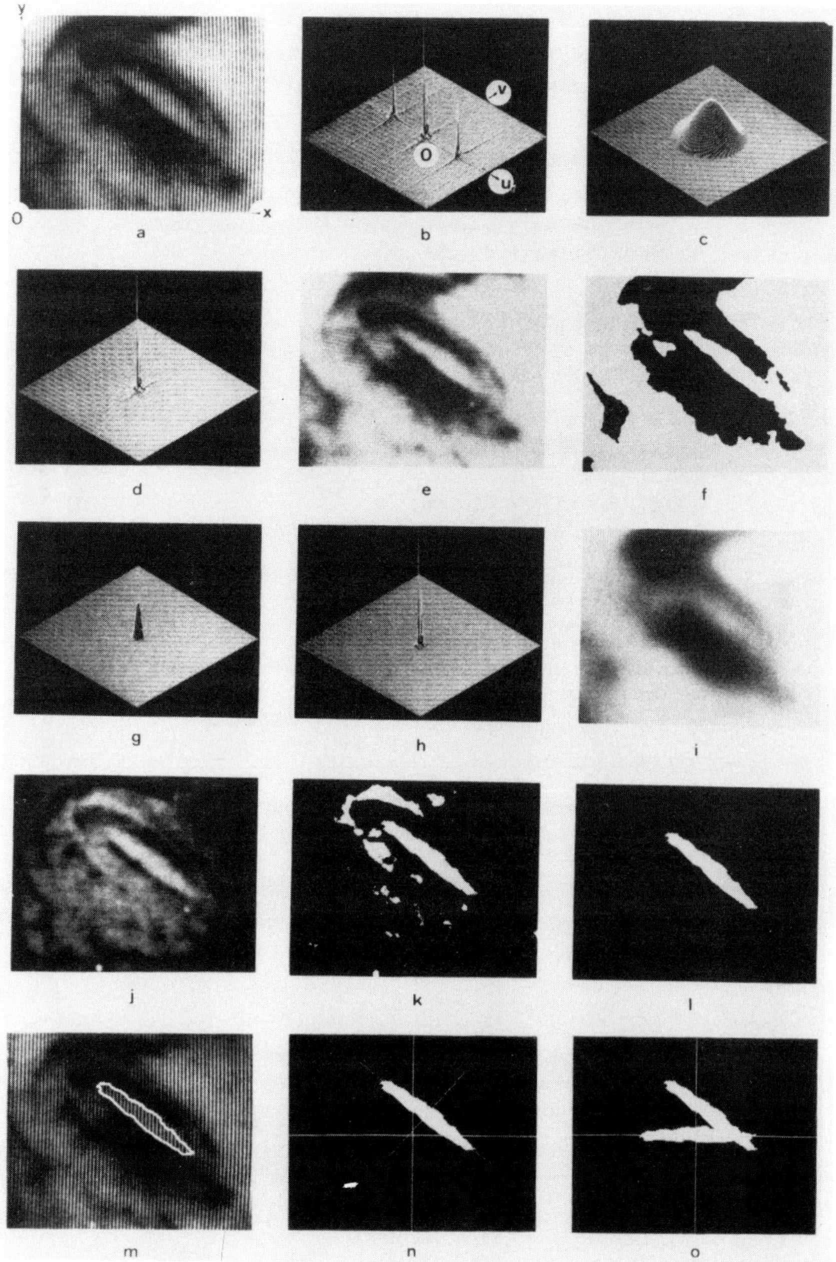

FIGURE 14. Sequence of digital image processing for evaluating stomatal aperture according to the flow diagram of Figure 12. (a) to (o) correspond to the processing steps in Figure 12. The coordinates are shown in (a) for (x,y) and (b) for (u,v). (From Omasa, K. and Onoe, M., *Plant Cell Physiol.*, 25, 1379, 1984. With permission.)

that the pore region had been exactly extracted even when the quality of the original image was poor, demonstrating the accuracy of the method.

3. Variety of Stomatal Movements

We examined the responses of neighboring stomata to SO_2 in a small leaf region of an intact growing plant. Figure 16 shows that six adjacent stomata differ in response to about

TABLE 1
Accuracy in the Evaluation of Area, Length, and Width of a Stomatal Pore

	Area (μm^2)	Length (μm)	Width (μm)
Mean value	433	34.1	16.4
Standard error	10	0.3	0.3

From Omasa, K. and Onoe, M., *Plant Cell Physiol.*, 25, 1379, 1984. With permission.

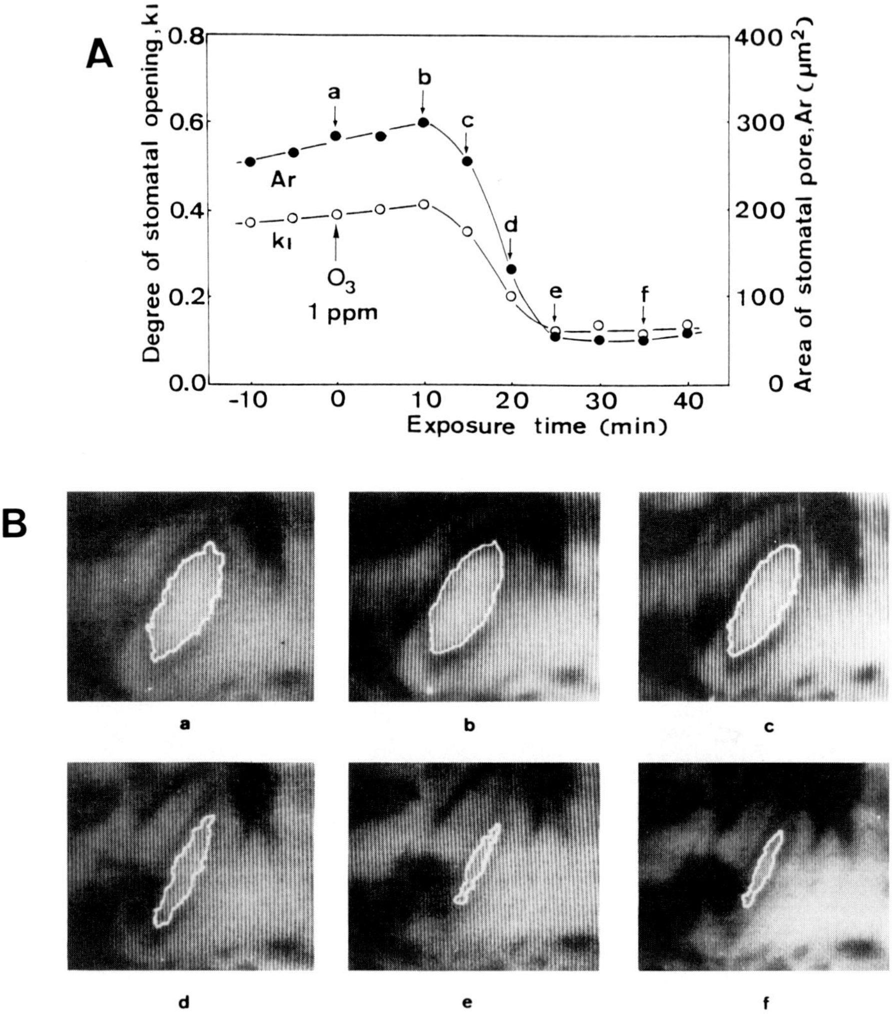

FIGURE 15. Responses of an intact stoma on an attached sunflower leaf to about 1 $\mu l\ l^{-1}$ O_3. (A) Changes in area of stomatal pore, A_r, and degree of stomatal opening, k_l; (B) original digital images superimposed with the pore edge at the time points (a to f) in A. Environmental conditions: air temperature = 25.0°C; relative humidity = 60%; light intensity = 600 μmol photons $m^{-2}\ s^{-1}$. (From Omasa, K. and Onoe, M., *Plant Cell Physiol.*, 25, 1379, 1984. With permission.)

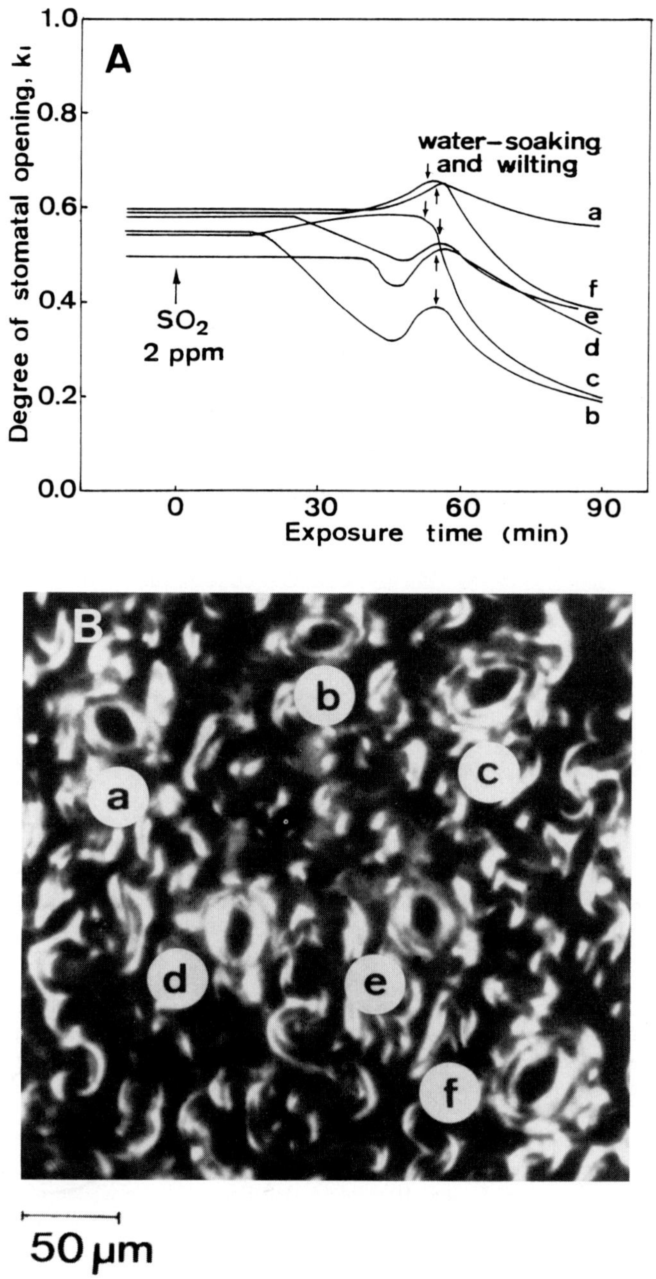

FIGURE 16. Response of six neighboring stomata on an attached sunflower leaf to about 2 $\mu l\ l^{-1}$ SO_2. (A) Changes in k_l of stomata (a to f) in B; (B) photomicrograph of a small region of the leaf. Environmental conditions are similar to those in Figure 15. (From Omasa, K. and Onoe, M., *Plant Cell Physiol.*, 25, 1379, 1984. With permission.)

2 $\mu l\ l^{-1}$ SO_2. These stomata showed almost uniform and constant k_l until about 20 min after the start of the exposure, and then a wide variety of stomatal movements began. The largest value in k_l was about twice as large as the smallest value at 45 min and became about three times as large at 90 min. Water soaking and wilting began to appear in the subsidiary cells at about 55 min, when k_l was at a maximum value, and then all the stomata

began to close. This phenomenon was assumed to be caused by increased water loss from the subsidiary cells because of SO_2, which affects the membrane and osmotic pressure, with a resulting difference in turgor between the guard cells and the subsidiary cells.[21,27]

Changes in stomatal conductance or stomatal resistance of a small leaf region to air pollutants have been investigated using a parameter and a thermal-image instrumentation system.[10-12,27] The latter system shows changes in the spatial distribution of stomatal resistance over the surface of an attached leaf, and differences have been seen in the resistance at each leaf site.[10-12] Transpiration, which is an indicator of the stomatal movements of a whole plant, or of one leaf exposed to air pollutants, is also known to vary by species, leaf and plant age, growth environment, air pollutant concentration, and other factors.[13,14,28,29] However, since stomatal resistance and transpiration indicate the average movement of many stomata, the values cannot give the response of each stoma to air pollutants. Therefore, it is important to observe individual stomata and their neighboring cells directly to study stomatal movements.[21] Also, since this system is effective for observing many intact stomata easily and rapidly, it can be used to analyze the relationship between stomatal aperture and conductance under the plant's growing conditions.[20]

REFERENCES

1. **Raschke, K.,** Heat transfer between the plant and the environment, *Annu. Rev. Plant Physiol.,* 11, 111, 1960.
2. **Monteith, J. L.,** *Principles of Environmental Physics,* Edward Arnold, London, 1973.
3. **Jones, H. G.,** *Plants and Microclimate,* Cambridge University Press, Cambridge, 1983.
4. **Omasa, K.,** Image instrumentation methods of plant analysis, in *Modern Methods of Plant Analysis, New Series,* Vol. 11, Linskens, H. F. and Jackson, J. F., Eds., Springer-Verlag, Berlin, 1990, 203.
5. **Omasa, K., Abo, F., Hashimoto, Y., and Aiga, I.,** Measurement of the thermal pattern of plant leaves under fumigation with air pollutant, *Res. Rep. Natl. Inst. Environ. Stud.,* 11, 239, 1980.
6. **Horler, D. N. H., Barber, J., and Barringer, A. R.,** Effects of cadmium and copper treatments and water stress on the thermal emission from peas (*Pisum sativum* L.): controlled evironment experiments, *Remote Sens. Environ.,* 10, 191, 1980.
7. **Hashimoto, Y., Ino, T., Kramer, P. J., Naylor, A. W., and Strain, B. R.,** Dynamic analysis of water stress of sunflower leaves by means of a thermal image processing system, *Plant Physiol.,* 76, 266, 1984.
8. **Omasa, K. and Aiga, I.,** Environmental measurement: image instrumentation for evaluating pollution effects on plants, in *Systems & Control Encyclopedia,* Vol. 2, Singh, M. G., Ed., Pergamon Press, Oxford, 1987, 1516.
9. **Omasa, K., Tajima, A., and Miyasaka, K.,** Diagnosis of street trees by thermography: Zelkova trees in Sendai City, *J. Agric. Meteorol.,* 45, 271, 1990.
10. **Omasa, K., Abo, F., Aiga, I., and Hashimoto, Y.,** Image instrumentation of plants exposed to air pollutants — quantification of physiological information included in thermal infrared images, *Trans. Soc. Instrum. Control Eng.,* 17, 657, 1981.
11. **Omasa, K., Hashimoto, Y., and Aiga, I.,** A quantitative analysis of the relationships between SO_2 or NO_2 sorption and their acute effects on plant leaves using image intrumentation, *Environ. Control Biol.,* 19, 59,1981.
12. **Omasa, K., Hashimoto, Y., and Aiga, I.,** A quantitative analysis of the relationships between O_3 sorption and its acute effects on plant leaves using image instrumentation, *Environ. Control Biol.* 19, 85, 1981.
13. **Omasa, K. and Abo, F.,** Studies of air pollutant sorption by plants. I. Relation between local SO_2 sorption and acute visible leaf injury, *J. Agric. Meteorol.,* 34, 51, 1978.
14. **Omasa, K., Abo, F., Natori, T., and Totsuka, T.,** Studies of air pollutant sorption by plants. II. Sorption under fumigation with NO_2, O_3 or NO_2 + O_3, *J. Agric. Meteorol.,* 35, 77, 1979.
15. **Omasa, K.,** Sorption of air pollutants by plant communities — analysis and modelling of phenomena, *Res. Rep. Natl. Inst. Environ. Stud. Jpn.,* 10, 367, 1979.
16. **Gates, D. M., Keegan, H. J., Schleter, J. C., and Weidner, V. R.,** Spectral properties of plants, *Appl. Opt.,* 4, 11, 1965.
17. **Fuchs, M. and Tanner, C. B.,** Infrared thermometry of vegetation, *Agron. J.,* 58, 597, 1966.

18. **Hashimoto, Y., Morimoto, T., and Funada, S.,** Image processing of plant information in the relation between leaf temperature and stomatal aperture, in *Technological and Methodological Advances in Measurement,* Vol. 3, Striker, G., Havrilla, K., Solt, J., and Kemeny, T., Eds., Elsevier/North-Holland, Amsterdam, 1983, 313.
19. **Omasa, K. and Onoe, M.,** Measurement of stomatal aperture by digital image processing, *Plant Cell Physiol.,* 25, 1379, 1984.
20. **Omasa, K., Hashimoto, Y., and Aiga, I.,** Observation of stomatal movements of intact plants using an image instrumentation system with a light microscope, *Plant Cell Physiol.,* 24, 281, 1983.
21. **Omasa, K., Hashimoto, Y., Kramer, P. J., Strain, B. R., Aiga, I., and Kondo, J.,** Direct observation of reversible and irreversible stomatal responses of attached sunflower leaves to SO_2, *Plant Physiol.,* 79, 153, 1985.
22. **Omasa, K.,** Study on changes in stomata and their surrounding cells using a nondestructive light microscope system: responses to air pollutants, *J. Agric. Meteorol.,* 45, 251, 1990.
23. **Omasa, K. and Maruyama, S.,** Study on changes in stomata and their surrounding cells using a nondestructive light microscope system: responses to changes in water absorption through roots, *J. Agric. Meteorol.,* 45, 259, 1990.
24. **Heath, O. V. S.,** The water relations of stomatal cells and the mechanisms of stomatal movements, in *Plant Physiology,* Vol. 2, Steward, F. C., Ed., Academic Press, New York, 1959, 193.
25. **Meidner, H. and Mansfield, T. A.,** *Physiology of Stomata,* McGraw-Hill, London, 1968.
26. **Kappen, L., Andresen, G., and Losch, R.,** *In situ* observations of stomatal movements, *J. Exp. Bot.,* 38, 126, 1987.
27. **Unsworth, M. H. and Black, V. J.,** Stomatal responses to pollutants, in *Stomatal Physiology,* Jarvis, P. G. and Mansfield, T. A., Cambridge University press, Cambridge, 1981, 187.
28. **Kondo, N. and Sugahara, K.,** Changes in transpiration rate of SO_2-resistant and -sensitive plants with SO_2 fumigation and the participation of abscisic acid, *Plant Cell Physiol.,* 19, 265, 1978.
29. **Furukawa, A., Isoda, O., Iwaki, H., and Totsuka, T.,** Interspecific differences in responses of transpiration to SO_2, *Environ. Control Biol.,* 17, 153, 1979.

Chapter 11

PROCESSING AND ANALYSIS OF ELECTRON IMAGES OF BIOMOLECULES*

George Harauz and Egbert J. Boekema

TABLE OF CONTENTS

* This review is dedicated to Professors E. Zeitler and E. F. J. van Bruggen.

I. INTRODUCTION

A. THE NECESSITY OF IMAGE ANALYSIS

Electron microscopy (EM) is a valuable technique for determining low-resolution structures of biological macromolecules and their complexes (biomolecules), especially larger ones with masses ranging from several hundred to several thousand kilodaltons. Usually electron images of isolated biomolecules contain structural information approaching a spatial resolution of 1 to 3 nm, at which important questions about shape and number of subunits can often be answered. However, it is very difficult to do so from a direct visual interpretation of the electron micrographs. The reasons for this lie in the complex processes that make electron images of biomolecules noisy.

First of all, the biological specimen lies on a support film such as carbon which provides a nonuniform background.[1] Adsorption to the support film and drying from aqueous solution may induce the fragile biomolecule to distort and flatten. Agents such as heavy atom salts are necessary to produce contrast and to stabilize the specimen against drying, and these may have a random distribution as well as limiting resolution.[2,3] During the recording process in the emulsion layer of the micrograph, additional noise is created since the electrons lose their energy in multiple interactions and thus hit many silver grains. The most serious effect of all, radiation damage, manifests itself in mass loss and random structural rearrangements.[4-7]

Careful and sophisticated experimental techniques such as low-dose cryo-electron microscopy of the specimen in frozen hydrated form can ameliorate many of these problems.[8-10] Other ways of enhancing biomolecular micrographs include instrumental techniques such as scanning transmission and dark-field EM.[11] In many circumstances, though, experimental approaches may be unavailable, inadequate, or inappropriate to yield electron images of improved quality. As stated before, structural information cannot then be easily deduced by visual inspection. One requires computational approaches using special processing, analysis, and reconstruction algorithms suited to the subject and the type of image.

Computer analysis of electron micrographs is frequently being reviewed in the literature.[11-19] Here, we restrict our discussion to the analysis of randomly oriented single biomolecules and their two-dimensional crystals. We first examine how relevant structural information can be extracted from projections of isolated biomolecules (''single particles'') using pattern recognition, followed by a general discussion of artificial intelligence to illustrate the current state of software development. The final section describes how structures of certain biomolecules such as membrane proteins can be determined to a higher resolution when they can be crystallized into monolayers; a different algorithmic approach than for single particles is required. Examples are presented to show the range of useful applications of these computational techniques.

B. IMAGE PROCESSING TOOLS

Before discussing electron image analysis in detail, it is worthwhile to say something about the hardware and software tools required. Until recently, the only computers powerful enough to do this sort of work were mainframes or large minicomputers, the latter being primarily of the DEC (Digital Equipment Corporation) variety. Good-quality image display devices interfaced to the computer were also expensive. Nonetheless, many software packages were developed in this environment for EM purposes: SEMPER,[20,21] EM,[22] PIC,[23] SPIDER,[24] and IMAGIC,[25] as well as others.[26-29] Today, larger centers have access to supercomputers and parallel processors, and the revolution in microcomputing technology has made EM image processing feasible even in modest laboratories. Much software is also commercially available. It remains, however, a significant financial and time investment to set up an image analysis facility. Important factors to be considered are portability, compatibility, and lifetime of software or hardware. Some way of digitizing the electron image, i.e., converting the optical density variations on a film into an array of numbers manipulable by a computer, is also required.[30]

The examples presented in this paper were mainly worked on using the IMAGIC image processing system.[25] Since its creation, IMAGIC has been extensively developed by van Heel and colleagues, especially in aspects of pattern recognition. This package consists of programs written in standard FORTRAN 77, which has somehow remained a workhorse of scientific computing despite the worst insults of informatics theorists. IMAGIC currently runs under the VAX-VMS, UNIX, and MS-DOS operating systems in a number of laboratories worldwide. A master program called IMAGIC interfaces to both the operating system and to the user. Commands (level 3) are typed by the user in a natural, near-English style, and then the appropriate programs (level 2) are run to perform specific functions (image rotations and shifts, Fourier transformation, reconstruction, etc.). Commonly used subroutines for scientific and arithmetic calculations, input/output, and buffering are found in specific level 1 libraries. Hardware and operating system-dependent routines are strictly segregated in level 0 libraries to facilitate transfer to new environments. Extensive user interaction is made possible by an interface to an image display device. IMAGIC has recently become commercially available with extensive user support.

II. SINGLE-IMAGE MANIPULATION, ENHANCEMENT, AND RESTORATION

A. RESTORATION OF DISTORTED AND NOISY IMAGES

Simple image manipulations such as scaling, thresholding, magnification, and arithmetic image combination are readily programmed and found in most software packages. Other techniques such as histogram equalization can be used to expand the dynamic range of the image and, in principle, increase object contrast (Figures 1a and b).[31]

Image enhancement and restoration require considerably more sophisticated mathematics.[31,32] Basically, an image is considered to be a combination of an object function and degradative functions, which represent blurring and distortion characteristic of the data collection mechanism as well as noise. The purpose of the restoration process is to extract the object function. The enhanced image then (hopefully) gives a better representation of the specimen structure. A specific example in EM is defocus correction which is applied, e.g., in studying virus structure,[9] in correcting for specimen tilting,[33] or for astigmatism or other distortion.[34,35] The implementation of such electron image enhancement is relatively straightforward, although on needs to know certain instrumental parameters.[11]

There are many mathematical transformations that can be applied to signals, including images.[32] The most important of these is the Fourier transform, which can be calculated for any image and displayed as what is called a diffraction pattern (Figure 2). Such patterns can also be generated optically.[11] The information contained in the diffraction pattern is essentially the amount of image information at all spatial frequencies, from large (near the origin) to small (at the edge of the transform). The Fourier transform is also a good indicator of image quality. For example, astigmatism or specimen charging are readily apparent.

When the noise is degradative in nature, the problem or restoring the image becomes more difficult. In principle, knowledge of the blurring and noise functions and the way that they interact with the object function (sometimes linearly) allow one to perform a "deconvolution", called Wiener filtering in some circumstances.[36] For the most part, however, the different types of noise are nonlinear and nonadditive, and consequently the degradative process is difficult to model analytically. Although it is impossibe to make a clear-cut distinction between signal and noise, the relative contribution of noise to the image can be reduced by techniques such as selective spatial filtering,[11,37] median filtering,[31,38] and maximum entropy and its variations.[39-43]

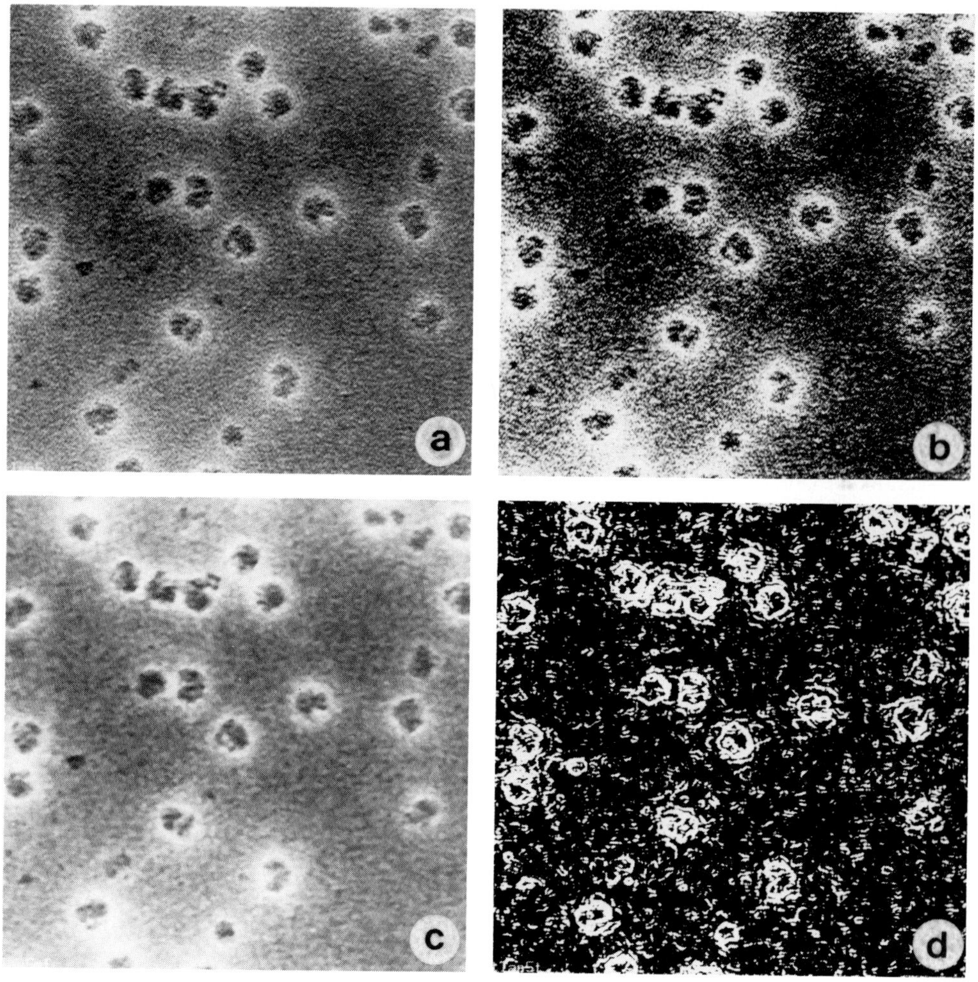

FIGURE 1. Various types of low- and high-level image processing with an electron micrograph of negatively stained ribosomes from the fungus *Thermomyces lanuginosus*.[38,77] (a) A portion of the original micrograph, digitized in 512 × 512 pixels and displayed on a computer monitor with 256 shades of gray; (b) the image in (a) after histogram equalization, a process which distributes pixels equally among all possible gray levels, thereby increasing image contrast; (c) the image in (a) after median filtering with a window size of 5 × 5 pixels, a process which smooths the noise due to the background carbon film and the negative stain which surrounds each complex; (d) the image in (c) after applying a hybrid edge-detection operator and binarizing the treated image. Edges of the ribosomes stand out clearly; ''edges'' due to fluctuations in the carbon support are also detected. (e) The image in (c) after more severe thresholding and binarization to distinguish ribosome edge points from those in the background; (f) individual ribosomes automatically detected by a component-labeling algorithm and bounded by boxes. Now specific characteristics of the complexes (sizes, shapes, and moments) can be calculated or individual biomolecules can be extracted for a single particle analysis.

B. SELECTIVE SPATIAL FILTERING

We shall illustrate selective spatial filtering using a micrograph of a eukaryotic transcription initiation complex (Figure 3).[44,45] The original images (not shown) differ in average brightness and some have low-level gradients of optical density, while fine detail is masked by a peppery, high spatial frequency background due primarily to the underlying support and the negative stain. These factors can be ameliorated by band-pass filtering, i.e., multiplying the Fourier transform of the image by a filter function which attenuates the low and high spatial frequency components to a small fraction of their original ''power'' (Figure 3). When a greater degree of filtering is applied, fine object details become blurred.

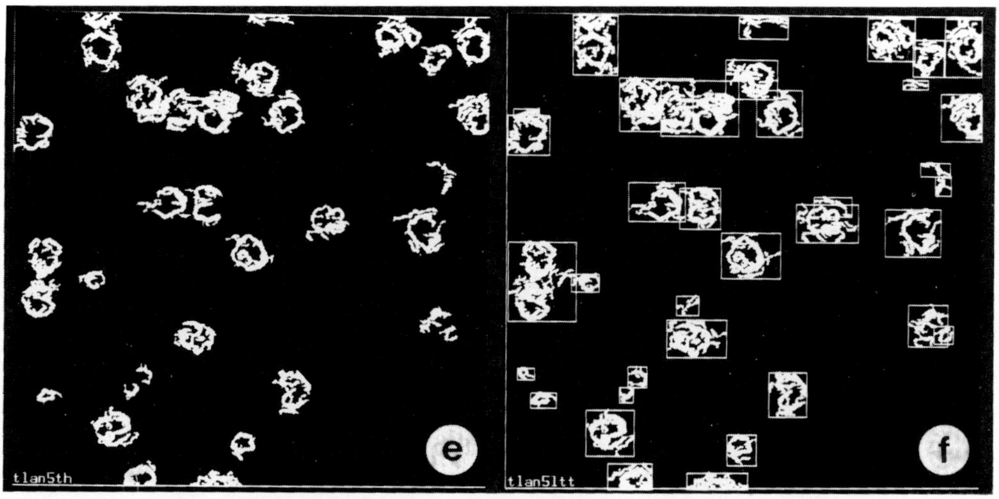

FIGURE 1e and f.

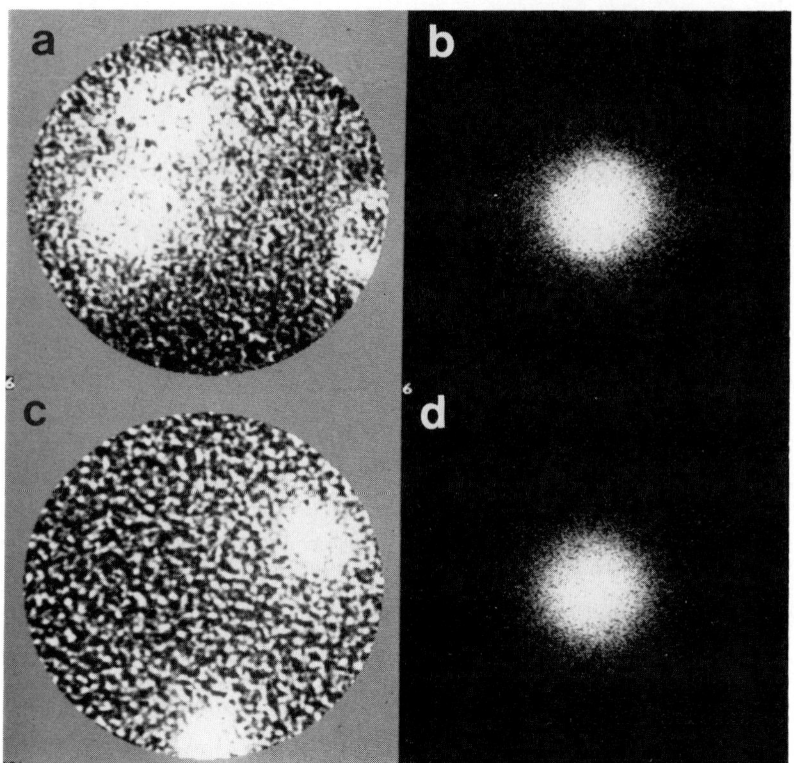

FIGURE 2. Two 512 × 512 portions of digitized micrographs (labeled 6 and 29) of negatively stained ribosomal subunits from the eubacterium *Thermus aquaticus* were treated with a circular mask (a, c) and Fourier transformed. The latter is represented by complex numbers whose magnitude is calculated to give a diffraction pattern which can be displayed like any other picture (b, d). The best-quality micrographs are those with a circularly symmetric diffraction pattern. Most of the information in these images is contained in the low to intermediate spatial frequencies. Distortion due to astigmatism or specimen charging would be evident by an elliptical or cross-shaped diffraction pattern.

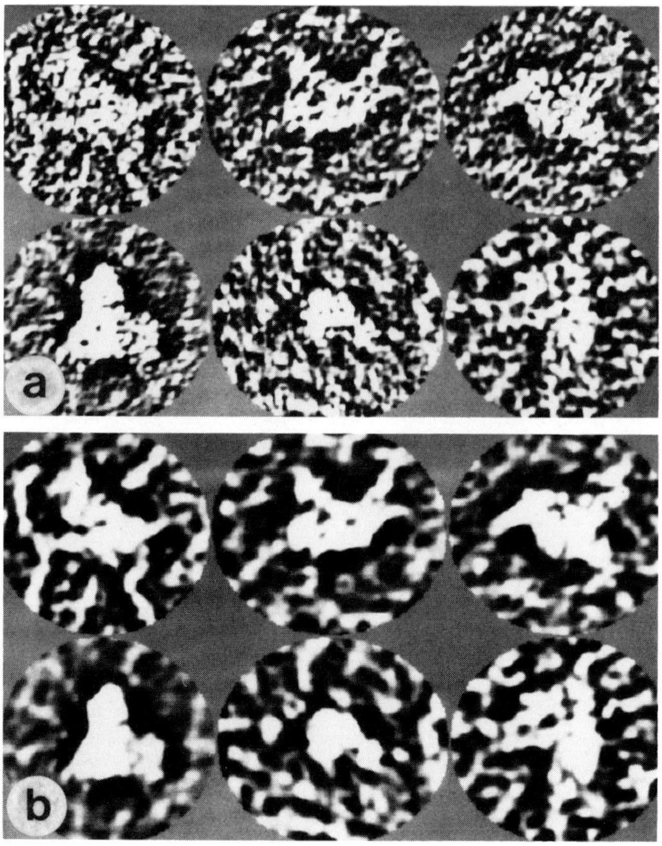

FIGURE 3. Selective spatial filtering of a set of six transcription initiation complexes present on the promoter of the murine ribosomal protein L32 gene.[45] The promoter was cloned into a p3AR2.8 plasmid. The nucleoprotein complex presents various projection views.[44] (a) Images of the complex after band-pass filtering to remove most low spatial frequency components; this results in an image which is ''floated'' to zero average density and has illumination gradients removed; (b) images of the complex after band-pass filtering with a more stringent high spatial frequency cutoff than in (a). The peppery texture of the background carbon support is now blurred out, and the shape of the complex is more readily apparent.

Selective spatial filtering (either ''sharpening'' or ''blurring'') can also be implemented directly in real space without recourse to Fourier transformation. A window of a certain size m × n pixels (e.g., 3 × 3, 5 × 5, etc.) is moved across the image and defines a neighborhood within it. At every position, the central point in the pixel in the neighborhood is replaced by a weighted average. For example, the 3 × 3 operator:

$$\begin{matrix} 1/18 & 1/9 & 1/18 \\ 1/9 & 1/3 & 1/9 \\ 1/18 & 1/9 & 1/18 \end{matrix} \qquad (1)$$

weights the density of the central pixel most heavily, followed by the pixels above, below, and to the sides, with the corner ones being weighted least. Sharpening filters have a high central peak with negative values at the periphery,[46,47] e.g.,

$$\begin{matrix} -1/8 & -1/4 & -1/8 \\ -1/4 & 5/2 & -1/4 \\ -1/8 & -1/4 & -1/8 \end{matrix} \qquad (2)$$

occasionally called a "Mexican hat" filter.

C. NONLINEAR FILTERING

The real-space blurring filters (e.g., Equation 1) smear out edges as well as suppress some noise. A nonlinear median filter has the advantageous properties of suppressing "spike" noise while retaining edges in the image[38] (Figure 1c) and can be defined formally as follows. Applying $MED_{m,n}$ (the median filter operator with a mask of size m × n) to the input image P gives the image $MED_{m,n} \cdot P$, where

$$\begin{aligned} MED_{m,n} \cdot P(x, y) &= \text{median } \{P(x - p, y - q)\} \\ \text{for } \lfloor (m - 1)/2 \rfloor &\leq p \leq \lceil (m + 1)/2 \rceil \\ \lfloor (n - 1)/2 \rfloor &\leq q \leq \lceil (n + 1)/2 \rceil \end{aligned} \qquad (3)$$

Either 3 × 3 or 5 × 5 masks are usually sufficient for electron micrographs (Figure 1c). Some very efficient implementations of the median filter which involve an overlapping moving window have been described (cf. references in Reference 38).

A recently developed nonlinear technique that has received great attention in many fields of data processing (e.g., X-ray crystallography, image restoration, and three-dimensional reconstruction) is maximum entropy, the basis of which has interesting parallels in statistical thermodynamics.[39,43] Briefly, an image can be considered as a set of probabilities p_i, where p_i is the ratio of the number of quanta (e.g., electrons) detected in pixel i to the total number of quanta detected in the entire image. The "entropy" of the image is defined as:

$$S = -\Sigma_j p_j \ln p_j \qquad (4)$$

which is simply the negative of what is called the "information content" as defined by Shannon.[31,39,43] The scene is treated as a collection of probabilities. Using statistical models of the noise and imposing a constraint on the distribution of the errors between the measured data and the reconstructed data, one can generate a set of images representing feasible object functions. The solution selected is the one with the maximum entropy. This choice is "safest" in that it yields the simplest possible result, containing the minimum of configurational information consistent with the recorded image data.

Farrow and Ottensmeyer have evaluated maximum entropy processing of dark-field-like images in which the primary sources of noise are the clutter of the support film and radiation-induced structural alterations.[40,41] A considerable increase in the signal-to-noise ratio was achieved with no loss in spatial resolution. However, it was also demonstrated that knowledge of the relative size of structure being imaged needed to be known, and that the final density values were biased either positively or negatively in different regions. These observations caution against imprudent application of maximum entropy reconstruction to electron images. Other image enhancement methods that have been suggested for use in EM include median window filtering,[42] maximum *a posteriori* reconstruction,[42] and "blind" deconvolution.[48] Each requires certain assumptions (e.g., image positivity) and has practical advantages and disadvantages which must be considered carefully before implementation.

D. EDGE DETECTION

Another example of low-level (iconic) image processing, preferably implemented after noise suppression, is edge detection. Some of the best computational edge detectors use adaptive or nonlinear operations.[49] We have implemented a hybrid edge detector where noise attenuation in one window and edge detection in a second window are combined in a single algorithm using the median operation.[38]

Here, we illustrate the operators using 3 × 3 masks. Define H_{1x} to be the 3 × 3 window operator:

$$\begin{matrix} -1.0 & 0.5 & 0.5 \\ -1.0 & 0.5 & 0.5 \\ -1.0 & 0.5 & 0.5 \end{matrix} \tag{5}$$

Similarly, H_{1y} is

$$\begin{matrix} 0.5 & 0.5 & 0.5 \\ 0.5 & 0.5 & 0.5 \\ -1.0 & -1.0 & -1.0 \end{matrix} \tag{6}$$

H_{2x} is a Prewitt operator:

$$\begin{matrix} 1 & 0 & -1 \\ 1 & 0 & -1 \\ 1 & 0 & -1 \end{matrix} \tag{7}$$

H_{2y} is the Prewitt operator in the orthogonal direction.

$$\begin{matrix} 1 & 1 & 1 \\ 0 & 0 & 0 \\ -1 & -1 & -1 \end{matrix} \tag{8}$$

H_3 gives the output 0 for all input values, i.e., is a 3 × 3 mask consisting entirely of zeroes.

The output of each filter H_i at position (x, y) is denoted by $g_i(x, y)$. The horizontal and vertical edge detector outputs, $g_x(x, y)$ and $g_y(x, y)$, respectively, are defined as follows:

$$g_x(x, y) = \text{median } \{1/3\ g_{1x}(x, y),\ 1/3\ g_{2x}(x, y),\ g_3(x, y)\} \tag{9}$$

$$g_y(x, y) = \text{median } \{1/3\ g_{1x}(x, y),\ 1/3\ g_{2y}(x, y),\ g_3(x, y)\} \tag{10}$$

The overall edge detector output g_{xy} can be defined as either

$$g_{xy}(x, y) = \sqrt{\{g_x^2(x, y) + g_y^2(x, y)\}} \tag{11}$$

or

$$g_{xy}(x, y) = ABS\{g_x(x, y)\} + ABS\{g_y(x, y)\} \quad (12)$$

The latter is computationally quicker, which is a consideration for large-scale, automated processing. On constant areas of the image, at least two of the subfilters should have as small an output signal as possible for good noise attenuation. On an edge, at least two of the subfilters should have a strong signal.

An example of edge detection is shown in Figure 1d. At this point, a simple thresholding method with a predetermined value is able to distinguish particle edge points from non-edge points caused by fluctuations in the background carbon film (Figure 1e). This is useful for particle extraction, which is a particular kind of image segmentation (Figure 1f).[38,50]

A reduction of computation time in image manipulation algorithms can usually be achieved at the cost of a programmer's efforts in devising and implementing more efficient algorithms. Programs can usually be rewritten to run very quickly on a parallel processing architecture.[51] IMAGIC itself is structured so that fast array or parallel processors can be used if they are available.

In practice in EM, the value of single-image restoration approaches exemplified here lies in accentuating certain features of electron images not for visual interpretation per se, but in preparation for further computational manipulations. These include image extraction and segmentation, alignment, sorting, and reconstruction.

III. PATTERN RECOGNITION OF MULTIPLE-IMAGE ENSEMBLES

The nature of EM is such that both the object and the image are degraded. We begin with the premise that all biomolecules from a biochemically pure preparation represent the same, average three-dimensional structure. It is thus more rational to consider a set of images of a large, biochemically homogeneous population of particulate biomolecules rather than each image individually. In a set of micrographs of numerous macromolecular complexes, the "noise" will vary in each; also, different individual particles lie in distinct orientations upon the support film and at diverse positions and angles within the image plane. Exhaustive visual surveys can be performed to sort images into common views and to build up a visual picture of the three-dimensional structure from these. The ingenious realization of averaging by photographic superposition[1] or optical filtering[52] of dark-field electron micrographs has enabled the visualization of single atoms and the elucidation of a number of small protein structures. The advent of computer image processing in the framework of "single-particle electron crystallography" has made this task easier and more flexible.[14,15,37,53-57] Many more images can be analyzed with increased objectivity, and quantification and a greater variety of image manipulations are made possible.

In typical EM preparations of single particles, a finite number of views are presented, i.e., the object has a number of preferred positions in which it lies on the support film. Much work has focused recently on developing algorithms for sorting these characteristic projections from a large (a few 100 to a few 1000) population of images of individual biomolecules from a biochemically pure sample.[54,56,58-62] In particular, a multivariate statistical technique called correspondence analysis is used to decompose a set of images into its principal components.[37,57] This then facilitates the next stage of the analysis, namely, the grouping together of similar images so that they can be summed, giving a "class average" with a better signal-to-noise ratio and improved reproducible spatial resolution than any of the original images.[53,63,64]

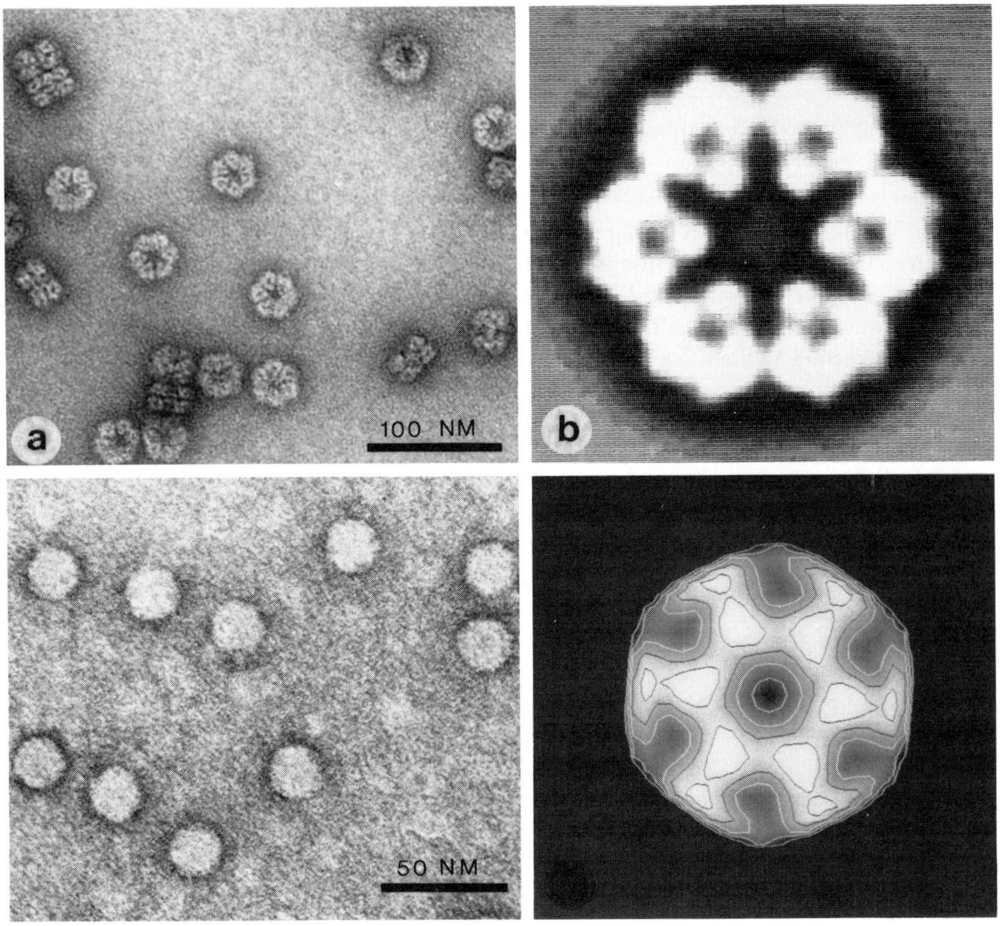

FIGURE 4. Alignment of projections by cross-correlation. (a) Hexagonal (top view) projections from worm (*Lumbricus terrestris*) hemoglobin, an extremely large biomolecule[75] having very clear features. Any well-preserved, single noisy top-view projection will perfectly align a data set of selected top-view projections. These can be added to give sums with an enhanced signal-to-noise ratio, as in (b). When this sum is taken as a next reference, the new rotational and translational shifts do not deviate significantly from those of the first alignment. In general, most objects give projections where subunit features are not as well recognizable as in hemoglobin, e.g., spherical viruses.[97] This means that the first (noisy) references need to be selected with care and that the alignments have to be repeated iteratively with different references, or that alternate methods making use of anticipated symmetry[85] need to be applied. In cases of quite undefined projections of objects smaller than 10 to 15 nm, the features of sums of aligned projections will always look somewhat like the reference (reference dependency).[15,65]

A. IMAGE ALIGNMENT BY CROSS-CORRELATION

The first step in comparing images of projections of biomolecules is to bring them into register in the plane: the alignment process (Figure 4). However, an alignment requires a reference image, which choice influences the outcome.[65] This leads to a circular ''bootstrapping'' problem: the ideal references are the best representations of the various projections of the macromolecular structure, which initially is unknown! An extensive visual survey of the data set should suffice to familiarize one with the highly repeating views, and representative images extracted. These are then arithmetically processed (contoured, thresholded, and centered) to remove the influence of the surrounding stain, and used as references. Each image in the data set is compared with each reference in turn, and rotational and translational cross-correlation functions are computed to determine the best angle and shifts to align the two images.[37,57] The parameters corresponding to the highest correlation coefficient are chosen. This iterative alignment process readies the data for the next step — comparison of different images.

There are alternative strategies to cross-correlation techniques for image registration. These include visual superposition using a high-quality display device for image subtraction in electron energy loss spectroscopy[66] and "robust" iterative algorithms in which extremes of some goodness of fit functions are found.[67-69] The former method is tedious, but surprisingly precise. In practice, we have found the correlation coefficient to be a poor indicator of image similarity and of the quality of the alignment in general. Unfortunately, we have no other alternative that can be computed quickly and hence be practical for large data sets.

B. MULTIVARIATE STATISTICAL ANALYSIS

Once an ensemble of micrographs has been suitably aligned, they can be compared in some numerical way. Pattern recognition approaches, namely, correspondence analysis and classification, were devised as tools to handle such data sets.[70] A set of "T" images digitized in "P" pixels can be envisaged as a cloud of T points in a P-dimensional space, where each axis represents the gray value of a particular pixel. A multivariate statistical technique called correspondence analysis effectively rotates and shifts this P-coordinate system. This is performed computationally by an eigenvalue-eigenvector decomposition:[58]

$$\mathbf{X}^T\mathbf{NXMu}_q = \lambda_q\mathbf{u}_q \tag{13}$$

under the orthonormalization constraint:

$$\mathbf{u}_q^T\mathbf{Mu}_{q'} = \boldsymbol{\delta}_{q,q'} \tag{14}$$

with

$$\mathbf{M} = \mathrm{diag}(m_j),\ m_j = 1/\Sigma_i x_{ij} \tag{15}$$

$$\mathbf{N} = \mathrm{diag}(n_i),\ n_i = 1/\Sigma_j x_{ij} \tag{16}$$

In these equations, X is the T $\times$ P matrix representing the full set of T images, each of P pixels (row i contains the densities of all P pixels of image i); u_q is the new unit vector adapted to the shape of the image cloud; M and N are metrics of the image and pixel hyperspaces, respectively; and λ_q is the q-th eigenvalue, ordered such that:

$$\lambda_1 > \lambda_2 > \lambda_3 > \ldots > \lambda_q > \ldots > \lambda_P$$

In the new representation of the hyperspace, the axes are ordered by decreasing eigenvalue, with the first axis representing the largest proportion of interimage variance, the second axis representing the largest proportion of remaining interimage variance, etc. The greatest sources of interimage variance are (ideally) due to orientational differences, but significant contributions arise from conformational flexibility of appendages, structural deformations, radiation-induced alterations, and the statistical distribution of negative stain around the particle.[14,15,37,57,62] Noise due to quantum imaging statistics and irregularities in the background support film is also a source of interimage variance, but is represented more by the higher-order axes defined by the correspondence analysis procedure.

By considering only the most significant axes, the set of images can be represented in a hyperspace of highly reduced dimensionality. It is now quicker to compute distances between images, and better since some of the noise is being ignored. The correspondence analysis step can thus be considered a special form of feature extraction or principal components analysis.[58]

With the generalized metrics used in correspondence analysis, the squared distance in hyperspace between two images is

$$d^2_{ii'} = \Sigma_j m_j (n_i x_{ij} - n_{i'} x_{i'j})^2 \quad (17)$$

The vector c_q containing all images relative to the new basis vector $\mathbf{u}_q$ is calculated by

$$\mathbf{c}_q = \mathbf{NXMu}_q \quad (18)$$

We can equivalently compute the Euclidean metric:

$$d^2_{ii'} = \Sigma_q (c_{iq} - c_{i'q})^2 \quad (19)$$

which is identical to the χ^2 distance if summed over all eigenvectors. Regardless of the distance used for comparing the data originally, a simple Euclidean metric can be used for the classification step.[54,58]

C. IMAGE CLUSTERING

A key to acquiring knowledge about an image data set is its reduction into manageable, coherent units, i.e., classification. Especially with large sets, this process may be difficult since the data may not be arranged such that (any) relationships are visible. The problem is compounded by extraneous noise. Some image data may be similar and obviously form groups. Other images may be difficult to categorize; these are ''outliers'' which obscure the natural boundaries between the clusters. The groupings may be distinct when the particles lie in a few preferential positions,[55] or may be slices of quasicontinuous orientational distributions.[15,61]

In general, the results of a classification depend on the choice of characteristics with which to represent entities and the measures used to compare similarities. One wishes to make as few assumptions about the data set as possible. An exploratory data analysis can be achieved by feature extraction, making the classification step less labor intensive since it is performed on fewer prominent data points. The multivariate statistical analysis described above is a reliable and efficient method for organizing the information extraction process. The purpose of clustering algorithms is to find the intrinsic structure of a data cloud, and initially some concept of an optimum partition must be defined. A ''cluster'' can be defined from different perspectives:

1. A set of entities which are alike, implying that those from different clusters are not alike — This definition emphasizes the similarities between objects in the data set as the basis for grouping.
2. An aggregation of points such that the distance between any two points in the cluster is less than the distance from any point in the cluster to any point outside — This is based on a geometric interpretation of the data points.
3. A connected region of a P-dimensional space containing a relatively high density of points, separated from other clusters by areas with a relatively low density of data points — This definition interprets clustering from a probabilistic perspective. Classification programs thus differ significantly because they are governed by the programmer's idea of what a cluster is![15,58,61,62]

There are two main types of clustering algorithms: hierarchical and partitional. Hierarchical methods result in a tree-structured output called a dendogram, which reflects hierarchically the similarities between data points, and are commonly employed in biological taxonomy. Partitional clustering algorithms are more frequently used in image processing. These techniques group data patterns by labeling each according to class membership, e.g., based on nearest-neighbor criteria. Regardless of which approach is used, the tendency of any algorithm to form artificial clusters and the validity of clusters found must be evaluated.

The different kinds of clustering algorithms that have been employed in electron image analysis include K-means, dynamic clouds, hierarchical ascendant classification (HAC), and favorite hybrids thereof.[15,58,61,62] The input required by these programs can be either a proximity matrix whose entries are the similarity measures between the data points, or a pattern matrix which is represented by rows identifying the patterns of the data and by columns which represent the features (the feature space being a set of orthogonal axes). Proximity matrices can be generated from pattern matrices by use of distance techniques; ordination performs the reverse transformation.

IMAGIC uses an automatic hierarchical ascendant (or agglomerative) classification (HAC) in which similar images are initially merged together to form classes; these are themselves grouped together to form larger classes. The classification procedure is usually stopped when a predetermined number (K) of classes (of the order of one-twentieth of the original number of images) is obtained. This cutoff, or proximity threshold, represents a tolerance for which dissimilar images will be grouped. A postprocessing step then moves class members around to optimize the partition still further.[54,58] Summing the aligned members of each class gives an "average" image with an enhanced signal-to-noise ratio (Figures 4 and 5).[53,71] These are then interpreted as representing different characteristic projections of the macromolecular complex under investigation.

The objective of the HAC is defined to be the minimization of the total sums of intraclass variances of K classes.[54] A naive approach of searching all possible combinations of clusters is far beyond the capabilities of any computer, even in a feature space of low dimensionality. The HAC algorithm represents one viable means of approximating the optimal partition of the data. The HAC relies on a principle due originally to Huygens which states that the sum of the total interclass and intraclass variances is a constant. In the current implementation of HAC, the added intraclass variance is chosen to be minimal at each step of merging. This results in an algorithm whose time and memory requirements increase only linearly according to the number of images.[58]

The correspondence analysis and classification approaches described above have been used by various groups in diverse ways.[15,55,58,61] Regardless of the form of implementation, these computational tools have proved invaluable in elucidating the structures of various macromolecular complexes, especially ribosomes.[37,55,57,61,72-77] One measure of improvement in image quality due to averaging is the reproducible spatial resolution (RSS) between two class sums.[63,64,78,79] In the case of negatively stained preparations of 30S ribosomal subunits from *Bacillus stearothermophilus,* an RSS of 1.7 nm was achieved.[55] The RSS between any two members of a class is only about 9.0 nm.[37,77]

In the remainder of this section on single-particle averaging, we address specific aspects of refining these techniques: the weighting of eigenvectors used in classification, position invariant classification, and artificial intelligence approaches in EM image analysis.

D. PROBABILISTIC INFERENCE AND EVENT COVERING

Regardless of which clustering approach one takes, the classification step is made computationally feasible only by the prior extraction by correspondence analysis of the (typically 8 to 24) principal components describing most of the interimage variation in the aligned population.[37,57] Some of these features may represent true structural variations (such as different orientations); others may be irrelevant.

Schiske[80] defined a limit for the highest eigenvalue that could be generated entirely by

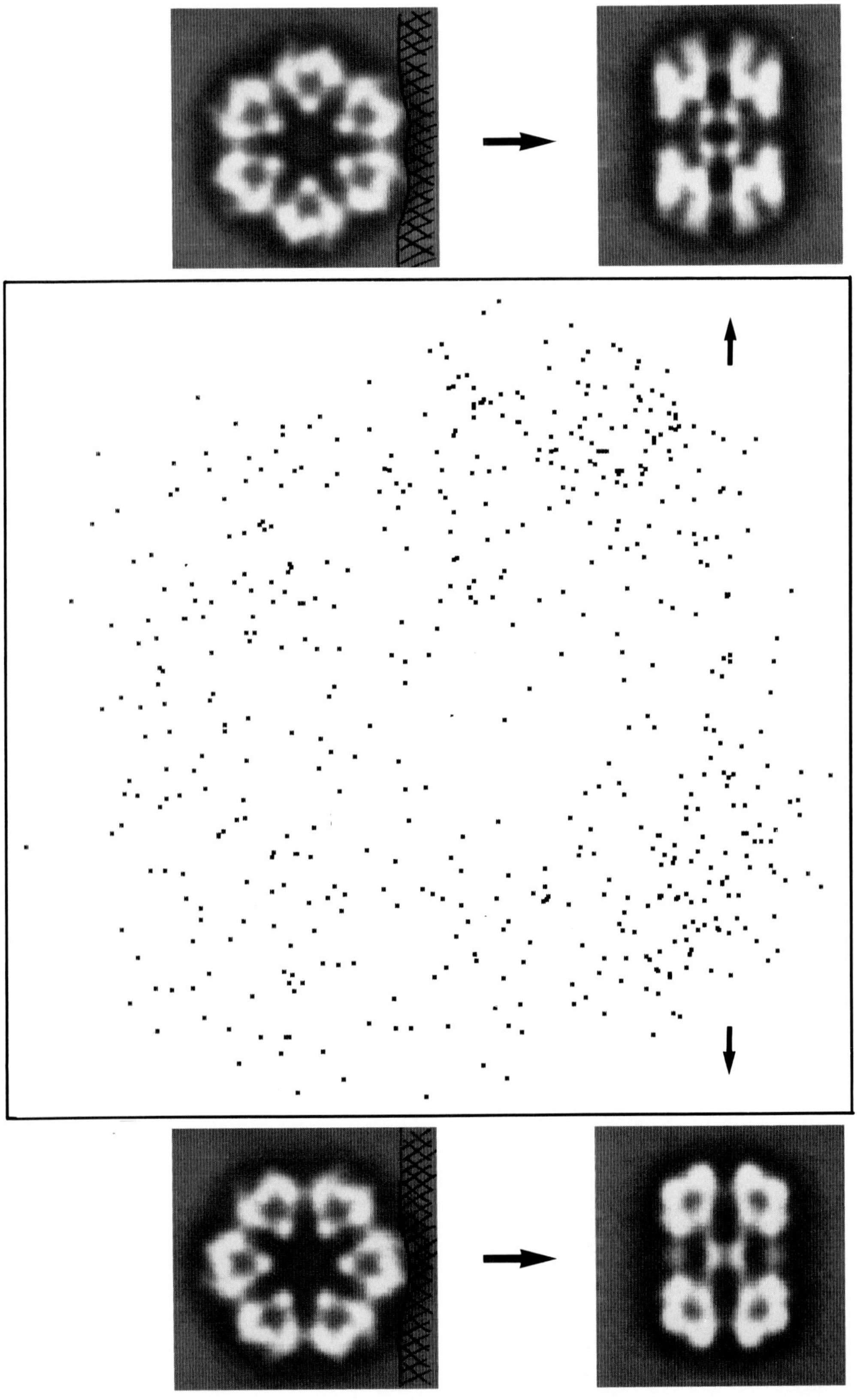

FIGURE 5. Example of a portion of a multivariate statistical analysis map and its interpretation. The map shows 546 side-view projections of *Lumbricus terrestris* hemoglobin molecules[75] plotted on a two-dimensional space defined by the first two eigenimages representing most of the interimage variance. Each point on this graph represents one projection, and the distances between the various points are an indication of similarity with respect to the major features defined by these two axes. There are two clusters in the lower half of the map: on the left and on the right. They represent the two most common types of side views, and result from two possible stable positions that the hexameric molecules can have on the carbon support, as indicated. Generally, separated clusters on MSA maps indicate differences in projections due to different molecular positions on the carbon support.

additive Gaussian noise of variance V:

$$\lambda_{noise} < V[\sqrt{T} + \sqrt{P}]^2 \tag{20}$$

where we have T images of P pixels. However, the noise in electron micrographs is not purely Poisson in nature, and thus a degree of uncertainty is introduced in deciding which eigenimages are significant and should be used.

It is possible to apply some sort of weighting scheme to the image coordinates $\mathbf{c}_{iq}$ relative to the eigenvector $\mathbf{u}_q$. For example, a weighting by $1/\sqrt{\lambda_q}$ leads to an equal appreciation of all significant eigenvectors irrespective of the total power described.[54] Frank and colleagues perform interpretational weightings, whereby they choose certain eigenvectors as being more important in describing orientational differences than others.[15,61,81] However, it may not be clear in other circumstances how to perform this. In practice, we[37,55,77] have been calculating and using the first 6 to 24 factorial coordinates, all weighted equally, in computing interimage distances for use in the HAC. The choice of number of coordinates has only small effects on the final classes obtained, e.g., different images end up in different classes which still represent the same projection. Until now, however, we have had no fundamentally sound way of selecting the most relevant eigenvectors.

Here, we mention some of our preliminary ideas on another strategy for choosing the best principal components upon which to base a clustering. Wong and Chiu have been developing the theory and applications of procedures that they call probabilistic inference and event covering.[82,83] To understand these, it is first necessary to define the concept of interdependency of events. Say there is a variable pair X_j and X_k with possible discrete outcomes $x_j = a_{jt}$, $t \in [1, L_j]$, and $x_k = a_{ks}$, $s \in [1, L_k]$. The *observed* marginal frequencies of these outcomes are $obs(a_{jt})$ and $obs(a_{ks})$, respectively. The *expected* frequency of the cooccurrence of these two events can be calculated as:

$$\exp(a_{ks}, a_{jt}) = obs(a_{ks})obs(a_{jt})/M(X_j, X_k) \tag{21}$$

where $M(X_j, X_k)$ is the total number of observations in which the outcomes of both X_j and X_k are observed. The degree that an observed frequency deviates from the expected frequency is

$$D_k^j = \sum_{t=1}^{L_j} (obs(a_{ks}, a_{jt}) - \exp(a_{ks}, a_{jt}))^2/\exp(a_{ks}, a_{jt}) \tag{22}$$

which has been shown to be asymptotically χ^2 with $(L_j - 1)$ degrees of freedom.[82,83] If this statistic D_k^j is greater than $\chi^2(L_j - 1, \alpha)$, one can state that the two events are interdependent at the α-level of significance.

In the single-particle EM application, the events whose interdependence is being examined are the coordinates of individual images in principal features space, and their belonging to a particular class. In an experiment[98] involving images of ribosomes from *Thermomyces lanuginosus*,[77] their coordinates were scaled and discretized using a maximum entropy approach.[84] Interestingly, it was discovered that the lower-order factors 4 and 8 (of the first 12 originally used for clustering) were not significant. A reclassification was then performed with these two coordinates weighted to zero. The results differed only very slightly from those of the classification without weighting, with some images falling in different classes representing a common projection view. This was not a profound improvement and it was not possible to assign meaning to these particular eigenvectors with this data set, primarily because of the globular nature of the ribosomes and the relatively low resolution of the images. Nonetheless, these preliminary results are intriguing and the approach the only rational one that has hitherto been proposed.

E. ORIENTATION-INVARIANT CLASSIFICATION

The problem of clustering is more difficult to achieve in situations where particles can have a range of orientations. From the primary images, one must calculate secondary images which have the properties of rotational and/or translational invariance.[85] One of these is the power spectrum, which is invariant to translations of a bounded object within an image (Figure 2). Another is the autocorrelation function, which has the advantage of a greater statistical homogeneity of pixel values than the power spectrum.[60] These alternative representations of the biomolecular images allow one to avoid initially the problem of aligning them by cross-correlation.

IV. ARTIFICIAL INTELLIGENCE IN ELECTRON IMAGE ANALYSIS

The single-particle analysis is computationally intensive and requires constant human intervention between individual steps. In principle, the expertise of a trained operator can be codified to automate the analysis further, thereby assisting nonspecialists. We have proposed the aim of developing an expert system for analysis of electron microscopical images.[38,86] To satisfy the unwritten requirement of a suitable acronym, we have chosen ELMAR (electron micrograph analyzer). ELMAR is written in the "C" programming language and makes extensive use of the windowing environments available on the newest generations of graphics workstations, a significant difference of implementation compared to previous systems. As a first step toward intelligent programming, we have proposed a new solution to the problem of extracting images of individual biomolecules from large electron micrographs. The process consists of (1) noise suppression and edge detection, (2) component labeling and feature computation to bridge the gap between the iconic (low-level) processing and the final phase of symbolic (high-level) processing, and (3) deriving simple symbolic objects (bounding boxes) from the edges, which are manipulated in a decision-making process. The main phases of the algorithm are summarized in Figure 1.[38] The hierarchical nature of the analysis embodies a reduction in the amount of data and a change in its nature. Initially, thousands of pixels of continuous gray levels must be dealt with. After component labeling, there are less than 100 bounding boxes whose objective manipulation can easily be defined and articulated by an expert. This can thus serve as a basis for applying artificial intelligence methodologies to analysis of electron micrographs.[70]

Recently, Russ[87] has described a system that embodies "learning" in the fashion of contextual object recognition. Statistical tools are employed and a simple knowledge structure defined. Given a "learning" set of images of different objects, which can represent here different projections of a biomolecule, certain characteristics are measured. These include size (area, perimeter, length, breadth, and width), shape (form factor, roundness, and aspect ratio), brightness (average brightness, contrast, and texture), and position. A multiple regression method is used to select the important parameters for discrimination of clusters defined by the human "expert". So-called "context lines" connecting the centers of the clusters are determined and then fixed. This is the first stage of the contextual learning process, i.e., studying the training population in which the identification is performed. The second step is to identify clusters to which unknowns belong by using Bayesian statistics. With a test set of images of nuts, the accuracy of discrimination ranged from 96 to 99%. Russ's method[87] is a significant complement to the component labeling that we have implemented.[38]

V. IMAGE ANALYSIS OF TWO-DIMENSIONAL CRYSTALS

A. REGULAR VS. IRREGULAR ASSEMBLIES

An underlying theme in EM image analysis is the use of the degree of order in the specimen to assist in picture restoration. With images of isolated biomolecules, the special pattern recognition techniques discussed above can be said to snatch "order" from the jaws of an apparently disordered collection. In this section, we describe computational techniques that are applied to micrographs of *regular* two-dimensional assemblies.[10,13,18,88-92]

As stated previously, there are many serious factors limiting the resolution of images of biomolecules. The most important one is radiation damage caused by the electron beam interacting with the fragile biological specimen.[4,6] Another one is the application of heavy metal stains, such as uranyl acetate, as contrast agents. By omission of staining agents and by recording images of the objects at an extremely low electron dose, the structural detail of biomolecules can, in principle, be preserved with a resolution well under 1 nm.[8,10,90] Under these conditions, the contrast and signal-to-noise ratio of biomolecular images are extremely low. The projections are almost lost in the background noise and, as a consequence, their positions cannot be determined accurately. If biomolecules have been crystallized into two-dimensional monolayers, however, the repeating distances within the crystal can be very precisely calculated. Therefore, the position and orientation of the individual biomolecules can be determined and their projections can be summed even though they are barely visible.

Image analysis is especially fruitful if the objective lens of the microscope and the sample have been cooled down to liquid helium temperature, i.e., cryo-electron microscopy.[90] Keeping the sample so cold minimizes radiation damage by a factor of 10 to 20.[7] Moreover, such cooling of the objective lens makes it superconductive, thus stabilizing the magnetic lens field. Recently, cryo-electron microscopy has been used to resolve the structure of the purple membrane protein from *Halobacterium halobium* with a resolution sufficiently high to define an atomic model where the folding of the polypeptide chain can be seen.[10] *In vivo*, bacteriorhodopsin pumps protons across the membrane. By resolving the structure, it has become clear where and how the various amino acid side chains involved in this process are located. The successful elucidation of the structure of purple membrane is a tremendous achievement that has been hitherto unparalleled in electron microscopy. The remarkable stability of this membrane protein has been an important factor since not all protein crystals are stable enough under the preparation conditions used for bacteriorhodopsin. However, several other membrane proteins have already been imaged with a resolution well under 1 nm in projection,[93] which indicates that structure determination by high-resolution electron microscopy is a very relevant technique. Such studies need unique image analysis programs since the recorded micrographs have to be corrected for some singular effects which are irrelevant at lower resolution.[10,88,90] Although the procedures for analyzing such high-resolution electron micrographs have been worked out, the two-dimensional crystallization process is not yet a routine operation. Presently, it is only in favorable cases that the biomolecules are stable enough and/or the crystals are well ordered and large enough to solve the structure with a resolution of better than 1 nm.[94] However, structure determination is a gradual process in which a lower-resolution analysis of negatively stained crystals is usually the first step to be taken. In this section, we describe briefly three important procedures in this process: (1) the determination of the packing of the crystal, (2) Fourier filtering, and (3) extracting structural information from the crystal by the so-called correlation averaging technique.

B. DETERMINATION OF CRYSTAL PACKING

The first stage is the calculation of the Fourier transform of a digitized crystal. Two-dimensional Fourier transformation is a mathematical operation that results in a diffraction pattern which can be viewed like any other image. The way the spots of the diffraction pattern are arranged tells us many things about the crystal (Figure 6). The lattice parameters can be determined from the distances between rows of spots. The crystal packing can be verified from the amplitudes and phases of the spots. This is very important: before we can

FIGURE 6. Analysis of a two-dimensional crystal of NADH dehydrogenase. The enzyme NADH dehydrogenase from beef heart (Complex I) can be crystallized into large two-dimensional sheets.[89] (A) A fragment of a digitized crystal image with a size of 128 × 128 pixels, equivalent to 61.5 × 61.5 nm, contains about 130 enzyme molecules. Noise limits the direct observation of the packing of these molecules within the crystal. (B) The result of correlation averaging. The digitized crystal image was divided into 60 nonoverlapping fragments, each of 256 × 256 pixels, which were aligned and summed. The central 128 × 128-pixel fragment of the sum is shown. A considerable improvement in the signal-to-noise ratio is evident when compared to the image in (A). (C) A Fourier transform of the sum in (B) results in an image of the crystal lattice in reciprocal (Fourier) space. The numbers and positions of the bright diffraction spots, as well as their intensity and sharpness, give us information about the crystal packing and the resolution of information present in the image. There are six to seven rows of spots on a square lattice. The most intense spots are on the first, second, and fourth rows. The distance from the center of the transform to the fist row gives the repeat of the crystal, 15.3 nm. The outer spots on the sixth and seventh rows are related to structural detail of about 2 nm in size. On the Fourier transform of 1536 × 1536 pixels of the whole crystal (not shown), most of these outer spots were not visible. This means that correlation averaging improved the resolution by correcting for small lattice imperfections. For instance, the average rotational correction of each of the 60 fragments bringing them into the exact position of the reference was 0.5°. (D) The repeating motif is called the unit cells of the crystal. Here, it measures 15.3 × 15.3 nm, indicated by the black lines forming a square. Usually, the unit cell is drawn with respect to special points of symmetry that the crystals might have. The NADH dehydrogenase crystals are highly symmetrical: there is four-fold rotational symmetry about the center of each unit cell, which contains four dimers, or eight individual molecules. Further improvement of the signal-to-noise ratio can be obtained by imposing the known symmetry on the sum obtained by correlation averaging, as shown in (D).

average the projected images of biomolecules forming the crystal, we must determine how they are arranged. In theory, there are 17 possibilities for the packing, called plane groups.[13,95] In the simplest cases, the repeating motif may consist of just one molecule. It is also possible that the crystal has a high form of symmetry, as in our example of Figure 6. In this case, the repeating motif consists of eight molecules in a special square arrangement with up- and down-oriented molecules.

C. FOURIER PEAK FILTERING

The simplest way to analyze images of two-dimensional crystals is by Fourier peak filtering. In this stage, a Fourier transform is first determined from the largest contiguous part of a crystal. Then, a mask is constructed with holes at the positions of peak diffraction spots, which contain the crystal information. A reverse Fourier transformation from the masked transform reconstructs the crystal image, but with the noise between the holes removed. The Fourier peak filtering is optimal when the locations of the spots have been precisely determined by least-squares fitting; from these positions, the reciprocal lattice vectors can be calculated. Even extremely noisy images can be treated in this way, although the best results are obtained from perfectly ordered crystals.

D. CORRELATION AVERAGING

Lattice imperfections and long-range distortions of a crystal significantly reduce the resolution of the Fourier-filtered image. These effects can be corrected by the so-called correlation averaging technique, developed by Saxton and Baumeister[92,96] and illustrated in Figure 6. A large crystal is divided into small parts, usually containing 10 to 20 biomolecules. The fragments are all shifted rotationally and translationally by correlation methods and summed thereafter. The first reference for this alignment is made from the center of a Fourier peak-filtered crystal. The (small) rotational and translational shifts correct for lattice imperfections very well, and all the biomolecular projections from a crystal can then be summed to yield a better-quality average. Moreover, information from several crystals can be easily combined. By correlation averaging, thousands of projections can be analyzed in the same time as hundreds using single-particle analysis. For small- to medium-sized objects (<500 kDa), correlation averaging is superior to single-particle averaging; the disadvantage is that usually only one type of projection view is obtained.

VI. CONCLUDING COMMENTS

The revolutionary advances in information technology of the past 2 decades have had profound effects in all scientific disciplines. In particular, the availability of increasingly more powerful computers, and the concomitant development of sophisticated software, have rendered electron image analysis an indispensable adjunct to structural studies of biological macromolecules and their complexes.

ACKNOWLEDGMENTS

This work was supported by a grant from the Natural Sciences and Engineering Research Council of Canada to G.H. and by a fellowship from the Royal Netherlands Academy of Arts and Sciences to E.J.B. We are grateful to Dr. Marin van Heel for his insight, generous assistance, and instruction in image processing techniques and also to our current collaborators, Drs. Wilko Keegstra, Amelia Fong-Lochovsky, and David Chiu. The micrographs of ribosomes from thermophiles and of the eukaryotic transcription initiation complex were made in G.H.'s laboratory by Derrick Flannigan and Elizabeth Welsh, respectively.

REFERENCES

1. **Ottensmeyer, F. P., Schmidt, E. E., Jack, T., and Powell, N.,** Molecular architecture: the optical treatment of dark field electron micrographs of atoms, *J. Ultrastuct. Res.*, 40, 546, 1972.
2. **Valentine, R. C., Shapiro, B. M., and Stadmann, E. R.,** Regulation of glutamine synthetase. XII. Electron microscopy of the enzyme from *Escherichia coli, Biochemistry,* 7, 2143, 1968.
3. **Steven, A. C. and Navia, M. A.,** Specificity of stain distribution in electron micrographs of protein molecules contrasted with uranyl acetate, *J. Microsc.*, 128, 145, 1982.
4. **Ottensmeyer, F. P., Bazett-Jones, D. P., Rust, H. P., Weiss, K., Zemlin, F., and Engel, A.,** Radiation exposure and recognition of electron microscopic images of protamine at high resolution, *Ultramicroscopy,* 3, 191, 1978.
5. **Egerton, R. G.,** Measurement of radiation damage by electron energy-loss spectroscopy, *J. Microsc,* 118, 389, 1980.
6. **Boudet, A. and Kubin, L. P.,** The limitations to resolution in the observation of radiation-sensitive specimens by electron microscopy, *Ultramicroscopy,* 8, 409, 1982.
7. **Lamvik, M. K., Kopf, D. A., and Robertson, J. D.,** Radiation damage in L-valine at liquid helium temperature, *Nature,* 301, 332, 1983.
8. **Rachel, R., Jakubowski, U., and Baumeister, W.,** Electron microscopy of unstained, freeze-dried macromolecular assemblies, *J. Microsc.*, 141, 179, 1985.
9. **Vogel, R. H., Provencher, S. W., von Bonsdorff, C.-H., Adrian, M., and Dubochet, J.,** Envelope structure of Semliki Forest virus reconstructed from cryo-electron micrographs, *Nature,* 320, 533, 1986.
10. **Henderson, R., Baldwin, J. M., Ceska, T. A., Zemlin, F., Beckmann, E., and Downing, K. H.,** Model for the structure of bacteriorhodopsin based on high-resolution electron cryomicroscopy, *J. Mol. Biol.*, 213, 899, 1990.
11. **Misell, D. L.,** Image analysis, enhancement, and interpretation, in *Practical Methods in Electron Microscopy,* Vol. 7, Glauert, A. M., Ed., Elsevier/North-Holland, Amsterdam, 1978.
12. **Aebi, U., Fowler, W. E., Buhle, E. L., and Smith, P. R.,** Electron microscopy and image processing applied to the study of protein structure and protein-protein interactions, *J. Ultrastruct. Res.*, 88, 143, 1984.
13. **Engelhardt, H.,** Correlation averaging and 3-D reconstruction of 2-D crystalline membranes and macromolecules, *Methods Microbiol.*, 20, 357, 1988.
14. **Frank, J.,** Image analysis of single macromolecules, *Electron Microsc. Rev.*, 2, 53, 1989.
15. **Frank, J.,** Classification of macromolecular assemblies studied as "single particles", *Q. Rev. Biophys.*, 23, 281, 1990.
16. **Hawkes, P. W.,** Subjects in search of an author: Speculations on the future of electron image processing, *J. Electron Microsc. Tech.*, 9, 421, 1988.
17. **Slump, C. H. and Ferwerda, H. A.,** Statistical aspects of image handling in low-dose electron microscopy of biological material, *Adv. Electr. Electron Phys.*, 66, 201, 1986.
18. **Stewart, M.,** Introduction to the computer image processing of electron micrographs of two-dimensionally ordered biological structures, *J. Electron Microsc. Tech.*, 9, 301, 1988.
19. **Stewart, M.,** Computer image processing of electron micrographs of biological structures with helical symmetry, *J. Electron Microsc. Tech.*, 9, 325, 1988.
20. **Saxton, W. O., Pitt, T. J., and Horner, M.,** Digital image processing: the SEMPER system, *Ultramicroscopy,* 4, 343, 1979.
21. **Saxton, W. O. and Koch, T. L.,** Interactive image processing with an off-line minicomputer: organization, performance and application, *J. Microsc.*, 127, 69, 1982.
22. **Smith, P. R.,** An integrated set of computer programs for processing electron micrographs of biological structures, *Ultramicroscopy,* 3, 153, 1978.
23. **Hegerl, R. M. and Altbauer, A.,** The "EM" program system, *Ultramicroscopy,* 9, 109, 1982.
24. **Frank, J., Shimkin, B., and Dowse, H.,** SPIDER — a modular software system for electron image processing, *Ultramicroscopy,* 6, 343, 1981.
25. **van Heel, M. and Keegstra, W.,** IMAGIC: a fast, flexible and friendly images analysis software system, *Ultramicroscopy,* 7, 113, 1981.
26. **Crewe, A. V. and Ohtsuki, M.,** A "stand alone" image processing system for STEM images, *Ultramicroscopy,* 9, 101, 1982.
27. **Russ, J. C. and Russ, J. C.,** Image processing in a general purpose microcomputer, *J. Microsc.*, 135, 89, 184.
28. **Squire, J. M., Luther, P. K., and Agnew, G. D.,** Averaging of periodic images using a microcomputer, *J. Microsc.*, 142, 289, 1986.
29. **Bright, D. S.,** A LISP-based image analysis system with applications to microscopy, *J. Microsc.*, 148, 51, 1987.

30. **Olofsson, A., Kavéus, U., and Hebert, H.,** Digitization of electron micrographs: a comparison of three different types of scanners, *J. Electron Microsc. Tech.,* 8, 381, 1988.
31. **Gonzalez, R. C. and Wintz, P.,** *Digital Image Processing,* Addison-Wesley, Reading, MA, 1977.
32. **Bracewell, R. N.,** Numerical transforms, *Science,* 248, 697, 1990.
33. **Schiske, P.,** A posteriori correction of object tilt for the CTEM, *Ultramiscoscopy,* 9, 17, 1982.
34. **Carragher, B., Bluemke, D. A., Potel, M. J., and Josephs, R.,** The restoration of electron micrographs blurred by drift and rotation, *Proteins: Struct. Funct. Genet.,* 1, 176, 1985.
35. **Krakow, W.,** The use of histograms for transmission electron microscopy: defocus and astigmatism correction at high resolution, *J. Electron Microsc. Tech.,* 2, 371, 1985.
36. **Dhawan, A. P., Rangayyan, R. M., and Gordon, R.,** Image restoration by Wiener deconvolution in limited-view computed tomography, *Appl. Opt.,* 24, 4013, 1985.
37. **Harauz, G., Boekema, E., and van Heel, M.,** Statistical image analysis of electron micrographs of ribosomal subunits, *Methods Enzymol.,* 164, 35, 1988.
38. **Harauz, G. and Fong-Lochovsky, A.,** Automatic selection of macromolecules from electron micrographs by component labeling and symbolic processing, *Ultramicroscopy,* 31, 333, 1989.
39. **Burch, S. F., Gull, S. F., and Skilling, J.,** Image restoration by a powerful maximum entropy method, *Comput. Vision Graph. Image Process.,* 23, 113, 1983.
40. **Farrow, N. A. and Ottensmeyer, F. P.,** Maximum entropy methods in dark field electron micrographs and elemental maps, in *Image and Signal Processing in Electron Microscopy,* Hawkes, P. W., Saxton, W. O., Ottensmeyer, F. P., and Rosenfeld, A., Eds., Scanning Microscopy International, AMF O'Hare, Chicago, 1988, 75.
41. **Farrow, N. A. and Ottensmeyer, F. P.,** Maximum entropy methods and dark field microscopy images, *Ultarmicroscopy,* 31, 275, 1989.
42. **Frieden, B. R.,** A comparison of the maximum entropy (ME), maximum a posteriori (MAP), and median window (MW) restoring algorithms, in *Image and Signal Processing in Electron Microscopy,* Hawkes, P. W., Saxton, W. O., Ottensmeyer, F. P., and Rosenfeld, A., Eds., Scanning Microscopy International, AMF O'Hare, Chicago, 1988, 107.
43. **Skilling, J.,** The maximum entropy method, *Nature,* 309, 748, 1984.
44. **Welsh, E. A., Zahradka, P., Larson, D. E., Sells, B. H., and Harauz, G.,** Structural analysis of a eukaryotic transcription initiation complex, paper presented at the 33rd Annu. Meet. Can. Fed. of Biological Societies, Halifax, June 14 to 16, 1990, 69.
45. **Zahradka, P., Larson, D. E., and Sells, B. H.,** Characterization of a mammalian ribosomal protein gene promoter, *Biochem. Cell Biol.,* 68, 949, 1990.
46. **Bronskill, M. J., Harauz, G., and Ege, G. N.,** Computerized internal mammary lymphoscintigraphy in radiation treatment planning of patients with breast carcinoma, *Int. J. Radiat. Oncol. Biol. Phys.,* 5, 573, 1979.
47. **Bright, D. S. and Steel, E. B.,** Two-dimensional top hat filter for extracting spots and spheres from digital images, *J. Microsc.,* 146, 191, 1987.
48. **Bates, R. H. T. and Lane, R. G.,** Deblurring should now be automatic, in *Image and Signal Processing in Electron Microscopy,* Hawkes, P. W., Saxton, W. O., Ottensmeyer, F. P., and Rosenfeld, A., Eds., Scanning Microscopy International, AMF O'Hare, Chicago, 1988, 149.
49. **Peli, T. and Malah, D.,** A study of edge detection algorithms, *Comput. Graph. Image Process.,* 20, 1, 1982.
50. **Ballard, D. H. and Brown, C. M.,** *Computer Vision,* Prentice-Hall, Englewood Cliffs, NJ, 1982.
51. **Fong-Lochovsky, A.,** The effect of window mask shape on algorithm design in pipeline architectures, in Proc. Vision Interface '89, London, Ontario, 1989.
52. **Ottensmeyer, F. P., Andrew, J. W., Bazett-Jones, D. P., Chan, A. S. K., and Hewitt, J.,** Signal to noise enhancement in dark field electron micrographs of vasopressin: filtering of arrays of images in reciprocal space, *J. Microsc.,* 109, 259, 1977.
53. **Frank, J., Verschoor, A., and Boublik, M.,** Computer averaging of electron micrographs of 40S ribosomal subunits, *Science,* 214, 1353, 1981.
54. **van Heel, M.,** Multivariate statistical classification of noisy images (randomly oriented biological macromolecules), *Ultramicroscopy,* 13, 165, 1984.
55. **van Heel, M. and Stöffler-Meilicke, M.,** Characteristic views of *E. coli* and *B. stearothermophilus* 30S ribosomal subunits in the electron microscope, *EMBO J.,* 4, 2389, 1985.
56. **Unser, M., Steven, A. C., and Trus, B. L.,** Odd men out: a quantitative objective procedure for identifying anomalous members of a set of noisy images of ostensibly identical specimens, *Ultramicroscopy,* 19, 337, 1986.
57. **Frank, J., Radermacher, M., Wagenknecht, T., and Verschoor, A.,** Studying ribosome structure by electron microscopy and computer-image processing, *Methods Enzymol.,* 164, 3, 1988.
58. **van Heel, M.,** Classification of very large electron microscopical image data sets, *Optik,* 82, 114, 1989.
59. **Borland, L. and van Heel, M.,** Classification of image data in conjugate representation spaces, *J. Optic. Soc. Am.,* A, 7, 601, 1990.

60. **Schatz, M. and van Heel, M.**, Invariant classification of molecular views in electron micrographs, *Ultramicroscopy,* 32, 255, 1990.
61. **Frank, J., Bretaudiere, J.-P., Carazo, J.-M., Verschoor, A., and Wagenknecht, T.**, Classification of images of biomolecular assemblies: a study of ribosomes and ribosomal subunits of *Escherichia coli, J. Microsc.* 150, 99, 1988.
62. **Unser, M., Trus, B. L., and Steven, A. C.**, Normalization procedures and factorial representations for classification of correlation-aligned images: a comparative study, *Ultramiscoscopy,* 30, 299, 1989.
63. **van Heel, M.**, Similarity measures between images, *Ultramicroscopy,* 21, 95, 1987.
64. **Unser, M., Trus, B. L., and Steven, A. C.**, A new resolution criterion based on spectral signal-to-noise ratios, *Ultramicroscopy,* 23, 39, 1987.
65. **Boekema, E. J., Berden, J. A., and van Heel, M. G.**, Structure of mitochondrial F1-ATPase studied by electron microscopy and image processing, *Biochim. Biophys. Acta,* 851, 353, 1986.
66. **Bazett-Jones, D. P. and Ottensmeyer, F. P.**, DNA organization in nucleosomes, *Can. J. Biochem.*, 60, 364, 1982.
67. **Appledorn, C. R., Oppenheim, B. E., and Wellman, H. N.**, An automated method for the alignment of image pairs, *J. Nucl. Med.*, 21, 165, 1980.
68. **Barber, D. C.**, Automatic alignment of radionuclide images, *Phys. Med. Biol.*, 27, 387, 1982.
69. **Bonnet, N. and Liehn, J.-C.**, Image registration in electron microscopy: application of a robust method, *J. Electron Microsc. Tech.*, 10, 27, 1988.
70. **Patrick, E. A. and Fattu, J. M.**, *Artificial Intelligence with Statistical Pattern Recognition,* Prentice-Hall, Englewood Cliffs, NJ, 1986.
71. **Unser, M. and Eden, M.**, Weighted averaging of a set of noisy images for maximum signal-to-noise ratio, *IEEE Trans. Acoust. Speech Signal Process.*, 38, 890, 1990.
72. **Bijlholt, M. M. C., van Heel, M., and van Bruggen, E. F. J.**, Comparison of 4 × 6-meric hemocyanins from three different arthropods using computer alignment and correspondence analysis, *J. Mol. Biol.*, 161, 139, 1982.
73. **Frank, J.**, The role of multivariate image analysis in solving the architecture of the *Limulus polyphemus* hemocyanin molecule, *Ultramicroscopy,* 13, 153, 1984.
74. **Woodcock, C. L. F. and Frank, J.**, Nucleosome mass distribution using image averaging, *J. Ultrastruct. Res.*, 89, 295, 1984.
75. **Boekema, E. J. and van Heel, M.**, Molecular shape of *Lumbricus terrestris* erythrocruorin studied by electron microscopy and image analysis, *Biochim. Biophys. Acta,* 957, 370, 1989.
76. **Boekema, E. J., Dekker, J. P., Rögner, M., Witt, I., Witt, H. T., and van Heel, M.**, Refined analysis of the trimeric structure of the isolated photosystem I complex from the thermophilic cyanobacterium *Synechococcus* sp., *Biochim. Biophys. Acta,* 974, 81, 1989.
77. **Harauz, G. and Flannigan, D.**, Structure of ribosomes from *Thermomyces lanuginosus* by electron microscopy and image processing, *Biochim. Biophys. Acta,* 1038, 260, 1990.
78. **Unser, M., Trus, B. L., Frank, J., and Steven, A. C.**, The spectral signal-to-noise ratio resolution criterion: computational efficiency and statistical precision, *Ultramicroscopy,* 30, 429, 1989.
79. **Sass, H. J., Büldt, G., Beckmann, E., Zemlin, F., van Heel, M., Zeitler, E., Rosenbusch, F. P., Dorset, D. L., and Massalski, A.**, Densely packed β-structure at the protein-lipid interface of porin is revealed by high-resolution cryo-electron microscopy, *J. Mol. Biol.*, 209, 171, 1989.
80. **Schiske, P.**, Signifikanzgrenzen der Eigenwerte bei der multivariaten Analyse von Bildreihen, *Optik,* 78, 144, 1988.
81. **Bretaudiere, J.-P. and Frank, J.**, Reconstitution of molecule images by correspondence analysis: a tool for structural interpretation, *J. Microsc.* 144, 1, 1986.
82. **Chiu, D. K. Y. and Wong, A. K. C.**, Synthesizing knowledge: a cluster analysis approach using event covering, *IEEE Trans. Syst. Man Cybern.*, 16, 251, 1986.
83. **Wong, A. K. C. and Chiu, D. K. Y.**, An event-covering method for effective probabilistic inference, *Pattern Recognition,* 20, 245, 1987.
84. **Chiu, D. K. Y., Cheung, B., and Wong, A. K. C.**, Information synthesis based on hierarchical maximum entropy discretization, *J. Exp. Theor. Artif. Intelligence,* 2, 117, 1990.
85. **Kunath, W. and Sack-Kongehl, H.**, Circular harmonic averaging of noisy single-molecule images, *Ultramicroscopy,* 27, 171, 1989.
86. **Matsuyama, T.**, Expert systems for image processing: knowledge-based composition of image analysis processes, *Comput. Vision Graph. Image Process.*, 48, 22, 1989.
87. **Russ, J. C.**, Computerized object recognition using contextual learning, *J. Comput. Assist. Microsc.*, 1, 105, 1989.
88. **Amos, L. A., Henderson, R., and Unwin, P. N. T.**, Three-dimensional structure determination by electron microscopy of two-dimensional crystals, *Prog. Biophys. Mol. Biol.*, 39, 183, 1982.
89 **Brink, J., Boekema, E. J., and van Bruggen, E. F. J.**, Electron microscopy and image analysis of the complexes I and V of the mitochondrial respiratory chain, *Electron Microsc. Rev.*, 1, 175, 1988.

90. **Henderson, R., Baldwin, J. M., Downing, K. H., Lepault, J., and Zemlin, F.,** Structure of purple membrane from *Halobacterium halobium:* recording, measurement and evaluation of electron micrographs at 3.5 Å resolution, *Ultramicroscopy,* 19, 147, 1986.
91. **Holzenburg, A., Mayer, F., Harauz, G., van Heel, M., Tokuoka, R., Harata, K., Ishida, T., Pal, G. P., and Saenger, W.,** Structure of the D-ribulose-1,5-bisphosphate carboxylase/oxygenase from *Alcilagenes eutrophus* H16, *Nature,* 325, 730, 1987.
92. **Saxton, W. O. and Baumeister, W.,** The correlation averaging of a bacterial cell envelope protein, *J. Microsc.,* 127, 127, 1982.
93. **Jap, B. K.,** The structure of PhoE porin in projection at 3.5 Å resolution, *J. Struct. Biol.,* 103, 57, 1990.
94. **Boekema, E. J.,** The present state of two-dimensional crystallization of membrane proteins, *Electron Microsc. Rev.,* 3, 87, 1990.
95. **Hovmöller, S.,** Three-dimensional structure of membrane proteins, in *Techniques for the Analysis of Membrane Proteins,* Ragan, C. I. and Cherry, R. J., Eds., Chapman and Hall, London, 1986, 315.
96. **Hegerl, R. and Baumeister, W.,** Correlation averaging of a badly distorted lattice: the surface protein of *Pyrodictium occultum, J. Electron Microsc. Tech.,* 9, 413, 1988.
97. **Tischer, I., Gelderblom, H., Vettermann, W., and Koch, M. A.,** A very small porcine virus with circular, single-stranded DNA, *Nature,* 295, 64, 1982.
98. **Chiu, D. and Harauz, G.,** unpublished data.

Chapter 12

THREE-DIMENSIONAL RECONSTRUCTION OF SINGLE PARTICLES IN ELECTRON MICROSCOPY

Michael Radermacher

TABLE OF CONTENTS

I. INTRODUCTION

For an understanding of the function of many biological particles, knowledge of their three-dimensional architecture is a prerequisite. For many smaller molecules, the answer often can be found using the techniques of X-ray crystallography, which is capable of resolving structure at the atomic level.[1,2] Electron microscopy of noncrystalline biological specimens, however, yields structural information at a much lower resolution, typically in the range of 2 to 3 nm in three dimensions, and although images of two-dimensional crystals have been recorded in the electron microscope with a resolution of 0.3 nm,[3] such fine details cannot be resolved yet in images of single particles. The advantage of electron microscopic structure determination of single particles lies in the fact that no crystallization is necessary and virtually all macromolecular structures are amendable to such an approach. A low-resolution, three-dimensional reconstruction of a macromolecular complex still can lead to an explanation of function if it allows the spatial relationships among its constituent proteins to be determined.

Throughout the development of three-dimensional reconstruction in electron microscopy, however, the goal has been to determine the architecture of biological particles with higher and higher resolution. There are two major effects that limit resolution. One limit is set by the artifacts that occur during preparation of macromolecules for electron microscopy. The effects of negative staining and subsequent drying have been discussed widely in the literature.[4] Only lately, with the advent of frozen hydrated preparation techniques,[5-8] can these artifacts be avoided. A second serious limitation of the resolution is the radiation sensitivity of biological objects.[9] Single atoms can be visualized in imaging some inorganic specimens, but the electron dose required for such imaging would reduce the biological material to ashes. The only solution to this problem is the use of extremely low electron doses. In fact, the higher the intended resolution, the lower the electron dose in the imaging process of biological specimens has to be. It is widely accepted that at resolutions on the order of 2 nm, electron doses of 10 el/Å^2 are acceptable. At this level, however, the signal-to-noise ratio of a single image has a magnitude of 1 meaning that noise and signal strength are about equal.

The first three-dimensional reconstruction in electron microscopy was done of a helical particle (the tail of bacteriophage T4).[10] Such particles provide in a single image all views necessary to determine the three-dimensional structure. Although the primary reason for choosing such a specimen was the simplicity of the data collection, this approach also minimized the effects of radiation damage.

For the first reconstructions of single asymmetrical particles, the different views were obtained from a tilt series, where the specimen is tilted by small increments and an image is recorded for each specimen orientation.[11-15] Even the use of extremely low doses for each single image cannot avoid a high cumulative electron exposure of the particle.

Whereas in the processing of two-dimensional images of asymmetrical single particles averaging techniques have been used for a long time to increase the signal-to-noise ratio while keeping the electron dose as low as possible,[16-18] such techniques have only recently been expanded successfully into three dimensions. In such averaging techniques, the different views needed for a three-dimensional reconstruction are obtained from images of identical particles in different orientations. The first attempts were made at the same time by Verschoor et al.[19,20] and van Heel[21] with the reconstruction of the 30S ribosomal subunit from *Escherichia coli;* yet in these studies the inaccuracy of the measurement of the viewing angles was great, which led to a low resolution in the final three-dimensional structure. Major progress, however, was achieved with the development of the reconstruction technique from a single-exposure, random conical tilt series,[22,24] which for the first time allowed the combination of images of many particles to form one common conical tilt series in which all projecting directions could be measured with great accuracy. This technique currently is the most reliable approach for the reconstruction of single asymmetrical particles, and a large number of structures have been determined with this technique in a comparably short period of time.[23]

The theory underlying all three-dimensional reconstruction techniques was originally developed by Radon in 1917,[26] who analyzed what now is called the Radon transform, an integral transform and its inversion. The same theory was developed independently much later by Cormack,[27,28] who at the time was not aware of the existence of Radon's theory. Other, equivalent approaches originate from reconstruction techniques used in X-ray crystallography, taking the point of view that a single particle is a crystal degenerated into a single unit cell. With some modifications, similar reconstruction methods can then be applied to those known in crystallography.[10,12] The basis of this approach is the projection theorem, which underlies Fourier-type reconstruction schemes. Mathematically, the inversion of the Radon transform and Fourier inversion are equivalent formulations of the same theory.

Based on the different theoretical approaches, there exist different reconstruction algorithms. In electron microscopy, the direct inversion of the Radon transform is seldom used and the most popular reconstruction schemes are based on Fourier transforms. These include the Fourier-Bessel reconstruction techniques,[27-30] Fourier interpolation techniques,[31] and weighted or deconvoluted back-projection algorithms.[32-36]

A second class of methods uses an algebraic approach and treats the reconstruction problem as a matrix inversion. Given a number of measurements — the points in the projections — and a number of unknowns — the points in the volume — a system of equations can be written and solved iteratively, provided sufficient measurements are available. The first iterative technique that was developed and applied to electron micrographs is the algebraic reconstruction technique (ART).[11,37] Claims about the performance of this method at that time created a controversy between the group that had developed the algorithm and groups using Fourier inversion-type algorithms.[38,39] However, iterative techniques allow a simple implementation of *a priori* knowledge about the object that is being reconstructed. Already, the first formulation of the algorithm contained a constraint postulating positive densities in the reconstruction. Many other constraints have been added later.[40] For non-iterative reconstruction techniques, the adding of *a priori* knowledge can be done in a second step after the reconstruction using a method known as projection onto convex sets (POCS).[41-43]

II. THE RADON TRANSFORM AND THE PROJECTION THEOREM

An object can be reconstructed in three dimensions if all its projections are known. This statement is based on the theory of Radon transforms or on the projection theorem, both of which are equivalent formulations of the same mathematical problem.

The Radon transform is an integral transform defined as follows.[26,44] Let $f(\mathbf{r})$ be a function of vector r in an n-dimensional Euclidean space, and let f obey certain regularity conditions. Then the integral transform,

$$\hat{f}(p, \boldsymbol{\xi}) = \int f(\mathbf{r})\delta(p - \boldsymbol{\xi} \cdot \mathbf{r})d\mathbf{r} \tag{1}$$

exists and has an inverse transform; $\boldsymbol{\xi}$ is a unit vector defining a direction in n dimensions, and p is a positive real number; δ is the delta function.

In two dimensions, $\boldsymbol{\xi}$ is a function of a single angle ϕ

$$\boldsymbol{\xi}(\phi) = (\cos\phi, \sin\phi) \tag{2}$$

and Equation 1 describes the set of line integrals in any direction in the plane. The integral extends over the whole plane and the δ-function selects a line in the direction of $\boldsymbol{\xi}$. The function $\hat{f}(p,\boldsymbol{\xi})$ is the set of all one-dimensional projections along the directions $\boldsymbol{\xi}(\phi)$.

In three dimensions, Equation 1 is an integral over planes, and in n dimensions, and integral over (n − 1)-dimensional hyperspaces. For three-dimensional reconstructions, only the situations n = 2 and n = 3 are of importance.

The inverse Radon transform for even dimensions (n ≥ 2) is

$$f(\mathbf{r}) = \frac{C_n}{i\pi} \Delta_{\mathbf{r}}^{(n-2)/2} \int_{|\boldsymbol{\xi}|=1} d\boldsymbol{\xi} \int_{\infty}^{+\infty} dp \frac{\hat{f}(p, \boldsymbol{\xi})}{p - \boldsymbol{\xi} \cdot \mathbf{x}} \tag{3}$$

The inverse of the Radon transform for odd dimensions (n ≥ 3) is

$$f(\mathbf{r}) = C_n \int_{|\boldsymbol{\xi}|=1} \left(\frac{\partial}{\partial p}\right)^{n-1} \hat{f}(p, \boldsymbol{\xi})\, d\boldsymbol{\xi} \tag{4}$$

$$\text{with } C_n = \frac{(-1)^{(n-1)/2}}{2(2\pi)^{(n-1)}} \tag{5}$$

Equation 3 states, for n = 2, that if all one-dimensional projections of an object f(r) are known, then the object can be recovered from its projections. Equation 4 states that if the sums over all planes in all directions of a three-dimensional object are known, then the object can be recovered. Although the projections obtained in the microscope are not sums over planes, but sums along lines (the projecting electron beam), the sum over all planes can be determined in theory by a subsequent two-dimensional Radon transform of the two-dimensional projections. The object then can be recovered using Equation 4.

An equivalent formulation of the mathematical problem is the projection theorem for Fourier transforms. It states that the Fourier transform of a projection is a central section through the Fourier transform of the object. If all projections are known, then all central sections through the object's Fourier transform are known, and by an appropriate inverse Fourier transform of the set of projection data, the object can be recovered. This statement is valid for all dimensions. In the following, we prove the projection theorem in three dimension.

Let f($\mathbf{r}$) be the three-dimensional object and

$$p(\mathbf{q}) = \int f(\mathbf{r})\, dz \tag{6}$$

be its projection, for simplicity assumed to be along the z-direction. Let

$$F(\mathbf{r}^*) = \int f(\mathbf{r})\, e^{-2\pi(xx^* + yy^* + zz^*)}\, dx\, dy\, dz \tag{7}$$

be the Fourier transform of f(r). The central section through the Fourier transform at $z^* = 0$ can be written as:

$$F(\mathbf{r}^*) = \int f(\mathbf{r})\, e^{-2\pi(xx^* + yy^* + z \cdot 0)}\, dx\, dy\, dz \tag{8}$$

$$= \int f(\mathbf{r})\, dz\, e^{-2\pi i(xx^* + yy^*)}\, dx\, dy \tag{9}$$

The first integral over z in Equation 9 is the projection operation, which then is Fourier transformed in two dimensions. Thus, the central section through the Fourier transform (Equation 8) is the Fourier transform of a projection (Equation 9).

The equivalence between the Radon transform and the central section theorem can most easily be seen for n = 2.

Let $f(\mathbf{r})$, $\mathbf{r} \in \mathbf{R}^2$ be a function in two dimensions. Its Radon transform is

$$\hat{f}(p, \boldsymbol{\xi}) = \int f(\mathbf{r})\, \delta(p - \boldsymbol{\xi} \cdot \mathbf{r})\, \mathbf{dr} \tag{10}$$

The Fourier transform of f(r) is

$$F(\mathbf{r}^*) = \int f(\mathbf{r})\, e^{-2\pi i \mathbf{r} \cdot \mathbf{r}^*}\, dr \tag{11}$$

which can be rewritten as

$$\int f(\mathbf{r})\, e^{-2\pi i t} \cdot \delta(t - \mathbf{r}^* \cdot \mathbf{r})\, dt\, \mathbf{dr} \tag{12}$$

Substitution of $\mathbf{r}^*$ by $s\boldsymbol{\xi}$ and t by $s \cdot p$ results in

$$F(s\boldsymbol{\xi}) = |s| \int_{-\infty}^{+\infty} \int f(\mathbf{r})\delta(sp - s\boldsymbol{\xi} \cdot \mathbf{r})\, e^{-2\pi i s p}\, \mathbf{dr}\, dp$$

$$= \int_{-\infty}^{+\infty} \int f(\mathbf{r})\delta(p - \boldsymbol{\xi} \cdot \mathbf{r})\, \mathbf{dr}\, e^{-2\pi i s p}\, dp \tag{13}$$

The inner integral over $\mathbf{r}$ is the Radon transform of $f(\mathbf{r})$ which describes the one-dimensional projection of f in the direction of the unit vector $\boldsymbol{\xi}$. The integral over p is the Fourier transform of the projections. The left-hand side $F(s \cdot \boldsymbol{\xi})$ is a section through the Fourier transform of f in direction $\boldsymbol{\xi}$. Equation 13 states that the one-dimensional Fourier transform along p of the two-dimensional Radon transform yields the two-dimensional Fourier transform of the original function $f(\mathbf{r})$, sampled along radii in directions $\boldsymbol{\xi}$ (ϕ).

III. TILT GEOMETRIES

For the determination of the three-dimensional structure, projections of the object from many different directions spanning a large angular range need to be available. Several different data collection schemes are used in electron microscopy, and the choice between them depends on the properties and size of the particle that is being analyzed as well as on the reconstruction algorithm used.

A. SINGLE-AXIS TILTING

The single-axis tilt geometry was the first one used in electron microscopy. Here, the specimen grid is mounted on a stage that can be rotated around one axis. For side-entry stages, this tilt axis is along the specimen rod. Normally, the tilt angles range from −60° to +60° and the images are taken at small angular intervals (Figure 1), typically between 1 and 10°, depending on the resolution desired in the reconstruction (see Section V of this chapter). The Fourier transforms of the projections provide measurements of the three-dimensional Fourier transform of the object. For the complete sampling of the three-

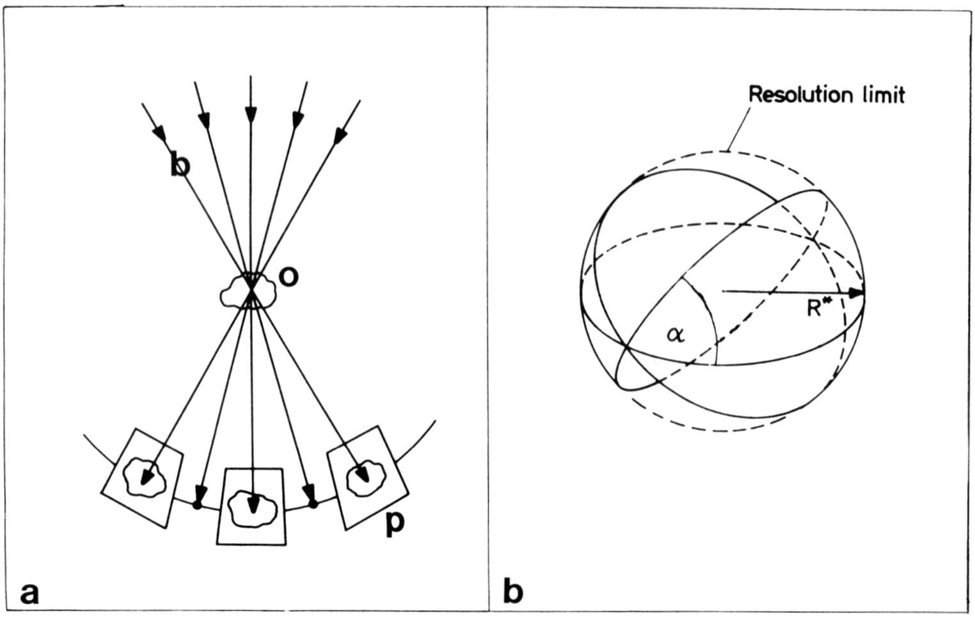

FIGURE 1. Single-axis tilt geometry. (a) Real space, where o = object, p = projection, and b = electron beam; (b) coverage of the Fourier space, where R* = radius of the resolution sphere, a maximum tilt angle. (From Radermacher, M., Dreidimensionale Rekonstruktion bei kegelförmiger Kippung im Elektronenmikroskop, Ph.D. thesis, Technische Universität München, FRG, 1980.)

dimensional Radon transform or Fourier transform, a tilt range of 180° would be needed; however, in most microscopes, only a range of 120° (±60°) can be achieved. Because of this limitation in a single-axis tilt series, there is a double wedge in the Fourier transform of the object that is not covered by measurements. If the tilt range is $\pm\alpha$ and the three-dimensional Fourier transform of the object is enclosed in a sphere of radius R*, then the volume of this "missing wedge" is[46]

$$V_w = \frac{8}{3} R^{*3} \left(\frac{\pi}{2} - \alpha\right), \quad \alpha \text{ in radians} \tag{14}$$

With the volume of the complete sphere being

$$V_{sp} = \frac{4}{3} \pi R^{*3} \tag{15}$$

the ratio between the volume of the missing wedge and the volume of the complete Fourier transform is

$$\frac{V_{sp}}{V_w} = \frac{\pi - 2\alpha}{\pi} \tag{16}$$

If the Fourier transform is assumed to be enclosed in a cylinder along the tilt axis with radius R*, instead of a sphere, the same volume ratio applies. With a tilt range of ±45°, 50% of the Fourier transform is undetermined, and for ±60°, still 33% of the Fourier coefficients are unknown. This missing wedge leads to artifacts in the reconstruction.

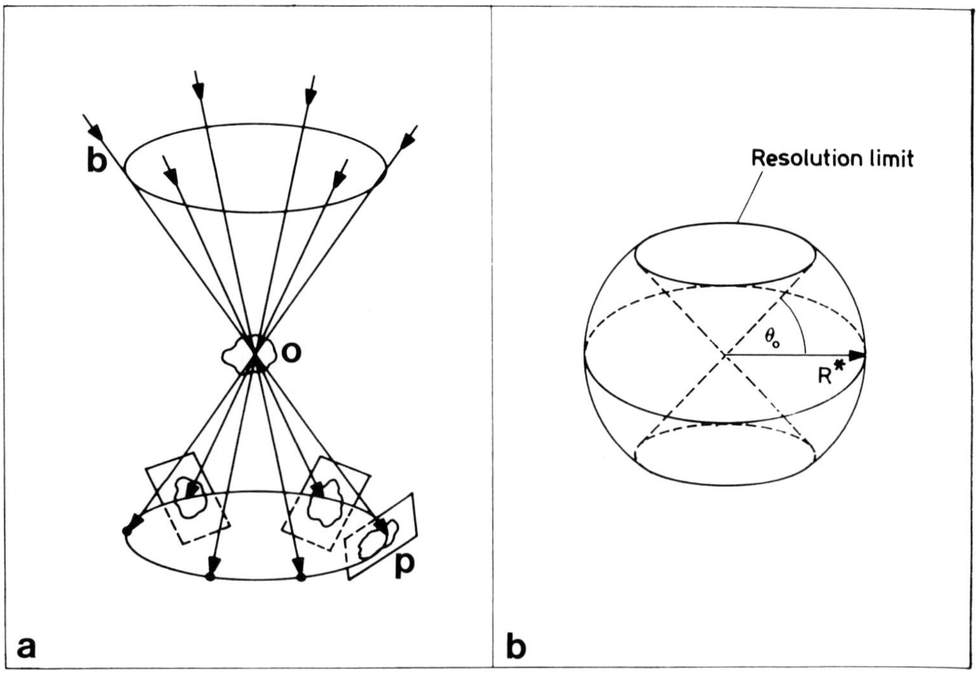

FIGURE 2. Conical tilt geometry. (a) Real space, where o = object, p = projection, and b = electron beam; (b) coverage of the Fourier space, where R* = radius of the resolution sphere and j = fixed tilt angle. (From Radermacher, M., Dreidimensionale Rekonstruktion bei kegelförmiger Kippung im Elektronenmikroskop, Ph.D. thesis, Technische Universität München, FRG, 1980.)

B. CONICAL TILTING

It was pointed out before that in a single-axis tilt experiment with limited tilt range, a double wedge in the three-dimensional Fourier transform of the object cannot be measured. The volume of the undetermined part of the Fourier transform can be decreased if a conical tilt geometry is employed (Figure 2). Here, the specimen grid is tilted by one fixed angle, which should be as large as possible. The tilted specimen then is rotated within its plane by small angular increments, and for each position an image is recorded. This procedure is equivalent to changing the illumination direction along the surface of a cone. The undetermined region in the Fourier transform of the object now is a double cone, the volume of which depends on the size of the fixed tilt angle θ_0.

Assuming again a spherical limitation of the three-dimensional Fourier transform of the object, the volume of the missing double cone is[46]

$$V_{dk} = \frac{4}{3}\pi R^{*3}(1 - \sin\theta_0) \tag{17}$$

and the volume ratio is now

$$\frac{V_{dk}}{V_{sp}} = (1 - \sin\theta_0) \tag{18}$$

As an example, the undetermined volume of the Fourier transform is 29.3% for θ_0 = 45° and only 13.4% for θ_0 = 60°. Because this volume is much smaller than in a data set obtained by single-axis tilting, the artifacts in the reconstruction are substantially reduced.

The disadvantage of this method is that, for a given object size and resolution, about twice as many images need to be available compared to a single-axis tilt geometry, and the problems of radiation damage are magnified.

C. RANDOM CONICAL TILTING

The random conical tilt method[16,22,24] has the advantage of the conical tilt geometry, namely, in that a larger part of the Fourier transform can be measured than with single-axis tilting, and at the same time it overcomes the problem of specimen damage by the large, cumulative electron doses that are encountered in the tilt experiments described above. The data collection scheme starts from the premise that many particles with identical structure are available. In addition, most macromolecular assemblies have the tendency to attach to the carbon support film in a specific, preferred orientation such that only one rotational degree of freedom is left, namely, a rotation within the specimen plane. If such a specimen is tilted, typically by 50° to 60°, then the set of all particle images forms a conical tilt series (Figure 3). Here, in constrast to the conical tilting scheme explained above, the rotation within the tilted specimen plane is not achieved by a mechanical rotation of the specimen, but by the random in-plane orientation of the different particles. As a result, the azimuthal angles in such a conical tilt series are distributed randomly. The reconstruction method implicity is an averaging procedure, because the tilt series is comprised of images of many different particles. As one important part of the evaluation of such a series, all images used in the reconstruction must show particles of identical shape and identical orientation relative to the specimen plane. This analysis is done with images extracted from a second micrograph of the same specimen area, but without tilt. The methods used are the same methods, multivariate statistical analysis and classification,[47-52] that are used in the two-dimensional averaging and alignment of single particles. The 0° micrograph is recorded after the image of the tilted specimen and thus does not add to the cumulative dose of the projections that are used for the acutal reconstruction. If the particle shows more than one preferred orientation, separate reconstructions can be performed. Their relative orientation in three dimensions can then be determined and subsequently all images can be used for a single, merged reconstruction. In this case, the projections no longer form a single random conical tilt series, but a set of such series. The geometry is referred to as a multicone geometry.[53] However, before a set of random conical tilt series can be merged for the calculation of a single reconstruction, the invariance of the particle structure among the separate reconstructions has to be ensured. Particle distortions can frequently be observed that depend on the orientation of the particle with respect to the support, and a combination of such data sets for the calculation of a single structure could lead to invalid results.[54]

The random conical tilt geometry is a technique that allows a reconstruction of asymmetrical particles under very low dose conditions. Exposures of less than 10 el/Å^2 can easily be achieved.

D. RANDOM TILTING

A number of experimental setups are conceivable that yield projections of particles in entirely random directions.

If surface effects are kept minimal, images of particles in random orientations can most easily be obtained from a specimen of particles suspended in vitreous ice.[8] The technique is being used extensively in the reconstruction of symmetric particles, especially viruses with icosahedral symmetry, where, because of the symmetry, the orientation of each particle in space can be found using the method of common lines.[55] Because of the symmetry, only a small number of images may be needed to provide sufficient data for a three-dimensional reconstruction, e.g., an image of a single icosahedral particle may correspond to 60 different views of this object. For asymmetrical particles, on the other hand, a reconstruction from

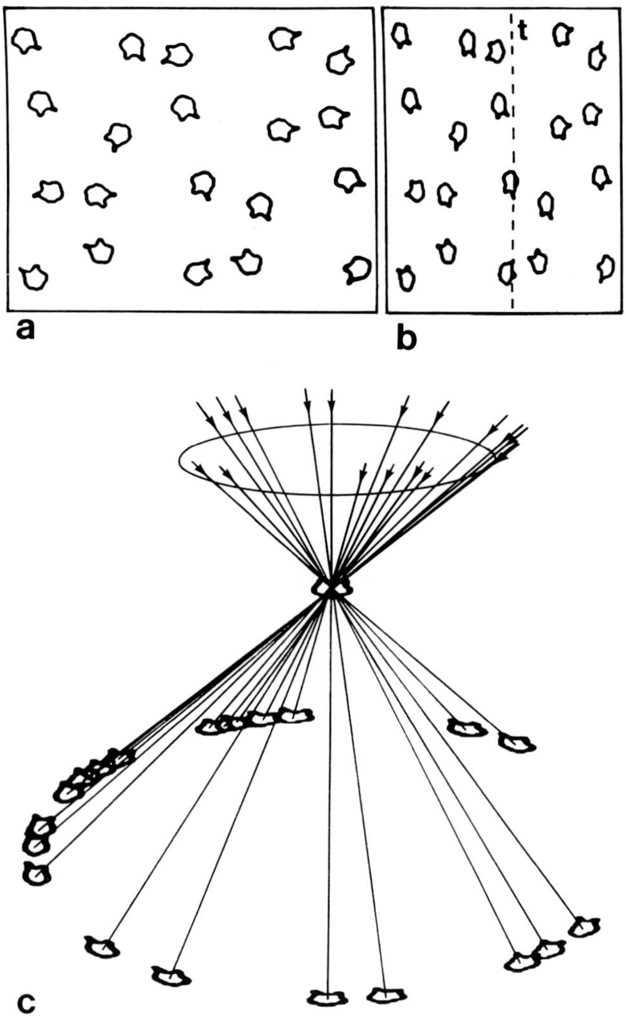

FIGURE 3. Geometry of the single-exposure random conical tilt collection scheme. (a) Specimen with identical particles in the same in-plane orientation without tilt, (b) same specimen area as in (a) tilted by 50°; (c) images extracted from (b) in their position in an equivalent conical tilt series. (From Radermacher, M., Wagenknecht, T., Verschoor, A., and Frank, J., *J. Microsc.*, 146, 113, 1987. With permission.)

a set of entirely randomly oriented projections poses a number of problems. First, the number of images needed is very large compared to the reconstruction of symmetrical particles. Second, because there is no preferred orientation, there exists as yet no technique for monitoring structural variations.

Early schemes to calculate a reconstruction from randomly oriented particles have been proposed by Kam.[56] A recent technique is the reconstruction method developed by Provencher and Vogel,[57,58] where, starting with approximate angular estimates, the correct tilt angle is fitted as part of the reconstruction algorithm. However, application of this method to particles without symmetry requires several micrographs of the same specimen area to be recorded with different tilt angles.

Other techniques for three-dimensional reconstruction from randomly oriented projections are being developed that use a common line approach to determine the orientation of particles in three dimensions. This can most easily be achieved by one-dimensional cross-correlation of the two-dimensional Radon transforms of the projections.[59,60] A two-dimensional Radon transform is often referred to as a sinogram. Each line in a two-dimensional Radon transform is a one-dimensional projection of the two-dimensional image. Any two two-dimensional projections of a three-dimensional object have at least one one-dimensional projection in common. Finding these common one-dimensional projections in a set of at least three noncoaxial projections uniquely determines their projecting directions, or, since the projections originate from different particles, these common lines determine uniquely the orientation of the three associated particles. Although the method is quite straightforward, it is extremely noise sensitive and does not produce the desired results when applied to single, low-dose images. Improvements can only be achieved by preaveraging a number of projections that show the same particle orientation, thus increasing the signal-to-noise ratio, or — as subaverages might be difficult to obtain from totally random-oriented particles — a technique needs to be found that is able to find all intersections between all projections simultaneously and use redundancies between them to increase the accuracy of the technique.

E. DEFINITION OF THE ROTATION MATRICES

Let $\mathbf{r}^j$ be a vector in the coordinate system of the projection p_j which is obtained by projecting the object tilted by the three angles ψ_j, θ_j, and ϕ_j along the direction z^j. The transformation between $\mathbf{r}^j$ and a vector $\mathbf{r}$ in the coordinate system fixed to the object then is

$$\mathbf{r}^j = \omega_j \mathbf{r}$$

$$\text{with} \quad \omega_j = D_{\psi_j} \cdot D_{\theta_j} \cdot D_{\phi_j} \tag{19}$$

$$D_{\psi_j} = \begin{pmatrix} \cos\psi_j & \sin\psi_j & 0 \\ -\sin\psi_j & \cos\psi_j & 0 \\ 0 & 0 & 1 \end{pmatrix} \tag{20}$$

$$D_{\theta_j} = \begin{pmatrix} \cos\theta_j & 0 & -\sin\theta_j \\ 0 & 1 & 0 \\ \sin\theta_j & 0 & \cos\theta_j \end{pmatrix} \tag{21}$$

$$D_{\phi_j} = \begin{pmatrix} \cos\phi_j & \sin\phi_j & 0 \\ -\sin\phi_j & \cos\phi_j & 0 \\ 0 & 0 & 1 \end{pmatrix} \tag{22}$$

For a conical and a random conical tilt series, $\psi \equiv 0$. The tilting is done by first rotating the object around its z-axis by the angle ϕ_j and then tilting it by θ_j around the new y-axis. Throughout a conical tilt series, θ_j = constant for all projections.

For single-axis tilting, $\psi \equiv 0$ and $\phi \equiv 0$; thus, the data collection is done by tilting the object around its y-axis by $-\theta_j$ and then projecting along z^j.

IV. RECONSTRUCTION ALGORITHMS

The reconstruction algorithms that are used in electron microscopy can be divided into three groups: Fourier reconstruction methods, weighted or deconvoluted back-projection methods, and iterative direct-space methods. The largest choice of algorithms is available for single-axis tilt geometry, in part because it was the first tilt method used and in part because the reconstruction problem can be separated into a set of reconstructions of two-dimensional slices perpendicular to the tilt axis from one-dimensional projections. This reduction to a lower dimensionality simplifies the reconstruction problem and reduces the amount of data that in some type of algorithms need to be handled simultaneously, thus allowing computer implementations of algorithms that could not be implemented for a truly three-dimensional problem because of the large volume of data involved.

A. FOURIER AND FOURIER-BESSEL RECONSTRUCTION ALGORITHMS

Fourier reconstruction algorithms are based on the projection theorem: *the Fourier transform of a projection is a central section through the Fourier transform of the object.* Thus, the three-dimensional Fourier transform can be measured from a set of projections.

The points in the Fourier transform that are known from the projections most often do not coincide with the sampling points needed to carry out the inverse Fourier transform, and an interpolation from the measured points onto the appropriate coordinate system is necessary.[29,31] For a Cartesian grid, the correct interpolation method is a Whittaker-Shannon interpolation.[61] The values in the Fourier transform of an image that is sampled optimally can change very rapidly, and thus, for example, a bilinear interpolation can lead to substantial errors.

The Whittaker-Shannon interpolation is based on Shannon's sampling theorem.[61] Assume that the object is enclosed in a rectangular box with edge lengths a, b, and c. The Fourier transform of the object then is completely determined, if the values on grid points spaced 1/a, 1/b, and 1/c apart are known. All points (x^*, y^*, z^*) not lying on the sampling grid, here the points known from the projections, are determined through the equation

$$F(x^*, y^*, z^*) = \sum_h \sum_k \sum_l F_{hkl}\, C_{hkl,h',k',l'} \tag{23}$$

with $x_j^* = h' \cdot 1/a$, $y_j^* = k' \cdot 1/b$, $z_j^* = l' \cdot 1/c$, where h', k', and l' are real numbers and

$$C_{hklj} = \frac{\sin\pi(ax_j^* - h) \cdot \sin\pi(by_j^* - k) \cdot \sin\pi(cz_j^* - l)}{\pi(ax_j^* - h)\pi(by_j^* - k)\pi(cz_j^* - l)} \tag{24}$$

$F_{hkl} = F(h \cdot 1/a, k \cdot 1/b, l \cdot 1/c)$ are the values of the Fourier transform on the Cartesian grid. To solve the interpolation problem, the matrix C_{hkl} needs to be inverted, and the Fourier values F_{hkl} can then be calculated.

$$F_{hkl} = \sum_{h'} \sum_{k'} \sum_{l'} C^{-1}_{h'k'l',h,k,l} F(x^*, y^*, z^*) \tag{25}$$

Already for an object with edge dimension 64, this matrix would have 4096^3 elements, and a solution can only be attempted if the problem can be reduced in dimensionality. As was pointed out before, a reduction is possible if the reconstruction is done from a single-axis tilt series because in this case the interpolation can be carried out across slices perpendicular to the tilt axis. Still, the inversion of a matrix of size 4096^2 is a formidable task.

For the two-dimensional reconstruction problem posed by the single-axis tilt geometry, a more appropriate transform to use is the Fourier-Bessel transform.[29] Such a transform requires that the measurement points lie on a polar grid which is, in fact, the grid obtained when the set of images is recorded with equal angular increments. In this case, no interpolation is necessary. According to the projection theorem, the Fourier transform of each projection is a one-dimensional central section through the Fourier transform of a slice. The Fourier coefficients of this slice then are known on a star-shaped polar grid. The radial sampling depends on the size and resolution of the projections, and the angular sampling on the tilt increment.

Let (r, θ, y) be cylindrical polar coordinates in real space, y being the coordinate along the tilt axis, and (r*, θ, y*) the corresponding coordinates in Fourier space. Let f(r, θ, y) be the object density and let F(r*, θ, y*) be its Fourier transform. The expansion of f and F into cylinder functions yields:

$$f(r, \theta, y) = \sum_n \int g_n(r, y^*)\, e^{in\theta} e^{2\pi i y y^*} dy^* \tag{26}$$

$$F(r^*, \theta, y^*) = \sum_n G_n(r^*, y^*)\, e^{in\theta + \pi/2} \tag{27}$$

where $G_n(r^*, y^*)$ is the Fourier-Bessel transform of $g_n(r, y^*)$.

$$G_n(r^*, y^*) = \int_{-\infty}^{+\infty} g_n(r, y^*) J_n(2\pi r^* r) 2\pi r\, dr \tag{28}$$

$$g_n(r, y^*) = \int_{-\infty}^{+\infty} G_n(r^*, y^*) J_n(2\pi r^* r) 2\pi r^*\, dr \tag{29}$$

In the application to a single-axis tilt series with equal angular increments, Equation 27 can be solved analytically for G_n. Using Equation 29, followed by Equation 26, the three-dimensional object can be recovered.

If the tilt series is not recorded with equal angular increments, then Equation 27 can be regarded as observational equations that may be solved by the method of least squares.[29]

B. SUMMATION OR SIMPLE BACK-PROJECTION

An algorithm that forms the central part of the modified back-projection methods and of the iterative methods to be explained later is the summation or simple back-projection algorithm. It is best explained as a heuristic approach that produces an approximation to the three-dimensional object. From each of the projections in a tilt series which may have been obtained through the use of any tilt geometry, back-projection bodies[13] are created by smearing each projection perpendicular to the projection plane into a three-dimensional cube. Mathematically, this operation can be described as a convolution of the two-dimensional projection with the three-dimensional function:

$$l(x, y, z) = \delta(x, y) \cdot c(z) \tag{30}$$

with

$$c(z) = \begin{cases} 1 & \text{for} \quad -a \leqslant z \leqslant +a \\ 0 & \text{otherwise} \end{cases} \tag{31}$$

2a is chosen larger than the diameter of the object in any direction. Usually, 2a is chosen equal to one of the dimensions of the projection. A convolution of the projection with the

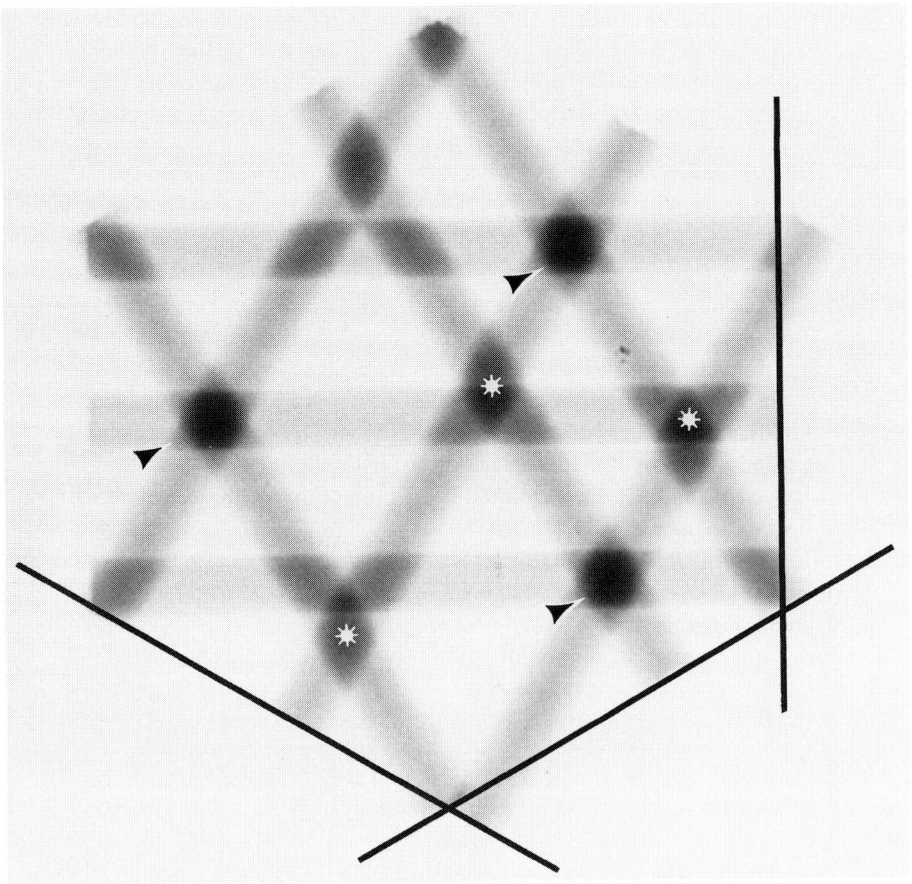

FIGURE 4. Simple back-projection from three projections. The three disks from the original object are marked by arrows. Note the rays around each reconstructed point and spurious densities that could be mistaken for additional disks, marked with an asterisk (*).

function l(x, y, z) (Equation 30) replaces each point in the projection by a line perpendicular to the projection plane where every point along this line has the density value of the projection. The three-dimensional object created in this way is the back-projection body. If p(x, y) is the projection, then the back-projection body $\hat{p}(x, y, z)$ is

$$\hat{p}(x, y, z) = \iint p(x, y) \cdot l(x - x', y - y', z)\, dx\, dy \tag{32}$$

In a simple back-projection procedure, these three-dimensional back-projection bodies are rotated in space using the tilt angles of the projections and then added up.[33] The result is a three-dimensional density distribution which is an approximation of the object's density. A reconstruction by a simple back-projection, however, contains systematic errors. Part of the problem is illustrated in Figure 4, which shows a plane object containing three disks reconstructed from just three projections. Obviously, there are density variations visible where no density had been in the original. The intersections of the rays created in smearing out the projections create a star-shaped structure around each disk and the intersections of these rays can form spurious high-density areas that can be mistaken for additional objects. However, if many projections are used, the situation improves considerably.

Even though simple back-projection results in only a crude approximation of the object, the technique is important to understand. Since it is a linear algorithm, the reconstruction that is created by its use can be analyzed using the concept of point-spread and transfer functions. The density distributions calculated can therefore be corrected by a deconvolution process to yield the accurate reconstruction. Simple back-projection is also used as the first approximation in many iterative reconstruction techniques.

C. WEIGHTED AND DECONVOLUTED BACK-PROJECTION METHODS

The simple back-projection algorithm represents a linear process, and the shape of each reconstructed point is independent of its location in the volume. Thus, according to the theory of linear systems,[62,63] the performance of this algorithm can be described by a single point-spread function, from which a convolution kernel can be determined to deconvolute the simple back-projection and obtain the correct reconstruction.

Two different types of algorithms can be found in the literature: (1) weighted back-projection methods, in which this deconvolution is achieved by dividing the Fourier transform of the reconstructed object by the appropriate transfer function (the Fourier transform of the point-spread function), and (2) deconvoluted back-projection methods, which carry out this deconvolution in real space.[24,32,35,36,64]

In the following, the transfer function of the simple back-projection is derived by making use of the projection theorem and by comparing the Fourier transforms of the original object with that of the object obtained by simple back-projection.

The Fourier transform $\hat{P}$ of a back-projection body $\hat{p}$ is

$$\hat{P}(x^*, y^*, z^*) = \int_{-\infty}^{+\infty} \hat{p}(x, y, z)e^{-2\pi i(xx^* + yy^* + zz^*)}\, dx\, dy\, dz \tag{33}$$

Substitution of Equations 32 and 30 into Equation 33 yields:

$$\begin{aligned}\hat{P}(x^*, y^*, z^*) &= \iiiint\!\!\int_{-\infty}^{+\infty} p(x', y')\delta(x' - x, y' - y) \\ &\quad \cdot c(z)\, dx'\, dy' \cdot e^{-2\pi i(xx^* + yy^* + zz^*)}\, dx\, dy\, dz \\ &= \int_{-\infty}^{+\infty}\int_{-\infty}^{+\infty}\int_{-a}^{+a} p(x, y)\, e^{-2\pi i(xx^* + yy^*)}\, dx\, dy\, e^{-2\pi i(zz^*)}\, dz \\ &= P(x^*, y^*) \cdot 2a \cdot \mathrm{sinc}(2a\pi z^*)\end{aligned} \tag{34}$$

with $\mathrm{sinc}(x) = \sin(x)/x$.

The back-projection algorithm consists of a summation of all back-projection bodies placed in their proper orientation:

$$b(x, y, z) = \sum_j \hat{p}(x^j, y^j, z^j) \tag{35}$$

Using Equation 34, the Fourier transform of the back-projected object becomes

$$B(x^*, y^*, z^*) = \sum_j P(x^{*j}, y^{*j}) \cdot 2a \cdot \mathrm{sinc}(2a\pi z^{*j}) \tag{36}$$

where $r^j = (x^j, y^j, z^j) = \omega^j r$, with ω as defined in Equation 19.

For determination of the point-spread function, the algorithm is applied to a set of projections of a single point. The Fourier transform of an image or a three-dimensional volume containing a single point in its origin is constant and, for simplicity, assumed to be 1. Setting $P(x^{*j}, y^{*j}) \equiv 1$ yields:

$$H(x^*, y^*, z^*) = \sum_j 2a \cdot \mathrm{sinc}(2a\pi z^{*j}) \tag{37}$$

where e is the Fourier transform of a single point reconstructed by simple back-projection. H in Equation 37 is the transfer function of a simple back-projection. As no assumptions were made about the specific tilt geometry, H is the transfer function for a back-projection from projections with arbitrary tilt geometry. To obtain the correct reconstruction, the Fourier transform of the back-projection has to be divided by H. H is real valued and thus equivalent to a weight applied to the Fourier transform. A reconstruction obtained by back-projection with subsequent division by H is called a weighted back-projection. According to the convolution theorem for Fourier transforms, a division of the three-dimensional Fourier transform of the back-projection by H is equivalent to a deconvolution of the object with the point-spread function of the back-projection. The division by H is possible as long as $H \neq 0$. However, wherever $H = 0$, no information is present, as is the case, for instance, within the missing wedge for a single-axis tilt series with limited angular range, and without the use of *a priori* knowledge, no information can be recovered within this range. The artifacts caused by the missing information are not specific for a weighted back-projection, but occur in any reconstruction algorithm that does not make use of *a priori* knowledge.

The division by H can either be applied to the Fourier transform of the three-dimensional volume after back-projection or to the Fourier transforms of the projections before back-projection. Because of the projection theorem, the two different modes of application are mathematically equivalent. The amount of calculations necessary and the accuracy of the reconstruction, however, may be quite different, depending on the data set and tilt geometry used. If only a few projections are available, then an application of the weighting function to the projections is more efficient; however, if the number of projections becomes very large, then the amount of calculations will be smaller when the weighting function is applied to the three-dimensional Fourier transform of the back-projected object. The application of the weighting function amounts to a specific high-pass filtration. If it is applied to the projections, then an appropriate interpolation has to be used during the summation of the back-projection bodies to limit the numerical errors.

From Equation 37, special analytical forms of the weighting function can easily be derived. If a is assumed to be infinite, then the sinc function becomes a δ-function. For a conical tilt series where $\psi = 0$ and $\theta_j = \theta_0 =$ constant, the right-hand side of Equation 37 is solely a function of ϕ_j. If, furthermore, the angle ϕ is considered to be continuous, the sum in Equation 37 can be replaced by an integral, resulting in:

$$H(x^*, y^*, z^*) = \int_{-\pi}^{+\pi} \delta(z'^*(\phi))\, d\phi \tag{38}$$

Expressing $z'^*(\phi)$ by the coordinates in the object coordinates system using the transformation Equation 19, Equation 38 becomes:

$$H(x^*, y^*, z^*) = \int_{-\pi}^{+\pi} \delta[x^*\sin(\theta_0)\cos(\phi) + y^*\sin(\theta_0)\sin(\phi) + z^*\cos(\theta_0)]d\phi \tag{39}$$

After x^*, y^*, and z^* are expressed in cylinder coordinates τ^* Γ, z^*, Equation 39 can be solved with the aid of the relation:

$$\delta(f(x)) = \sum_n \frac{1}{|f'(x_n)|} \delta(x - x_n), \quad \text{with } f'(x_n) = \left.\frac{df(x)}{dx}\right|_{x=x_n} \tag{40}$$

Equation 40 is valid for $f'(x_n) \neq 0$. x_n are the zeros of f(x) within the integration range.

The resulting transfer function of a back-projection for conical geometry is

$$H(r^*, \Gamma, z^*) = \frac{2}{\sqrt{r^{*2}\sin^2(\Theta_0) - z^{*2}\cos^2(\Theta_0)}} \qquad r^* > z^*\cot\Theta_0 \tag{41}$$

and the three-dimensional weighting function W_c for conical tilting is

$$W_c = 1/H(r^*, \Gamma, z^*) = 1/2 \cdot \sqrt{r^{*2}\sin^2(\Theta_0) - z^{*2}\cos^2(\Theta_0)} \tag{42}$$

In an analogous fashion, the transfer function for single-axis tilting with equal angular increments can be determined. In a single-axis tilt series, ψ and ϕ in the matrix ω (Equation 19) are 0 and θ is assumed to be continuous. Again, the a in Equation 37 is assumed to be ∞.

The transfer function of a simple back-projection from a single-axis tilt series therefore becomes

$$H(r^*, \Gamma, z^*) = 1/r^* \tag{43}$$

and the weighting function W_s for single-axis tilting is simply

$$W_s = r^* \tag{44}$$

which leads to the so-called r*-weighted back-projection.[32,33,35] Because the single-axis tilt geometry was the first used in electron microscopy, the r*-weighted back-projection was also the first to be applied. The weighting function for conical tilting was derived much later,[36] and the weighting function in Equation 37 for arbitrary tilt geometry was first applied in 1986[24] in connection with the reconstruction from a single-exposure random conical tilt series.

For single-axis tilt geometries, the deconvolution has not only been done by a multiplication of the Fourier transform with r*, but also by the equivalent convolution applied to the real-space data, the convolution kernel being the inverse Fourier transform of r*. However, as r* is not square integrable, as is required for the Fourier transform to exist, assumptions about a band limit of the function have to be made. The convolution kernels found in the literature differ because of different band-limiting functions used.

One of the earliest deconvolution kernels was derived in connection with three-dimensional reconstruction in radio astronomy by Bracewell and Riddle.[32] Let F(x*, y*) be the Fourier transform of f(x, y).

$$f(x, y) = \int_{-\infty}^{+\infty} \int_{-\infty}^{+\infty} f(x^*, y^*) e^{2\pi i(x^*x + y^*y)}\, dx^*\, dy^* \tag{45}$$

By converting the spatial frequency variables to polar coordinates and assuming a maximum Fourier radius M as a band limit, Equation 45 becomes:

$$= \int_0^{\pi} \int_{-M}^{M} F(r^*, \Theta)\, e^{2\pi i r^*(x\cos\Theta + y\sin\Theta)}\, r^*\, dr^*\, d\Theta \tag{46}$$

The integral over θ in Equation 46 can be interpreted as the continuous form of the back-projection, $F(r^*, \theta)$ as the set of Fourier transforms of the one-dimensional projections at angles θ, and r^* as the weighting function for single-axis tilting. In the article by Bracewell and Riddle,[32] the r^* is replaced by a band-limited function

$$K(r^*) = \Pi\left(\frac{r^*}{2M}\right) - \Lambda\left(\frac{r^*}{M}\right)$$

with

$$\Pi(x) = \begin{cases} 1 & \text{for} \quad |r^*| < 1/2 \\ 0 & \text{for} \quad |r^*| > 1/2 \end{cases}$$

and (47)

$$\Lambda(r^*) = \begin{cases} 1 - |r^*| & \text{for} \quad |r^*| < 1 \\ 0 & \text{for} \quad |r^*| > 1 \end{cases}$$

The inverse Fourier transform of $K(r^*)$ is the real-space convolution kernel of a weighted back-projection from a single-axis tilt series.

$$\int_{-\infty}^{+\infty} K(r^*)\, e^{2\pi i r^* r}\, dr = 2M \operatorname{sinc}(2\pi M r^*) - M \operatorname{sinc}^2(M\pi r) = k(r) \tag{48}$$

While for the derivation of Equation 48 a cutoff filter has been used to limit the Fourier transform, a number of other convolution kernels can be found that are derived using other band-limiting functions.[65,66]

D. ITERATIVE RECONSTRUCTION METHODS

Another class of methods are the iterative reconstruction algorithms. These were introduced by Gordon et al.[37] in application to a single-axis tilt geometry and modified later by Goitein,[67] by Gilbert and by Budinger and Gullberg.[70,73] They have been extended to conical tilt geometry by Colsher,[68,69] where an extensive review of these techniques also can be found.

The three major types of iterative algorithms are ART (Arithmetic Reconstruction Technique),[37] SIRT (Simultaneous Iterative Reconstruction Technique),[70] and ILST (Iterative Least-Squares' Technique).[67]

All three techniques start with an estimate of the object, which can either be a uniform density distribution or a simple back-projection of the object. The calculated three-dimensional density is projected in the direction of the original projections and the differences between the experimental projections and the calculated images are determined. From these differences, corrections to the points in three dimensions are calculated. In ART, these calculations are carried out sequentially for one projection at a time, whereas in SIRT the corrections from all projecting directions are applied simultaneously. In ART and SIRT, the corrections are based on the difference between the calculated and measured projections. In ILST, the error is the squared difference divided by an estimate of the variance of the noise along each ray.

In a conical geometry, the direction of a projection is determined by the angles θ_0 and ϕ_1. The value of a point at location (m, n) in the projection at angles (θ_0, ϕ_1) is calculated using the ray sum

$$R^q(\Theta_0, \phi_1, m, n) = \sum_{\substack{\text{all points contained}\\ \text{in ray } (\Theta_0,\, \phi_1,\, m,\, n)}} \sum \sum D^q(i, j, k) \tag{49}$$

$D^q(i, j, k)$ is the three-dimensional density. The index q indicates the q-th iteration. Let $P(\theta_0, \phi_1, m,n)$ be the value of the experimental projection at point m,n; then in ART and SIRT, the error at this point is calculated as

$$E_q(\Theta_0, \phi_1, m, n) = P(\Theta_0, \phi_1, m, n) - R^q(\Theta_0, \phi_1, m, n) \tag{50}$$

In additive ART, the error is distributed over the projecting ray, and the density at each voxel is changed according to

$$D^{q+1}(i, j, k) = D^q(i, j, k) + E^q(\Theta_0, \phi_1, m, n)/N_{lmn} \tag{51}$$

for (i, j, k) in ray (l, m, n). N_{lmn} is the number of points within the ray.

In multiplicative ART, the voxel value is changed according to:

$$D^{q+1}(i, j, k) = D^q(i, j, k) \cdot \frac{P(\Theta_0, \phi_1, m, n)}{R^q(\Theta_0, \phi_1, m, n)} \tag{52}$$

again for voxel (i, j, k) within the ray (l, m, n).

Equations 51 and 52 are applied for each projection, one at a time. The result is that after one iteration, the differences between the last ray sums and the values of the points in the last-used projection are minimal.

In SIRT, the errors for all projections are calculated simultaneously, and the corrections that are applied to each point in the three-dimensional density are calculated from the combined error of all projections.

In the third iterative method, ILST,[67] the residual between all experimental and all calculated projections, defined as

$$\Delta^q = \sum_l \sum_m \sum_n [P(\Theta_0, \phi_1, m, n) - R^q(\Theta_0, \phi_1, m, n)]^2/\sigma^2(l, m, n) \tag{53}$$

is minimized. σ is the variance of the noise in the ray at location (m,n) on the l-th projection. The solution to Equation 53 is found by an iterative relaxation technique.

V. RESOLUTION OF A THREE-DIMENSIONAL RECONSTRUCTION

Before a three-dimensional structure determination is attempted, the resolution that is required to provide an answer to the functional problem being studied should be estimated. For example, if the existence of a channel with a diameter of 2 nm in expected, then the resolution of the reconstruction should be in that range. The required resolution determines aspects of the reconstruction process from the very beginning, the image recording in the microscopy, and sometimes even the biochemical preparation. As shown below, the final resolution in three dimensions is dependent on the ratio between the size of the complete object being reconstructed and the number of projections available. Given a fixed number of projections, the resolution deteriorates when the size of the object gets larger. If the detail in question is smaller than the predicted resolution, two options are available. Either the number of images can be increased or a biochemical preparation should be tried that contains only a fraction of the particle. For example, if a particle consists of a number of identical subunits, then their relative location can be found in a reconstruction with rather low resolution. If details at higher resolution are required, the subunits should be isolated and analyzed separately.

A. RESOLUTION AS A FUNCTION OF THE NUMBER OF PROJECTIONS

Although the number of projections and the object size are the primary limitations of the resolution, other factors contribute. The resolution of each image in a tilt series should be better than the resolution expected in the reconstruction. Further limitations of resolution can result from inaccuracies in the numerical procedures used for the alignment of the projections toward a common origin or inaccuracies in the measured tilt angles and, in reconstructions from random or random conical tilt series, from structural variabilities among particles that are assumed to be identical.

The relation between the number of projections and the size of the object is determined by Shannon's sampling theorem,[61] which states that for an image with a resolution d, at least two measurements per resolution element need to be available. Assuming equally spaced measurements, they should not be separated by more than d/2. For the Fourier transform, Shannon's sampling theorem states that for an object of diameter D, the measurements points in the Fourier transform should not be further than 1/D apart.

The number of projections needed for a specific resolution in a reconstruction from a single-axis tilt series is[29,32]

$$N = \pi \frac{D}{d} \tag{54}$$

or, if expressed in terms of the angular increment $\Delta\Theta = \pi/N$ between tilts, the resolution can be determined as

$$d = \Delta\Theta \cdot D \tag{55}$$

For a conical tilt series with equal azimuthal angular increments, the relation between the number of projections, the resolution, and the diameter of the reconstructed object is[25,45,46]

$$d = 2\pi \cdot D/N \cdot \sin\Theta_0 \qquad \text{for} \qquad N \text{ even} \tag{56}$$

and

$$N = \frac{\pi D}{2d} \tan\Theta_0 \cdot \sqrt{16\cos^2\Theta_0 - (d/D)^2} \qquad \text{for} \qquad N \text{ odd}, \quad \Theta_0 < \pi/2 - \Delta\phi/4 \tag{57}$$

$\Delta\phi$ being the azimuthal angular increment. For large N, both formulas result in approximately the same value for N. If θ_0 in Equation 56 is set to $\pi/2$, which would make the conical tilting scheme degenerate into a single-axis tilt geometry, then the number of projections resulting is twice as large as found with Equation 54. The reason is that the azimuthal angle ϕ in a conical tilt series ranges over 360° instead of just 180°, and thus half of the projections contain redundant information.

Because the resolution formulas above do not take into account the missing wedge in a limited-angle, single-axis tilt series or the missing cone in a conical tilt series, the resolution d is the resolution in the direction of the x-axis or within the x-y plane, respectively.

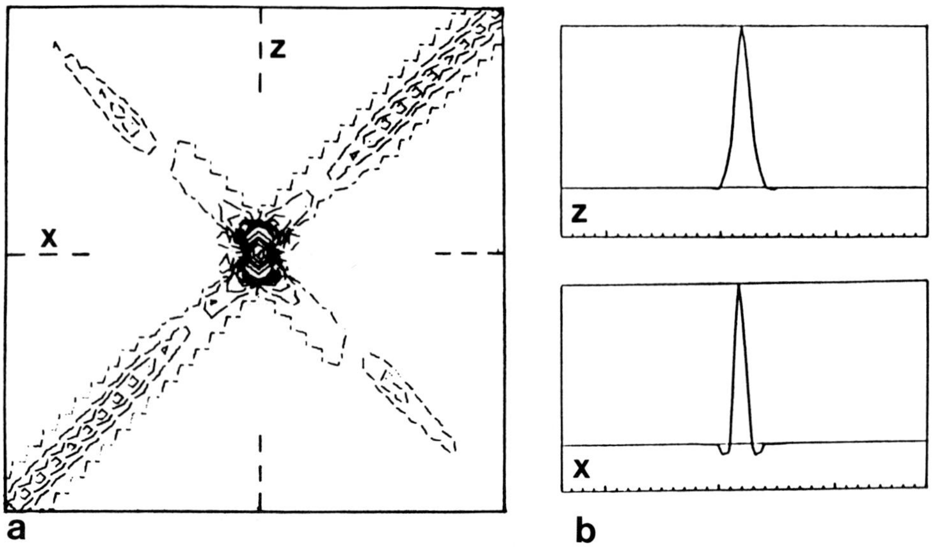

FIGURE 5. (a) Section through a three-dimensional image point reconstructed by weighted back-projection from a conical tilt series. Tilt angle 45°, point reconstructed from 25 projections. Because of the missing cone, the point is elongated and shows side minima. Contour levels relative to image maximum: ----- 1%<0; — — — — 1%>0; —·—— 0; ——— at levels of 10%, 20%, 40% . . . 100%. (b) Cross sections through (a), as indicated. (From Radermacher, M., Dreidimensionale Rekonstruktion bei kegelförmiger Kippung im Elektronenmikroskop, Ph.D. thesis, Technische Universität München, FRG, 1980.)

B. EFFECTS OF THE MISSING DATA

Because of the missing wedge in the Fourier transform of a reconstruction from single-axis tilt series and the missing cone in a reconstruction from a conical tilt series, the point-spread function does not have a circular or spherical symmetry. Figure 5 shows such a point-spread function for a reconstruction from a conical tilt series. It can be seen that the image point is elongated in the z-direction, the direction of the electron beam in the microscope. A closer analysis shows that side minima of this point can also be seen, as well as traces of the projecting rays. A complete analytical description of this point-spread function has only be obtained for a single-axis tilt geometry.[71] For conical geometry, such a description was obtained by numerical integration.[72] Some aspects of this function can also be found by using a quadratic approximation of the image point, essentially describing it as an ellipsoid.[25,45,46] From the ratio between the principal axes of this ellipsoid, an estimate for the resolution in the z-direction can be obtained. The resolution in the z-direction then can be expressed as an elongation factor — the ratio between the long principal axis divided by the short principal axis — multiplied with the resolution in the x-y plane (for conical tilting) or in the x-direction for single-axis tilting around y. However, because this elongation factor is based on an approximation of the point-spread function, it only describes the shape of isolated points, separated sufficiently such that other effects of the side minima and projecting rays (Figure 5) can be neglected.

The elongation factor for single-axis tilting is

$$e_{xz} = \sqrt{\frac{\alpha + \sin\alpha\cos\alpha}{\alpha - \sin\alpha\cos\alpha}} \tag{58}$$

with α being the maximum tilt angle, and for conical tilting

$$e_{xz} = \sqrt{\frac{3 - \sin^2\Theta_0}{2\sin^2\Theta_0}} \tag{59}$$

with θ_0 being the fixed cone angle.

VI. RESTORATION

The distortions caused by the missing data range in the three-dimensional Fourier transform of a three-dimensional reconstruction from a single-axis tilt series with limited tilt range or from a conical tilt series cannot be corrected by deconvoluting the reconstruction with a convolution kernel inverse to the point-spread function. This can be understood most easily by looking at the equivalent problem of dividing the Fourier transform of the reconstruction by the transfer function. This function is equal to 0 in the areas of the missing wedge or missing cone, precluding such a division.

The distortions can be reduced, however, if additional information is available that can be used to provide values for the Fourier transform in the missing region. *A priori* information can be built into the reconstruction process, as is done in some iterative procedures, by including this information in the error criterion that is being minimized, or it can be applied to the finished reconstruction using the method of projections onto convex sets (POCS).[41-43] Because a detailed description of this method is beyond the scope of this chapter, only the basic ideas will be outlined here.

Normally, in a reconstruction from projections that do not cover the complete angular range, all points in regions of the Fourier transform where no data are available are set to 0. This choice selects one specific solution out of many possible ones. Any combination of values chosen for the Fourier transform in the nonmeasured range would result in a three-dimensional volume whose calculated projections would be equally consistent with the measured projection. Thus, for a set of given projections, an infinite set of valid solutions can be found. This solution set can be reduced by the introduction of *a priori* knowledge, such as the limited volume that the structure occupies — in practice, defining a boundary around a reconstructed particle outside of which the values in the volume are assumed to be 0 — or the condition that the range of densities in the reconstruction is limited. Both conditions are applied to the real-space data and will affect every point in the Fourier transform of the object, thus creating inconsistencies with the measured projections. To obtain an object that again is consistent with the projection data, the three-dimensional density distribution is now Fourier transformed, and every point in the Fourier transform for which measured data are available is replaced by its original value. The result is a real-space density distribution that again is in agreement with the projections, but now satisfies the imposed conditions only approximately. The procedure is iterated until the changes to the object become minimal. The final result is a structure that is consistent with the projections as well as with the imposed conditions. The theory of the method of POCS shows that a unique solution can be found if the solution sets defined by *a priori* knowledge are convex and form a common intersection.

VII. PRACTICAL ASPECTS OF THREE-DIMENSIONAL RECONSTRUCTION

In practice, the three-dimensional reconstruction of single particles requires more than just a choice of the reconstruction algorithm. In the following, an outline will be given of the single steps required to carry out a reconstruction, without, however a detailed analysis of the underlying theories. More detailed analysis of the mathematical aspects can be found in the cited literature. We assume as an example a single particle, approximately globular, with a diameter of 30 nm that is to be reconstructed with a resolution better than 3 nm.

A. RECONSTRUCTION FROM SINGLE-AXIS AND CONICAL TILT SERIES

As most of the steps are the same for a single-axis and a conical tilt series, both with equal angular increments, the two geometries will be treated in the same section.

According to Equation 54, N = 31 projections are needed to obtain 3-nm resolution from a single-axis tilt series, which is equivalent to an angular increment of 5.8°. The next smaller integer step is a 5° increment. Assuming a goniometer tilt range of ±60°, a total of 25 images need to be recorded in a single-axis series. This will result in a reconstruction that has details up to a size of 2.6 nm visible in the direction parallel to the specimen plane and approximately 4 nm in the perpendicular direction (calculated from Equation 58).

If the object is to be reconstructed from a conical tilt series using the maximum tilt angle the microscope allows, assumed to be 60°, then according to Equation 57, 55 images are needed corresponding to an azimuthal angular increment of 6.5°. Choosing 6° instead leads to 60 images and, in turn, to a resolution of 2.7 nm within planes parallel to x-y. Using Equation 59, the resolution in the z-direction can be estimated to be 3.3 nm.

Low-dose techniques should be used during microscopy.[74,75] A dose of about 10 el/Å^2 can easily be achieved in a single exposure, so that the cumulative exposure of the specimen is 250 el/Å^2 for a single-axis tilt series and 600 el/Å^2 for a conical tilt series.

As a next step, the images are analyzed in an optical diffractometer. The micrograph is illuminated with a coherent parallel laser beam and the diffraction pattern is observed. All images in the tilt series should be in the underfocus range,[76] and the first 0 transition of the microscope's transfer function should be at a radius that corresponds to a resolution better than 3 nm, the resolution aimed for in the reconstruction. If a higher resolution is anticipated, then this first 0 transition has to be at a larger radius or the transfer function has to be corrected in the computer.[77]

Subsequently, the images are digitized. For a 3-nm resolution in the reconstruction, each pixel in the image should correspond to less than 1.5 nm on the object scale. The digitization should be done with a pixel size that is again smaller by a factor of about 2, since in the course of image processing, the image quality may deteriorate due to interpolations used in rotating and shifting the images. Oversampling by a factor of 2 ensures that the final resolution will not be limited by these operations.

The direction of the tilt axis, which in most microscopes is not parallel to either of the micrograph axes, must be separately determined for each image. This is done by using markers,[78] either gold particles that are applied to the specimen during specimen preparation or other small features that can be followed through the whole tilt series.

The images in the tilt series then are aligned to a common origin in three dimensions.[79] Although this alignment can in many cases be achieved with the aid of the marker coordinates, a higher accuracy can be achieved using cross-correlation techniques. As analyzed in detail in Guckenberger,[79] the images are aligned in a series relative to each other, starting with the 0° image, centered in such a way that its center of gravity coincides with the image center. If the angular increment between the projections in a single-axis tilt series is $\Delta\phi = \pi/N$, then the projection p_i with angle $i \cdot \Delta\phi$ is stretched perpendicular to the tilt axis by a factor of $1/\cos(i \cdot \Delta\phi)$ and the images are successively cross-correlated with each other and shifted such that the center of gravity of the cross-correlation peak coincides with the center of the image. This procedure can be modified by changing the order in which images are aligned to minimize the accumulation of alignment errors.[80] Analogously, for a conical tilt series, each image is stretched by $1/\cos(\theta_0)$, rotated by the azimuthal angular difference to its predecessor, and successively cross-correlated, again such that the center of gravity of the cross-correlation peak coincides with the image center after alignment. The aligned tilt series then are ready to be used in one of the reconstruction algorithms described in this chapter.

B. RECONSTRUCTION FROM A RANDOM CONICAL TILT SERIES

A reconstruction from a single-exposure random conical tilt series to date is the technique that allows the determination of the structure of asymmetrical single particles with minimal radiation damage. Because only one image of each particle is used in the reconstruction, the cumulative dose can easily be kept to less than 10 el/Å^2 compared to 250 to several thousand el/Å^2, which even under low-dose imaging conditions are accumulated in the collection of a regular tilt series.

During microscopy, one image is recorded under low-dose conditions with the specimen tilted by a large angle (>45°). A second 0° image of the same specimen area is taken afterwards and is only used for the determination of particle orientation and particle variability.

The parts that apply to the diffractometry and digitization of the regular tilt series described above also apply for a reconstruction from a single-exposure random conical tilt series. The tilt image is first analyzed in the diffractometer. Usually, it contains a whole range of defocus values, depending on the location of each particle in the micrograph and ranging from underfocus to overfocus. Again, only that area in underfocus is used which also shows the desired resolution. The direction of the tilt axis and the tilt angle again is determined from marker coordinates or small features seen in the image that can serve as markers in a comparison of the tilt image and the 0° image.

Besides the fixed tilt angle, the azimuthal angle of each particle in the image field also has to be determined. This angle is the rotation angle of each particle within the specimen plane and is determined from images extracted from the 0° micrograph, using rotational and translational cross-correlation procedures.[22] Because a reconstruction from a random conical tilt series inherently is an averaging procedure, all particles used in the reconstruction should be of identical shape. Furthermore, they should have the same orientation relative to the specimen support film, because the same tilt angle θ_0 is assumed for all particles. As a pattern recognition problem, this translates into the condition that the images of all particles in the 0° projection must be identical after rotational alignment. In general, this will not be the case, and the subset of particles that fulfills this condition has to be selected. Multivariate statistical analysis, correspondence analysis, nonlinear mapping, and classification techniques[47-52] can be used to achieve this selection.

All angles are known after the 0° images are aligned, and the selected particles now form a conical tilt series with known, random azimuthal angles. The alignment of the tilt images to a common origin is best achieved by a cross-correlation of each tilt image with its 0° counterpart, again after stretching the tilt image by $1/\cos(\theta_0)$. To increase the significance of the cross-correlation peak, a large area surrounding the particle can be used in this cross-correlation, since not only do the particles in the 0° image and the tilted image have the same shape, but the surrounding areas also are projections of the same part of the support film.

The aligned tilt images then can be used in a reconstruction using the weighted back-projection algorithm for arbitrary geometry, or any of the iterative methods.

VIII. EXAMPLES OF APPLICATIONS

The reconstruction from a single-axis tilt series was originally developed for the reconstruction of single macromolecules,[13] but because of the problem of radiation damage, macromolecules now are mostly reconstructed using the random conical reconstruction technique. However, reconstructions from single-axis tilt series are still important for particles that do not occur in multiple identical copies, precluding the application of averaging techniques. Its main application is therefore found in the study of subcellular structures in the high-voltage microscope.

A. RECONSTRUCTION OF THICK SECTIONS IN HIGH-VOLTAGE ELECTRON MICROSCOPY

The main application of the single-axis tilt geometry can be found in electron microscopy of sectioned material.[81,82] Here, the structures analyzed differ from one preparation to the next, and only the principal building plan may be conserved. Secondly, because the structures are larger and the required resolutions are lower, radiation damage is less of a problem.

One of the first objects studied by high-voltage electron microscopy and reconstruction was the cilium of newt lung tissue, embedded in a 0.25 μm plastic section.[82] Here, 53 projection were recorded within an angular range of -54 to $+50°$ in 2° increments (Figures 6 and 7). With a diameter of the reconstruction volume of 320 nm, the final resolution was 12 nm. Clearly visible are the central microtubules, the outer double microtubules, and the dynein arms.

B. RECONSTRUCTION OF MACROMOLECULES

The first reconstruction of a macromolecular assembly was the reconstruction of the bacteriophage T4,[10] followed later by the reconstruction of fatty acid synthetase,[13-15] which was done using a single-axis tilt series. Further reconstructions from single-axis tilt series were the reconstruction of the 30S and 50S ribosomal subunit from *E. coli,*[84,85] the 30-nm chromatin fiber,[86] and the Balbiani ring ribonuceloprotein particles.[87]

The only macromolecular structure reconstructed from a regular conical tilt series was the reconstruction of fatty acid synthetase.[45] Because of the limited availability of truly eucentric tilt-rotation stages for electron microscopy, and more so because of the high cumulative dose in collecting such a tilt series, the technique has not been used further for the reconstruction of macromolecules.

Most reconstructions of macromolecular assemblies have been carried out using the single-exposure random conical tilt technique, because of the very low radiation damage of the specimen and also because of the ease of data collection (very few tilt pairs are needed). The first such reconstruction was of the 50S ribosomal subunit from *E. coli* (Figure 8).[22,88] Later, this reconstruction was followed by the structure determination of the 50S subunit depleted of protein L7/L12,[89] the 70S ribosome from *E. coli*,[90] the 40S ribosomal subunit from rabbit reticulocytes,[91] the 80S ribosome from rabbit reticulocytes,[92] the calcium-release channel from sarcoplasmic reticulum,[93] and the reconstruction of the *Androctonus australis* hemocyanin.[94]

The first successful reconstruction of a single asymmetrical particle from random tilts was the reconstruction of the 50S ribosomal subunit from *E. coli* in a frozen, hydrated preparation.[58]

IX. CONCLUSION

Within the last 20 years, three-dimensional reconstruction of macromolecular assemblies from electron micrographs has increasingly gained importance. A large number of reconstruction algorithms are available, some specialized for specific geometries or symmetries, others having general applicability. The quality of both Fourier methods and weighted back-projection methods can be described by a point-spread function, whereas for iterative reconstruction methods, the quality of the result is less predictable and can depend on the object. Originally, only iterative methods had allowed an implementation of *a priori*

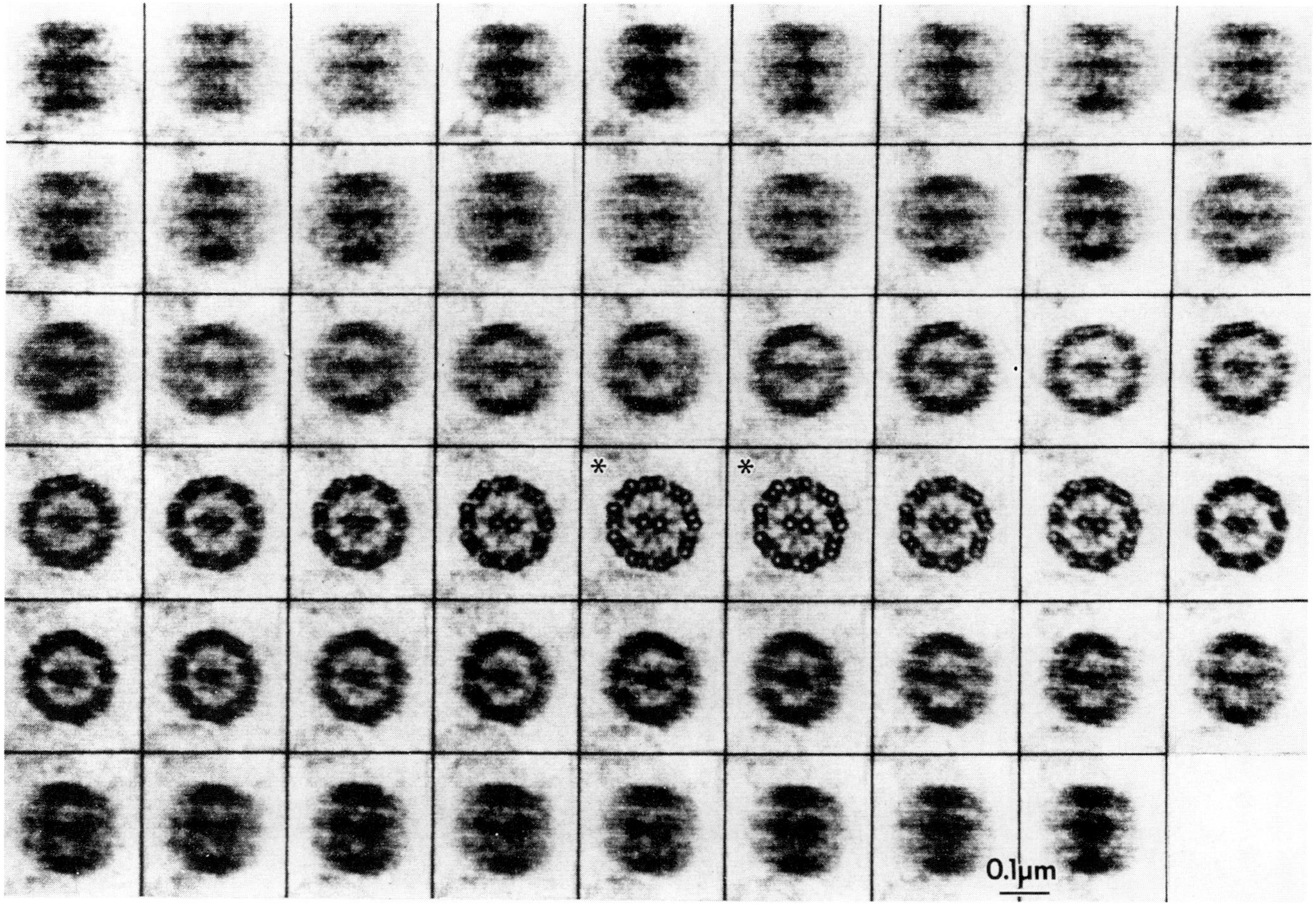

FIGURE 6. Single-axis tilt series of the newt lung cilium. (From McEwen, B., Radermacher, M., Rieder, C. L., and Frank, J., *Proc. Natl. Acad. Sci. U.S.A.*, 83, 9040, 1986. With permission.)

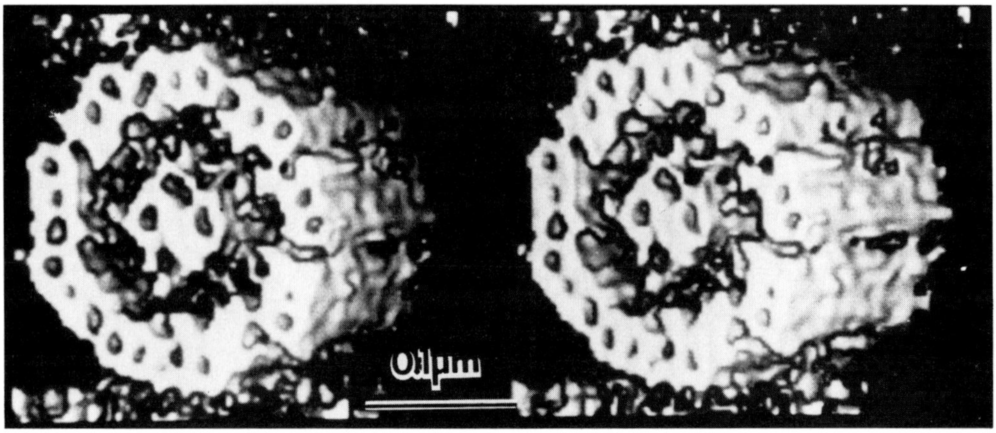

FIGURE 7. Three-dimensional reconstruction of the newt lung cilium. (From Frank, E., McEwen, B., Radermacher, M., and Rieder, C. L., in *Proc. 44th Annu. Meet. EMSA,* Bailey G. W., Ed., San Francisco Press, San Francisco, CA, 1986, 18. © 1990 by San Francisco Press, Inc. and produced by permission.)

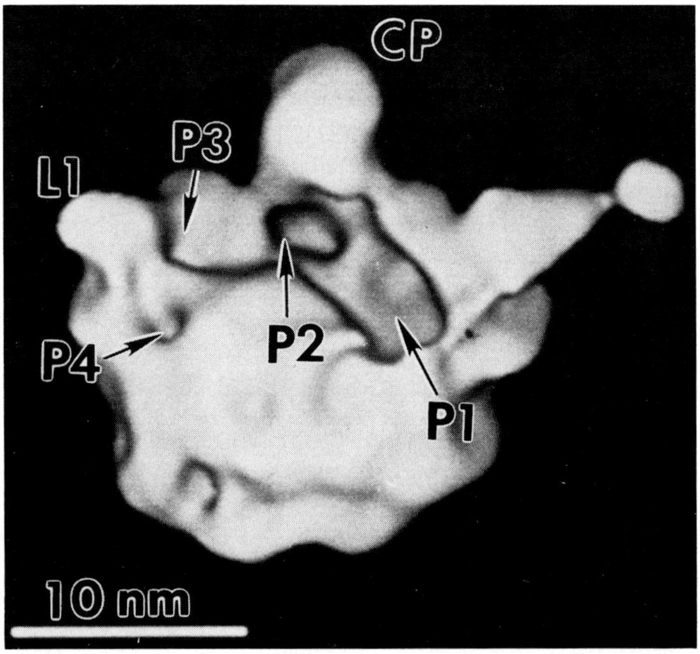

FIGURE 8. Reconstruction of the 50S ribosomal subunit from *Escherichia coli* showing clearly the central protuberance (CP), a large interface canyon with several deep pockets (P1, P2, and P3). P1 and P3 are in the location where antibodies had been mapped that affect peptidyl transferase activity.[95-98] P2 is connected to a channel that leads to the back of the subunit[99,100] and, consistent with the immunoelectron microscopical findings by Ryabova et al.,[101] was concluded to be the location of the peptidyl transferase center. (From Frank, J., Verschoor, A., Radermacher, M., and Wagenknecht, T., in *The Ribosome, Structure, Function & Evaluation,* Hill, W. E., Moore, P. B., Dahlberg, A., Schlessinger, D., Garret, R. A., and Warner, J. R., Eds., American Society for Microbiology, Washington, D.C., 1990, 107. With permission.)

information, which later also became possible for noniterative reconstruction methods through use of the POCS method. Because, however — in contrast to reconstructions in medical X-ray tomography — *a priori* knowledge about macromolecules in electron microscope preparations is very limited, an ideal reconstruction technique would rely solely on measured data. One such technique that allows a complete determination of the structure is the multirandom conical tilt scheme, where several random conical tilt series of particles with different orientations relative to the specimen support are combined into a single reconstruction. Much work is currently being done to allow reconstruction from projections of totally randomly oriented particles, which could provide a set of projections that covers the complete angular range and would not lead to a missing region in the Fourier transform of the reconstruction. One of the most successful approaches was the reconstruction technique developed by Provencher and Vogel,[57] which thus far is the only one that has produced interpretable results.

A number of problems concerning reconstruction of single asymmetrical particles still need to be solved. One of the major problems in reconstruction from random projections is the determination of the relative particle orientations. Solutions have been proposed by van Heel[59] and by Goncharov,[60] yet these techniques are extremely noise sensitive, and their applicability to practical problems still needs to be proven. Except for the multicone reconstruction method, where volumes from separate random conical reconstructions can be compared before the data are combined into a single reconstruction, there currently is no method available that can ensure that the images used in an entirely random set of projections represent particles with identical structure.

For the future of reconstruction from single-particle specimens, it will be necessary to improve the moderate resolution achieved to date. Resolution limitations caused by staining, drying, and the associated particle deformations have been overcome by the development of frozen, hydrated preparation techniques, and macromolecules now can be preserved in their native physiological state.

Full advantage of these preparation techniques, however, can only be taken if the resolution in the three-dimensional reconstruction is improved. One limitation is the lack of stability of currently available cryo-holders, which causes blurring in the micrographs. Also, high resolution can only be obtained if the specimen is embedded in very thin ice, but the reproducibility of thin-ice preparations is still limited. In images of frozen, hydrated preparations taken at tilt angles above 45°, large blurred areas are often found that may be attributed to charging of the specimen embedded in the nonconductive ice layer.

Solutions to the above problems are being pursued, and it can be expected that cryopreparation techniques combined with three-dimensional reconstruction methods will lead to the determination of structures within the 0.5-nm resolution range in the near future.

ACKNOWLEDGMENTS

I acknowledge support by National Institutes of Health (NIH) Grants 1R01 GM29169, 1S10 RR03998, and 1RO1 GM40165, and National Science Foundation (NSF) Grant 8313405. Further, I thank J. Frank, C. Mannella, and A. Verschoor for many discussions and critical reading of this chapter.

REFERENCES

1. **Blundell, T. L. and Johnson, L. N.,** *Protein Crystallography,* Academic Press, New York, 1976, 381.
2. **Smith, J. L., Corfield, P. W. R., Hendrickson, W. A., and Low, B. W.,** Refinement at 1.4 Å resolution of a model of erabutoxin $β. Treatment of ordered solvent and discrete disorder, *Acta Crystallogr. Sect. A,* 44, 357, 1988.
3. **Henderson, R., Baldwin, J. M., Ceska, T. A., Zemlin, F., Beckmann, E., and Downing, K. H.,** Model for the structure of bacterioodopsin based on high resolution cryomicroscopy, *J. Mol. Biol.,* 213, 899, 1990.
4. **Moody, M. F.,** Structure of the sheath of bacteriophage T4, *J. Mol. Biol.,* 25, 167, 1967.
5. **Taylor, K. A. and Glaeser, R. M.,** Electron microscopy of frozen hydrated biological specimens, *J. Ultrastruct. Res.,* 55, 448, 1976.
6. **Dubochet, J., Lepault, J., Freeman, R., Berriman, J. A., and Homo, J.-C.,** Electron microscopy of frozen water and aqueous solutions, *J. Microsc.,* 128, 219, 1982.
7. **Lepault, J., Booy, F. P., and Dubochet, J.,** Electron microscopy of frozen biological suspensions, *J. Microsc.,* 129, 89, 1983.
8. **Dubochet, J., Adrian, M., Chang, J.-J., Homo, J.-C., Lepault, J., McDowall, A. W., and Schultz, P.,** Cryo-electron microscopy of virtified specimens, *Q. Rev. Biophys.,* 21, 129, 1988.
9. **Glaeser, R. M.,** Limitations to significant information in biological electron microscopy as a result of radiation damage, *J. Ultrastruct. Res.,* 36, 466, 1971.
10. **DeRosier, D. J. and Klug, A.,** Reconstruction of three dimensional stuctures from electron micrographs, *Nature,* 217, 130, 1968.
11. **Bender, R., Bellman, S. H., and Gordon, R.,** ART and the ribosome, *J. Theor. Biol.,* 29, 483, 1970.
12. **Hoppe, W., Langer, R., Knech, G., and Poppe, C.,** Proteinkristallstrukturanalyse mit Elektronenstrahlen, *Naturwissenschaften,* 55, 333, 1968.
13. **Hoppe, W., Schramm, H. J., Sturm, M., Hunsmann, N., and Gaßmann, J.,** Three-dimensional electron microscopy of individual biological objects. I. Methods, *Z. Naturforsch. Teil A,* 31, 645, 1976.
14. **Hoppe, W., Schramm, H. J., Sturm, M., Hunsmann, N., and Gaßmann, J.,** Three-dimensional electron microscopy of individual biological objects. II. Test calculations, *Z. Naturforsch. Teil A,* 31, 645, 1976.
15. **Hoppe, W., Schramm, H. J., Sturm, M., Hunsmann, N., and Gaßmann, J.,** Three-dimensional electron microscopy of individual biological objects. III. Experimental results on yeast fatty acid synthetase, *Z. Naturforsch. Teil A,* 31, 645, 1976.
16. **Frank, J., Goldfarb, W., Eisenberg, D., and Baker, T. S.,** Reconstruction of glutamine synthetase using computer averaging, *Ultramicroscopy,* 3, 283, 1978.
17. **Frank, J. and Goldfarb, W.,** Methods for averaging of single molecules and lattice fragments, in *Electron Microscopy at Molecular Dimensions,* Baumeister, W. and Vogell, W., Eds., Springer-Verlag, Berlin, 1980, 261.
18. **Frank, J., Verschoor, A., and Boublik, M.,** Computer averaging of electron micrographs of 40S ribosomal subunits, *Science,* 214, 1353, 1981.
19. **Verschoor, A., Frank, J., Radermacher, M., Wagenknecht, T., and Boublik, M.,** Three-dimensional reconstruction of the 30S ribosomal subunit from randomly oriented particles, in *Proc. 41st Annu. Meet. EMSA,* Bailey, G. W., Ed., San Francisco Press, San Francisco, 1983, 758.
20. **Vershoor, A., Frank, J., Radermacher, M., Wagenknecht, T., and Boublik, M.,** Three-dimensional reconstruction of the 30S ribosomal subunit from randomly oriented particles, *J. Mol. Biol.,* 178, 677, 1984.
21. **van Heel, M.,** Three-dimensional reconstruction of the 30S ribosomal subunit, in *Proc. 41st Annu. Meet. EMSA,* Bailey, G. W., Ed., San Francisco Press, San Francisco, 1983, 460.
22. **Radermacher, M., Wagenknecht, T., Verschoor, A., and Frank, J.,** Three-dimensional reconstruction from a single-exposure random conical tilt series applied to the 50S ribosomal subunit of *Escherichia coli, J. Microsc.,* 146, 113, 1987.
23. **Frank, J., Verschoor, A., Wagenknecht, T., Radermacher, M., and Carazo, J.-M.,** A new non-crystallographic image-processing technique reveals the architecture of the ribosome, *Trends Biochem. Sci.,* 13, 123, 1988.
24. **Radermacher, M., Wagenknecht, T., Verschoor, A., and Frank, J.,** A new 3-D reconstruction scheme applied to the 50S ribosomal subunit of *E. coli, J. Miscosc.,* 141, RP1, 1986.
25. **Radermacher, M.,** Three-dimensional reconstruction of single particles from random and nonrandom tilt series, *J. Electron Microsc. Tech.,* 9, 359, 1988.
26. **Radon, J.,** Über die Bestimmung von Funktionen durch ihre Integralwerte längs gewisser Mannigfaltigkeiten, *Ber. Verh. König. Sächs. Ges. Wiss. Leipzig, Math. Phys. Kl.* 69, 262, 1917.
27. **Cormack, A. M.,** Representation of a function by its line integrals, with some radiological applications, *J. Appl. Phys.,* 34, 2722, 1963.

28. **Cormack, A. M.,** Representation of a function by its line integrals, with some radiological applications. II, *J. Appl. Phys.,* 35, 2908, 1964.
29. **Crowther, R. A., DeRosier, D. J., and Klug, A.,** The reconstruction of a three-dimensional structure from projections and its application to electron microscopy, *Proc. R. Soc. London, Ser. A,* 317, 319, 1970.
30. **Smith, P. R., Peter, T. M., and Bates, R. H. T.,** Image reconstruction from a finite number of projections, *J. Phys. A Math. Nucl. Gen.,* 6, 361, 1973.
31. **Hoppe, W.,** Das Endlichkeitspostulat und das Interpolationstheorem der dreidimensionalen elektronenmikroskopischen Analyse aperiodischer Strukturen, *Optik,* 29, 617, 1969.
32. **Bracewell, R. N. and Riddle, A. C.,** Inversion of fan-beam scans in radio astronomy, *Astrophys. J.,* 150, 427, 1967.
33. **Vainshtein, B. K.,** Finding the structure of objects from projections, *Sov. Phys. Crystallogr.,* 15, 781, 198.
34. **Vainshtein, B. K. and Orlov, S. S.,** Theory of the recovery of functions from their projections, *Sov. Phys. Crystallogr.,* 17, 253, 1972.
35. **Gilbert, P. F. C.,** The reconstruction of a three-dimensional structure from projections and its application to electron microscopy. II. Direct methods, *Proc. R. Soc. London Ser. B,* 182, 89, 1972.
36. **Radermacher, M. and Hoppe, W.,** 3-D reconstruction from conically tilted projections, *Proc. 9th. Int. Congr. Electron Microsc.,* 1, 218, 1978.
37. **Gordon, R., Bender, R., and Herman, G. T.,** Algebraic reconstruction techniques (ART) for three-dimensional electron microscopy and X-ray photography, *J. Theor. Biol.,* 29, 471, 1970.
38. **Crowther, R. A. and Klug, A.,** ART and science, or, conditions for 3-D reconstruction from electron microscope images, *J. Theor.Biol.,* 32, 199, 1971.
39. **Bellman, S. H., Bender, R., Gordon, R., and Rowe, J. E.,** ART is science being a defense of algebraic reconstruction techniques for three-dimensional electron microscopy, *J. Theor. Biol.,* 32, 205, 1971.
40. **Trussel, H. J.,** *A priori* knowledge in algebraic reconstruction methods, in *Advances in Computer Vision and Image Processing,* Vol. 1, JAI Press, Greenwich, CT, 1984, 265.
41. **Tam, K. C. and Perez-Mendez, V.,** Tomographical imaging with limited-angle input, *J. Opt. Soc. Am.,* 71, 582, 1981.
42. **Sesan, M. I. and Stark, H.,** Image restoration by the method of convex projections. II. Applications and numerical results, *IEEE Trans. Med. Imag.,* 1, 95, 1982.
43. **Carazo, J.-M. and Carracosa, J. L.,** Information recovery in missing angular data cases: an approach by the convex projections method in three dimensions, *J. Microsc.,* 145, 23.
44. **Deans, S. R.,** *The Radon Transform and some of its Applications,* John Wiley & Sons, New York, 1983.
45. **Radermacher, M.,** Dreidimensionale Rekonstruktion bei kegelförmiger Kippung im Elektronenmikroskop, Ph.D. thesis, Technische Universität München, FRG, 1980.
46. **Radermacher, M. and Hoppe, W.,** Properties of 3-D reconstruction from projections by conical tilting compared to single axis tilting, in *Proc. 7th Eur. Congr. Electron Microscopy,* Vol. 1, The Hague, 132, 1980.
47. **van Heel, M. and Frank, J.,** Use of multivariate statistics in analyzing the images of biological macromolecules, *Ultramicroscopy,* 6, 187, 1981.
48. **Frank, J. and van Heel, M.,** Correspondence analysis of aligned images of biological particles, *J. Mol. Biol.,* 161, 134, 1982.
49. **van Heel, M.,** Multivariate statistical of noisy images (randomly oriented biological macromolecules), *Ultramicroscopy,* 13, 165, 1984.
50. **Radermacher, M. and Frank, J.,** Use of nonlinear mapping in multivariate image analysis of molecule projection, *Ultramicroscopy,* 17, 117, 1986; Erratum, *Ultramicroscopy* 19, 75, 1986.
51. **Frank, J.,** Classification of macromolecular assemblies studied as single particles, *Q. Rev. Biophys.,* 23, 281, 1990.
52. **Carazo, J.-M., Rivera, F. F., Zapata, E. L., Radermacher, M., and Frank, J.,** Fuzzy sets-based classification of electron microscopy images of biological macromolecules with an application to ribosomal particles, *J. Microsc.,* 157, 187, 1990.
53. **Frank, J., Carazo, J.-M., and Radermacher, M.,** Refinement of the random conical reconstruction technique using multivariate statistical analysis and classification, *Eur. J. Cell Biol. Suppl.,* 25, 48, 143, 1988.
54. **Carazo, J.-M., Wagenknecht, T., and Frank, J.,** Variations of the three-dimensional structure of the *Escherichia coli* ribosome in the range of overlap views, *Biophys. J.,* 55, 465, 1989.
55. **Crowther, R. A.,** Procedures for three-dimensional reconstruction of spherical viruses by Fourier synthesis from electron micrographs, *Philos. Trans. R. Soc. London Ser. B,* 261, 221, 1971.
56. **Kam, Z.,** The reconstruction of structure from electron micrographs of randomly oriented particles, *J. Theor. Biol.,* 82, 15, 1980.
57. **Provencher, S. W. and Vogel, R.,** Three-dimensional reconstruction from electron micrographs of disordered specimens. I. Method, *Ultramicroscopy,* 25, 209, 1988.

58. **Vogel, R. and Provencher, S. W.,** Three-dimensional reconstruction from electron micrographs of disordered specimens. II. Implementation and results, *Ultramicroscopy,* 25, 223, 1988.
59. **van Heel, M.,** Angular reconstitution: *a posteriori* assignment of projection directions for 3D reconstruction, *Ultramicroscopy,* 21, 111, 1987.
60. **Goncharov, A. B. and Gelfand, M. S.,** Determination of mutual orientation of identical particles from their projections by the moments method, *Ultramicroscopy,* 25, 317, 1988.
61. **Shannon, C. E.,** Communication in the presence of noise, *Proc. IRE,* 37, 10, 1949.
62. **Goodman, J. W.,** *Introduction to Fourier Optics,* McGraw-Hill, New York, 1968.
63. **Papoulis, A.,** *System and Transforms with Applications in Optics,* Robert E. Krieger, FL, 1968; McGraw-Hill, New York, 1986 (reprint).
64. **Ramachandran, G. N. and Lakshminarayanan, A. V.,** Three-dimensional reconstruction from radiographs and electron micrographs: application of convolution instead of Fourier transforms, *Proc. Natl. Acad. Sci. U.S.A.,* 68, 2236, 1971.
65. **Kwok, Y. S., Reed, I. S., and Truong, T. K.,** A generalized $|\omega|$-filter for 3D-reconstruction, *IEEE Trans. Nucl. Sci.,* 24, 1990, 1977.
66. **Budinger, T. F, Gullberg, G. T., and Huesman, R. H.,** Emission computed tomography, in *Image Reconstruction from Projections, Topics in Applied Physics,* Vol. 32, Herman, G. T., Ed., Springer-Verlag, Berlin, FRG, 1979, 147.
67. **Goitein, M.,** Three-dimensional density reconstruction from a series of two-dimensional projections, *Nucl. Instr., Methods,* 101, 509, 1972.
68. **Colsher, J. G.,** Iterative Three-Dimensional Image Reconstruction from Projections: Applications in Electron Microscopy, Ph.D. thesis, Lawrence Livermore Laboratory, University of California, Livermore, 1976.
69. **Colsher, J. G.,** Iterative three-dimensional image reconstruction from tomographic projections, *Comput. Graphics Image Process,* 6, 513, 1977.
70. **Gilbert, P. F. C.,** Iterative methods for the three-dimensional reconstruction of an object from projections, *J. Theor. Biol.,* 36, 105, 1972.
71. **Tam, K. C. and Perez-Mendez, V.,** Limited-angle 3-D reconstruction using Fourier transform iterations and Radon transform iterations, *Opt. Eng.,* 20, 586, 1981.
72. **Chiu, M. Y., Barrett, H. H., Simpson, R. G., Chou, C., Arendt, J. W., and Gindi, G. R.,** Three-dimensional radiographic imaging with a restricted view angle, *J. Opt. Soc. Am.,* 69, 1323, 1979.
73. **Budinger, T. F. and Gullberg, G. T.,** Three-dimensional reconstruction in nuclear medicine emission imaging, *IEEE Trans. Nucl. Sci.,* 21, 2, 1974.
74. **Williams, R. C. and Fischer, H. W.,** Electron microscopy of tobacco mosaic virus under conditions of minimal beam exposure, *J. Mol. Biol.,* 52, 121, 1970.
75. **Knauer, V., Schramm, H. J., and Hoppe, W.,** A minimal dose technique in 3-D electron microscopy, in *Electron Microscopy,* Vol. 2, Sturgess, J. M., Ed., Microscopical Society of Canada, Toronto, 1978, 4.
76. **Typke, D. and Radermacher, M.,** Determination of the phase of complex atomic scattering amplitudes from light optical diffractograms of electron microscope images, *Ultramicoscopy,* 9, 131, 1982.
77. **Hoppe, W.,** Use of zone correction plates and other techniques for structure determination of aperiodic objects at atomic resolution using a conventional electron microscope, *Philos. Trans. R. Soc. Ser. B,* 261, 71, 1971.
78. **Luther, P. K., Lawrence, M. C., and Crowther, R. A.,** A method for monitoring the collapse of plastic sections as a function of electron dose, *Ultramicroscopy,* 24, 7, 1988.
79. **Guckenberger, R.,** Determination of a common origin in the micrographs of tilt series in three-dimensional electron microscopy, *Ultramicroscopy,* 9, 167, 1982.
80. **Frank, J., McEwen, B. F., Radermacher, M., Turner, J. N., and Rieder, C. L.,** Three-dimensional tomographic reconstruction in high-voltage electron microscopy, *J. Electron Microsc. Tech.,* 6, 193, 1987.
81. **Olins, A. L., Olins, D. E., Levy, H. A., Durfee, R. C., Margle, S. M., Tinnel, E. P., Hingerty, B. E., Dover, S. D., and Fuchs, H.,** Modeling balbiani ring gene transcription with electron microscope tomography, *Eur. J. Cell Biol.,* 35, 129, 1984.
82. **McEwen, B. F., Radermacher, M., Rieder, C. L., and Frank, J.,** Tomographic three-dimensional reconstruction of cilia ultrastructure from thick sections, *Proc. Natl. Acad. Sci. U.S.A.,* 83, 9040, 1986.
83. **Frank, J., McEwen, B., Radermacher, M., and Rieder, C. L.,** Methods for three-dimensional reconstruction of cellular components within a thick section, in *Proc. 44th Annu. Meet. EMSA,* Bailey, G. W., Ed., San Francisco Press, San Francisco, CA, 1986, 18.
84. **Knauer, V., Hegerl, R., and Hoppe, W.,** Three-dimensional reconstruction and averaging of 30S ribosomal subunits of *Escherichia coli* from electron micrographs, *J. Mol. Biol.,* 163, 409, 1983.
85. **Oettl, H., Hegerl, R., and Hoppe, W.,** Three-dimensional reconstruction and averaging of 50S ribosomal subunits of *Escherichia coli* from electron micrographs, *J. Mol. Biol.,* 163, 431, 1983.
86. **Subirana, J. A., Munoz-Guerra, S., Radermacher, M., and Frank, J.,** Three-dimensional reconstruction of chromatin fibers, *J. Biomol. Stereodyn.,* 1, 705, 1983.

87. **Skoglund, U., Anderson, K., Standberg, B., and Daneholt, B.,** Three-dimensional stucture of a specific pre-messenger RNP particle established by electron microscope tomography, *Nature,* 319, 560, 1986.
88. **Radermacher, M., Wagenknecht, T., Verschoor, A., and Frank, J.,** Three-dimensional stucture of the large ribosomal subunit from *Escherichia coli, EMBO J.,* 6, 1107, 1987.
89. **Carazo, J.-M., Wagenknecht, T., Radermacher, M., Mandiyan, V., Boublik, M., and Frank, J.,** Three-dimensional structure of 50S *Escherichia coli* ribosomal subunits depleted of protein L7/L12, *J. Mol. Biol.,* 201, 393, 1988.
90. **Wagenknecht, T., Carazo, J.-M., Radermacher, M, and Frank, J.,** Three-dimensional reconstruction of the ribosome from *Escherichia coli, Biophys. J.,* 55, 455, 1989.
91. **Verschoor, A., Zhang, N. Y., Wagenknecht, T., Obrig, T., Radermacher, M., and Frank, J.,** Three-dimensional reconstruction of mammalian 40S ribosomal subunit, *J. Mol. Biol.,* 209, 115, 1989.
92. **Verschoor, A. and Frank, J.,** Three-dimensional structure of the mammalian cytoplasmic ribosome, *J. Mol. Biol.,* 214, 737, 1990.
93. **Wagenknecht, T., Grassucci, R., Frank, J., Saito, A., Inui, M., and Fleischer, S.,** Three-dimensional architecture of the calcium channel/foot structure of sarcoplasmic reticulum, *Nature,* 338, 167, 1989.
94. **Boisset, N., Taveau, J.-C., Lamy, J., Wagenknecht, T., Radermacher, M., and Frank, J.,** Three-dimensional reconstruction of native *Androctonus australis, J. Mol. Biol.,* 216, 743, 1990.
95. **Stöffler, G., Bald, R., Kastner, B., Lührman, R., Stöffler-Meilicke, M., and Tischendorf, G.,** Structural organization of the *Escherichia coli* ribosome and localization of functional domains, in *Ribosome: Structure, Function and Genetics,* Chambliss, G., Craven, G. R., Davies, J., Davis, K., Kahan, L., and Nomura, M., Eds., University Park Press, Baltimore, 1980, 171.
96. **Lührman, R., Bald, R., Stöffler-Meilicke, M., and Stöffler, G.,** Localization of the puromycin binding site on the large ribosomal subunit of *Escherichia coli* by immunoelectron microscopy, *Proc. Natl. Acad. Sci. U.S.A.,* 78, 7276, 1981.
97. **Olson, H. M., Grant, P. D., Cooperman, B. S., and Glitz, D. G.,** Immunoelectron microscopic localization of puromycin binding on the large subunit of the *Escherichia coli* ribosome, *J. Biol. Chem.,* 257, 2649, 1982.
98. **Olson, H. M., Nichoolson, A. W., Cooperman, B. S., and Glitz, D. G.,** Localization of sites of photoaffinity labeling of the large subunit of *Escherichia coli* by arylazide derivative of puromycin, *J. Biol. Chem.,* 260, 10236, 1985.
99. **Radermacher, M., Frank, J., and Wagenknecht, T.,** The probable exit site of the polypeptide in the ribosome: analysis of densities in three-dimensional reconstruction, in *Proc. 9th Eur. Congr. on Electron Microscopy,* Vol. 3, Inst. Physics Conf. Ser. 93, Dickinson, H. G. and Goodhew, P. J., Eds., Institute of Physics, Bristol, 1988, 323.
100. **Frank, J., Vershoor, A., Radermacher, M., and Wagenknecht, T.,** Morphologies of eubacterial and eucariotic ribosomes as determined by three-dimensional electron microscopy, in *The Ribosome, Structure, Function & Evolution,* Hill, W. E., Moore, P. B., Dahlberg, A., Schlessinger, D., Garret, R. A., and Warner, J. R., Eds., American Society for Microbiology, Washington D.C., 1990, 107.
101. **Ryabova, L. A., Selivanova, O. M., Baranov, V. I., Vasiliev, V. D., and Spirin, A. S.,** Does the channel for the nascent polypeptide exist in the ribosome? Immune electron microscopy study, *FEBS Lett.,* 226, 255, 1988.

Chapter 13

COMPUTER-AIDED THREE-DIMENSIONAL RECONSTRUCTION FROM SERIAL SECTION IMAGES

Norio Baba

TABLE OF CONTENTS

I. INTRODUCTION

The serial section reconstruction technique has been utilized for a long time as a basic method in biological fields in which a model is built by stacking materials like balsa wood sheets or plastic plates cut along contour lines of objects. Thus, the object is reconstructed from section images. Since the 1970s, computer processing and computer graphics were introduced to assist this work.[1-20] At present, commercial image processors and software packages are provided for the reconstruction, although sometimes investigators may regard these as insufficient, depending on the level of their research work.[3] With advances in the technology of microcomputers and electronics, useful systems based on personal computers were also developed.[21-30] Furthermore, with improvements in the quality of on-line image detectors, even in electron microscopy, a total reconstruction system including an electron microscope for on-line image acquisition was developed.[31] Recently, computer-aided techniques have become more convenient for investigators, although the preparation of serial sections and their images is still cumbersome. Modern systems are capable of visualizing reconstructed models, including shading, from any three-dimensional (3-D) viewpoint with smooth surfaces using colors to easily identify specific parts of the object. Stereo pair images are possible to observe the model in three dimensions,[32-41] which is useful for morphological interpretations.

This chapter describes the method of computer-aided serial section reconstruction from a technological point of view. Most of the recent systems are based on contour line acquisition and polygon-based surface or solid model reconstruction. This chapter mainly focuses on the contour line-based method. The author would like to stress two points. One is the importance of the analysis of relations between serial contour lines. An original set of contour lines should be divided into subsets; the contour lines belonging to each subset reconstruct individual objects, each defined as a physical or biological entity. Of course, fully automatic computer analysis is impossible, but a computer is a powerful aid to analyze contour lines and can select, for example a subset of contour lines reconstructing a single isolated object based on the calculation of the degree of similarity between contour lines.[41] The importance of this analysis increases with the complexity of the structure under investigation. The second point is the necessity of a technique for smoothing the reconstructed object itself by changing the 3-D coordinates of polygons to reconstruct the surface.[40,41] This technique is not based on smooth shading, which does not change the coordinates of the polygons.[43] The proposed smoothing technique is useful when — even though the smooth shading is performed — the roughness of the surface still disturbs visualization of the complex reconstructed model; the details are described in Section IV.

The last part of this chapter describes a second method (using no contour lines), based on a voxel representation of the object, in which all section image data are kept available. By analogy to the picture element (pixel), the volume raster element is termed a voxel. In a computer, the voxels are stored as a stack of section images. This method frees the user from the cumbersome task of tracing contour lines. This technique has similarities to computer tomography (CT) and 3-D confocal microscopy concerning visualization and 3-D image processing.

II. CONTOUR LINE ACQUISITION

First of all, a computer must receive contour line data extracted from serial section images. Several acquisition techniques have been developed: an operator can input contour lines, tracing them from a section image on paper by means of a tablet or with a mouse on a computer screen displaying the section image recorded by a TV camera. In the latter case, computer image processing can assist the tracing. If the contrast of the contour image is sufficient, the image processor can semiautomatically trace the contour line. The operator indicates a point on the contour line with a mouse and the algorithm searches in a direction along which the slope of the contrast is smallest.[35] Furthermore, if the object is distinguishable from the background by a threshold, the processor can obtain the contour lines by using binary image processing. Recent image processors operate interactively: the alignment of the section image series is performed by human visual adjustment of every neighboring section pair on a computer display. If possible, it is preferable to nick a specimen with a laser or other suitable device before cutting to facilitate the alignment.[31] In the procedure using no section micrographs, the investigator traces contour lines in each section micrograph with a black pen on a transparent sheet placed over a photograph. The copied contour lines are input into a computer using a TV camera and binary processing. Each contour line image is stored as a binary image. It is advantageous to produce a set of aligned serial contour line images with transparent sheets in order to roughly envision the object before the final analysis.

The fundamental problem for the imaging of serial sections is the expansion or contraction of section images or local deformations of sections arising during specimen preparation or observation with a microscope. Helpful software programs can be written to correct simple cases, e.g., a global expansion or contraction of the image, or if the nature of the deformation is known. There is an interesting method for correcting local deformations using an elastic model where the image plane is modeled on a material such as a rubber sheet.[44] In this approach, the local distortions were practically corrected by comparing pairs of neighboring section images. The method seems to be a useful tool for serial sectioning. Of course, it is necessary to make a system interactive to manually correct defective contour line images. Figure 1b shows an example of a contour line image extracted from a section image of Figure 1a using an image processor (the specimen is a yeast cell; for details see Section V).

Probably, investigators still fear the difficulty and time involved in preparing and working with long series of serial sections, even though better computer technology could be developed for the reconstruction. A partial solution to this problem is the use of thicker sections, limited, of course, by the specific research project. Marko et al.[34] developed a useful reconstruction system for thick serial section images taken with a high-voltage electron microscope using stereo pair images. The principle is based on the stereo photogrammetry for making terrain maps. Once the stereo pair image has been displayed, contour lines are drawn and superimposed on the stereo view at a preselected number of parallel planes within the thickness of the section. The lines are drawn using a cursor viewed in stereo on the graphics terminal.

Each contour line input produced as mentioned above is subdivided into a set of straight line segments. Therefore, the lines is represented by a set of points which are the connecting points of the straight line segments defined in three dimensions by their x-y-, and z-coordinates. The z-coordinate is determined by estimating the section thickness. Further, an identification label or number is added to the set to distinguish it from other structural parts belonging to a different biological object. Since the biological objects to be reconstructed consist of several structural parts, input contour lines corresponding to such parts should be sorted into groups in order to analyze them individually and display them in a distinguishable manner in the final presentation. A computer can assist in the sorting process if each group is distinguishable from a morphological point of view.

III. ANALYSIS OF RELATIONS BETWEEN SERIAL CONTOUR LINES

Each contour line is linked to one or more neighboring contour lines of the sections above and below. The computer needs the information on which pair of contour lines (or

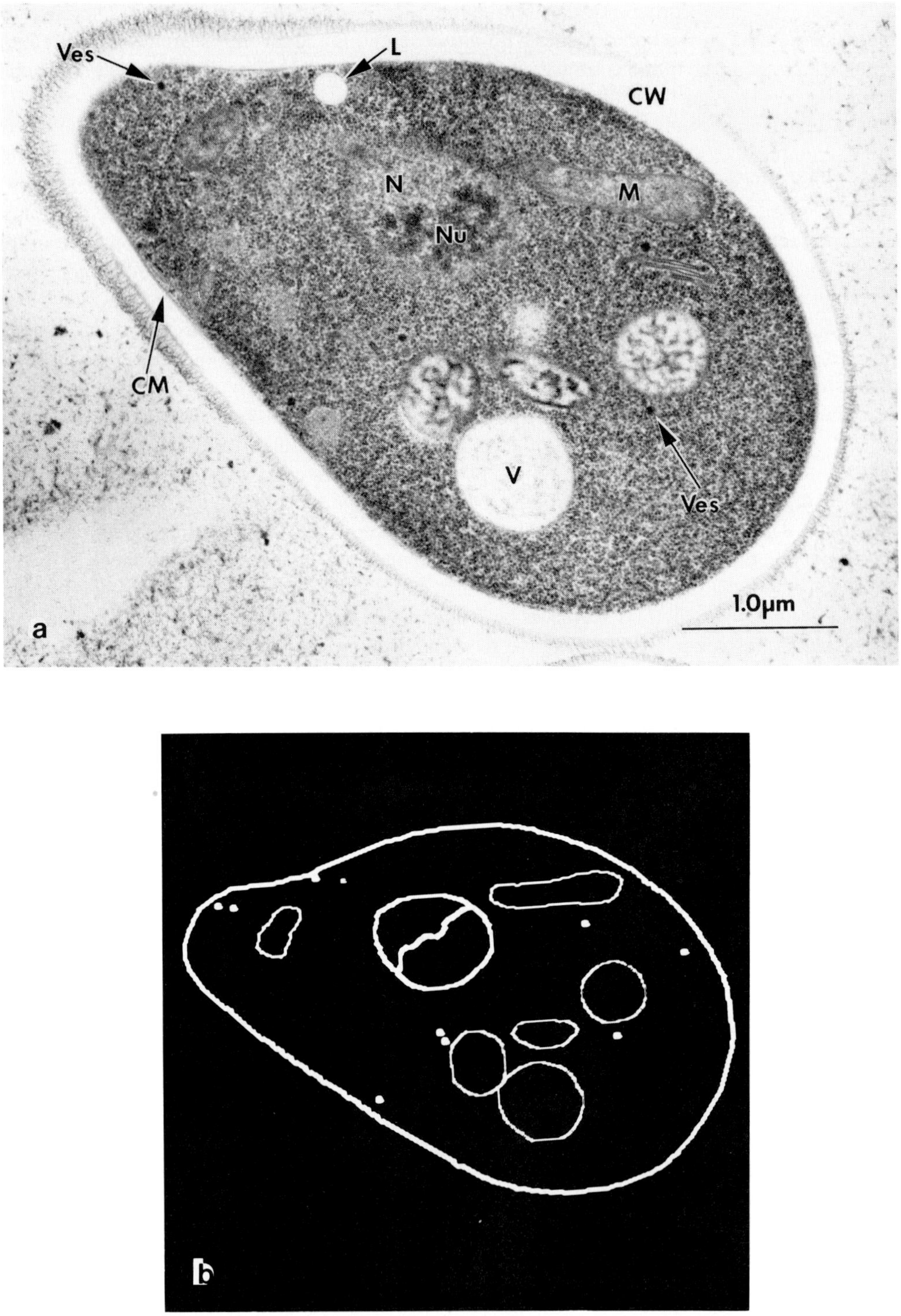

FIGURE 1. (a) Section of a yeast target cell induced by the mating pheromone (α-factor) taken with a 200-kV electron microscope. This section is a typical one among the serial section micrographs. (b) Contour lines of organelles and positions of lipid droplets and small vesicles. Abbreviations: CM, cell membrane; CW, cell wall; L, lipid droplet; M, mitochondrion; N, nucleus; Nu, nucleolus; V, vauole; Ves, vesicle. (From Baba, N. and Kanaya, K., *Scanning Microsc. Suppl.*, 2, 303, 1988. With permission; Baba, N., Baba, M., Imamura, M., Koga, M., Ohsumi, Y., and Kanaya, K., *J. Electron Microsc. Tech.*, 11, 16, 1989. With permission.)

multiple pairs of contour lines) should be linked between neighboring sections. In reconstructions with many contour lines and sections, the operator is probably reluctant to define the relations between all contour lines. The author proposed a simple method to obtain the relations from the overlapping rate between each neighboring pair of contour lines.[42] Figures 2b and 2c illustrate the method using a pair of neighboring contour line images as a typical example. Figure 2a shows a section image taken with a scanning electron microscope (SEM) corresponding to the upper contour lines used in the reconstruction shown in Figure 7. The method is described below. In Figures 2b and 2c, the upper contour lines are drawn as heavy lines and the lower ones are drawn as thin lines.

Every pair of contour lines between the upper and lower sections is listed in a table by the computer. For example, if there are n contour lines in the upper section and m contour lines in the lower one, a total of n × m pairs are listed in the table. The table consists of n rows and m columns. Each pair is examined for a possible linkage according to the criterion described below, and the result is put into the table. This is exemplified by one pair which consists of two contour lines in the upper and lower sections, respectively (see A and B in Figure 2c). Subsequently, binary images are created, and each area is determined for the upper (S_u) and the lower contour line (S_l), respectively: for a closed contour line, the inside is defined as 1 and the outside as 0. In the case of an open contour line, the binary processor dilates the line homogeneously (this dilation processing is similar to that seen in R1 to R5 in Figures 2b and 2c, as described below) and sets the dilated region to 1 inside and 0 outside. For example, in Figure 2c, S_u is an area surrounded by line A and S_l is an area surrounded by line B. An overlapping part (common part) between two binary images is generated by a logical operation, and the area of the common part, S_c, is also calculated. For example, S_c is the area indicated by C in Figure 2c. Finally, the ratios S_c/S_u and S_c/S_l are compared, and the greater ratio is used for further comparison using a predefined threshold. If the measured ratio is greater than the threshold, the pair of contour lines is judged to be linked.

If a single contour line has been found to be related to several contour lines in the upper or lower section, using the method described above, the structure is branched, and the single contour line is appropriately divided into several lines. For instance, contour line D shown in Figure 2c (upper section) is judged to be branched to three contour lines, E, F, and G (lower section). Contour line D is therefore divided into three parts according to the boundary and three isolated regions (R1, R2, and R3) are generated by "dilation processing", as shown in the figure. The "dilation processing" extends the three regions surrounded by the E, F, and G lines homogeneously, but every boundary line between every isolated region is preserved. A similar binary processing is performed for the lower section contour line H in Figure 2b. Line H is divided according to the two regions, R4 and R5, generated by dilating areas I and J, respectively.

This processing is executed for all contour lines through all serial sections to produce a large, final table which shows all relations between the serial contour lines. Therefore, the computer can output the numbers of subsets of contour lines which define isolated solids and branch points, etc., together with the measured surface areas and volumes of every subset or solid, as described below. An application of this automatic subdivision processing of contour lines is shown below. Further development of such analytical methods is desirable so that they can be extended to other types of 3-D analyses and allow easier 3-D data manipulation.

IV. SURFACE AND SOLID MODELINGS

A. SMOOTH TILING

At this point, a 3-D model has been constructed by establishing links between contour lines of the same group. To define such connections, fitting triangular patches between

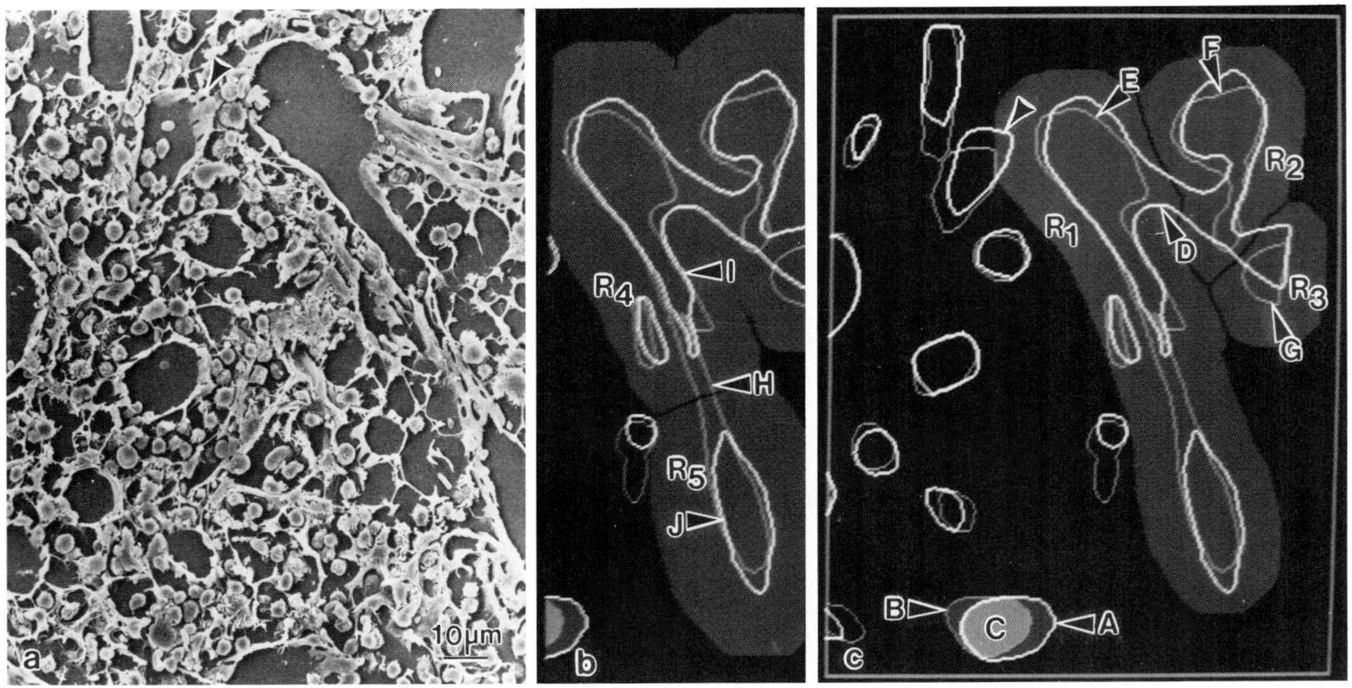

FIGURE 2. Explanatory illustration for automatic analysis of relations between contour lines and automatic branching processing using a practical case. (b, c) Neighboring contour lines in which the upper section contours are drawn in heavy lines and the lower section contours in thin lines; (a) section of the red pulp in the splenomegaly in rat taken with SEM corresponding to the upper contours. For example, a pair of contour lines A and B in (c) is examined to determine whether these should be linked, based on the degree of overlapping between the pair. For instance, contour line D in (c) is judged to be branched to three contour lines, E, F, and G, and automatically divided into three parts. Similar binary processing is performed for contour line H in (b). H is divided according to two regions, R4 and R5, generated by dilating areas I and J, respectively (see text for details).

neighboring contour lines is generally used. This fitting is also referred to as "smooth tiling".[37] The algorithm for the automatic connection used here is based on the following basic idea. A contour line consists of contour points; for each contour point, the algorithm tries to locate corresponding points on the adjacent contour lines which would be the most proper ones for connection. Furthermore, the algorithm examines whether the tiling is continuous or smooth.[40] Figure 3a shows an example of a "wire frame" model of a yeast cell membrane constructed by this algorithm, a contour line within the frame is shown in Figure 1b.

There is a point that must be treated carefully: if the size of an object is comparable to the thickness of the section or a tube-like object, which is curved three dimensionally, neighboring contour lines to be lined frequently deviate considerably, and reasonable links cannot be generated by using the current method. A simple solution is presented in the following.[42] Figure 4 shows a comparison between two methods for the connection in the above case using a pair of elliptical contour lines as an example, in which A in Figure 4 describes the result of the current method and C that of the present, improved one. The present method is performed by the following procedure. First, two binary images are created filling the inside of each contour line with a 1 and the outside with a 0. Second, a cross-correlation function of these binary images is calculated by simple binary processing, and the two contour lines are aligned at the position indicating the maximum correlation, as shown in B in Figure 4. Finally, the two aligned contour lines are linked by the current method and, preserving the generated links, the moved contour line returns to the original position. The improved technique can perform "smooth tiling".

B. CUTTING THROUGH A RECONSTRUCTED MODEL

Some techniques are available to cut through a reconstructed model. The system used here cuts all contour lines for every section along an arbitrary 3-D boundary plane chosen by the user (see Figure 6). The boundary plane is defined by serial two-dimensional boundary lines for all serial sections. Then, all data of the cut contour lines are input into the system again, and the cut 3-D model is reconstructed automatically, except for some minor editing of contour lines. As shown in Figures 6c and 6d, the cut 3-D image not only shows the cut end, but also the whole structure remaining after cutting.[45]

C. MODELING OF SIMPLE STRUCTURAL OBJECTS IN SIMPLE FORMS

In some cases, certain structural objects can be displayed in the final 3-D image in a simplified form such as a sphere and a rod, if this is adequate for the research purpose or from the viewpoint of the resolution. Computer graphics is suitable for this purpose. In the application of intracellular structures in yeast target cells reported here, vesicles and lipid droplets were modeled as spheres, and their sizes were approximated roughly. Shading in the final image is simple because the unit vector normal to any point on the surface of the model can be solved mathematically with ease. Similarly, in a previous paper, microtubules were displayed as lines.[40,41]

If the modeled simple object extends over several sections, the simple form is divided into some section parts, i.e., the model must be produced by stacking those parts in order. The data for serial sections of the modeled object that are needed for drawing the final image are input into the system together with the contour line data tracing the genuine object as precisely as possible. This technique is used especially in displaying lipid droplets and vesicles as well as other intracellular organelles of a yeast cell (see Figure 5). For this purpose, a spherical model was employed, and the sphere was divided into several thin, circular planes that completely restore the sphere by stacking.

D. MEASUREMENT OF VOLUMES AND SURFACE AREAS

As a byproduct of computer-aided reconstruction, the volume and surface area of the

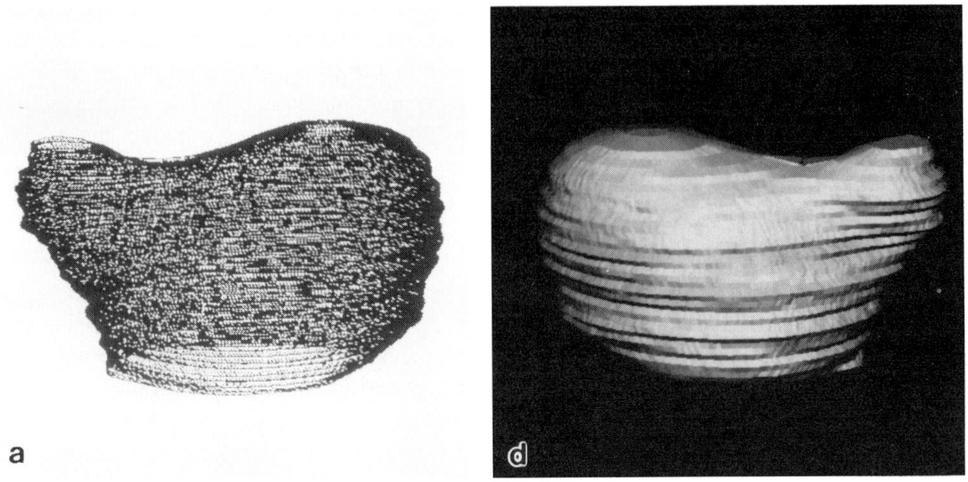

FIGURE 3. Effects of smoothing methods. (a and d) The original "wire frame" model and shaded image, respectively; (b and e) the result of smoothing the coordinates of 3-D surface points (contour points) which are seen in the "wire frame" and solid model, respectively; (f) the result of the smooth shading performed on the result of (b) or (e). (c) The surface reconstruction of a set of small triangular planes and explains the technique of the proposed smoothing. Coordinates of a contour point j are averaged over neighboring points (for example, i, j − 1, j, j + 1, k). This reconstruction is for a cell membrane of a yeast target cell shown in Figure 1. (From Baba, N. and Kanaya, K., Scanning Microsc. Suppl., 2, 303, 1988. With permission; Baba, N., Baba, M., Imamura, M., Koga, M., Ohsumi, Y., and Kanaya, K., *J. Electron Microsc. Tech.*, 11, 16, 1989. With permission.)

reconstructed object can be determined by simple methods.[46] The volume of the object is calculated by summing up every volume of planar plates formed between neighboring contour lines. In recent image processors, a dedicated high-speed binary processing hardware unit, which calculates areas surrounded by arbitrary curved lines, is utilized. The volume is easily obtained at high speed.

The surface area of the reconstructed object is obtained by summing up all areas of the triangular patches constructing the 3-D contour of the object (see Figure 3c). If **ai** and **bi** are vectors of two sides of the i-th triangle patch, $|\mathbf{ai} \times \mathbf{bi}|/2$ is the area of the triangle plane. For the shading processing, every unit vector, **ni**, normal to the small triangle plane is calculated ($\mathbf{ni} = (\mathbf{ai} \times \mathbf{bi})/|\mathbf{ai} \times \mathbf{bi}|$; × means vector product). Thus, by summing up all areas of triangle patches, the surface area is obtained. The accuracy of these calculations depends on the accuracy of the estimation of section thickness and on errors in the lowest and the highest sections. The latter errors may arise from having a boundary of the object in the lowest or highest section not cut perpendicular, but more or less parallel to its long axis. As an example, the calculated volumes for the yeast target cells are shown in Table 1.

E. SMOOTHING SURFACE

It is necessary to generate a smooth surface image of the reconstructed model because surface roughness is an obstacle in the visualization of the 3-D model. Surface roughness is caused by insufficient numbers of sections with respect to the complexity of the structure to be investigated or by artifacts arising during specimen preparation or photography with a microscope. Smooth shading is generally used to calculate the smooth brightness distribution on the displayed 3-D surface.[43] To further ameliorate the effect of smooth shading, it is possible to smooth the surface of the reconstructed model itself.[40,41] There are various methods for this purpose. Although it is possible to employ established methods in the field of computer graphics,[43,47] the author applied a simple method to genuine samples.[40,41] When

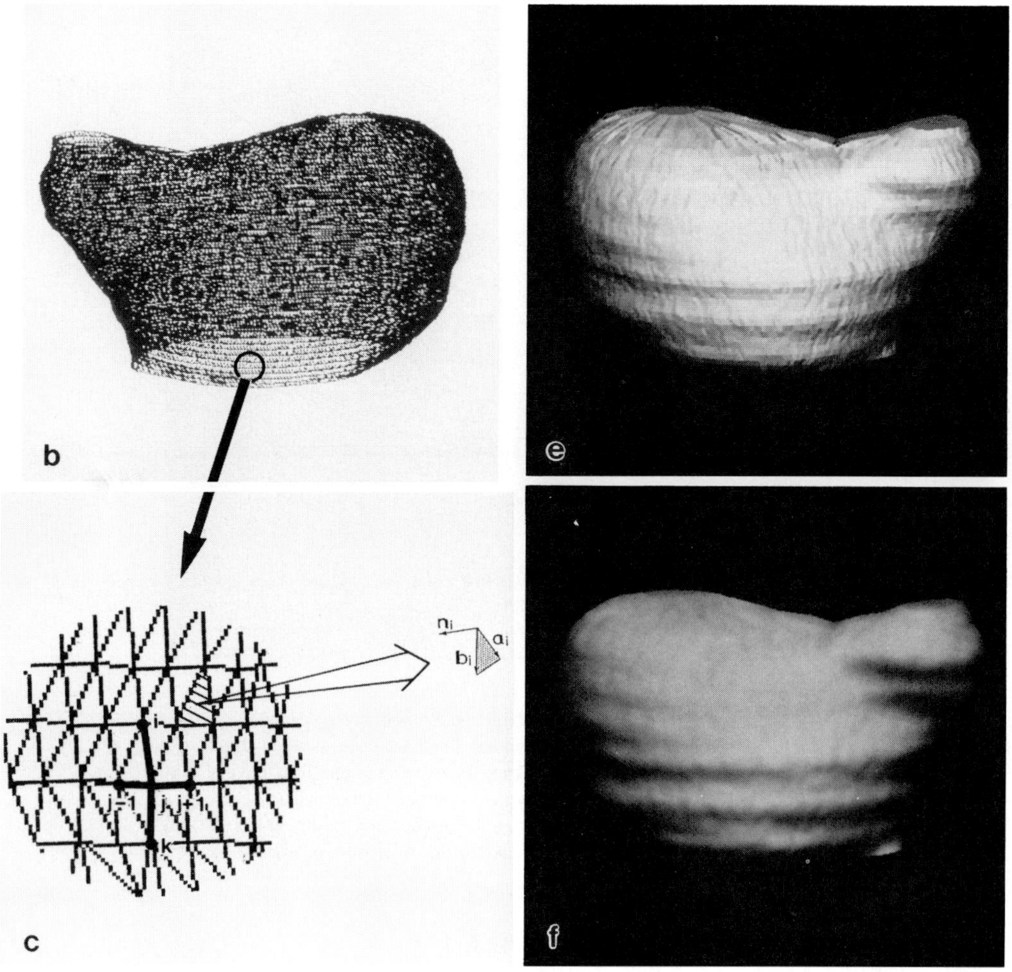

FIGURE 3 (continued).

such a smoothing is performed, the smooth shading may not necessarily be optimal. The proposed smoothing averages the x- and y-coordinates over neighboring 3-D contour points located around a certain point. For instance, as shown in Figure 3c, the coordinates of xj and yj are obtained by averaging five coordinates of (xi, xj − 1, xj, xj + 1, and xk) and (yi, yj − 1, yj, yj + 1, and yk).[40,41] Figure 3d shows a shaded image for the original "wire frame" model (Figure 3b), and Figure 3e shows a shaded image for the smoothed surface by the above. Figure 3f shows the model after smooth shading.

Perhaps, in practice, techniques that create a model with a smooth surface are necessary to reconstruct more complex structures. Depending on the complexity of a reconstructed model, using only smooth shading may be insufficient to visualize the structure in the final image. For a complex model, it is necessary to smooth the model itself using a more sophisticated technique. Recently, we tried to reconstruct splenic sinuses of splenomegaly in portal hypertensive rats. This morphological study was started to clarify the relationship between the development of splenomegaly and portal hypertension, mainly by scanning electron microscopy.[48] Figure 2a shows a section of the red pulp of the splenomegaly in rat taken with a SEM. Endothelial cells and ring fibers compose sinuses. An example of the first reconstructed result of a sinus is shown in Figure 7a. In the figure, a single solid model

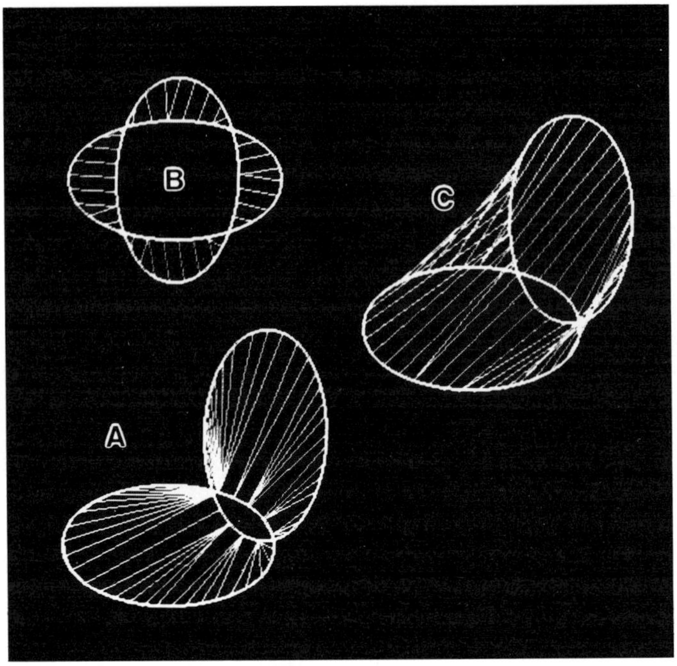

FIGURE 4. Effect of the proposed linkage method using alignment of a pair of contour lines in which ellipse contour lines are used as an example. (A) Typical example of unreasonable linkage where two contour lines deviate from each other; (C) preferable linkage by the improved method where two contour lines are aligned, as shown in (B) before the linkage processing, and preserving the generated links when the moved contour line returns to its original position, as shown in (C).

TABLE 1
Volume Measurement of Cell Organelles During Projection Formation in Yeast Target Cells Shown in Figure 5

			Volume in μm³ (% of total cell volume)			
Stage	Number of vesicles	Number of filasomes	Total cell	Nucleus	Mitochondria	Vacuoles
I	137	4	33.3	2.6 (7.7)	0.49 (1.5)	5.3 (15.9)
II	235	8	31.8	2.7 (8.4)	0.62 (1.9)	8.4 (26.4)
III	284	13	39.1	3.1 (7.9)	0.60 (1.5)	6.2 (15.8)
IV	179	16	36.7	2.9 (7.8)	0.76 (2.1)	3.6 (9.7)

From Baba, M., Baba, N., Ohsumi, Y., Kanaya, K., and Osumi, M., *J. Cell Sci.*, 94, 207, 1989. With permission.

is shown, i.e., all parts link through branching points, which was automatically analyzed by our system according to the method described previously and automatically reconstructed except for the contour line acquisition. The surface roughness due to the large variation in the ratio between neighboring contours was a problem to be solved. This was remarkable since the structure is very complex. However, the surface roughness was smoothed by the following, improved manner. First, interpolated contour lines were generated and inserted into the original contour line data to increase the outward resolution.[49,50] In this case, the number of 36 original sections was increased to 71 sections. Second, a simple binary voxel processing was introduced to smooth the surface. All planar plates formed by contour lines were located in a 3-D binary memory space, i.e., the original model was constructed in space, and the 3-D averaging processing was performed. Finally, new contour line data converted from the smoothed model by the averaging were input into the system and the final image was visualized (Figures 7b to d). By smoothing, the quality of the model image was increased remarkably. However, it should be pointed out that the investigator must pay attention to excess smoothing because a fine structure may be deformed or disappear.

V. APPLICATIONS

Three practical applications are reported here which include intracellular details of yeast target cells during a projection formation, a periodontal structure of a dog tooth moved orthodontically, and a 3-D structure of splenic sinuses in portal hypertensive rats. The system used for these applications consists of a LUZEX 5000 image processor (Nireco Corporation) and a TV image digitizer. A digitizer tablet was used to input menu choices into the computer. The TV image digitizer was also used to input serial contour line images copied on transparent sheets with a black pen. The processor provides a high-speed hardware processing unit that is exclusively dedicated to binary image processing, which is of advantage in editing contour line images. The tablet was also used for correcting contour lines and for entering interstitial contour lines.[40,41]

A. RECONSTRUCTION OF YEAST TARGET CELLS[51]

Figure 1a shows an electron micrograph of a sectioned yeast target cell (*Saccharomyces cerevisiae* X2180-1A (*MATa*))[52] induced by a mating pheromone, called the α-factor. Specimens were prepared for transmission electron microscopy by freeze substitution. This fixation method preserves well the intracellular organelles and the membrane systems.[53] The thickness of the section was estimated to be 70 to 100 nm as judged from the interference colors and by determining the number of sections that covered a spherical body of a given diameter. As shown in the micrograph, after α-factor treatment, cells developed a pointed projection, which is a unique pattern of oriented cell surface growth. To understand the process of morphological changes in the cells, serial thin sections of a total of 30 cells were prepared. The cells were classified into four groups by the shape of the projection. A typical cell from each stage was chosen and reconstructed (Figure 5). The photos show intracellular organelles clearly observable through the transparent cell membrane. To display the intracellular structure as clearly as possible, the reconstructed outer cell membrane was illuminated orthogonal to the direction of the viewpoint in order to illuminate mainly the external membrane. (The direction of illumination for the cell membrane was, of course, different from that for the intracellular structural objects.) It may be hard to distinguish each organelle by observing the photos; the following organelles are contained: filasome, Golgi body, lipid granule shown as a spherical model, mitochondrion, nucleus, nucleolus, rough endoplasmic reticulum, and membrane vesicle in a particle model (for more results and details, see previous papers.[41,51]) The four reconstructed views in Figure 5 represent cells in different stages, respectively. At stage I, cells were spherical without a projection. At stage II, cells that had just started projection formation contained more membrane vesicles than at the earlier stage. Only a small number of vesicles was located near the projection. The nucleolus was located at the side opposite the projection. At stage III, cells had a short projection and more membrane vesicles, which were notably concentrated at the tip of the projection. At stage IV, cells having an elongated projection showed apparently concentrated membrane vesicles in the projection. The nucleus moved close to the neck portion of the projection. Since the transparent display method used here enabled us to visualize the cell membrane

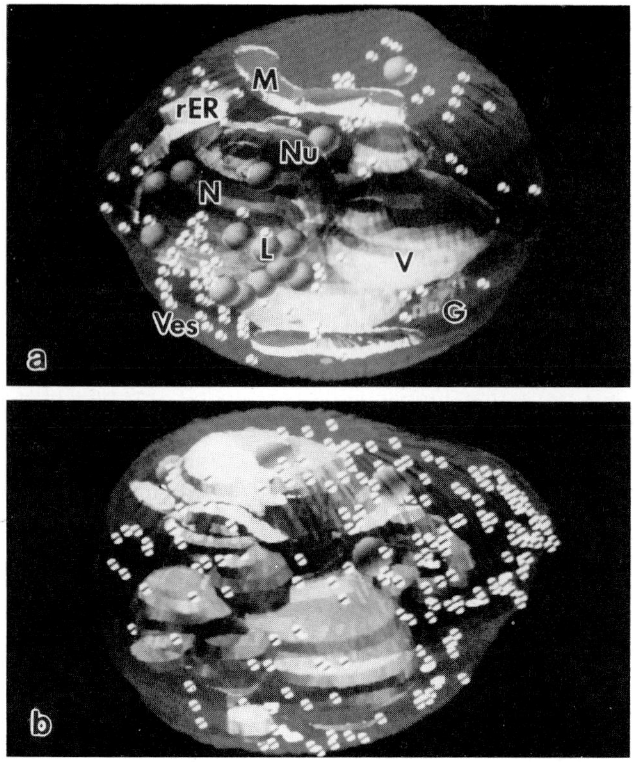

FIGURE 5. Reconstructed images of α-factor-treated cells. Four typical cells corresponding to four stages in projection formation are shown from top to bottom. The following cell organelles are seen through the cell membranes: lipid droplet (L), mitochondrion (M), nucleus (N), nucleolus (Nu), vacuole (V), vesicle (Ves), Golgi body (G), rough endoplasmic reticulum (rER), and filasome (F), in which vesicles and lipid droplets are represented by simple, small spheres in the computer graphics. (From Baba, M., Baba, N., Ohsumi, Y., Kanaya, K., and Osumi, M., *J. Cell. Sci.*, 94, 207, 1989. With permission.)

and intracellular organelles simultaneously, some interesting results were obtained for the topological relationship between various organelles in the process of projection formation. The results are described in detail by Baba et al.[51] As a byproduct of computer-aided reconstruction, the volume and surface area of the reconstructed object can be determined as described previously. Organelle volumes and their ratios to total cell volumes were analyzed quantitatively during projection formation. The results are shown in Table 1.

B. RECONSTRUCTION OF AN ORTHODONTICALLY MOVED TOOTH AND ITS PERIODONTAL TISSUES[45]

Figure 6a shows a light microscope image of a section of an orthodontically moved dog tooth and its periodontal tissues. After extraction of adjacent teeth, the upper central incisor of a dog was inclined by the application of an orthodontic force of about 80 g initially and 48 g finally (see Imamura[45] for details of the material and the specimen preparation, the measurement of the distance through which the tooth moved, and the X-ray data). After 7 weeks of movement, the tooth and its periodontal tissue were fixed, embedded in celloidin, and cut into 232 serial sections about 50 μm thick. In this experiment, celloidin was used for embedding because the convenient method of embedding in paraffin is apt to cause a

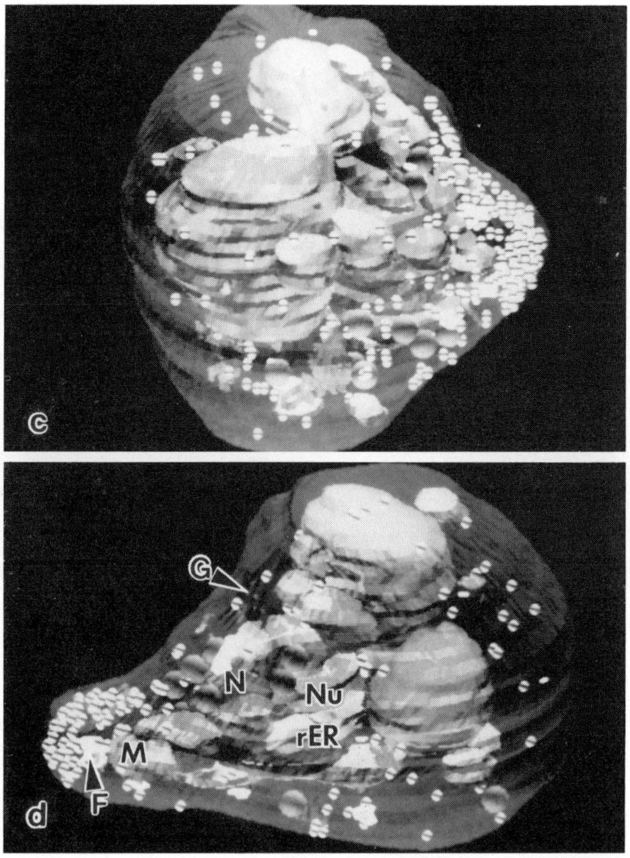

FIGURE 5c and d.

section to shrink and, frequently, to be partially destroyed or deformed. Successful serial sections were obtained that can be regarded as not deformed. The thickness of the section was checked every ten sections with a micrometer; 50 ± 2 to 3 μm was obtained as a mean value for one section. All 232 sections were observed with a light microscope and used for reconstruction. The arrow in Figure 6a indicates the direction of the tooth movement, and the asterisk points to the area of root resorption. Figure 6b shows the cut contour lines generated by a computer following a defined boundary plane. In this example, two cutting planes were used to observe labio-lingual and mesio-distal cut surfaces of the tooth and of the adjacent alveolar bone. Figure 6c shows the labio-lingual cut surface. This cut surface is orthogonal to the other mesio-distal cut surface. Four sets of cut contour lines for all 232 sections were reconstructed. In this text, only three typical views are shown. Other views are reported by Imamura[45] and the author.[40,41] Figures 6c to e show results for the labio-lingual cut surfaces and the tooth viewed from the mesio-side. Figure 6c shows both the cut tooth and the bone, 6d shows only the bone. The tooth surface in Figure 6e faces the inner surface in Figure 6d. Labio and lingual sides are indicated by axes inserted in Figure 6c and marked with La and Li, respectively. Since these images are reconstructed from many sections, the parts of the reconstructed images at every interval of 20 sections (indicated by lines) are distinguished from one another. Part of the 60th section (Figure 6a) is indicated in Figure 6c. These reconstructed images together with other ones reported by Imamura[45] revealed, for example, root and bone resorption and massive bone formation. From a comparison of the reconstructed images with histological sections, various important results

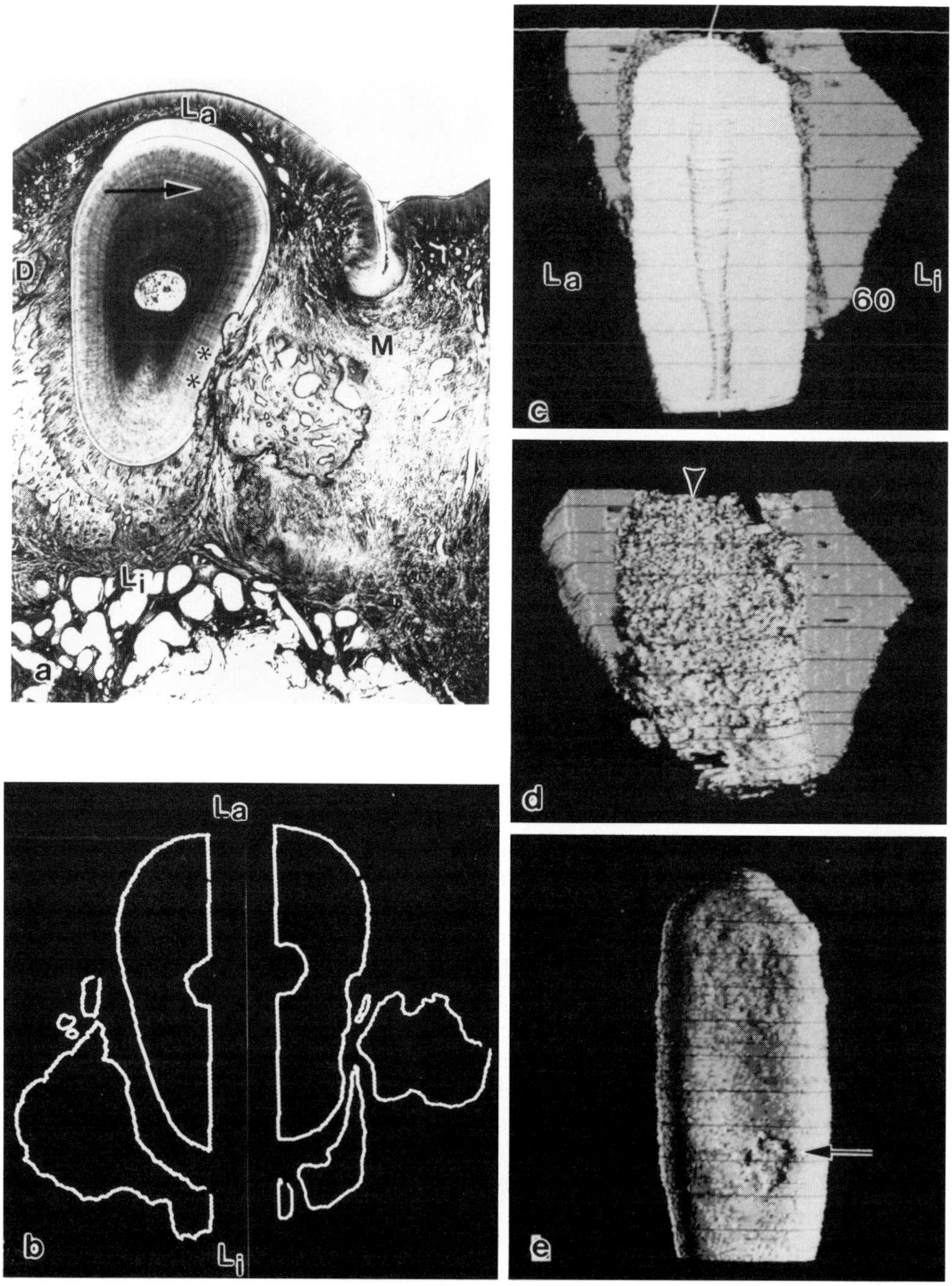

FIGURE 6. Reconstructions of the labio-lingual cut surfaces of an orthodontically moved tooth and of adjacent alveolar bone. (a) Typical section used in the reconstruction where the arrow indicates the direction of orthodontical movement and the asterisk points to the root resorption; (b) the cut contour lines automatically generated which are used in the reconstruction of the 3-D-cut model. Labio and lingual sides are indicated by La and Li, respectively. (c) The cut tooth and bone; (d) only bone; (e) the reconstructed tooth. The arrow in (e) shows root resorption; the arrow in (d) shows massive bone formation. (From Baba, N., and Kanaya, K., Scanning Microsc. Suppl. 2, 303, 1988. With permission; Baba, N., Baba, M., Imamura, M., Koga, M., Ohsumi, Y., Osumi, M, and Kanaya, K., *J. Electron Microsc. Tech.*, 11, 16, 1989. With permission.)

could be obtained on histological details of periodontal structures caused by the tooth movement.[45] One of the results was that the external profile of the root resorption site has a diamond shape as shown in Figure 6e. From both the section images and the reconstructed image, it was found that the resorption was slow at the center and rapid at the periphery. This view was very close to that observed in scanning electron micrographs.[54] It should be emphasized that the present method, serial sectioning and computer-aided reconstruction, makes it possible to investigate the reconstructed result, comparing and identifying each section image.

C. RECONSTRUCTION OF SPLENIC SINUSES OF SPLENOMEGALY IN PORTAL HYPERTENSIVE RATS[48]

Figure 2a shows a section of the red pulp of splenomegaly in portal hypertensive rat taken with a SEM. Endothelial cells and ring fibers compose splenic sinuses. We tried to reconstruct a structure of the splenic sinuses. This morphological study was started to clarify the relationship between the development of splenomegaly and portal hypertension.[48] The study was performed mainly by scanning electron microscopy in liver-cirrhotic rats. Liver cirrhosis was produced by intraperitoneal injections of 4% thioacetamide (TAA) in saline. The details for material and specimen preparation are described by Misawa et al.[48] Aliquots of fixed splenic tissue were alcohol-freeze-fractured and routinely processed for the SEM. Other aliquots were embedded in paraffin and serial sections 6 μm in thickness were prepared for the SEM. Figure 7 shows the reconstructed result of the splenic sinuses in rat from animals with highly elevated portal pressure. Figures 7c and d represent stereo pair images. As described above, our system first analyzed the relations between input serial contour lines and then divided them into several sets of contour lines. One set corresponded to a single isolated solid. This analysis was performed by investigating every relationship between neighboring contour lines. Figure 7a shows the largest isolate found by computer, and this had no connection with any other parts in the reconstructed whole model (cf. other photos in the same figure). Next, the system performed the smoothing of the model and reduced the surface roughness. Figures 7b to 7d show the result, and in Figure 7b, the same part is displayed as in Figure 7a in a distinguishable manner: the bright part in Figure 7b corresponds to the model in Figure 7a, and the dark parts in Figure 7b indicated by arrows correspond to other parts isolated from the bright part. Because the smooth shading processing was not finished, the surface roughness was not completely removed, but the stereo pair images (Figures 7c and d) revealed the complex structure well resolved in three dimensions. Reconstruction for various stages in the development of the splenomegaly is now in progress.

VI. SERIAL SECTION RECONSTRUCTION BY VOXEL PROCESSING

Finally, the second method is introduced in which the system does not need contour lines, but, rather, live section images. This method frees the user from the cumbersome task of tracing contour lines. It generates a realistic image because of a voxel representation of the object based on live image data. Since all section image data are kept available, an interactive parameter setting is possible in many cases. It is expected to be able to manipulate 3-D-reconstructed data more easily with the advance of 3-D image processing software and computer technology for high-speed processing, and expansion of memory capacity. This also stimulates the application of 3-D computer analysis in morphology. The reconstruction method is based on the same principle as that in 3-D confocal microscopy (e.g., see Reference 55) and that from serial sectional images obtained by CT. Furthermore, voxel processing needed for serial section reconstruction is closely related to that in the other types of reconstructions mentioned above and in postprocessings used in 3-D CT for visualization.[56]

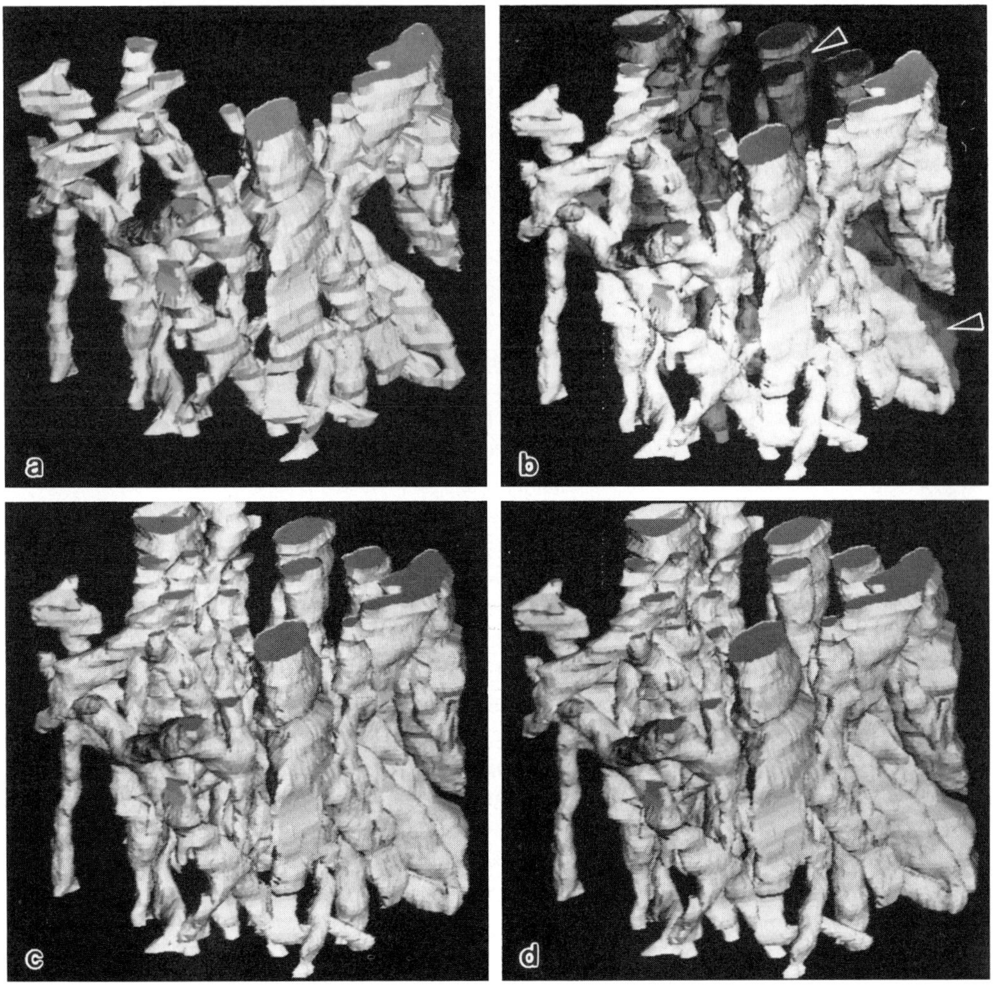

FIGURE 7. Reconstructed images of splenic sinuses of splenomegaly in portal hypertensive rats. (a) The result before the smoothing; (b, c, and d) the results of smoothing. In (b), the same part as (a) is displayed distinguishing the bright part, which corresponds to the model in (a), and the dark part, indicated by arrows, which correspond to the other parts isolated from the bright part. A stereo image pair can be seen in (c) and (d).

Figure 8 shows an example of the present reconstruction. The sample was minute colonic adenomas in human tissues from familial adenomatosis coli.[57] A few dozen section images taken with a light microscope were reconstructed by voxel processing. For the 3-D image processing, a LUZEX III image processor (Nireco Corporation) was used, which is an enhanced version of the LUZEX-5000 used in earlier work; it has 16 Mbyte of internal memory of user area. In addition, the processor has various high-speed processing units for 1024 × 1024 image areas and a high-quality display unit for a 1024 × 1024 image. The section images were input into the 3-D memory space (in this case, 256 × 256 × 256 × 8-bit memory space) by a TV image digitizer after alignment. Depending on the thickness of the section, interpolated section images were generated and inserted, filling gaps between original section images in the 3-D space. In this step, the reconstructed result was considered to be a 3-D intensity distribution. To visualize this result, a display principle based on surface shading was used. First, the 3-D intensity distribution was changed to a binary distribution generated by a simple intensity threshold. Because a surface of the object was determined, the surface normals could be calculated and the surface shading could be performed, as

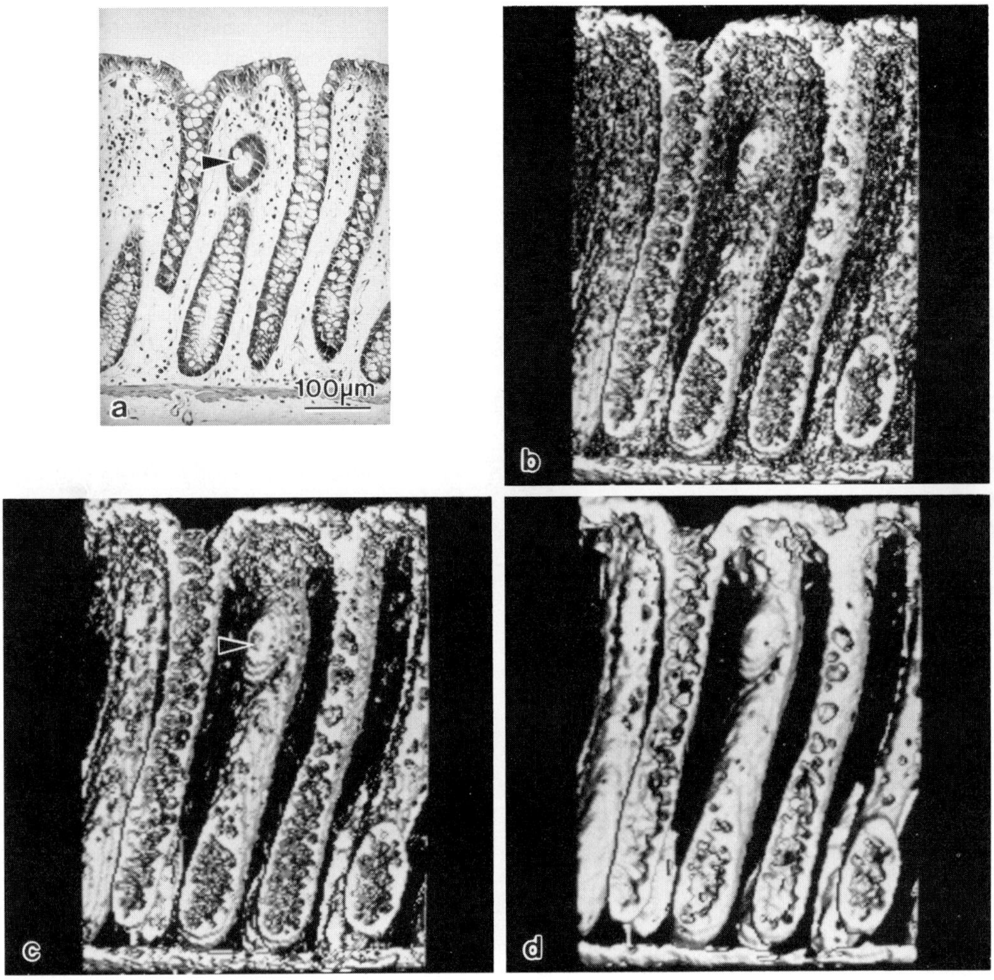

FIGURE 8. Reconstruction by voxel processing. (a) A typical section of a minute colonic adenoma (arrow) in human tissues from familial adenomatosis coli; (b, c, and d) reconstructed results; (c) the result of eliminating noise due to the removal of very small-scale objects; (d), the result of smoothing or filtering the 3-D intensity distribution with a 5 × 5 × 5 Gaussian filter.

shown in Figure 8b. Two examples for voxel processings are shown in Figures 8c and d. One is for eliminating noise by removing very-small-scale objects. Because the objects were based on binary values and the boundary surfaces were very clear, it was easy for the computer to recognize individual objects and measure their volumes. Therefore, by determining a threshold on the volumes, small objects could be eliminated, as shown in Figure 8c. The other example is smoothing or filtering the 3-D intensity distribution. Of course, it is possible to expand 2-D image processing techniques to 3-D processing. In this example (Figure 8d) the 3-D intensity distribution was smoothed with a 5 × 5 × 5 Gaussian filter.

VII. CONCLUSION

The method of computer-aided serial section reconstruction is described from a technological point of view which is applied to complex structures. Most of the sections in this chapter refer to the contour line-based method. In the last section, the voxel processing method is mentioned. It seems that if a satisfactory serial section image is prepared, the latter method is preferable because it is not necessary to trace the many contour lines. A result more faithful to the originally recorded image can be reconstructed as compared with the former method. However, the contour line-based method is also a very significant tool because of the following reasons. The voxel processing method has advantages only if the investigated 3-D object is reconstructed from a 3-D density distribution or section images are taken with a transmission microscope. As shown in the third application described in the text, it is impossible to use the voxel processing method if the section image is an SEM image. Furthermore, the stacking of serial section images produces very large amounts of data which are not easy to understand or display. To trace contours of objects means to select information from a larger amount of data. It should be stressed that the reconstruction methods described above are based on different viewpoints.

ACKNOWLEDGMENTS

The author thanks Professor M. Osumi of Japan Women's University, Associate Professor Y. Ohsumi of the University of Tokyo, Dr. M. Baba of Kogakuin University, Associate Professor M. Koga and Dr. M. Imamura of Tokyo Dental College, Professor T. Suzuki and Dr. T. Misawa of The Jikei University School of Medicine, and Associate Professor S. Nakamura of Hamamatsu University School of Medicine for much helpful advice and preparing the materials and serial sections.

REFERENCES

1. **Capowski, J. J.,** Computer-aided reconstruction of neuron trees from several sections, *Comput. Biomed. Res.,* 10, 617, 1977.
2. **Groves, P. M. and Linder, J. C.,** Dendro-dendritic synapses in substantia nigra: descriptions based on analysis of serial sections, *Exp. Brain Res.,* 49, 209, 1983.
3. **Huijsmans, D. P., Lamers, W. H., Los, J. A., and Strackee, J.,** Toward computerized morphometric facilities: a review of 58 software packages for computer-aided three-dimensional reconstruction, quantification, and picture generation from parallel serial sections, *Anat. Rec.,* 216, 449, 1986.
4. **Johnson, E. M. and Capowski, J. J.,** A system for the three-dimensional reconstruction of biological structures, *Comput. Biomed. Res.,* 16, 79, 1983.
5. **Johnson, E. M. and Capowski, J. J.,** Principles of reconstruction and three-dimensional display of serial sections using a computer, in *The Microcomputer in Cell and Neurobiology Research,* Mize, R. R., Ed., Elsevier, New York, 1985, 249.
6. **Kropf, N., Sobel, I., and Levinthal, C.,** Serial section reconstruction using CARTOS, in *The Microcomputer in Cell and Neurobiology Research,* Mize, R. R., Ed., Elsevier, New York, 1985, p 265.
7. **Levinthal, C. and Ware, R.,** Three-dimensional reconstruction from serial sections, *Nature,* 236, 207, 1972.
8. **Macagno, E. R., Levinthal, C., and Sobel, I.,** Three-dimensional computer reconstruction of neurons and neuronal assemblies, *Annu. Rev. Biophys. Bioeng.,* 8, 323, 1979.
9. **Moens, P. B. and Moens, T.,** Computer measurements and graphics of three-dimensional cellular ultrastructure, *J. Ultrastruct. Res.,* 75, 131, 1981.
10. **Peachey, L. D., Damsky, C. H., and Veen, A.,** Computer assisted three-dimensional reconstructions from high-voltage electron micro-graphs of serial slices of biological material, in *Proc. 8th Int. Congr. Electron Microscopy,* Vol. 1, Australian Academy of Science, Canberra, Australia, 1974, 330.
11. **Perkins, W. J. and Green, R. J.,** Three-dimensional reconstruction of biological sections, *J. Biomed. Eng.,* 4, 37, 1982.
12. **Rakic, P., Stensas, L. J., Sayre, E. P., and Sidman, R. L.,** Computer-aided three-dimensional reconstructions and quantitative analysis of cells from serial electron microscopic montages of foetal monkey brain, *Nature,* 250, 31, 1974.
13. **Sobel, I., Levinthal, C., and Macagno, E. R.,** Special techniques for the automatic computer reconstruction of neuronal structures, *Annu. Rev. Biophys. Bioeng.,* 9, 347, 1980.

14. **Stevens, B. J.,** Variation in number and volume of the mitochondria in yeast according to growth conditions. A study based on serial sectioning and computer graphics representation, *Biol. Cell,* 28, 37, 1977.
15. **Stevens, J. K., Davis, T. L., Friedman, N., and Sterling, P.,** A systematic approach to reconstructing microcircuitry by electron microscopy of serial sections, *Brain Res. Rev.,* 2, 265, 1980.
16. **Veen, A. and Peachy, L. D.,** Trots: a computer graphics system for three-dimensional reconstructions from serial sections, *Comput. Graphics,* 2, 135, 1977.
17. **Willey, T. J., Schultz, R. L., and Gott, A. H.,** Computer graphics in three dimensions for perspective reconstruction of brain ultrastructure, *IEEE Trans. Biomed. Eng.,* 20, 288, 1973.
18. **Wong, Y.-M. M., Thompson, R. P., Cobb, L., and Fitzharris, T. P.,** Computer reconstruction of serial sections, *Comput. Biomed. Res.,* 16, 580, 1983.
19. **Young, S. J., Fram, E. K., and Craig, B. A.,** Three-dimensional reconstruction and quantitative analysis of rat lung type II cells: a computer-based study, *Am. J. Anat.,* 174, 1, 1985.
20. **Coombs, G. H., Tetley, L., Moss, V. A., and Vickerman, K.,** Three-dimensional structure of the Leishmania amastigote as revealed by computer-aided reconstruction from serial sections, *Parasitology,* 92, 13, 1986.
21. **Braverman, M. S. and Braverman, I. M.,** Three-dimensional reconstruction of objects from serial sections using a microcomputer graphics system, *J. Invest. Dermatol.,* 86, 290, 1986.
22. **Briarty, L. G. and Jenkins, P. H.,** GRIDSS: an integrated suite of microcomputer programs for three-dimensional graphical reconstruction from serial sections, *J. Microsc.,* 134, 121, 1984.
23. **Baba, N., Naka, M., Muranaka, Y., Nakamura, S., Kino, I., and Kanaya, K.,** Computer-aided stereographic representation of an object reconstructed from micrographs of serial thin sections, *Micron Microsc. Acta,* 15, 221, 1984.
24. **Gras, H. and Killman, F.,** NEUREC — a program package for 3-D reconstruction from serial sections using a microcomputer, *Comput. Programs Biomed.,* 17, 145, 1983.
25. **Kinnamon, J. C.,** Three-dimensional reconstructions from serial sections using the IBM PC, *Eur. J. Cell Biol.,* 48 (Suppl. 25), 65, 1989.
26. **Moss, V. A.,** The computation of 3-dimensional morphology from serial sections, *Eur. J. Cell Biol.,* 48 (Suppl. 25), 57, 1989.
27. **Prothero, J. and Prothero, J.,** Three-dimensional reconstruction from serial sections. I. A portable microcomputer-based software package in Fortran, *Comput. Biomed. Res.,* 15, 598, 1982.
28. **Street, C. H. and Mize, R. R.,** A simple microcomputer-based three-dimensional serial section reconstruction system (MICROS), *J. Neurosci. Methods,* 7, 359, 1983.
29. **Sundsten, J. W. and Prothero, J. W.,** Three-dimensional reconstruction from serial sections. II. A microcomputer-based facility for rapid data collection, *Anat. Rec.,* 207, 665, 1983.
30. **Young, S. J., Royer, S. M., Groves, P. M., and Kinnamon, J. C.,** Three-dimensional reconstructions from serial micrographs using the IBM PC, *J. Electron Microsc. Tech.,* 6, 207, 1987.
31. **Bron, C., Gremillet, P., Launay, D., Jourlin, M., Gautschi, H. P., Bächi, T., and Schüpbach, J.,** Three-dimensional electron microscopy of entire cells, *J. Microsc.,* 157, 115, 1990.
32. **Cookson, M. J., Dykes, E., Holman, J. G., and Gray, A.,** A microcomputer based system for generating realistic 3D shaded images reconstructed from serial sections, *Eur. J. Cell Biol.,* 48 (Suppl. 25), 69, 1989.
33. **Lockhausen, J., Kristen, U., Menhardt, W., and Dallas, W. J.,** Three-dimensional reconstruction of a plant dictyosome from a series of ultrathin sections using computer image processing, *J. Microsc.,* 158, 197, 1989.
34. **Marko, M., Leith, A., and Parsons, D.,** Three-dimensional reconstruction of cells from serial sections and whole-cell mounts using multilevel contouring of stereo micrographs, *J. Electron Microsc. Tech.,* 9, 395, 1988.
35. **Menhardt, W., Lockhausen, J., Dallas, W. J., and Kristen, U.,** An environment for three-dimensional shaded perspective display of cell components, *Micron Microsc. Acta,* 17, 349, 1986.
36. **Nakamae, E., Harada, K., Kaneda, K., Yasuda, M., and Sato, A.,** Reconstruction and semi-transparent stereographic display of an object consisting of multi-surfaces, *Trans. Inf. Process. Soc. Jpn,* 26, 181, 1985.
37. **Yaegashi, H., Takahashi, T., and Kawasaki, M.,** Microcomputer-aided reconstruction: a system designed for the study of 3-D microstructure in histology and histopathology, *J. Microsc.,* 146, 55, 1987.
38. **Baba, N., Nakamura, S., Kino, I., and Kanaya, K.,** Three-dimensional reconstruction from serial section images by computer graphics, *J. Electron Microsc. Tech.,* 3, 401, 1986.
39. **Baba, N., Baba, M., Osumi, M., Nakamura, S., Imamura, M., Koga, M., and Kanaya, K.,** Computer graphics representation for a three-dimensional reconstruction from serial sections, *Sci. Form,* 2, 49, 1986.
40. **Baba, N. and Kanaya, K.,** Three-dimensional reconstruction from serial section images by computer graphics, *Scanning Microsc. Suppl.,* 2, 303, 1988.
41. **Baba, N., Baba, M., Imamura, M., Koga, M., Ohsumi, Y., Osumi, M., and Kanaya, K.,** Serial section reconstruction using a computer graphics system: applications to intracellular structures in yeast cells and the periodontal structure of dog's teeth, *J. Electron Microsc. Tech.,* 11, 16, 1989.

42. **Baba, N., Baba, M., Osumi, M., Nakamura, S., and Kanaya, K.,** Improvements of a serial section reconstruction system to be applicable to more complicated structures, in *Proc. 9th Eur. Congr. Electron Microscopy,* Vol. 3, Dickinson, H. G. and Goodhew, P. J., Eds., IOP Publishing, York, England, 1988, 373.
43. **Newman, W. M. and Sproull, R. F.,** *Principles of Interactive Computer Graphics,* McGraw-Hill, New York, 1975.
44. **Peterhans, E. and von der Heydt, R.,** Correction of distorted image pairs with elastic models, *Eur. J. Cell Biol.,* 48 (Suppl. 25), 85, 1989.
45. **Imamura, M.,** Histological study on the periodontal structures incident to experimental tooth movement in dog, *Shikwa Gakuho,* 86, 1319, 1986.
46. **Chawla, S. D., Glass, L., Friewald, S., and Proctor, J. W.,** An interactive computer graphics system for 3-D stereoscopic reconstruction from serial sections: analysis of metastatic growth, *Comput. Biol. Med.,* 7, 223, 1982.
47. **Rogers, D. F.,** *Procedural Elements for Computer Graphics,* McGraw-Hill, New York, 1985.
48. **Misawa, T., Hataba, Y., Baba, N., Kanaya, K., Sakurai, K., and Suzuki, T.,** Ultrastructural characteristics of experimental splenomegaly in portal hypertensive rats — a scanning electron microscopic study of the red pulp, in *Proc. 12th Int. Congr. Electron Microscopy,* San Francisco Press, San Francisco, 1990, 270.
49. **Otten, E. and van Leeuwen, J. L.,** MacReco, a 3D reconstruction package froms serial sections and images, *Eur. J. Cell Biol.,* 48 (Suppls. 25), 73, 1989.
50. **Zsuppan, F. and Rethelyi, M.,** Computer reconstruction of serial EM pictures: approximation of missing sections, in *Proc. 8th Eur. Congr. Electron Microscopy,* Vol. 2, Budapest, 1984, 1361.
51. **Baba, M., Baba, N., Ohsumi, Y., Kanaya, K., and Osumi, M.,** Three-dimensional analysis of morphogenesis induced by mating pheromone α factor in *Saccharomyces cerevisiae, J. Cell Sci.,* 94, 207, 1989.
52. **Ohsumi, Y. and Anraku, Y.,** Specific induction of Ca^{2+} transport activity in MATa cells of *Saccharomyces cerevisiae* by a mating pheromone, α factor, *J. Biol. Chem.,* 169, 10482, 1985.
53. **Baba, M. and Osumi, M.,** Transmission and scanning electron microscopic examination of intracellular organelles in freeze-substituted *Kloeckera* and *Saccharomyces cerevisiae* yeast cells, *J. Electron Microsc. Tech.,* 5, 249, 1984.
54. **Kvam, E.,** Scanning electron microscopy of tissue changes on the pressure surface of human premolars following tooth movement, *Scan. J. Dent. Res.,* 80, 357, 1972.
55. **Forsgren, P.-O.,** Visualization and coding in three-dimensional image processing, *J. Microsc.,* 159, 195, 1990.
56. **Goldwasser, S. M. and Reynolds, R. A.,** Real-time display and manipulation of 3-D medical objects: the voxel processor architecture, *Comput. Vision Graphics Image Process.,* 39, 1, 1987.
57. **Nakamura, S. and Kino, I.,** Morphogenesis of minute adenomas in familial polyposis coli, *J. Natl. Cancer Inst.,* 73, 41, 1984.

Chapter 14

AUTOMATIC GRAIN COUNTING IN AUTORADIOGRAPHS BY COMPUTERIZED PATTERN RECOGNITION

Nico A. M. Schellart

TABLE OF CONTENTS

I. INTRODUCTION

In experimental biology and medical research, autoradiography is a well-known technique to study, e.g., DNA replication, RNA and protein synthesis, and glucose metabolism. Autoradiography is the visualization, usually on a photographic film, of the pattern of distribution of radiation emitted by the studied specimen itself. In biology and biomedicine, the specimen is mostly a light-microscopic or electron-microscopic (microautoradiography) section with incorporated radioactive material. The image on the film reflects in varying degree, dependent on autoabsorption and scatter, the distribution of the radioactivity in the section. All aspects of the autoradiographic method, concerning technique as well as interpretation, are discussed in the monograph by Rogers,[1] to which the reader is referred for further information.

In the early times of this technique, autoradiographic data were at most quantified manually, either by manually counting the autoradiographic grains or by manually controlled densitometry. During the last decades, computer processing speed and memories have been increased rapidly, which brought numerical, automatized processing of autoradiographs within reach.

Densitometry is suitable in those cases where the grain density is so high that a considerable part of the grains, say more than 10%, are touching or overlapping. By focusing the (monochromatic) measuring beam, currently often a laser beam, within the plane of the grains and by automatic scanning, e.g., by computer-controlled driving of the microscope stage, the intensity value of the transmitted beam can be stored and processed by the computer together with the coordinates of the stage. In this chapter, I will hardly deal with such techniques, but, rather, with automatized counting of individual silver grains by image analysis, mostly established with the light microscope (LM). Counting by computer, presently nearly 200 grains per second,[2] has two advantages compared to manual counting: (1) the speed of counting is several orders of magnitude higher than manual counting and (2) a higher reliability and reproducibility is attained since subjective (criteria) and unstable (fatigue, tedium) human factors are banished or at least strongly diminished, as discussed by Rogers (in Chapter 10 of Reference 1) and others.[3,4]

In short, the setup for grain counting consists of a research microscope (LM or electron microscope, EM), a high-quality video camera mounted on the phototube of the microscope, a video interface, and a computer with at least 0.5 Mbyte memory and a harddisk, a photocamera, or a high-quality hard-copy apparatus and a printer. Various complete setups or assemblies of parts are (were) commercially available, generally with image-handling software packages included, e.g., Quantimet,[5-7] Leitz TAS,[8] Lusex 450 Particle Analyzer,[9] and Classimat.[10] In some cases, the customer has to write the specific application software, sometimes in a dedicated language.[8] In recent years, it has become more and more easy for the investigator to assemble the hardware himself in order to explore a setup optimized for his specific application.[2,4,11] The basic strategy of grain-counting algorithms has a broader application than that in autoradiography. It can, for example, be applied to count densities of microorganisms or other small particles of biological material. Also, there are nonbiological applications, e.g., in astronomy (counting of stars).

Columns 1 and 2 of Table 1 summarize the types of applications together with the hardware and software of the various grain-counting systems described in the literature. Various commercially available systems for (microscopic) image analysis have a grain count capability. In addition to the various current versions of Quantimet (500 series, Cambridge Instruments, Cambridge, U.K.) and Leitz CBA 8000 (the successor of Leitz TAS, Wild Leitz, Wetzlar, FRG), during the last few years new systems have become available, such as IBAS (Kontron Elektronik, Zeiss, Eching/Munich, FRG) and the CUE series (Olympus Optical Co., Tokyo, Japan). Most of them are only suitable for grain counting at high magnification.

TABLE 1
Characteristics of Grain Counting Systems

Year	Appl	Hard- and software	Focus	Strategy	Pixel size	Grain diam (μm)	Area ratio (μm)	Reliability[b]		Ref.
								p	Error (%)	
1971	Cells	Quantimet		dc	0.3		Large	0.96		5
1974	Nuclei	Quantimet/PDP8e	aut			0.16				6
1974	chro	Classimat		dc			Large	(0.83)	4.2 (7.4)	10
1982	Nuclei	Leitz TAS		dc				0.97	10	8
1982	Nuclei	Quantimet					Large	0.92	0.6	7
1986	Tissue	LSI 11/23, ITEX/Q	s-aut	gr_d	0.68	0.9_e	1.4		5	4
1988	Tissue	PC-AT	Hand	gr_d	0.89	2.3	3.5		4 (9)	2
1990	Tissue	IBAS/Southern Micro		gr	0.17	0.56	9	0.98		11

Note: Abbreviations: appl, application; aut, automatized; chro, chromosomes; $_d$, dark field; dc, direct counting by static thresholding; diam, diameter; $_e$, extrapolated dark-field diameter; gr, grain recognition; s-aut, semiautomatized.

[a] Grain over pixel area ratio. [b]*p* is the correlation coefficient between manual and automatized counting and Error is the deviation between these two methods. Values in parentheses concern nonoptimal conditions.

The literature on computer-aided grain counting before 1979 has been reviewed by Rogers (in Chapter 10 of Reference 1). In this chapter, I will emphasize grain-recognizing strategies developed in recent years. Another main theme of this chapter is the probability theory of grain counting, a subject that has hardly been discussed in the literature.

II. TYPES OF APPLICATIONS

In medical biology, zoology, and botany, there are three major fields of application for automatic processing of autoradiographs: (1) grain counting or grain area determination in chromosome[10] or cell suspensions, (2) grain counting or grain area determination in nuclei of tissue sections,[6,7] and (3) grain counting in anatomical sections studied at low magnification.[2,4]

These three fields have their specific requirements and constraints. Applications 1 and 2 demand high magnification, achieved by a 100× objective, often with oil immersion and bright-field illumination. The image frame of interest is mostly selected manually under the microscope. However, for application 1, this is done such that a complete set of chromosomes (or the cell of interest) is within the frame. For application 2, a special procedure is needed to define the contour of the cells to be investigated. This can be done by drawing the contour with a light pen on a writing tablet or with a mouse on the screen. A software procedure also can be used to determine the contour automatically with the possibility of manual correction,[8] after which the number of grains or the area occupied by the grains can be determined. For applications 1 and 2, the area of interest is on the order of 100 to 1000 μm^2, but for application 3, it is 10^4 to 10^5 μm^2. Consequently, the latter applications need low magnification, usually with a 20, 25, or, at most, a 40× objective. Where a substantial number of grains are (nearly) touching or are more or less overlapping, the area of, for example, a chromosome or a nucleus occupied by grains is scored.[10] Of course, this makes scoring of individual grains impossible. Software for the determination of total grain area is based on a completely different strategy[5,10] than is software to count individual grains.[2,4,10]

A parameter which is inherent to the biological preparation and which is crucial for the choice of the counting strategy is the density of the grains. Very high densities, i.e., when substantial and multiple overlap between the grains occurs, such that more than 10% of the grains are fused, make microdenistometry most suitable. This is the normal practice in ^{14}C and ^{35}S autoradiography of, for example, brain tissue. Two types of radiographs with moderate density, suitable for grain counting, can be distinguished. One is an autoradiograph with local moderate densities, so that fusion occurs fairly frequently. This may lead to areas with single grains and with clustered grains. The other one is an autoradiograph with spatially gradually changing densities of the grains with interspersed fused grain. In these cases, grains should be recognized and counted individually with a correction for coinciding grains, in the same way as it is done for low densities (<5 grains per 100 μm^2). In addition, a correction for grain clusters, based on cluster size, will improve the accuracy of the grain count.

III. PROBABILITY THEORY OF GRAIN COUNTING

A. FACTORS INFLUENCING COUNTING STATISTICS

The interpretation of the results of grain counting has several problems of a probabilistic nature. Obvious and important pitfalls biasing the results are

1. Two or more grains can occlude each other.
2. A grain partly covers two or more pixels such that the pixel intenstities become subthreshold.

Methods will be described to correct for these complications. However, the impact of these procedures may be influenced by the variability in grain size and shape.

B. COINCIDING GRAINS

To obtain statistical insight into the problem of coincidence, some restrictions have to be made before mathematically formulating this problem. It is assumed that all grains are identical and ideal and that each grain occupies only one pixel. However, a pixel can contain several grains. The problem is to calculate the probability that a pixel contains 0, 1, 2, . . . , m grains when n grains are randomly distributed over N pixels. The probability, that the fraction of pixels contains m grains, p(n, m), is

$$p(n, m) = \frac{n!}{(n - m)!m!} \left(\frac{1}{N}\right)^{m} \left(1 - \frac{1}{N}\right)^{n-m} \qquad (1)$$

with $n = 0, 1, 2, \ldots, \infty$
$m = 0, 1, 2, \ldots, n$
$N = 2, 3, \ldots, \infty$

By multiplying p with N, the number of pixels with m grains are found. To calculate the values of the probability p(n, m), the simple recursion:

$$p(n, m) = p(n, m - 1) \cdot (n - [m - 1])/(m[N - 1]) \qquad (2)$$

is useful.

Figure 1 gives p(n, m) for $N = 1024$ (32×32 pixels). Along the horizontal axis, the grain density increases from low and moderate (the domain of grain counting) to intermediate (the domain of measuring the area covered by grains and grain clusters) to high (the domain of densitometry). The left part of the curves have slopes approximately equal to m. For grain counting, the most important part of the function space p(n, m) can be approximated by the Poisson distribution. The parts of the curves above the dashed, horizontal line deviate less than 1% from the Poisson distribution. (This also holds for $n/N \leq 0.2$ as long as N is large.)

For $n << N$, the number of empty pixels is nearly $N - n$. When $n = 0.281N$ (arrow), the curve $(N - n)/N$ (at most, 1 grain per pixel) deviates 5% from p(n, 0) due to multiple occupations.

Grain counts with high grain densities have to be corrected for multiple occupations, as shown by Schuette et al.[12] The actual grain density D_c, i.e., the density of radioactive disintegrations, is directly related to the measured density D_m, according to $D_c = C \times D_m$, where C is a correction factor which is dependent on the density and which can be calculated from p(n, m). Figure 2 gives $(1 - C)$ as a function of D_m with the number of frame pixels as parameter. The figure shows that for $D_m < 0.1$, correction only plays a minor role and that for lower values of D_m, correction is irrelevant. In contrast, the densities generally practiced in autoradiographic densitometry unequivocally need correction, especially since C is clearly nonlinearly related to D_m.

C. INFLUENCE OF GRAIN SIZE

The silver bromide crystals of the photographic emulsions in general use in autoradiography range from 0.06 to 0.34 μm (Rogers, Reference 1, appendix). After development of the crystals, the grains measure on the average from 0.2 to 0.7 μm, as found by various authors (see Reep and Greegan[2] for references). The size and shape of the grains are strongly dependent on the developer and developing conditions (time, temperature). The grain size

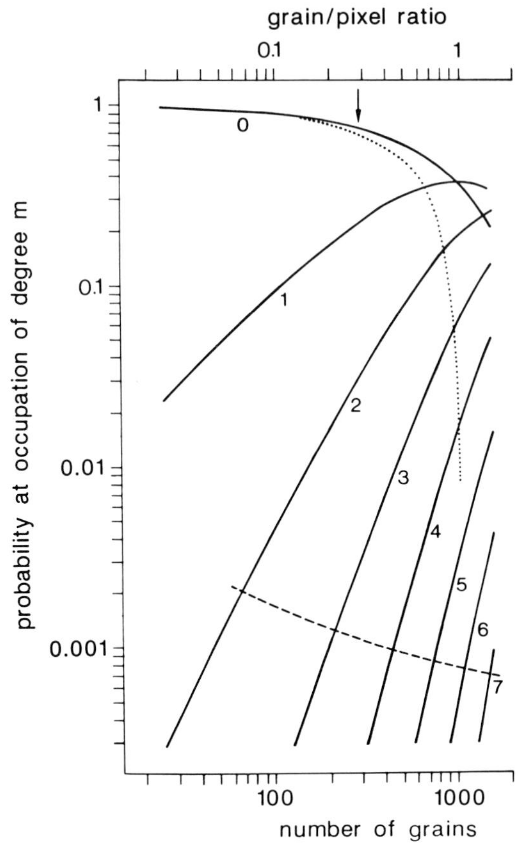

FIGURE 1. Probability of multiple occupations as a function of the number of grains in a 32^2 pixel matrix. The drawn curves give the probability at an m-fold occupation, indicated by the number at the curve. The part of the graph above the dashed line gives the probabilities deviating less than 1% from the Poisson distribution. The stippled curve represents $(N - n)/N$. The arrow indicates the point where this curve deviates 5% from $p(n, 0)$.

is below or at the border of LM resolution with bright-field (transmission) illumination, but in dark-field, the grains look two to four times larger.

An intriguing problem in the theory of grain counting is to optimize the size of the pixel with respect to that of the grain. Obviously, the ratio between grain size and pixel size strongly determines the counting strategy and also influences the reliability of the results. For grains considerably smaller than the pixel size, many grains will be indistinguishable from random local intensity fluctuations. For the reversed situation, grains much larger than the pixel size, the grain will be detected easily and reliably, but at the cost of processing speed and computer memory. In some, especially older, applications, high grain/pixel ratios have been used, mostly to quantify the labeled area of nuclei[7,8] or chromosomes[10] (Table 1). However, in more recent applications, concerning tissue examination, the grain/pixel area ratio is generally smaller and ranges from 1.5^2 to 9.[11]

Intuitively, the grain size seems to be optimal when, almost always, only one pixel is (nearly) completely covered. The problem can be solved when the distribution of the area of the pixel covered by a grain is known. When the grain covers more than one pixel, only the pixel with the largest covered area (at most 100%) is relevant since this pixel has the

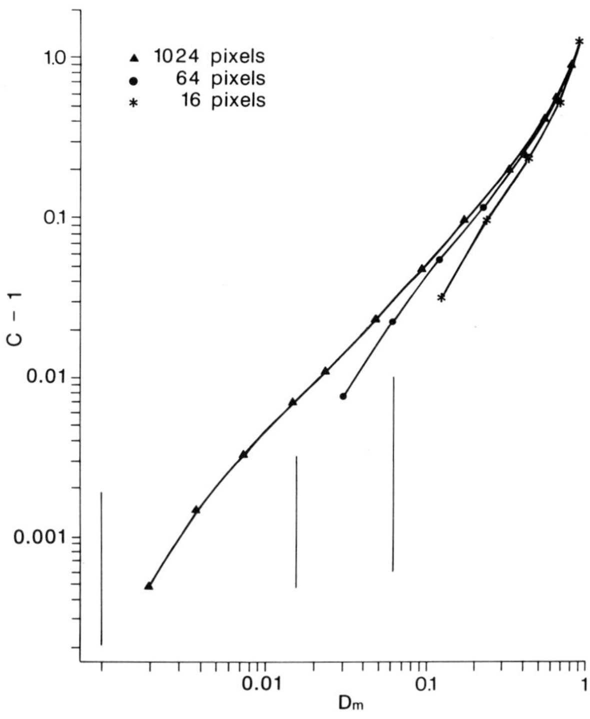

FIGURE 2. Correction of grain density for multiple occupations. The calculations have been made for a frame of 32^2, 8^2, and 4^2 pixels. Along the ordinate, the correction factor C minus one (C − 1) is plotted as a function of the measured grain density D_m (= occupied pixels/N). The lower end of the curves has a D_m value corresponding to two grains within the frame. The three vertical lines indicate the "asymptotes" (one grain).

highest chance to be above threshold. As far as is known, this problem cannot be solved analytically in geometrical probability theory. However, the distribution can easily be approximated by calculating the area of overlap for a finite number of pixel-to-grain positions. Here, the actual calculations were made for perfect circular grains with unit radius and a two-dimensional matrix with perfect square pixels with a width w of 1.0, 1.2, 1.4, or 1.6 units. To approximate the distribution, a step size of 0.1 or 0.05 was used in two orthogonal directions. Figure 3a shows the distribution of the area of overlap of the 576 grain-to-pixel positions with a pixel width of 1.2. The large peak to the right is due to grains which completely or nearly completely cover a pixel. Decreasing the pixel width results in a further increase of this peak and an attenuation of the peak at 0.8 (abscissa). For a pixel width $\leq \frac{1}{2}\sqrt{2}$, the distribution degenerates to a single narrow peak (δ-function) at w^2, since then at least one pixel always is completely covered. Therefore, the grain/pixel area ratio of 2π represents the transition between meaningful ratios yielding distributions of final width and distributions degenerated to a δ-function (ratio $> 2\pi$). For increasing pixel widths, the slender peak to the right diminishes and the broad peak to the left becomes larger and dominant. For $w \geq 2$, the maximally covered pixel area amounts to π. From here, increasing w again yields a slender peak to the right, but now at π, with a rapid decrease of the area of the distribution to the left of this peak. Since for these values of w the ratio of overlap is always smaller than $\frac{1}{4}\pi$, w should be chosen between $\frac{1}{2}\sqrt{2}$ and 2. Figure 3b is helpful when estimating an optimal value. Along the ordinate, the cumulative chance is plotted that at least 50 or 75% of the area of a pixel is covered by a grain, with the area ratio of the

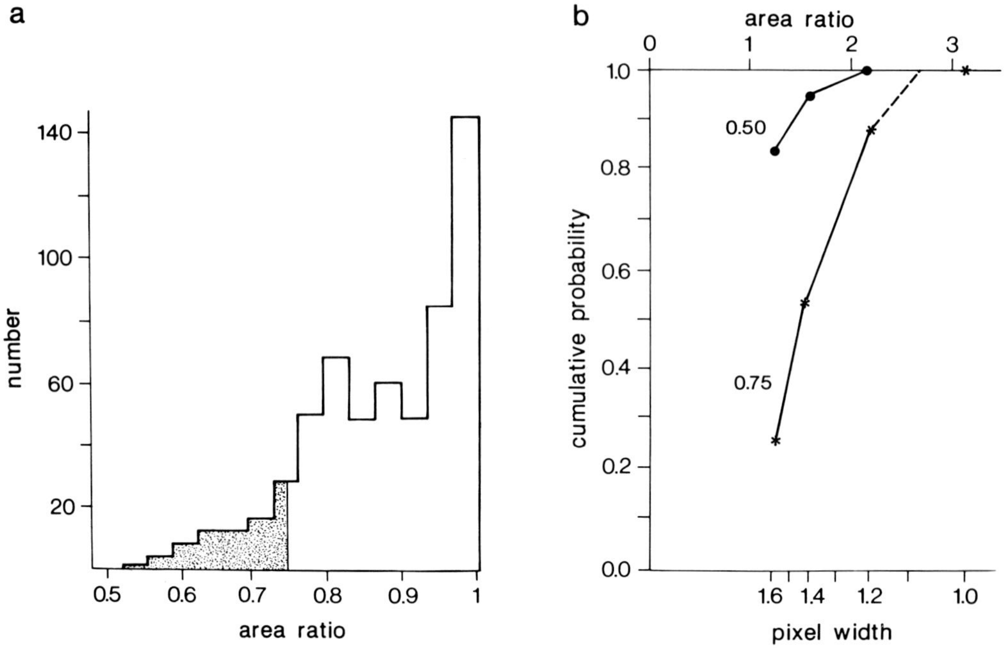

FIGURE 3. The influence of pixel size. (a) Distribution of the area of the pixel (1.2 × 1.2) covered by the grain calculated for 576 grain-to-pixel positions with a step size of 0.05 in both directions. The nonstippled area, expressed as a fraction of the total distribution, is depicted as one of the data points in b (abscissa 1.2, ordinate 0.87), (b) Cumulative probability of a coverage of at least 50 and 75%, calculated from the approximated distributions for pixel widths of 1.0, 1.2, 1.4, and 1.6. Along the abscissa, the area ratio (grain/pixel) as well as the pixel width (nonlinear scale) are indicated.

grain over the pixel along the abscissa. When one demands that at least 50% of the pixel should be covered by the grain, and that only 5% of the grain-to-pixel positions will not fulfill this condition, then the grain/pixel area ratio should be 1.6. For the 75% criterion, this ratio is 2.5.

Up to now, variability in grain size has not been taken into account. As long as the variability is small, it does not affect the results described above. However, for a large variability, a higher area ratio should be chosen to prevent smaller grains from not scoring. In the literature, substantial data about grain size variability are practically lacking. Small (coefficient of variation of 0.19)[4] as well as large (range of a factor of 2.4)[2] variabilities have been reported. Data about the shape of grains, besides some qualitative data (Rogers, in Chapter 2 of Reference 1), are also lacking. Studied at high magnification (EM), their shape often appears to be very irregular and variable. However, under the LM, they look fairly circular. To meet the demand for some quantitative data about size and shape, grains of ^{3}H-methylacrylate test plates made with Kodak NTB-2 emulsion and developed with Kodak D-19 were measured. It appears that the shape can be approximated by an ellipse. In bright-field, the ratio of the short vs. long axis is 0.69 ± 0.11 (N = 168), whereas in dark-field, 0.74 ± 0.085 (m ± SD, N = 165) is found. Thus, dark-field grains look more circular and vary less in size than bright-field grains. The grain size (defined as the diameter of the circle with the same area) is 0.46 ± 0.076 μm and 1.95 ± 0.31 μm, respectively.

In conclusion, taking into account the moderate variability in size and shape of the roughly circular grain, the mean grain area should be twice that of a pixel.

IV. PRECOUNTING PROCEDURES AND BASIC APPROACH

The use of dark-field illumination has several advantages over bright-field illumination:

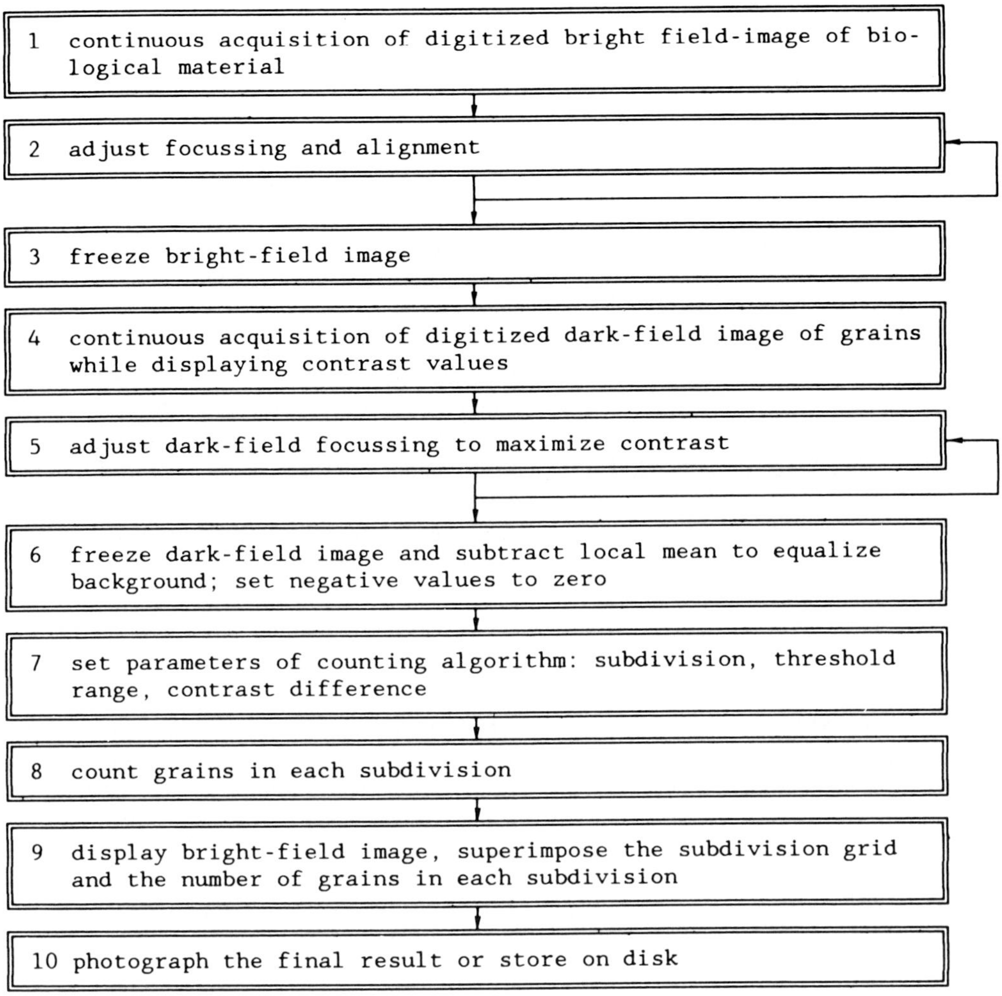

FIGURE 4. Flowchart of principal grain counting procedures of a grain-recognizing program. (Modified from Schellart, N. A. M., Zweijpfenning, R. C. J. V., van Marle, J., and Huysmans, D. P., *Comput. Methods Programs Biomed.*, 23, 103, 1986. With permission.)

1. The contrast between the grains, which show up as bright specks, and the background is considerably larger.
2. The small size and precise shape of the grain are less critical.
3. The magnification of the LM can be smaller than 100×.

Computerized grain counting enables separate storage of the (bright-field) image of the biological section and the (dark-field) image of the grains. After the counting procedure, the grain densities can be superimposed numerically or with graphic symbols in a matrix arrangement upon the section image.[4] In tissue with inevitable differences in grain densities, the identified grains can be superimposed on the section image as bright pixels (without counting them) in a way similar to that used in autoradiographic research on chromosomes.[10] Figure 4 gives a flow chart of the principal precount and count procedures as described by Schellart et al.[4]

V. STRATEGY OF GRAIN COUNTING

A. THE CENTRAL PIXEL TRUE APPROACH

1. Basic Procedure

In autoradiography, especially with ^3H, the grain density is often low to moderate, so that in general only 1 to 10% of the pixels are covered by the silver grains caused by radioactive decay. When a distribution of gray values of the pixels in the digitized image is determined, the distribution does not show a peak or shoulder which can be attributed to the grains, due to the large local variability of the background intensity. This background is caused by the underlying section (see Section VII.A), imperfect optics (see Section VII.B.2), and artifacts such as fine dust particles. The lack of a distinct and predominant contribution of the grains in the pixel intensity histogram rules out the use of straightforward or more sophisticated counting (or densitometry) for low (or moderate) grain densities. Owing to the often high variability of background intensity, grain discrimination by exceeding a static criterion generally fails. However, with a dynamic criterion, being a preset difference in gray value with respect to the immediate surrounding of the pixel to be tested (the central pixel), the pixels with a definite higher intensity can be scored as silver grains. Although this approach can be used under bright-field as well as dark-field conditions, reliable results can be obtained more easily with dark-field.[1,2,4]

The central pixel true approach has been described by Schellart et al.[4] and Reep and Greegan.[2] The procedure of Reep and Greegan[2] can be seen basically as an improvement and refinement of that of Schellart et al.[4] The method of Schellart et al.[4] is discussed first.

The digitized image is scanned from left to right and from top to bottom until a pixel is found which has an intensity value between a preset, fixed lower and upper level (to ignore noise and bright artifacts, respectively). Such a pixel, called true (T pixel), forms the central pixel of a 3 × 3 matrix. The eight pixels surrounding the central pixel are all connected by one side or by one corner with the central pixel (8-connected). The gray value of the central pixel is compared with the gray values of the 8-connected pixels. An 8-connected pixel is T when its intensity is larger than that of the central pixel minus a preset intensity difference. In this way, single grains of size 1 to 9 can be detected. However, the grain can only be scored when a 1-pixel-wide margin around the 3 × 3 matrix (this is a 5 × 5 margin) does not contain a T pixel (i.e., a false [F] pixel). If the 5 × 5 margin is not empty, the grain candidate is considered to be an artifact (too large or touching another grain). However, by this (too limited) approach grains consisting of a single (central) pixel with an empty 3 × 3 margin and a not-empty 5 × 5 margin will be rejected. Therefore, after finding a T pixel, the first test the grain candidate has to pass is whether the central pixel is a one-pixel grain. Thus, the steps in the grain recognition procedure are

1. Jump to the next pixel and determine whether the pixel is T. If the pixel is F, reenter step 1; if not, proceed to step 2.
2. Collect the 3 × 3 margin of the central T pixel and convert the intensity values to T and F values. Count the number of T pixels (n_p) in the 3 × 3 margin. If this number is zero, then increment the number of 1-pixel grains (size 1), set the 3 × 3 margin to zero intensity, jump to the next pixel, and go to step 1. If not zero, proceed to step 3.
3. Read the 5 × 5 margin and scan this margin for a T pixel. If a T pixel is found, return to step 1, otherwise go to step 4.
4. Increment the number of grains (size n_p). Set the 5 × 5 margin to zero intensity, jump two pixels to the right, and proceed with step 1.

The visual control of the counting process can be improved by adjusting the intensity of the T pixels above the upper threshold.

	configuration			%		configuration			%
	F	F	F			F	F	F	
(1)	F	T	F	44	(6)	F	T	F	3
	F	F	F			T	T	F	
	F	F	F			F	F	F	
(2)	F	T	T	25	(7)	T	T	T	2
	F	F	F			F	F	F	
	F	F	F			F	F	F	
(3)	F	T	F	6	(8)	F	T	F	2
	F	T	F			F	T	T	
	F	F	F			F	F	F	
(4)	F	T	T	6	(9)	F	T	T	1.5
	F	T	T			F	F	T	
	F	F	F			F	F	F	
(5)	F	T	T	3	(10)	T	T	F	1
	F	T	F			T	T	T	

FIGURE 5. The ten most frequently occurring configurations of grains in a 3 × 3 margin with the central pixel true approach. F indicates a false; T indicates a true pixel. (Modified from Schellart, N. A. M., Zweijpfenning, R. C. J. V., van Marle, J., and Huysmans, D. P., *Comput. Methods Programs Biomed.*, 23, 103, 1986. With permission.)

2. Evaluation of the Basic Procedure

In dark-field images, the ten most frequently occurring configurations represented the vast majority (about 95%)[4] of configurations, as illustrated in Figure 5. Seven of these ten configurations could have been detected with a 2 × 2 inner matrix, and all with a 3 × 2 matrix. Obviously, scanning from left to right and from top to bottom seldom yields a T pixel in the upper row and left column of the 3 × 3 matrix. When a grain density smaller than 5% (of the pixels covered) occurs, it can safely be concluded that only a minority of the 2-pixel grains are due to a pairing of two grains. Most of them will be caused by (slightly larger) grains sufficiently covering two pixels. In Figure 5, statistically half of the 3-pixel grains and most larger grains may be considered as artifacts.

3. Refinements

In a grain field, it is hard to say whether clusters of three to nine pixels within the 3 × 3 matrix are clustered grains or artifacts. The procedure of Reep and Greegan[2] has been

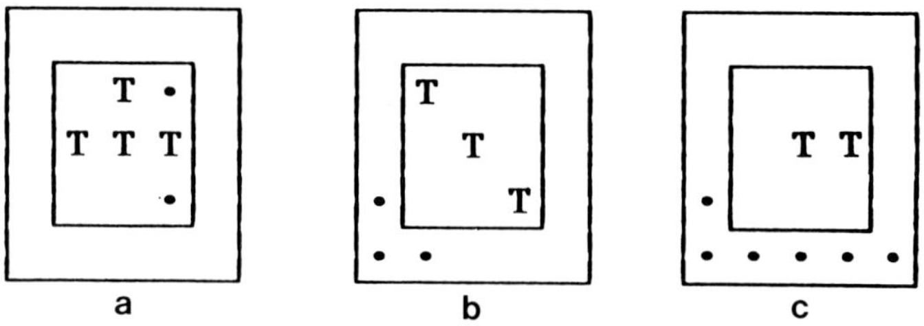

FIGURE 6. Examples of missed grains in complex configurations. The three configurations show a grain in the center which may be missed with the procedure of Reep and Greegan,[2] depending on the values of the two tests screening small (a) or large artifacts (b and c). In b and c, the grain candidate may be missed when in the 5 × 5 margin too many pixels indicated by dots are T. The middle and right configurations show a grain in the center which will be missed by the procedure of Schellart et al.[4] The dots in the pixels indicate pixels of which one or more are true.

designed for low grain densities. For low densities, clustered grains should be very rare. Therefore, they developed two tests to exclude such fake clusters. A contrast test checks whether the mean intensity value of the 3 × 3 margin is a preset intensity difference lower than the intensity of the central pixel. When this test is passed, the number of T pixels in the 3 × 3 margin having the same intensity as the central pixel is determined. (The very first test examines whether the central pixel has a higher intensity than any of the 3 × 3 margin pixels.) The chance for T pixels of equal intensity is substantial since here the grains are twice as wide as the pixels and the actual number of T intensity levels is only about 35.[2] When more than a preset number of pixels attains this intensity, the central pixel is considered to be part of an artifact and is rejected as grain center. The grain candidate in the example depicted in Figure 6a may be rejected by one of the tests, depending on the setting of the parameters.

To examine the 5 × 5 margin for rejecting large artifacts, Reep and Greegan[2] introduced (nearly) the same two tests as mentioned above for the 3 × 3 margin. These two tests are performed before the two small artifact tests described above are carried out. The two tests for the large artifacts are a less rigorous approach compared to the 5 × 5 empty-margin criterion of Schellart et al.[4] The tests overcome the problem that some combinations of grains with equal and/or different sizes are not scored (generally 1- and 2-pixel grains). With the plain 5 × 5 empty-margin criterion of Schellart et al.,[4] multipixel grains separated by one F pixel will be rejected, and multipixel grain candidates separated by one F pixel from an isolated T pixel will also be rejected by the empty 5 × 5 margin test (Figure 6b and c). Since with the procedure of Schellart et al.[4] 2-pixel grains may occur frequently (Figure 5), 2-pixel grain candidates were tested for an empty 4 × 3 or 3 × 4 margin,[4] a procedure severely reducing the number of false negatives.

One of the basic differences between the methods of Schellart et al.[4] and Reep and Greegan[2] is that the latter allows one or some (depending on the parameter settings) T pixels in the 5 × 5 margin. This means that a T pixel in the 5 × 5 margin may be adjacent to a T pixel in the 7 × 7 margin. However, this is not tested,[2] so there is a chance, depending on the setting of the parameters, that long thread-like structures are false positives. In contrast, a 3-pixel-long (double) grain with its central pixel at the end may be scored as a false negative. Figure 6b and c give some examples of configurations which yield false negatives in the empty 5 × 5 margin[4] and the nearly empty 5 × 5 margin[2] method. The small inaccuracies of both procedures can be overcome simply by determining whether each T pixel in the 3 × 3 margin is connected to a T pixel in the 5 × 5 margin. Obviously, this test increases analysis time substantially, so that it is hardly feasible with PC-AT computers or too-small minicomputers. A final refinement concerns a test on the size of the grains and grain clusters, which is illustrated by the following example. For densities smaller than 100 grains per 64^2 pixels and a mean grain size of 2, the maximal grain size can safely be limited to 4, irrespective of the counting strategy, since statistically it is unlikely ($p = 0.008$) that an occupied 3×3 matrix contains more than two grains. Grains of size 3 and 4 should obtain a weight of two (counted twice). In addition to the dependency on grain density and grain size, the size correction may also be influenced by characteristics of the biological material, e.g., the occurrence of local high grain densities. This may hamper the implementation of a simple test on grain size.

B. STATIC THRESHOLDING AND GRADIENT THRESHOLDING

The versions of the central pixel true (T) approach, based on grain size and shape, constitute a true grain-recognition procedure. A clearly distinct approach of grain counting is to decide exclusively on pixel intensity or intensity gradient, regardless of whether it is covered by a grain. Thus, the decision whether a pixel is T or F is solely based on thresholding. This approach is mostly chosen when the grains are heavily clustered. It is often used in labeling studies of chromosomes and cell nuclei. In these applications, the grains cover many pixels and the actual goal is to determine the area of the soma, nucleus, or chromosome covered by grains. This can be done directly with the raw digitized image,[5,7,10] although sometimes a preprocessed image is used. To remove coarse variations of background intensity in order to use a constant threshold, the neighborhood of every pixel can be filtered. This can be done by low-pass filtering of the image and subtracting the result of the original image,[11] which result in a kind of spatial high-pass filtering. This filter method to remove the background is promising for large field analysis of low-density autoradiographs with few artifacts, digitized so that the pixel size is slightly less than the grain size. The method has recently been applied to high-magnification grain counting (many pixels per grain) in tissue with single grains as well as clusters of grains.[11] Since this filtering method ignores grain clusters, the image with the removed background was added to a modified original image. The latter was obtained after subtracting a constant intensity, so that only solitary grains and clusters remained. With this method, there is a risk of false positives (large artifacts) and false negatives with respect to the grain clusters.

Older procedures apply thresholding straightforwardly in order to estimate the field area (often a nucleus or chromosome) occupied by grains. Here, the field area was often determined by the experimenter himself, e.g., with a light pen.

C. RELIABILITY AND SENSITIVITY

The reliability of counting algorithms can be tested by (1) comparing the grain counts obtained by manual counting and automated counting[2,5,7-10] and (2) comparing grain counts of the same frame, obtained by computer with different parameter settings.[2,4] The high correlation (often >0.95; Table 1) found by various authors[2,5,7-10] when comparing both ways of counting clearly shows that computerized counting is a very good and fast substitute for human counting. Remarkably, for moderate grain densities obtained at high LM magnification, the computer often yields lower counts than those of manual counting, whereas for low densities they are the same.[7,9,12] At low magnification, this difference was not found.[2] This effect is obviously due to the higher occurrence of adjacent (or overlapping) grains, which are scored as one grain by the computer, but double in manual counting. Reproducibility of the computer countings, which is better than for manual countings,[12] also shows high correlations, even when countings with optimal and suboptimal parameter settings are compared.[2,4]

Reep and Greegan[2] conducted the most extensive study of the sensitivity of a grain-recognizing program. Their algorithm has five parameters. The first three are the minimal intensity value of the potential grain center (Imin), the 5 × 5 margin mean contrast parameter (Pcon), and the number of pixels in the 5 × 5 margin not exceeding a small contrast difference (P5 × 5). The last two parameters concern the 3 × 3 margin. The sensitivity of the parameters was tested by varying the setting of the first one with relaxed (not critical) values of the other. After optimizing the first one, the second parameter was optimized, and so on. It appeared that changing Imin by 2 points on a 64-intensity scale changes the grain counts at most by 5%. The same change in the setting of Pcon results in a slightly higher error, whereas a change of P5 × 5 of only 1 pixel nearly doubles the error. The 4th and 5th parameters are less critical and contribute only marginally to an increase in accuracy.

D. COMPARISON OF COUNTING SPEED

A fair comparison of the counting speeds of the various dedicated systems and the commercial image processing systems is hard to make. This is due to large differences in computer systems, including image handling hardware, and differences in software, e.g., the choice of the computer language (machine language or one of the common higher-level languages). Detailed data about hard- and software are generally not available in the current literature, and precise information about counting speed is often lacking. Nevertheless, generally speaking, the multipuropse commercial systems seem to have lower speeds than dedicated systems. For example, the counting speed of the VIDAS system, a subsystem of IBAS, is about 7 grains per second for grains with an average size of 8 pixels (the minimal allowed size is 4). The dedicated system of Reep and Greegan,[2] based at a PC-AT with image processing board has a speed of 180 grains per second (mean grain size $3^1/_2$), and the system of Schellart et al.,[4] based on an LSI 11/23, has a speed of 110 grains per second (mean grain size 1.4).

VI. GRAIN BACKGROUND NOISE

In autoradiographs, grains show up which are not due to radiation of the incorporated radioactive material. The causes of the occurrence of these grains have been discussed extensively by Rogers (in Chapter 6 of Reference 1). Two types of background noise can be distinguished. The first one concerns spontaneously developed grains, induced thermically or photically. An (approximate) estimate of the amount of this Poisson-distributed background noise is given by the grain density well separated from the labeled biological material. It is generally less than 0.5 grains per 100 μm^2,[8,11] so that correction is necessary only for low grain densities. A mathematical procedure to correct the mean grain count per area of variable size (e.g., nuclei) for this background has been described by Korr and Schmidt.[13,14] For low grain numbers per nucleus (<5), the result of this correction deviates substantially from the difference between the mean grain count with and without labeling.

The second type of background consists of grains which are not homogeneously distributed over the section. These grains are caused by mechanical stress, chemography (due to biological or other agents), environmental radiation, etc. Standard methods to correct for these artifacts do not exist.

VII. DISTURBANCES AND LIMITATIONS

A. TISSUE EFFECTS

Large differences in the optical density of the biological material at a large spatial scale can give rise to false positives and negatives, depending on the algorithm. A remedy is to divide the frame into a number of fields and calculate the mean pixel intensity of each field. This value can be used to eliminate most of the background by subtracting it from the gray value of the individual pixels in the field (negative values are set to zero). A more sophisticated (no border effects between fields), but more complex and time-consuming method is to filter out the low spatial frequencies. Background effects can often be reduced further by applying monochromatic light.

B. OTHER DISTURBANCES

1. Focusing

Focusing is generally done manually, but sometimes is computer assisted[4] or automatized.[6] The effect of focusing is thoroughly examined by Reep and Greegan.[2] By determining the total grain count vs. focusing, they found an optimal range of 6 μm (error 1.5%). By calculating the spatial correlation of the grain count in the 256 squares of an image frame, Schellart et al.[4] obtained a range of 2 μm. For high-magnification images, the range goes down to 1.6 μm.[10] Thus, optimal focusing is a prerequisite for reliable counting.

2. Optical Limitations

A serious limitation of many microscope/video systems is that an homogeneous illumination across the microscope field (object field) can hardly be obtained.[4,15] In large field images, especially obtained with 18, 25, or 40 × objectives, illumination is at its best circular symmetric, with a slight decrease of intensity toward the edge of the field. This effect is generally strengthened by the optics of the camera. Homogeneity of the illumination can be improved optically by adjusting the illumination during display of the digitized image, without a biological object and with a reduced intensity. Again, the performance of the optics is improved by the use of monochromatic light, resulting in sharper grain images.

Another optical flaw is cushion-shaped distortion, especially at the edges, which even occurs with plan-apo objectives. The influence of the optical imperfections at the edges can be quantified by determining the grain density in the fields at the edge compared to fields in the center when using an autoradiographic test plate. By repeating this procedure several times, the averaged found differences, being some percent, are used to make a correction table.

3. Imperfections of Preparation

Counting inaccuracies can also be caused by fine dust particles and inhomogeneities in the layer thickness of the photographic emulsion. These and other imperfections of autoradiographic material are discussed by Rogers (in Chapter 7 of Reference 1).

VIII. CONCLUSIONS

The chosen strategy of grain counting depends on the kind of biological material (chromosomes, nuclei, or tissue) and the density of the labeling.

For chromosome and nuclear studies, mostly performed with high LM magnification in semithin sections, the static threshold approach generally suffices, as concluded from the results of reliability tests. Due to the often heavy labeling of nuclear material, the algorithm will detect isolated grains as well as clustered grains. This means that false positive (small artifacts) cannot be avoided.

In general, typical tissue studies demand a grain-recognizing algorithm. In order to have a sufficient resolution and depth of focus with a tissue area as large as possible, the LM magnification should be between 45 and 80 × . Counting in dark-field is more efficient than in bright-field. The grain/pixel area ratio should be about 2.0. Due to intensity fluctuations of the background (caused by the section), occurring over large areas, some type of spatial high-pass filtering is necessary. Strictly local photic background fluctuations demand dynamic thresholding (contrast criterion). Artifact rejection tests are a necessity. The central pixel true approach appears to be a successful and reliable strategy. The performance of this approach for moderate grain densities is expected to improve when the regular 5 × 5 margin is replaced by an irregular margin dependent on the pattern of the directly adjacent true pixels, and forming within the 5 × 5 matrix a 1-pixel-wide contour around the grain. The future increase in computational speed should make this refinement feasible.

For all applications, but especially those based on static thresholding, a rigorous standardization of the various steps of the procedure is helpful to improve the performance. This concerns the settings of the (stabilized) light source, the microscope, and the video camera. If the procedures of preparing and developing the autoradiographic material are also standardized, it is profitable to standardize also the values of the parameters of the algorithm.

ACKNOWLEDGMENTS

The author would like to thank Erwin Furth and André Linnenbank for their help in presenting the problem of multiple occupations. Professor J. Strackee is acknowledged for the stimulating discussions about probability theory and for carefully reading the manuscript. Thanks is due to Maarten Prins who provided the statistical data on grain size and shape.

REFERENCES

1. **Rogers, A. W.,** *Technique of Autoradiography,* 3rd ed., Elsevier/North-Holland, Amsterdam, 1979.
2. **Reep, R. L. and Greegan, W. L.,** An accurate method for automated counting of silver grains in autoradiographs, *Comp. Biomed. Res.,* 21, 244, 1988.
3. **Fry, D. L., Tousimis, A. J., Talbot, T. L., and Lewis, S. J.,** Methods to quantify silver in autoradiographs, *Am. J. Physiol.,* 238, H414, 1980.
4. **Schellart, N. A. M., Zweijpfenning, R. C. J. V., van Marle, J., and Huysmans, D. P.,** Computerized pattern recognition used for grain counting in high resolution autoradiographs with low grain densities, *Comput. Methods Programs Biomed.,* 23, 103, 1986.
5. **Prensky, W.,** Automatic image analysis in autoradiography, *Exp. Cell Res.,* 68, 388, 1971.
6. **Lipkin, L. E., Lemkin, P., and Carman, G.,** Automated autoradiographic grain counting in human determined context, *J. Histochem. Cytochem.,* 22, 755, 1974.
7. **Sklarew, R. J.,** Simultaneous Feulgen densitometry and autoradiographic grain counting with Quantimet 720D image-analysis system. II. Automatic grain counting, *J. Histochem. Cytochem.,* 30, 49, 1982.
8. **Kempf, J. and Kempf, C.,** Autoradiographic grain counting by fully automatic image analysis, *Microsc. Acta,* 86, 215, 1982.
9. **Kekki, M., Santamäki, T., Talanti, S., and Vesikari, E.,** Accuracy of an automatic counting method by television-computer equipment in the quantitative autoradiography, *Acta Histochem.,* 81, 171, 1987.
10. **Stengel-Rutkowski, S., Gundlach, H., and Zang, K. D.,** Quantitative electronic analysis of chromosome autoradiographs using a television image analyzing computer, *Exp. Cell Res.,* 87, 313, 1974.
11. **Parkkinen, J. J., Paukkonen, K., Pesonen, E., Lammi, M. J., and Markkanen, S.,** Quantitation of autoradiographic grains in different zones of articular cartilage with image analyzer, *Histochemistry,* 93, 241, 1990.
12. **Schuette, W. H., Chen, S. S., Occhipinti, S. J., Mujagic, H. S., and Shackney, S. E.,** Automated radioautographic grain counting. Correction for grain overlap, *Cell Tissue Kinet.,* 16, 221, 1983.
13. **Korr, H. and Schmidt, H.,** An improved procedure for background correction in autoradiography, *Histochemistry,* 88, 407, 1988.
14. **Korr, H. and Schmidt, H.,** A new procedure for correcting background in quantitative autoradiographic studies, *Acta Histochem.,* 37, S149, 1989.
15. **McCasland, J. S. and Woolsey, T. A.,** New high-resolution 2-deoxyglucose method featuring double labeling and automatic data collection, *J. Comp. Neurol.,* 278, 543, 1988.

Movement Analysis

Chapter 15

REAL-TIME TRACKING OF MICROORGANISMS

Donat-P. Häder and Kurt Vogel

TABLE OF CONTENTS

I. INTRODUCTION

The development of digital image analysis as a tool for automatic analysis of movement has found wide acceptance in many different fields in biology, medicine, and other sciences, facilitated by technological development in video and computing technology.[1] The applications range from the analysis of cloud movements recorded by satellites[2,3] to the tracking of microorganisms.[4,5] In medical research, automatic tracking by image analysis facilitates following the movements of lymphocytes[6] and sperm cells,[7] and in zoology it is utilized to track crawling nematodes or fish[8] (see also Chapter 17 in this volume). Generally, the techniques developed are independent of the size and nature of the moving objects, which can range from microscopic particles to celestial objects. The size of the tracked objects only affects the recording techniques. From an image analysis point of view, other parameters are more relevant, such as the relative speed in the image (calculated as pixels per time unit), the relative size and form parameters,[9-11] areas of a distinct color or gray shade,[12,13] or movement of individual elements in a sequence of images.[14-16]

Vision by man and machine differs in many respects, and certainly the human visual apparatus and brain are still far superior to machine vision.[17] Both the spatial and gray or color shade resolutions of the human eye surpass those of high-resolution video cameras and the image processing of the human brain far exceeds the speed and complexity of even the fastest computers.[18] The aims of introducing automatic tracking systems based on image analysis are high objectivity in contrast to manual assessment performed by a human operator, who cannot free himself from a biased analysis even in double blind experiments. Furthermore, high-speed, computer-controlled image analysis using modern equipment enhances the velocity of the measurements, thus increasing the statistical relevance of the data. Image analysis allows the extraction of various parameters from the sequence of frames:[19] in addition to form and size parameters of the objects as well as a gray scale or color distribution, motility can be analyzed in terms of the percentage of motile objects in the population visible on screen or in terms of the velocity distribution of the objects. Furthermore, movement vectors can be calculated by comparing the position of individual objects in subsequent frames, which allows one to determine the distribution of the direction of movement.

A. MOVEMENT AND ITS DETECTION IN IMAGE ANALYSIS

Movement of objects relative to the preceiving device, which can be a camera or the human eye, in a sequence of images is recorded as a change in the gray level or color at a given position with time. In the following discussion, we will concentrate on gray images; however, the same reasoning applies to color images. The analog image is usually digitized at a defined spatial resolution (e.g., 1024 × 1024 image elements, called pixels). Each pixel is assigned a unique gray (or color) value during digitization, and the gray-level range, which depends on the hardware analog-to-digital (A/D) converter, defines the dynamic range within the digitized image.[20] The gray value in a sequence of images is thus a function of time and space:

$$f(x,y,t) = f(x + dx,y + dy,t + dt) \tag{1}$$

Movement can also be regarded as the stream of gray-scale patterns through the image, which has been termed optical flow.[21] However, the image is only a two-dimensional representation of three-dimensional objects. Usually, only the projection of the objects onto a plane is analyzed since approaches to reconstructing the path of a motile particle in space increases the amount of data considerably. However, there are methods to track objects in three dimensions, e.g., by the technique of defocusing.[22,23] If we disregard rotational and deformational changes, we are left with translational movements which can be detected in the two-dimensional projection by the change in the position of surface points relative to each other.[24,25] For example, the movement of a particle in a microscope image can be detected as a dilation of the area when it moves in and out of focus.

Computer-controlled image analysis of moving objects is usually based on a sequence of frames, which are recorded and processed at discrete time intervals defined by the camera frequency (e.g., 50 half-frames in Europe and 60 half-frames in the U.S. video standard). The movement of the individual objects is therefore seen as a jump between subsequent frames. However, the movement can be too slow to be detected between subsequent frames, which is a problem, for instance, in the cell tracking of gliding microorganisms such as diatoms, desmids, or amoebae. This problem can be circumvented by introducing defined time intervals between subsequently recorded frames.

In very fast movements, analysis can be difficult since the translocation can be so far that the identification of the object fails, especially at high densities of moving objects, a problem known as the correspondence problem. If the objects cannot be distinguished by form or size parameters, a correspondence can only be determined if the movement vectors are smaller than the mean distance between the objects.[26] If the object distances are statistically distributed, misinterpretations cannot be excluded.

Another problem occurs when objects meet or partially or totally occlude each other. A related problem occurs if touching objects move apart; in this case, no correlation can be found to previous frames. These difficulties can be aggravated by shadow images, non-uniform irradiation in the image, or movement perpendicular to the plane of projection.

Two different methods can be distinguished for following the movement of objects in a sequence of images:[27] (1) determination of the optical flux by analysis of the difference between subsequent images and (2) the calculation of corresponding image segments.

In the first approach, which has been used to study, e.g., street scenes,[28] an image is digitized and stored in the video memory. When the next frame comes in, a difference image is calculated between the two images by comparing all individual pixels[29] (see also Chapter 18 in this volume). Provided no changes have occurred in the irradiation, the gray values in the static areas of the images do not change between the two images and therefore the subtraction results in a low value, which in the difference image occurs as a dark pixel.[30] In contrast, areas in which a change in gray levels occurs due to the movement of an object show up in bright pixels. In practice, this procedure is modified by initializing the whole image matrix with 0 values. An element in the matrix is incremented when the gray values of corresponding pixels differ between subsequent frames. The calculated first-order difference picture (FODP) shows the traces of the moving objects in the full width of the objects. The tracks can be extracted by the technique of skeletoning, during which shells of 1-pixel width are removed iteratively until only a line of 1-pixel width remains. From the FODP, a second-order difference picture (SODP) can be calculated by defining those elements in the pixel array as 1 where elements of different values are found in the FODP. The SODP can therefore be used to determine the start and end positions of tracks. Due to the large number of pixels in the array, these calculations which involve all picture elements in the image are rather time consuming and can hardly be performed in real time.

The second method is based on the determination of corresponding image segments in subsequent frames. The position and form of each object (or a limited number of objects for faster calculation) is determined in each frame of the sequence and stored for the following analysis. Using complex search algorithms allows one to identify each object in subsequent frames so that the movement vectors can be determined. One of the methods for identifying the individual objects is the cross-correlation method by Smith and Phillips[3] which, however, has the disadvantage of being rather time consuming. Takahashi and Kobatake,[5] in contrast, use a reduced correlation matrix of 32×32 pixels between subsequent frames, centered around the object found in the first image. The maximum of the correlation matrix is equal

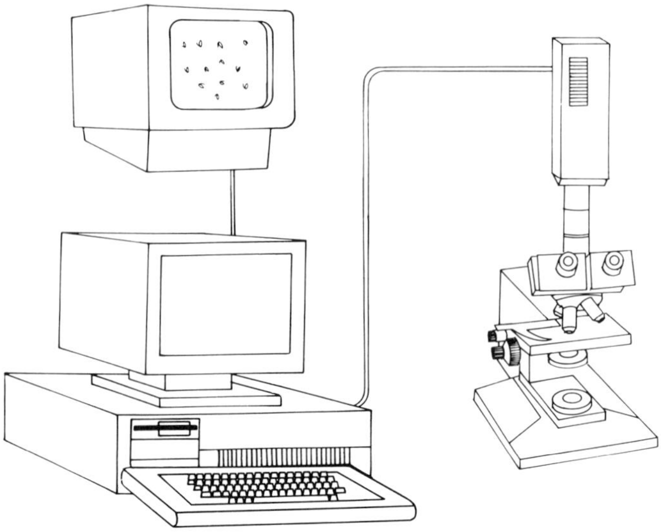

FIGURE 1. Set-up to track paths of microorganisms from a microscope. The image of the moving cells is recorded by a CCD camera on top of a microscope and digitized in a digitizer card plugged into a slot of an IBM AT-compatible 80386 microcomputer, which has access to the digitized image in memory.

to the dislocation of the object between two images. In a method developed by Noble and Levine,[31] the positions of objects in subsequent frames are stored in an array. Then the movement vectors are calculated by the nearest neighbor algorithm. Aggrawal and Duda[32] describe a method for the movement analysis of polygons which can even overlap partially, based on the detection of corresponding vortices. An algorithm to determine the translation, rotation, and dilation of moving objects has been developed by Schalkoff and McVey.[33]

II. HARDWARE REQUIREMENTS

The source of the image can be a video recorder or a camera (tube or charge coupled device, CCD), which, for tracking microorganisms, is usually mounted on top of a conventional light microscope or a stereo microscope (Figure 1). In order to increase the optical contrast of the organisms against the background, dark-field microscopy is used, in which the organisms appear bright on a dark background.[34-36] Since many organisms are sensitive to light and move, e.g., toward a bright spot, infrared (IR) irradiation often is used, which neither activates the orientation mechanisms of the organisms nor their photosynthetic apparatus.[37] In these cases, an IR sensitive camera needs to be employed, which can be a Newvicon or a CCD camera, the sensitivity of which usually extends well into the IR region.[38] In the case of larger organisms, such as large ciliates or multicellular organisms, even a macro lens can be sufficient to obtain the necessary magnification to track the organisms. In most cases, cellular details are not of interest, and only the position of the cells needs to be determined. If the organisms are to be tracked in three dimensions, two approaches are possible: (1) the stage of the microscope is moved in x, y-, and z-directions, so that an individual object is kept in the center of the image and in focus by suitable sensors and servo motors, the positions of which are recorded constantly;[22] (2) the alternative is to observe the moving object using two cameras mounted orthogonally to each other.[39] The tracking algorithm then needs to extract the displacement in all three directions (Figure 2).

Whatever the source of the video image is, it is digitized in a suitable hardware A/D

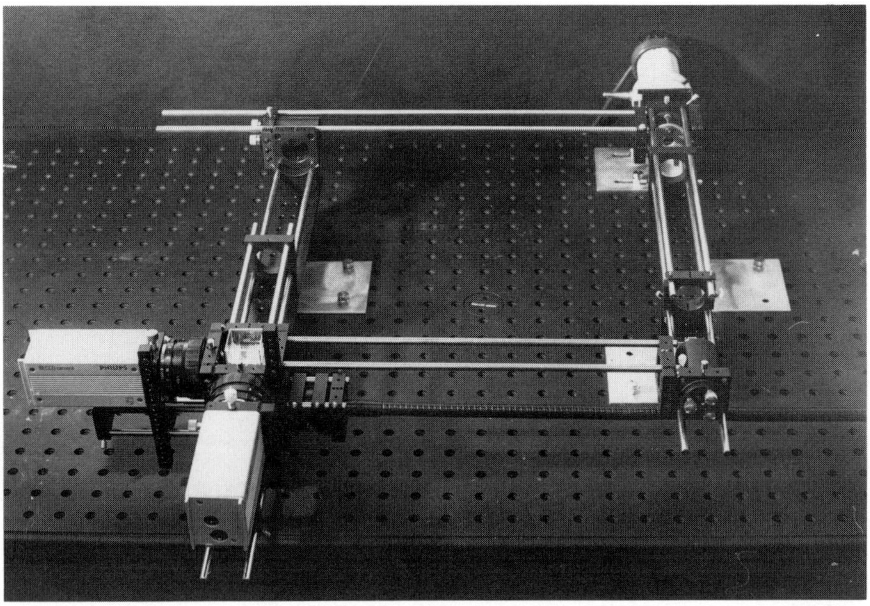

FIGURE 2. Set-up used by the authors to track large ciliates (*Stentor coeruleus*) in three dimensions. The moving cells are recorded by two CCD cameras, the images of which are combined into one using a video wiper. The combined image is then analyzed using standard equipment and software to extract the movement vectors in three dimensions.

converter. In the past, these devices used to be fairly slow and consumed up to several minutes to perform the digitization of one frame. This procedure, of course, did not allow for real-time image analysis. The next generation of instrumentation used so-called flash converters, which digitized the whole image at the same frequency as the frames were recorded (50 half-frames in the European CCIR standard and 60 half-frames in the American standard). While these machines were stand-alone instruments, most recent digitizers are reduced to the size of a card which can be inserted into the slot of a microcoputer.[40-42] There are many boards on the commercial market, also known as frame grabbers,[43,44] such as the PIP 1024 and MVP-AT (Matrox, Dorval, Quebec, Canada), VS-100 (Imaging Technology Inc., Woburn, MA, U.S., Figure 3) or GC1 (Delta-T Devices, Cambridge, England), to name only a few. Some digitizer cards allow color digitization, when using a color video input, such as Rivage 512 (Tecnom, Medon-la-Forêt, France), CR3000i (Bio-Rad, Düsseldorf, FRG) or MicroScale 1C PC-AT (Digihurst Ltd, Newark Close, Royston, England).

Often, the contrast and the brightness of the video image (gain and offset) can be manipulated before digitization by software (Figure 4). The digitizer hardware defines the spatial resolution of the digitized image, e.g., digitizing the video image at 10 MHz allows a resolution of 512 × 512 pixels, and an increase to 15 MHz produces an image of 768 × 512 pixels. Since the video image is not square, a pixel in a 512 × 512 matrix is not the representation of a square picture element, which needs to be taken into account during quantitative analysis, for instance, of tracks. Each pixel is defined by a specific gray (or color) value, which can, for example, be in the range of 0 (darkest black) to 255 (brightest white).

Before being stored in memory, the gray values of individual pixels can be manipulated.[45,46] For this purpose, a look-up table (LUT) can be defined in which the gray value of each pixel is looked up and replaced by the entry at this position, which has been defined by software beforehand (Table 1). Using this technique, it is possible to, e.g., invert an

FIGURE 3. The VS-100 (Imaging Technology Inc., Woburn, MA, U.S.) digitizer board. Stemmer, Puchheim, FRG. (With permission.)

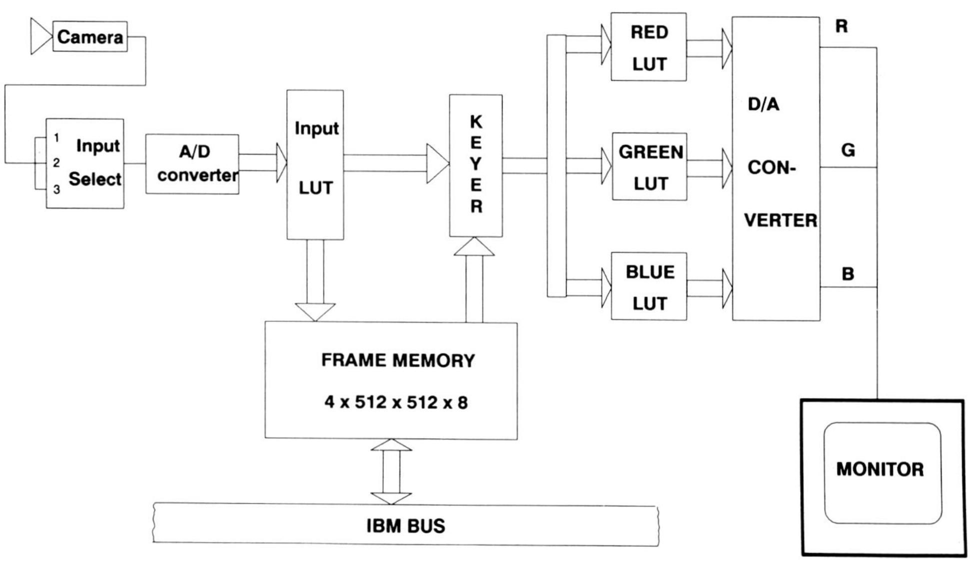

FIGURE 4. Block diagram of the hardware manipulation and digitization in a modern frame grabber (Matrox PIP 1024, Dorval, Quebec, Canada).

image into its negative by replacing a gray value of 0 by 255, a 1 by 254, etc. Likewise, every value between 0 and 127 (medium gray) can be mapped to 0, and every value between 128 and 255 can be mapped to 255; by this procedure, the contrast is boosted to its maximum, represented by only black or white (Table 1, B). LUTs can also be used to map the data in an exponential way or to expand part of the gray scale (Table 1, C).

The digitized image is usually displayed on-line on a suitable monitor (often in parallel to the camera input image before processing). For this purpose, the digital image is converted into an analog signal (D/A conversion) in order to display an analog image; there are also monitors which accept digital input. Before the D/A process and display, the image can be manipulated again using one or more output LUTs (Table 1). Pseudocolor representation of

TABLE 1
Use of Look-Up Tables (LUTs)

	A	B	C	D	
0	255	0	0	0	
1	254	0	0	3	
2	253	0	0	6	Blue
3	252	0	0	9	
4	251	0	0	12	
124	131	0	31	118	
125	130	0	63	121	
126	129	0	95	124	
127	128	255	137	127	Green
128	127	255	169	130	
129	126	255	201	133	
130	125	255	233	136	
131	124	255	255	139	
251	4	255	255	243	
252	3	255	255	246	
253	2	255	255	249	Red
254	1	255	255	252	
255	0	255	255	255	

Note: LUTs can be used to invert the image into its negative (A), increase the contrast to two levels (B), or expand the gray level of a small band (124 to 130) over the whole possible range (0 to 255) (C). Mapping certain gray level ranges to three colors results in pseudocolor representation (D).

black and white images uses three LUTs in parallel in which defined ranges of gray levels are mapped to specific colors (Table 1, D). With 256 possible gray levels in the stored image, all gray numbers ranging from 0 to 85 can be spread over the full range on the blue LUT, all between 86 and 170 to the green LUT, and the remainder to the red LUT. By this procedure, used arbitrarily as an example, all dark pixels are represented by various shades of blue, all bright hues by red colors with increasing brightness, and intermediate gray values by green. Any other combination or overlap of ranges can be used for a total of over 16 million colors. Real color processing, in contrast, requires separate digitization of each of the three color channels on the input side.

The digitization described above uses 8-bit planes (256 possible gray levels). Some frame grabbers provide additional bit planes for graphical overlay (video keying) or allow one to block and use some of the 8-bit planes for this purpose. The size of memory used by an image of 512 × 512 pixels is 262,144 bytes. Often the dedicated image memory allows one to store more than one image at a time;[47] roughly 1 Mbyte is required to store four frames. If the full gray scale is not required, the incoming images can be converted into binary images and stored in single-bit planes, which extends the storage capabilities of a 1-Mbyte board to 32 frames at maximal spatial resolution, which, of course, can also be decreased to further increase the number of individual frames. The host microcomputer has access to the image in memory and can both read the gray value of a randomly selected pixel or alter it by writing a new value into the memory cell.[48-50]

III. SOFTWARE APPROACHES

Due to the enormous numbers of pixels in a high-resolution image, fast algorithms need to be employed if analysis time is critical. In order to speed calculations up, the computer language C often is used, which requires less overhead time than other higher-level languages. Procedures and subroutines which are performed repetitively (such as pixel operations) are often even written in assembler language in order to optimize the time requirements. Algorithms have been developed to reduce the large numbers of bytes by specific coding techniques; however, this usually does not reduce the required time for processing.[51]

A. IMAGE MANIPULATIONS

Before detecting on organism in a frame, the image can be enhanced, which is facilitated by the numerical representation of the image in memory (Table 1, A). In addition to the LUT mapping described in the last section, the content of each individual memory cell, and thus the image, can easily be manipulated. The brightness can be increased or decreased by simply adding or subtracting a constant number to or from each pixel, respectively. Similarly, a horizontal and/or vertical ramp can be defined to compensate a linear spatial offset in the image. Multiplication or division of each pixel with a constant allows one to enhance or reduce the contrast. Thresholding is another useful technique to suppress the background below or above a predefined value which removes the noise.

Laplace or Sobel algorithms are more advanced mathematical techniques based on calculations involving 3 × 3 (or higher) matrices of pixels gliding over the image.[52] Each pixel in the derived image is calculated from its original value and that of its neighbors. During Laplace filtering, each pixel is multiplied by a specific number (kernel), which is eight for the most common algorithm, and the values of all its surrounding eight neighbors are subtracted from this value (Table 2).

$$\begin{matrix} -1 & -1 & -1 \\ -1 & 8 & -1 \\ -1 & -1 & -1 \end{matrix} \tag{2}$$

Laplace filtering results in a reduction in brightness of all areas with similar gray levels, while sharp boundaries show up as bright lines. Likewise, averaging can be achieved by replacing each pixel P by the mean value P′ calculated from the original value and that of its eight neighbors, which smoothes sharp boundaries and reduces noise.[53]

$$\begin{matrix} A & B & C \\ D & P & E \\ F & G & H \end{matrix} \tag{3}$$

$$P' = (P + A + B + C + D + E + F + G + H)/9 \tag{4}$$

Zooming can be achieved by spreading out each pixel over several ones, which magnifies the image, but does not increase the resolution.[54] Gradual zooming requires interpolation between pixels, but most systems allow only zooming by integer numbers. Panning in horizontal and vertical directions is useful in systems which display only part of the image on screen.

Other image enhancement techniques are discussed in other chapters in this volume. In real-time applications, only hardware image enhancement manipulations can be used since all software approaches are too time consuming, which is evident considering the large amount of calculations required.[55,56]

VI. CELL RECOGNITION AND QUANTIFICATION

Object recognition in an image can be performed by using one of several approaches developed in the past. A simple technique is based on a filling algorithm: the image in

TABLE 2a
Original Pixel Values in Part of a Digitized Image

60	52	61	63	75	71	68	72	78	81	79	84	85
61	64	140	72	75	71	79	80	85	81	84	89	92
58	61	64	67	80	70	72	75	78	80	78	82	85
64	71	72	75	71	69	220	231	225	228	229	231	230
65	70	58	73	75	230	229	228	234	227	230	232	218
69	72	75	73	225	247	230	231	235	237	228	229	240
71	75	72	74	68	242	250	231	232	232	228	237	239
79	69	72	75	76	71	228	229	235	235	236	228	227
60	63	65	71	72	72	228	229	230	225	230	230	225
63	70	68	72	74	73	230	229	231	234	234	230	228

TABLE 2b
Noise Reduction and Smoothing by Averaging over a 3 × 3 Matrix in Table 2a

60	52	61	63	75	71	68	72	78	81	79	84	85
61	69	72	77	72	73	73	76	79	80	82	84	92
58	73	76	80	72	90	107	127	129	130	131	133	85
64	65	68	71	90	124	158	177	178	179	180	179	230
65	68	71	89	126	177	213	229	231	230	230	230	218
69	70	71	88	145	200	235	233	232	231	231	231	240
71	73	73	90	128	182	218	233	233	233	232	232	239
79	70	71	72	91	145	198	232	231	231	231	231	227
60	68	69	72	73	125	177	230	231	232	231	230	225
63	70	68	72	74	73	230	229	231	234	234	230	228

TABLE 2c
Result of Laplace Filtering with a Kernel of 8 Using Values in Table 2a

60	52	61	63	75	71	68	72	78	81	79	84	85
61	0	255	0	31	0	53	33	55	5	18	43	92
58	0	0	0	70	0	0	0	0	0	0	0	85
64	56	37	40	0	0	255	255	255	255	255	255	230
65	14	0	0	0	255	146	0	30	0	0	21	218
69	21	33	0	255	255	0	0	28	50	0	0	240
71	21	0	0	0	255	255	0	0	0	0	41	239
79	0	12	30	0	0	255	0	37	32	43	0	227
60	0	0	0	0	0	255	0	0	0	0	2	225
63	70	68	72	74	73	230	229	231	234	234	230	228

memory is scanned line by line until a pixel is found which differs from the background by a predefined threshold value. Once a potential member of an organism is detected, all its neighbors are analyzed for being connected object pixels, which would then belong to the same organism. All pixels found to be object pixels are set to a specific gray level in order not to be scanned again, and all their respective neighbors are analyzed subsequently until the extent of the whole object has been found. Subsequently, the search is continued from the first found object pixel throughout the whole image until all objects are found. A variant of this method, which relies on line by line scanning, is described by Gualtieri in Chapter 18 of this volume.

This method requires the knowledge of the threshold value which separates objects from background. This value can be determined experimentally by the user or found automatically using a pixel-oriented thresholding to segment the image. All pixels P(x, y), the gray values

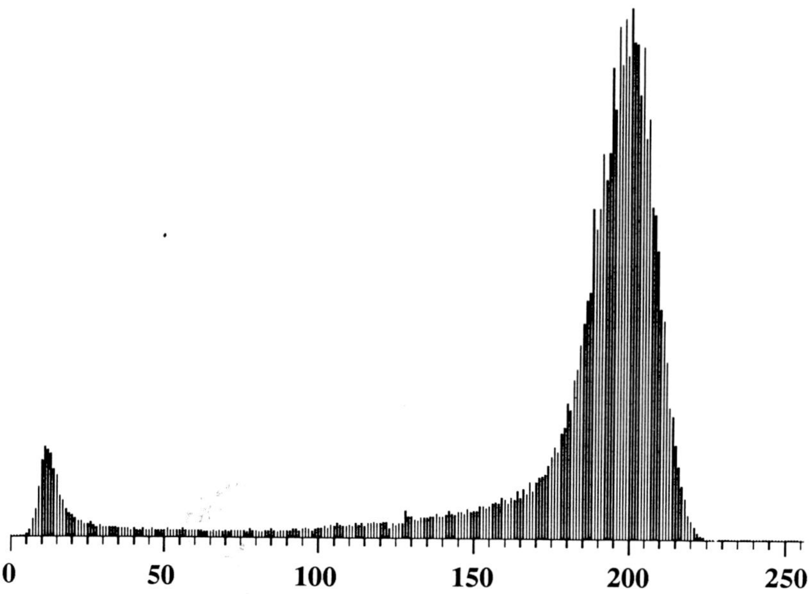

FIGURE 5. Histogram of the gray-level distribution in an image of dark organisms on a bright background.

of which are below a defined threshold T, are regarded as object pixels O, while all others are defined as background pixels B.

$$P(x,y) = \begin{cases} O \text{ for } P(x,y) < T \\ B \text{ else} \end{cases} \tag{5}$$

This definition holds for bright-light microscopy in which the organisms appear dark on a bright background; in dark-field microscopy, the situation is reversed. Histograms of the gray-level distribution in the image are calculated which, under optimal conditions, show a bimodal distribution, one range for the gray levels of the organisms and the other one for the background values (Figure 5). The threshold is defined by the valley between the two maxima. If no clear minimum is found in the histogram, which can be due to a noisy background, a modified technique is used[57] in which only those object pixels are considered which are not in the transition zone between background and object.

Alternative methods utilize edge detection algorithms.[58-60] A simple technique also starts by scanning the image line by line until an organism pixel has been found. The x- and y-coordinates of this edge pixel are stored and the neighboring edge pixel is searched in a clockwise direction by analyzing the eight neighbors of this first one, also in a clockwise direction. An edge pixel is defined by the fact that in a number of connected pixels there is at least one neighbor which does not belong to the object. A transition from a background pixel to an object pixel signals that the next edge pixel has been found. This procedure is repeated until the first edge pixel is found again, and thus the outline has been determined. During calculation, the extreme x- and y-coordinates in the outline are established from which the geometric center of gravity can be calculated.

A faster variant of this approach is the chain code algorithm.[61,62] A chain code is a sequence of short contour elements coded in numerical form which uniquely describe the

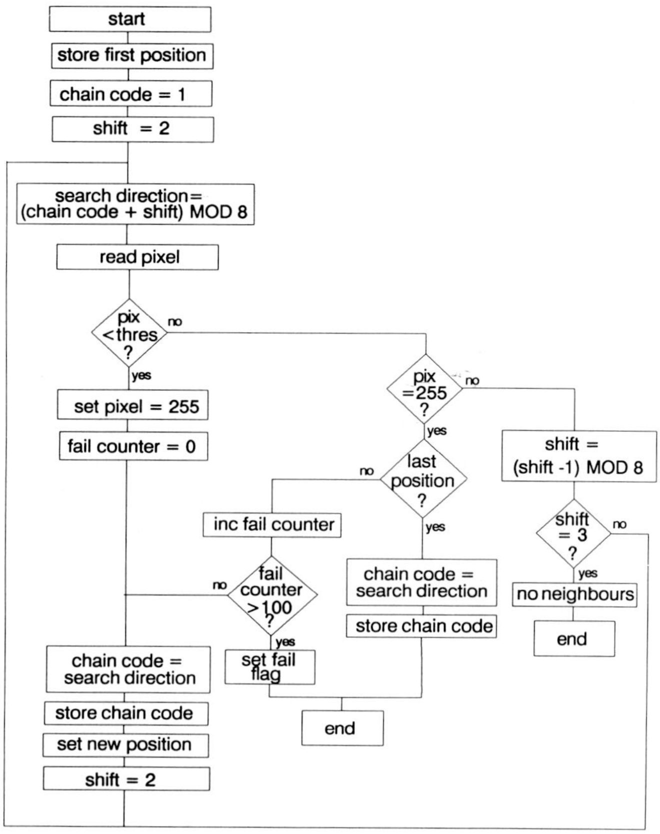

FIGURE 6. Flow chart of the outline detection routine based on the chain code algorithm.

outline of an object in a digital image. The directions from the central pixel P to its eight neighbors is defined by the numbers 0 through 7 in a counterclockwise fashion, starting to the right of the pixel.

The technique also starts with a line by line search until an edge pixel P has been detected (Figure 6). Starting from the current scan direction rotated two positions counterclockwise, the neighboring pixel is tested for being an edge pixel in a clockwise direction (Figure 7). This is performed by a modulo 8 addition of 2. Let us assume the first chain code element was defined as 1; thus, the next tested (possible) neighbor is searched in the direction 3. If the tested neighbor is not an organism pixel, the shift is decremented (modulo 8) and the search is continued in a clockwise fashion until an organism pixel is found. Now the new chain code element is calculated by adding the shift (modulo 8) to the previous value. In a clockwise direction, the maximal direction change is a shift of 4 (180°). If the shift value falls below 4, there are no neighbors. The whole outline is scanned until the first edge point is found again, with this routine performed iteratively. The determined directions to the next edge points are stored in the growing chain of elements.[61,62] The advantage of this robust technique is that it is very fast and enables one to reduce the amount of information drastically.

A. AREA AND POSITION DETERMINATION

From the chain code, the length of the outline, the area of the enclosed object, and the coordinates of the baricenter can be calculated using equations developed for this purpose. The length of the outline is given by

$$L = T(n_e + n_o\sqrt{2}) \tag{6}$$

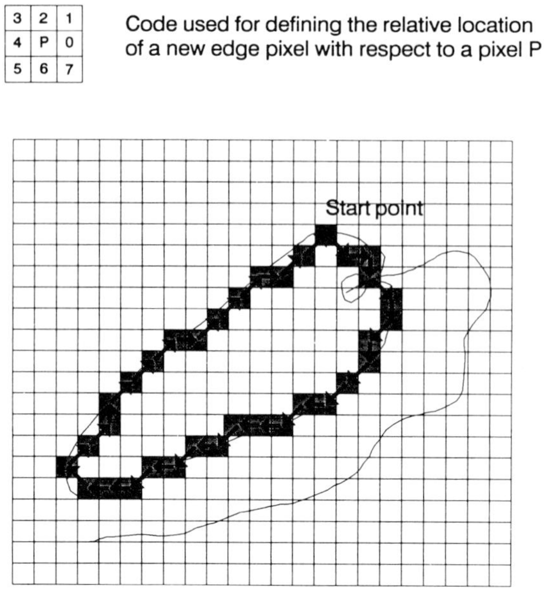

Traced contour of a cell

Chain code:

$A = a_1, a_2, \ldots a_n$

A = 7, 0, 6, 7, 6, 5, 6, 5, 5, 4, 5, 4, 4, 5, 4, 5
5, 4, 4, 1, 1, 2, 1, 1, 1, 0, 1, 1, 1, 0, 1, 1

FIGURE 7. Sketch to exemplify the chain code technique to analyze the outline of an organism.

where n_e is the number of even links, n_o is the number of odd links, and T is a scaling factor which depends on the optical magnification. The area A of the two-dimensional projection of the organism can be determined from the integration of the chain elements with respect to the x-axis:

$$A = \sum_{i=1}^{n} a_{ix}(y_{i-1} + {}^{1}/_{2}\, a_{iy}) \tag{7}$$

In order to calculate the baricenter, the first moments about the x- and y-axes are needed; about the x-axis:

$$M_1^x = \sum_{i=1}^{n} {}^{1}/_{2}\, a_{ix}[(y_{i-1})^2 + a_{iy}(y_{i-1} + {}^{1}/_{3}\, a_{iy})] \tag{8}$$

The first moment about the y-axis is determined by rotating the chain by $\pi/2$ about the origin and then employing Equation 10. The centroid with the coordinates x_c and y_c is calculated by using the first moments and the area:

$$x_c = -M_1^y/A \tag{9}$$

$$y_c = M_1^x/A \tag{10}$$

All calculations can be performed as integer operations, which facilitates computing in assembler language, which is important for real-time analysis. All edge pixels are arbitrarily set to 255 (bright white) after they have been identified. Thus, they serve as markers for already found organisms. Starting from the first found edge pixel of the previously calculated organism, the search is continued until the whole frame has been analyzed and all organisms are found. The centroids and areas of all organisms are stored in an array.

B. CELL COUNTING

Cell counting is routinely performed in many medical and biological laboratories to count colonies, microorganisms, or macroscopical objects such as leaves.[63] A software system based on the hardware described above has been developed to determine cell densities.[64] The cell suspension is transferred into a cuvette of known depth. Motile cells do not need to be fixed or immobilized for this purpose since real-time digitization is used. The image is scanned line by line until an object is found; then the algorithm continues to fill the object in shells around the initial point until all connected pixels have been identified as described above. The number of pixels found to belong to the object represent the area of the cell.[46,65] This technique is slower than the chain code algorithm, but has the advantage of detecting and correcting for ''holes'' in the found objects, which cannot be determined easily with edge detection algorithms.

When objects are in direct contact, errors in counting occur.[66] When the average size of the objects is known, bigger ones can be rejected or the area can be divided when it is outside predefined boundaries, which may still be inaccurate, especially in dense suspensions. In order to solve this problem, an erosion technique is applied: shells 1 pixel wide are removed and set to a specific gray value around the circumference of each object. This process is repeated until connecting areas are separated at the isthmus. Subsequently, the shells are blown up (dilated) again to their original value. However, a gap at least 1 pixel wide is left between adjacent objects. Objects can also be separated manually before analysis using the cursor controls, a mouse or digitizing tablet, or a touch-sensitive screen. For averaging purposes, the process of counting and area determination can be repeated on the same sample after renewed digitization or on a different sample. The results are used to calculate histograms of the area distribution and to analyze the largest diameters or finding the center of gravity.

Cell counting was recently applied in an ecological approach to determine the densities in samples of *Euglena* populations filled in a column submerged in a lake[67] (Figure 8). A peristaltic pump which could handle 18 samples in parallel from 18 outlets evenly spaced along the length of the column was used to take samples at regular time intervals in order to determine the distribution of the population in dependence of light and gravity. The data were documented in linear histograms (Figure 9).

V. ORGANISM TRACKING

Using subsequent frames recorded at frequent and regular time intervals allows one to follow movement vectors of organisms in the time domain.[68-76] Likewise, the growth of an organism or organelle can be followed.[77-81] The objects can range from individual cells such as flagellates or ciliates[37,82-86] to multicellular organisms.[8,87] Parameters such as directionality, velocity, or percentage of motile organisms in a population can be recorded in a population. After an object (or all objects in the frame) has been identified by using either of the techniques described above, it is wise to determine whether or not it is likely to be an organism by using a number of comparisons. Upper and lower limits for the area allow one to discriminate a valid organism from debris. Also, the contour length or the ratio of the contour length to the area can be employed as additional classification factors.

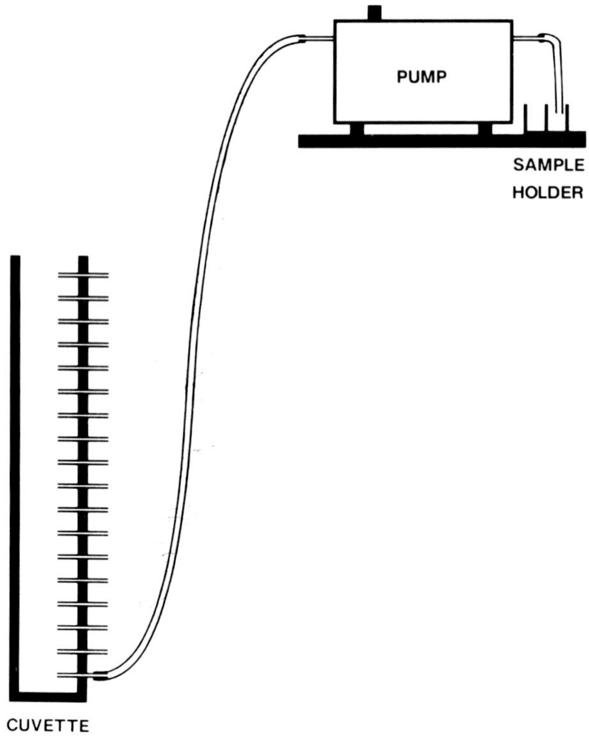

FIGURE 8. Schematic diagram of a Plexiglas column with 18 outlets from which 18 samples can be drawn in parallel using a peristaltic pump. (After Häder, D.-P. and Griebenow, K., *FEMS Microbiol. Ecol.*, 53, 159, 1988.)

A. QUANTIFICATION OF MOTILITY

In this section, we will concentrate on the movements of whole organisms and disregard the analysis of ciliary and flagellar beat patterns,[88-91] which are covered in other chapters in this volume. In an earlier approach to tracking microorganisms,[92] one organism was identified in the image using a random number algorithm and the coordinates of its baricenter were stored. After digitizing the next frame from the video sequence, the new position of this organism was determined starting from the baricenter in the previous frame. If this was not found to be in the new outline of the cell, it was regarded as lost.

In a new approach,[93] a different strategy was used: four frames were digitized and stored in the four quadrants of the digitizer memory without further analysis. These images are recorded at 80 ms intervals each; thus, the track segments are 240 ms long. Then the positions of all cells within the first frame are analyzed using the chain code algorithm. All baricenters and the chain codes are stored in a first array. Then the positions of the cells are determined in the second image. However, in this case, the image is not scanned line by line; rather, the organisms are looked for starting at the baricenters stored in the first array. If no organism is found at the previous position, the neighborhood is searched using a dynamically adjusted area, the size of which depends on the mean path length of the organism within the time interval of 80 ms. If there is only one object in the surrounding, the new area is compared to the old one. This object is accepted provided there are no large discrepencies in the areas. A considerable increase in the area is a sign that two organisms meet during the process of tracking, and a sudden decrease indicates that the object has moved out of the focus. If there is no more than one object in the search area, the organism is regarded as lost and not

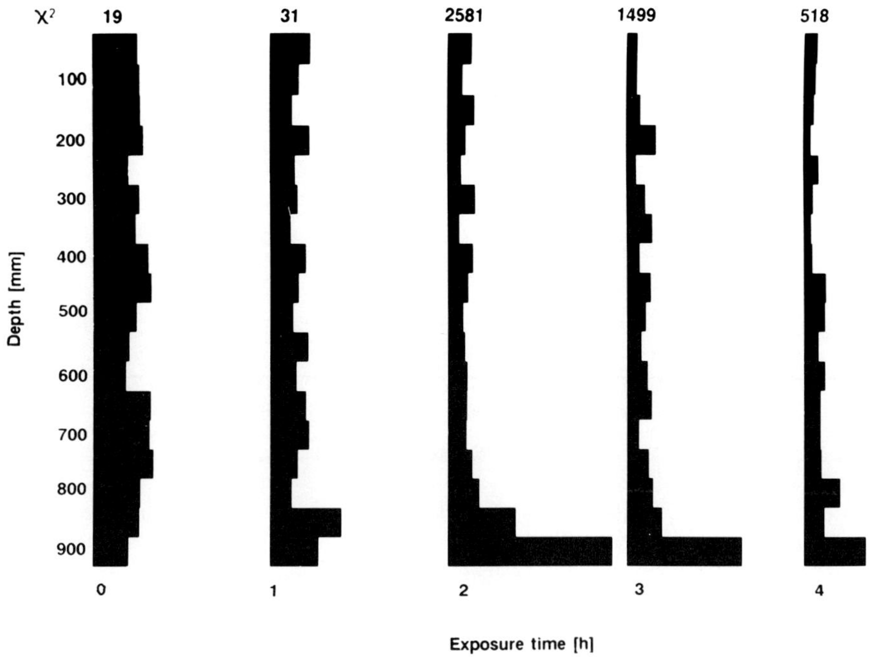

FIGURE 9. Linear histograms of the cell density distribution in the column after increasing periods of exposure to solar radiation. The numbers above the histograms indicate the chi-square values. (From Häder, D.-P. and Griebenow, K., *FEMS Microbiol. Ecol.*, 53, 159, 1988. With permission.)

considered for the rest of the calculation. The coordinates and the chain codes of the successfully identified organisms in the second frame are stored in a second array, and the procedure is repeated for the third and fourth frame, each time overwriting the second array.

From the data stored in the two arrays, the movement vectors can be calculated. In order to determine the velocity, the distance the organism moved between the first and the last frames is extracted:

$$s = \sqrt{(y_2 - y_1)^2 + (x_2 - x_1)^2} \tag{11}$$

The time can be read from the hardware clock of the computer, so that the velocity can be determined in pixels per unit time; in order to calculate the real velocity, the magnification needs to be considered.

Using a minimal distance an organism has to move to be regarded as motile allows one to determine the percentage of motile organisms in a population (Figure 10). Using only the motile organisms allows one to construct histograms of the velocity distribution in a population (Figure 11).

B. VECTOR ANALYSIS

Microorganisms use a number of external factors to orient in their environment.[94,95] While some motile microorganisms have been found to orient in the water column with the aid to thermal[96,97] and chemical[98,99] gradients, the magnetic field of the earth,[100,101] and even electrical currents,[102] the flagellate *Euglena gracilis* and other photosynthetic and nonphotosynthetic organisms mainly orient with respect to light[84,103-106] and gravity.[34,107]

From the baricenters of the beginning and the endpoints of the track segments defined in two arrays (see above), not only the velocity, but also the movement vectors can be

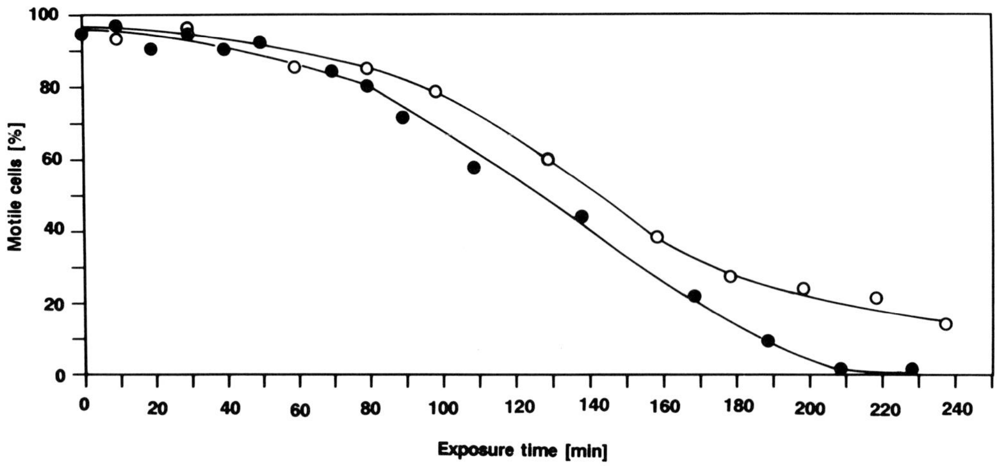

FIGURE 10. Percentage of motile microorganisms (*Peridinium gatunense*) in a population exposed to unfiltered solar radiation (closed circles) and radiation filtered through a cuvette filled with air enriched in ozone (open circles). (From Häder, D.-P., et. al., *BioSystems,* 23, 335, 1990. With permission.)

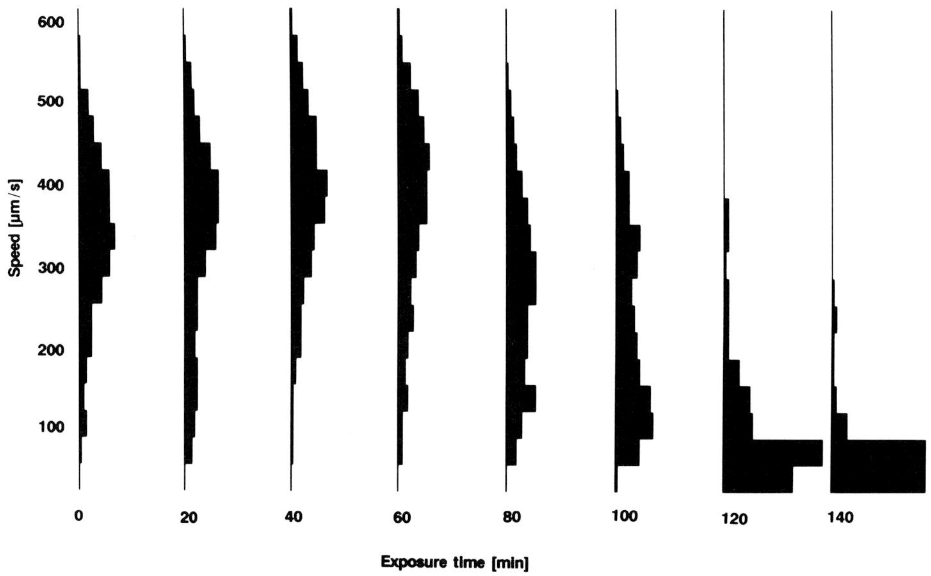

FIGURE 11. Velocity distribution in a population of *Peridinium gatunense* dependent on increasing exposure to solar radiation. (From Häder, D.-P., et. al., *BioSystems,* 23, 335, 1990. With permission.)

calculated:

$$\alpha = \mathrm{tn}^{-1} \frac{y_2 - y_1}{x_2 - x_1} \tag{12}$$

Thus, it is possible to determine the direction of movement of all organisms during a time segment (240 ms) and construct histograms of their direction distribution (Figure 12).

The precision of orientation can be calculated using the Rayleigh test:[108-110]

$$r = \frac{\sqrt{(\Sigma \sin\alpha)^2 + (\Sigma \cos\alpha)^2}}{n} \tag{13}$$

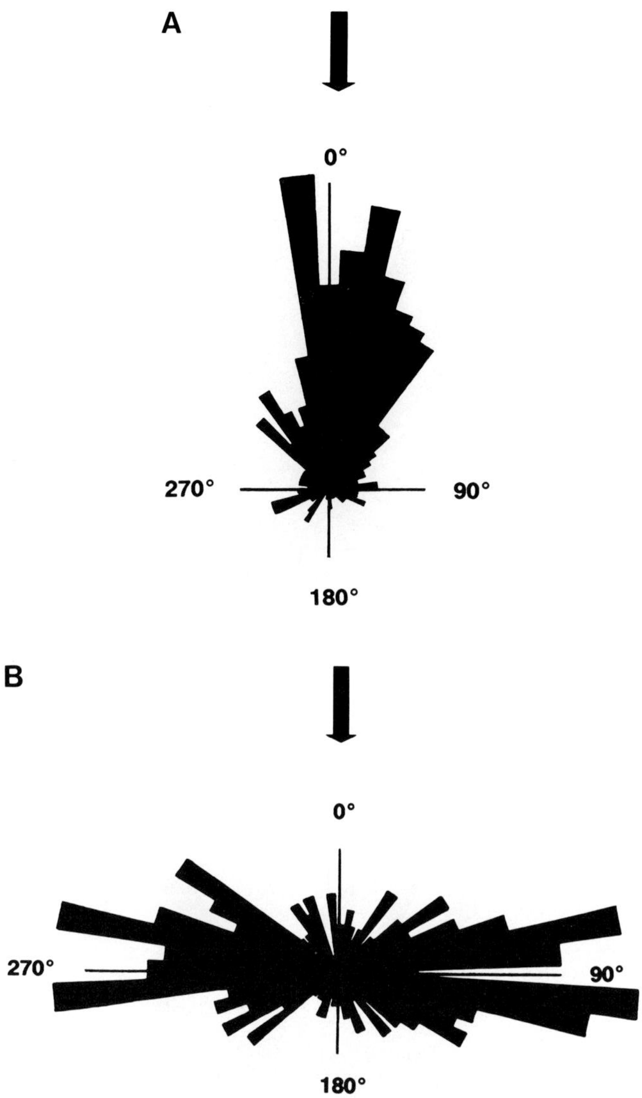

FIGURE 12. Histograms of the positive (A) and transverse (B) phototaxis in the flagellate *Peridinium gatunense* using 64 sectors of 5.6° each to bin the angular deviation from the light direction. (From Liu, S.-M., et. al., *FEMS Microbiol. Ecol.*, 73, 91, 1990. With permission.)

A value near 0 indicates a close to random orientation, while a value near 1 describes a very precise orientation.

This technique has been utilized to study the effect of microgravity conditions on microorganisms in a space flight.[93] In addition to phototaxis, the flagellate *E. gracillis* orients precisely in the gravity field of the earth (Figure 13a). During a TEXUS flight, the movement of the cells was recorded and the video signal was sent to a terrestrial station by video downlink. Under microgravity conditions, the cells showed an almost perfect random orientation (Figure 13b), which proved the assumption that gravity is the signal for orientation under terrestrial conditions.

C. LONG-TERM TRACKING

Often it is of interest to follow organisms over considerable time spans rather than only short time segments. This type of analysis is required, e.g., for the determination of the

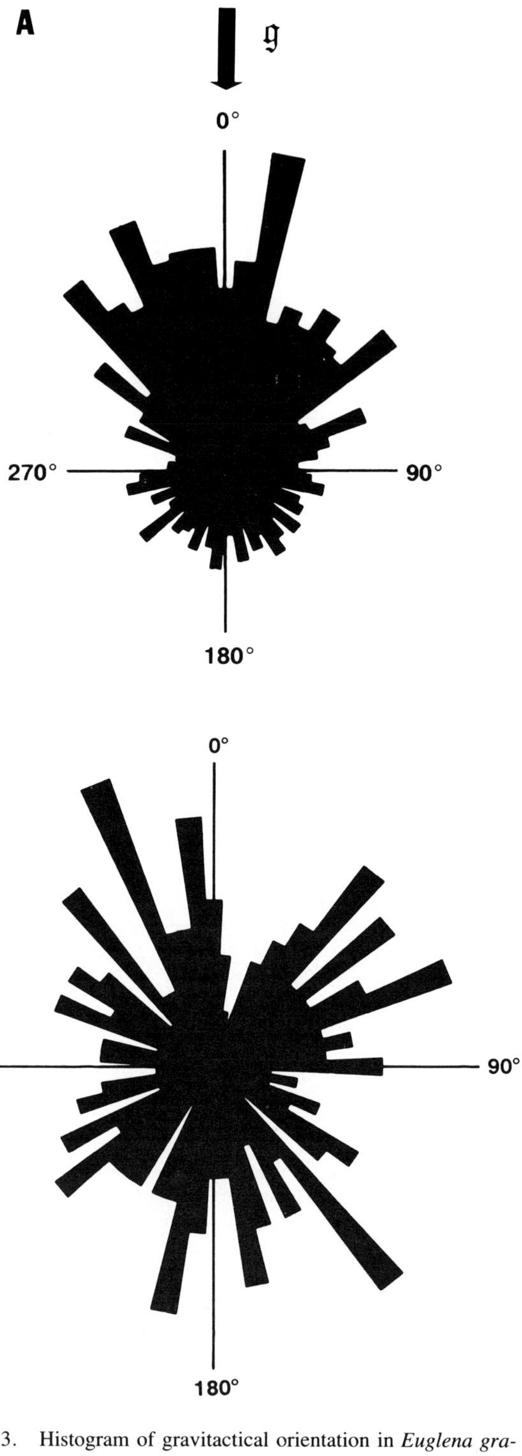

FIGURE 13. Histogram of gravitactical orientation in *Euglena gracilis* under terrestrial conditions (A) and under microgravity conditions (B). (From Hader, D.-P., et. al., *Micrograv. Sci. Technol.*, 3, 110, 1990. With permission.)

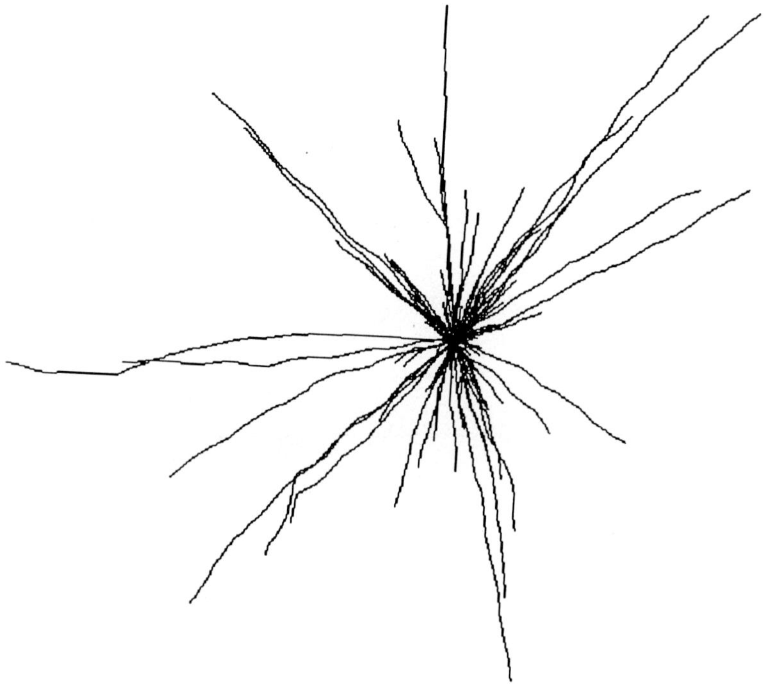

FIGURE 14. Tracks of microorganisms followed by long-term tracking and drawn from an arbitrarily chosen center.

free path length of the organisms.[111] For this purpose, a true real-time analysis is inevitable since only a limited number of frames can be stored in memory. One possible way around this problem is to record the data on video and then do the tracking in a frame by frame analysis during which each frame is replayed and digitized after the previous one has been analyzed.[4,112,113]

A random access to a pixel in the memory of a Matrox PIP 1024 requires about 9 μs. If we assume that an object has 200 edge points, a total of about 1000 pixel operations is necessary, which requires 9 ms. Therefore, during the interval between two video frames (40 ms), a maximum of four objects can be tracked, not even counting the calculations for area and baricenters.

In the program developed for long-term tracking, four objects are selected in the center of the field of view and followed using the chain code algorithm. The coordinates of the baricenters and the time interval to the last frame are written into a disk file for future analysis. If, during tracking, one of the organisms is lost (because it moved out of focus, left the field of view, or collided with another object), a zero is written into the file instead, while the tracking of the remaining organisms is followed until all are lost. Then the cycle is started again. The tracks of the organisms can be reconstructed on a graphic monitor or plotted starting from their true position in the analyzed frame[114] or from an arbitrarily choosen center (Figure 14).

The likelihood that organisms collide or tracks cross increases with the duration of tracking. In order to continue tracking even in the case of collisions, several techniques are used. Assuming that the organisms do not change their direction of movement drastically, the directions of two organisms can be predicted from their previous paths. A search sector is defined in front of the moving organism in which it is expected in the next image, which allows one to discriminate its track from that of another colliding organism.

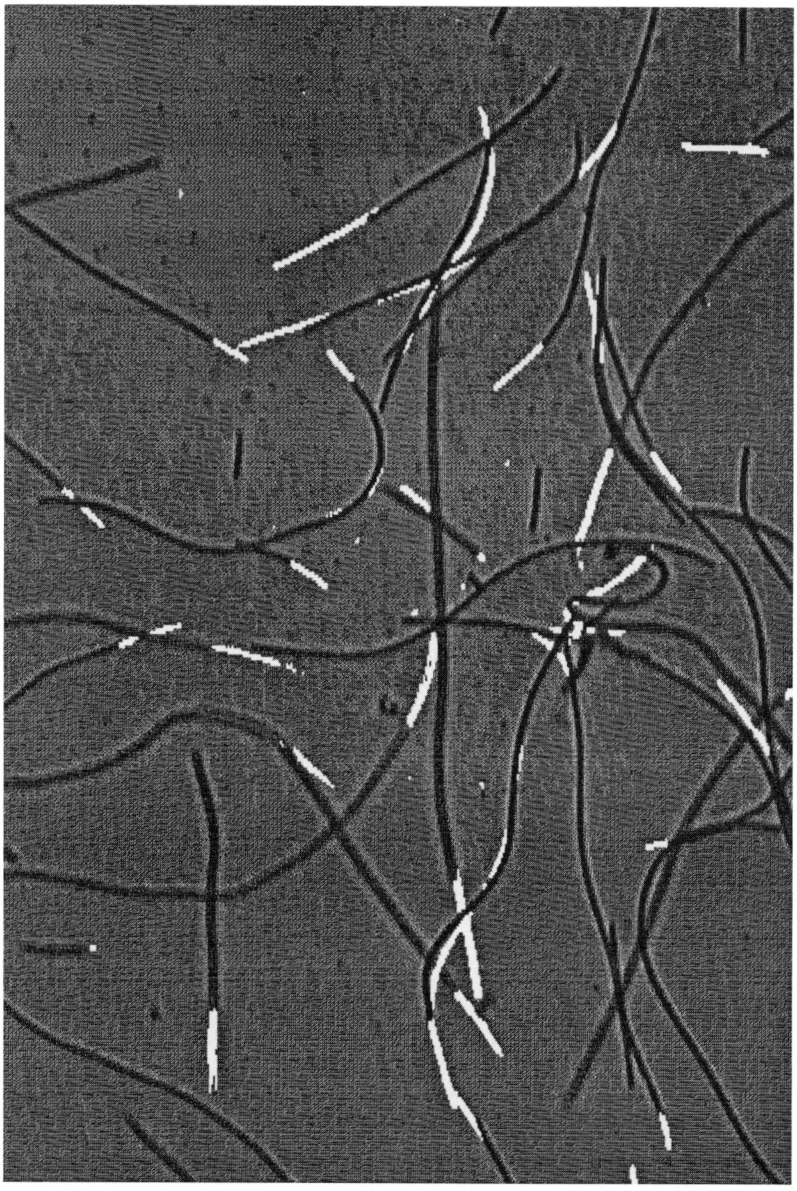

FIGURE 15. Analysis of the movement of cyanobacteria filaments using subtraction between subsequent frames which shows the longitudinal displacement of the trichomes.

D. TRACKING OF ORGANISMS BY INTERACTIVE PROCESSING

In some cases, automatic tracking of microorganisms is extremely difficult or even impossible because of the optical conditions or the unfavorable form (for machine vision) of the organisms. For example, trichomes of gliding cyanobacteria move at low speed along the long axis of the filament.[115] In a population, these filaments overlap, which makes it difficult to determine the outline of the individuals. Using the technique of image subtraction produces short track segments which indicate the longitudinal displacement of the filaments. This difference image is overlayed over the original image in order to distinguish the movement along the long axis from lateral movements due to interactions between organisms (Figure 15). The length of the individual track elements is calculated by inputting the beginning and end positions using a crosshair cursor controlled by the mouse. With the exception of the laterally moved trichomes, the displacement vectors also could have been analyzed automatically be determining the form parameters or using a skeletoning technique.

VI. CONCLUSIONS

Strategies to find and track motile microorganisms in a sequence of video frames are discussed in this chapter. The hardware requirements are considered and possible track algorithms explained. The applications of this technology range from cell recognition and quantification to organism counting and tracking. From the tracks, important features can be extracted such as the percentage of motile cells, the individual velocity of each organism, and the movement vector. Tracking of short segments is possible for a few hundred cells on the screen within a few hundred milliseconds, and real-time tracking can be performed on up to four cells at a time. In addition, interactive tracking is described, which requires the input of a human operator.

ACKNOWLEDGMENTS

This work was facilitated by a grant from the Bundesminister für Forschung und Technologie (KBF 57) and financial aid by the Deutsche Forschungsgemeinschaft (SFB 305) to D.-P. H. The skillful technical assistance of J. Schäfer, U. Schleier, and H. Vieten is gratefully acknowledged.

REFERENCES

1. **Shipton, H. W.,** The microprocessor, a new tool for the biosciences, *Annu. Rev. Biophys. Bioeng.,* 8, 269, 1979.
2. **Endlich, R. M., Wolf, D. E., Hall, D. J., and Brain, A. E.,** Use of pattern recognition technique for determining cloud motions from sequences of satellite photographs, *J. Appl. Meteorol.* 10, 105, 1971.
3. **Smith, E. A. and Phillips, D. R.,** Automated cloud tracking using precisely aligned digital ATS pictures, *IEEE Trans. Comput.,* 21, 715, 1972.
4. **Gualtieri, P., Colombetti, G., and Lenci, F.,** Automatic analysis of the motion of microorganisms, *J. Microsc.,* 139, 57, 1985.
5. **Takahashi, T. and Kobatake, Y.,** Computer-linked automated method for measurement of the reversal frequency in phototaxis of *Halobacterium halobium, Cell Struct. Funct.,* 7, 183, 1982.
6. **Levine, M. D., Youssef, Y. M., Noble, P. B., and Boyarsky, A.,** The quantification of blood cell motion by a method of automatic digital picture processing, *IEEE Trans. Pattern Anal. Mach. Intelligence,* 2, 444, 1980.
7. **Leidl, W., Riemke, P., and Schröppel, I.,** Computergesteuerte Videomikrographie-Auswertung zur Bestimmung der Spermienmotilität am Modell des Bullen, *Dtcsh. Tierärztl. Wochenschr.,* 8, 461, 1987.
8. **Dusenbery, D. B.,** Video camera-computer tracking of nematode *Caenorhabditis elegans* to record behavioral response, *J. Chem. Ecol.,* 11, 1239, 1985a.
9. **Hawkes, P. W.,** Digital image processing, *Nature,* 285, 174, 1980.
10. **Serra, J.,** Digitalization, *Mikroskopie,* 37 (Suppl.), 109, 1980.
11. **Kokubo, Y. and Hardy, W. H.,** Digital image processing, a path to better pictures, *Ultramicroscopy,* 8, 277, 1982.
12. **McMillan, P. J., Yakush, A., Frykman, G., Nava, P. B., and Ras, V. R.** Minima equalization, a useful strategy in automatic processing of microscopic images, *J. Microsc.,* 148, 253, 1987.
13. **Russ, J. C. and Russ, J. C.,** Automatic discrimination of features in grey-scale images, *J. Microsc.,* 148, 263, 1987.
14. **Lewandowska, K., Doroszewski, J., Haemmerli, G., and Struli, P.,** An attempt to analyze locomotion of leukemia cells by computer image processing, *Comput. Biol. Med.,* 9, 331, 1979.

15. **Turano, T. A., D'Arpa, P, Clark, W. L., and Williams, J. R.,** A time-lapse, image digitization videomicroscope system based on a mini computer with large peripheral memory, *Comput. Biol. Med.,* 15, 177, 1985.
16. **Thurston, G., Jaggi, B., and Palcic, B.,** Cell motility measurements with an automated microscope system, *Exp. Cell Res.,* 165, 380, 1986.
17. **Poggio, T.,** Vision by man and machine, *Sci. Am.,* 250, 68, 1984.
18. **Blake, A.,** Real-time seeing machines?, *Nature,* 328, 759, 1987.
19. **Coltelli, P. and Gualtieri, P.,** A procedure for the extraction of object features in microscope images, *Int. J. Biomed. Comput.,* 25, 169, 1990.
20. **Castillo, X., Yorkgitis, D., and Preston, K., Jr.,** A study of multidimensional multicolor images, *IEEE Trans. Biomed. Eng.,* 29, 111, 1982.
21. **Gibson, J. J.,** *The Perception of the Visual World,* Houghton Mifflin, Boston, 1950.
22. **Berg, H. C.,** How bacteria swim, *Sci. Am.,* 233, 36, 1975.
23. **Bianco, B. and Diaspro, A.,** Analysis of three-dimensional cell imaging obtained with optical microscopy techniques based on defocusing, *Cell Biophys.,* 15, 189, 1989.
24. **Roach, J. W. and Aggarwal, J. K.,** Computer tracking of objects moving in space, *IEEE Trans. Pattern Anal. Mach. Intelligence,* 1, 127, 1979.
25. **Roach, J. W. and Aggarwal, J. K.,** Determining the movement of objects from a sequence of images, *IEEE TRans. Pattern Anal. Mach. Intelligence,* 2, 554, 1980.
26. **Jähne, B.,** *Digitale Bildverarbeitung,* Springer-Verlag, Heidelberg 1989.
27. **Jain, R.,** Extraction of motion information from peripheral processes, *IEEE Trans. Pattern Anal. Mach. Intelligence,* 3, 489, 1981.
28. **Jain, R. and Nagel, H. H.,** On the analysis of accumulative difference pictures from image sequences or real world scenes, *IEEE Trans. Pattern Anal. Mach. Intelligence,* 1, 206, 1979.
29. **Kenny, P. A., Dowsett, D. J., Vernon, D., and Ennis, J. T.,** A technique for digital image registration used prior to subtraction of lung images in nuclear medicine, *Phys. Med. Biol.,* 35, 679, 1990.
30. **Gualtieri, P. and Coltelli, P.,** A digital microscope for real time detection of moving microorganisms, *Micron Microsc. Acta,* 20, 99, 1989.
31. **Noble, P. B. and Levine, M. D.,** *Computer Assisted Analysis of Cell Locomotion and Chemotaxis,* CRC Press, Boca Raton, FL, 1986.
32. **Aggarwal, J. K. and Duda, R. O.,** Computer analysis of moving polygonal images, *IEEE Trans. Comput.,* 24, 966, 1975.
33. **Schalkoff, R. J. and McVey, E. S.,** A model and tracking algorithm for a class of video targets, *IEE Trans. Pattern Anal. Mach. Intelligence,* 4, 2, 1982.
34. **Häder, D.-P.,** Polarotaxis, gravitaxis and vertical phototaxis in the green flagellate, *Euglena gracilis, Arch. Microbiol.,* 147, 179, 1987.
35. **Häder, D.-P. and Häder, M.,** Ultraviolet-B inhibition of motility in green and dark bleached *Euglena gracilis, Curr. Microbiol.,* 17, 215, 1988.
36. **Häder, D.-P. and Häder, M.,** Effects of solar radiation on photoorientation, motility and pigmentation in a freshwater *Cryptomonas, Bot. Acta,* 102, 236, 1989.
37. **Häder, D.-P.,** Effects of solar and artificial UV irradiation on motility and phototaxis in the flagellate, *Euglena gracilis, Photochem. Photobiol.,* 44, 651, 1986.
38. **Häder, D.-P.,** Electrical and proton gradients in the sensory transduction of photophobic responses in the blue-green alga, *Phormidium uncinatum, Arch. Microbiol.,* 130, 83, 1981.
39. **Crenshaw, H. C.,** Kinematics of helical motion of microorganisms capable of motion with four degrees of freedom, *Biophys. J.,* 56, 1029, 1989.
40. **Danielsson, P.-E. and Kruse, B.,** PICAP II — a second generation picture processing system, *Mikroskopie,* 37 (Suppl.) 425, 1980.
41. **Steinbach, T., Unland, F., and Müller, K.-M.,** Kostengünstiges Mikroprozessorsystem zur Ergänzung eines Quantimet 720-Bildanalysegerätes, *Microsc. Acta,* 86, 139, 1982.
42. **Allen, R. D. and Allen, N. S.,** Video-enhanced microscopy with a computer frame memory, *J. Microsc.,* 129, 3, 1983.
43. **Preston, K., Jr.,** Gray level image processing by cellular logic transforms, *IEEE Trans. Pattern Anal. Mach. Intelligence,* 5, 55, 1983.
44. **Desai, V. and Reimer, L.,** Digital image recording and processing using an Apple II microcomputer, *Scanning,* 7, 185, 1985.
45. **Bryan, S. R., Woodward, W. S., Griffis, D. P., and Linton, R. W.,** A microcomputer based digital imaging system for ion microanalysis, *J. Microsc.,* 138, 15, 1985.
46. **Häder, D.-P.,** Computer-assisted image analysis in biological sciences, *Proc. Indian Acad. Sci. (Plant Sci.),* 98, 227, 1988a.

47. **Eichler, H., Purz, H. J, and Phillip, B.,** Digitale Bildverarbeitung und Bildauswertung mikroskopischer Aufnahmen mit stark schwankender Untergrundintensität. II. Ermittlung von Strukturparametern und Anwendung der Methode zur Auswertung elektronenmikroskopischer Aufnahmen von Celluloseacetatmembranen, *Mikroskopie,* 42, 183, 1985.
48. **Kemnitz, H.-D. and Hougardy, H. P.,** Design of an interface for digital image storage, *Mikroskopie,* 7, 415, 1980.
49. **Russ, J. C. and Russ, J. C.,** Image processing in a general purpose microcomputer, *J. Microsc.,* 135, 89, 1984.
50. **Mayfield, C. I.,** A simple computer-based video image analysis system and potential applications to microbiology, *J. Microbiol. Methods,* 3, 61, 1984.
51. **Kunt, M.,** Image coding, *Scanning Microsc.,* S2, 1, 1988.
52. **Voss, K.,** Differentialgeometrie und digitale Bildverarbeitung, *Bild Ton,* 43, 165, 1990.
53. **Julez, B. and Harmon, L. D.,** Noise and recognizability of coarse quantized images, *Nature,* 308, 211, 1984.
54. **Hainfeld, J. F., Wall, J. S., and Desmond, E. J.,** A small computer system for micrograph analysis, *Ultramicroscopy,* 8, 263, 1982.
55. **Katz, D. F., Davis, R. O., Delandmeter, B. A., and Overstreet, J. W.,** Real-time analysis of sperm motion using automatic video image digitization, *Comput. Methods Prog. Biomed.,* 21, 173, 1985.
56. **Poole, P. S., Sinclair, D. R., and Armitage, J. P.,** Real time computer tracking of free-swimming and tethered rotating cells, *Anal. Biochem.,* 175, 52, 1988.
57. **Wezka, J. S. and Rosenfeld, A.,** Histogram modification for threshold selection, *IEEE Trans. Syst. Man Cybern.,* 9, 38, 1979.
58. **Grant, G. and Reid, A. F.,** An efficient algorithm for boundary tracing and feature extraction, *Comput. Graph. Imag. Process,* 17, 225, 1981.
59. **Pavlidis, T.,** *Graphics and Image Processing,* Springer-Verlag, Berlin, 1982.
60. **Freeman, H.,** On the encoding of arbitrary geometric configurations, *IRE Trans. Electron. Comput.,* 10 (2), 260, 1961.
61. **Freeman, H.,** Computer processing of line-drawing images, *Comput. Surv.,* 6, 57, 1974.
62. **Freeman, H.,** Analysis and manipulation of lineal map data, in *Map Data Processing,* Academic Press, New York, 1980, 151.
63. **Kincaid, D. T. and Schneider, R. B.,** Quantification of leaf shape with a microcomputer and Fourier transform, *Can. J. Bot.,* 61, 2333, 1983.
64. **Häder, D.-P. and Griebenow, K.,** Versatile digital image analysis by microcomputer to count microorganisms, *EDV Med. Biol.,* 18, 37, 1987.
65. **Kim, N. H., Wysocki, A. B., Bovik, A. C., and Diller, K. R.,** A microcomputer-based vision system for area measurement, *Comput. Biol. Med.,* 17, 173, 1987.
66. **Erhardt, R., Reinhardt, E. R., Schlipf, W., and Bloss, W. H.,** FAZYTAN A system for fast automated cell segmentation, cell image analysis and feature extraction based on TV-image pickup and parallel processing, *Anal. Quant. Cytol. J.,* 2, 25, 1980.
67. **Häder, D.-P. and Griebenow, K.,** Orientation of the green flagellate, *Euglena gracilis,* in a vertical column of water, *FEMS Microbiol. Ecol.,* 53, 159, 1988.
68. **Mineyuki, Y., Yamada, M., Takagi, M., Wada, M, and Furuya, M.,** A digital image processing technique for the analysis of particle movements, its application to organelle movements during mitosis in *Adiantum protonemata, Plant Cell Physiol.,* 24, 255, 1983.
69. **Allen, R. D.,** New directions and refinements in video-enhanced microscopy applied to problems in cell motility, in *Advances in Microscopy,* Alan R. Liss, New York, 1985, 3.
70. **Dow, J. A. T., Lackie, J. M., and Crocket, K. V.,** A simple microcomputer-based system for real-time analysis of cell behavior, *J. Cell Sci.,* 87, 171, 1987.
71. **Amos, L.,** Movements made visible by microchip technology, *Nature,* 330, 211, 1987.
72. **Davenport, D., Culler, G. J., Greaves, J. O. B., Forward, R. B., and Hand, W. G.,** The investigation of the behavior of microorganisms by computerized television, *IEEE Trans. Bio-Med. Eng.,* 17, 230, 1970.
73. **Davenport, D.,** Studies in microorganismal behavior by computerized television, in *Behavior of Microorganisms,* Perez-Miravete, A., Ed., Plenum Press, New York, 1973, 106.
74. **Coates, T. D., Harman, J. T., and McGuire, W. A.,** A microcomputer-based program for video analysis of chemotaxis under agarose, *Comput. Methods Prog. Biomed.,* 21, 195, 1985.
75. **Rikmenspoel, R. and Isles, C. A.,** Digitized precision measurements of the movement of sea urchin sperm flagella, *Biophys. J.,* 47, 395, 1985.
76. **Burton, J. L., Law, P., and Bank, H. L.,** Video analysis of chemotactic locomotion of stored human polymorphonuclear leukocytes, *Cell Motil. Cytoskel.,* 6, 485, 1986.
77. **Gordon, D. C., MacDonald, I. R., Hart, J. W., and Berg, A.,** Image analysis of geo-induced inhibition, compression, and promotion of growth in an inverted *Helianthus annuus* L. seedling, *Plant Physiol.,* 76, 589, 1984.

78. **Omasa, K. and Onoe, M.**, Measurement of stomatal aperture by digital image processing, *Plant Cell Physiol.* 25, 1379, 1984.
79. **Jaffe, M. J., Wakefield, A. H., Telewski, F., Gulley, E., and Biro, R.**, Computer-assisted image analysis of plant growth, thigmomorphogenesis, and gravitropism, *Plant Physiol.*, 77, 722, 1985.
80. **Omasa, K. and Aiga, I.**, Environmental measurement, image instrumentation for evaluating pollution effects on plants, in *System & Control Encyclopedia,* Vol. 2, Singh, M. G., Ed., Pergamon Press, New York, 1987, 1516.
81. **Popescu, T., Zängler, F., Sturm, B., and Fukshansky, L.**, Image analyzer used for data acquisition in phototropism studies, *Photochem. Photobiol.*, 50, 701, 1989.
82. **Mikolajczyk, E., Häder, D.-P., and Nultsch, W.**, Photodynamically induced chemoresponses of the colorless flagellate, *Astasia longa,* in the presence of riboflavin, *Arch. Microbiol.*, 142, 397, 1985.
83. **Mikolajczyk, E., Nultsch, W., and Häder, D.-P.**, Chemoaccumulation of the colorless flagellate, *Astasia longa,* in the presence of the photosensitizer methylene blue, *Acta Protozool.*, 25, 179, 1986.
84. **Häder, D.-P., Lebert, M.,and DiLena, M. R.**, New evidence for the mechanism of phototactic orientation of *Euglena gracilis, Curr. Microbiol.*, 14, 157, 1986.
85. **Häder, D.-P., Rhiel, E., and Wehrmeyer, W.**, Phototaxis in the marine flagellate *Cryptomonas maculata, J. Photochem. Photobiol.*, 1, 115, 1987.
86. **Häder, D.-P., Rhiel, E., and Wehrmeyer, W.**, Ecological consequences of photomovement and photobleaching in the marine flagellate, *Cryptomonas maculata, FEMS Microbiol. Ecol.*, 53, 9, 1988.
87. **Dusenbery, D. B.**, Using a microcomputer and videocamera to simultaneously track 25 animals, *Comput. Biol. Med.*, 15, 169, 1985b.
88. **Sanderson, M. J. and Dirksen, E. R.** A versatile and quantitative computer-assisted photoelectronic technique used for the analysis of ciliary beat cycles, *Cell Motil.*, 5, 267, 1985.
89. **Baba, S. A. and Mogami, Y.**, An approach to digital image analysis of bending shapes of eukaryotic flagella and cilia, *Cell Motil.*, 5, 475, 1985.
90. **Omoto, C. K. and Brokaw, C. J.**, Bending patterns of *Chlamydomonas flagella.* II. Calcium effects on reactivated *Chlamydomonas* flagella, *Cell Motil.*, 5, 53, 1985.
91. **Cantatore, G., Ascoli, C., Colombetti, G., and Frediani, C.**, Doppler velocimetry measurements of phototactic response in flagellated algae, *Biosci. Rep.*, 9, 475, 1989.
92. **Häder, D.-P. and Lebert, M.**, Real time computer controlled tracking of motile microorganisms, *Photochem. Photobiol.*, 42, 509, 1985.
93. **Häder, D.-P., Vogel, K., and Schäfer, J.**, Responses of the photosynthetic flagellate, *Euglena gracilis,* to microgravity, *Micrograv. Sci. Technol.*, 3, 110, 1990.
94. **Nultsch, W. and Häder, D.-P.**, Photomovement in motile microorganisms. II, *Photochem. Photobiol.*, 47, 837, 1988.
95. **Häder, D.-P.**, Ecological consequences of photomovement in microorganisms, *J. Photochem. Photobiol. B,* 1, 385, 1988b.
96. **Mizuno, T., Maeda, K., and Imae, Y.**, Thermosensory transduction in *Escherichia coli,* in *Transmembrane Signaling and Sensation,* Oosawa, F., Yoshioka, T., and Hayashi, H., Eds., Japan, Science Society Press, Tokyo, and VNU Science Press BV, Netherlands, 1984, 147.
97. **Poff, K. L.**, Temperature sensing in microorganisms, in *Sensory Perception and Transduction in Aneural Organisms,* Colombetti, G., Lenci, F., and Song. P.-S., Eds., Plenum Press, New York, 1985, 299.
98. **Berg, H. C.**, Physics of bacterial chemotaxis, in *Sensory Perception and Transduction in Aneural Organisms,* Colombetti, G., Lenci, F., and Song, P.-S., Eds., Plenum Press, New York, 1985, 19.
99. **MacNab, R. M.**, Biochemistry of sensory transduction in bacteria, in *Sensory Perception and Transduction in Aneural Organisms,* Colombetti, G., Lenci, F., and Song, P.-S., Eds., Plenum Press, New York, 1985, 31.
100. **Frankel, R. B.**, Magnetic guidance of organisms, *Annu. Rev. Biophys. Bioeng.*, 13, 85, 1984.
101. **Esquivel, D. M. A. and de Barros, H. G. P. L.**, Motion of magnetotactic microorganisms, *J. Exp. Biol.*, 121, 153, 1986.
102. **Mast, S. O.**, *Light and Behavior of Organisms,* John Wiley & Sons, New York, 1911.
103. **Diehn, B., Feinleib, M., Haupt, W., Hildebrand, E., Lenci, F., and Nultsch, W.**, Terminology of behavioral responses of motile microorganisms, *Photochem. Photobiol.*, 26, 559, 1977.
104. **Häder, D.-P., Colombetti, G., Lenci, F., and Quaglia, M.**, Phototaxis in the flagellates, *Euglena gracilis* and *Ochromonas danica, Arch. Microbiol.*, 130, 78, 1981.
105. **Colombetti, G., Häder, D.-P., Lenci, F., and Quaglia, M.**, Phototaxis in *Euglena gracilis:* effect of sodium azide and triphenylmethyl phosphonium ion on the photosensory transduction chain, *Curr. Microbiol.*, 7, 281, 1982.
106. **Lenci, F., Colombetti, G., and Häder, D.-P.**, Role of flavin quenchers and inhibitors in the sensory transduction of the negative phototaxis in the flagellate, *Euglena gracilis, Curr. Microbiol.*, 9, 285, 1983.
107. **Brinkmann, K.**, Keine Geotaxis bei *Euglena, Z. Pflanzenphysiol.*, 59, 12, 1968.

108. **Batschelet, E.,** Statistical methods for the analysis of problems in animal orientation and certain biological rhythms, in *Animal orientation and navigation,* Galles, S. R., Schmidt-Koenig, K., Jacobs, G. J., and Belleville, R. F., Eds., National Aeronautics and Space Administration, Washington, D.C., 1965, 61.
109. **Batschelet, E.,** *Circular Statistics in Biology,* Academic Press, London, 1981.
110. **Mardia, K. V.,** *Statistics of Directional Data,* Academic Press, London, 1972.
111. **Kessler, J. O.,** The external dynamics of swimming microorganisms, in *Progress in Phycological Research 4,* Round and Chapman, Eds., Biopress, 1986, 258.
112. **Glazzard, A. N., Hirons, M. R., Mellor, J. S., and Holwill, M. E. J.,** The computer assisted analysis of television images as applied to the study of cell motility, *J. Submicrosc. Cytol.,* 15, 305, 1983.
113. **Herman, B. and Albertini, D. F.,** A time-lapse video image intensification analysis of cytoplasmic organelle movements during endosome translocation, *J. Cell Biol.,* 98, 565, 1984.
114. **Kondo, T., Kubota, M., Aono, Y., and Watanabe, M.,** A computerized video system to automatically analyze movements of individual cells and its application to the study of circadian rhythms in phototaxis and motility in *Chlamydomonas reinhardtii, Protoplasma,* S1, 185, 1988.
115. **Häder, D.-P. and Hoiczyk, E.,** Gliding motility, in *Algal Cell Motility,* Melkonian, M., Ed., Chapman & Hall, New York, in press.

Chapter 16

AUTOMATED MEASUREMENT OF MOVEMENT RESPONSES IN HALOBACTERIA

Tetsuo Takahashi

TABLE OF CONTENTS

I. INTRODUCTION

The extremely halophilic bacterium *Halobacterium halobium* is well known for its remarkably simple light energy-harvesting apparatus, bacteriorhodopsin (bR).[1-4] This membrane protein is inducible and becomes a major membrane component, forming distinct patches (purple membrane) in the cell membrane when the basically aerobic *H. halobium* is subjected to low oxygen tension.[5] The function of bR is to transport protons from the inside of the cell to the external medium upon illumination. The transmembrane electrochemical gradient thus formed is utilized as an energy source for various cellular functions such as ATP synthesis, regulation of ionic composition inside the cell, and transport of nutrients. Therefore, *H. halobium* can survive like a phototroph under semiaerobic or anaerobic conditions. Under these circumstances, the cell also acquires the ability to be attracted by light and accumulates in areas where illumination suitable for bR function is available.[6,7]

H. halobium is probably the first bacterial species whose behavioral responses were extensively studied by computer-linked automated methods.[8,9] This is partly because the behavior is controlled by photostimuli which are easily maneuverable by an electronic shutter interfaced to the computer. Also, *H. halobium* swims slowly enough to be tracked with early laboratory computer techniques, presumably due to the fact that the behavior has to be observed in an almost saturating salt solution, as in the natural habitat of the organism. In addition, the photobehavioral study was facilitated by scientists' attention to the receptor molecule(s) as a model system or a system evolutionarily related to our visual reception mechanism, because both the visual pigment and bR use a common chromophore, retinal (vitamin A aldehyde).

In addition to bR, *H. halobium* and related species of halophilic bacteria possess a family of retinal-containing proteins with similar molecular weights of 23 to 27 kDa.[3,10-16] One is called halorhodopsin (hR), a light-driven, inwardly directed chloride pump,[17] in which also generates inside-negative membrane potential like bR. Others are two electrically neutral photosensory pigments which control the photobehavior of the cell. They are now called sensory rhodopsin (or sensory rhodopsin I) and phoborhodopsin (or sensory rhodopsin II). In this review, I will briefly introduce the methodology of the objective and less time-consuming measurement of the halobacterial photobehavior and its application.

II. BEHAVIOR OF *H. HALOBIUM*

H. halobium is a rod-shaped, polarly flagellated bacterium, 0.5 to 1.0 μm wide by 2 to 4 μm in length. The cell size varies with both the bacterial strain and the growth conditions. During the exponential phase of growth, the cells are short in length. Strains containing light-driven ion pumps look generally thin under a phase-contrast microscope. They swim parallel to their long axis in both directions at 2 to 6 μm/s. The swimming speed strongly depends on the temperature. Their flagella are observable under a high-magnification light microscope with bright dark-field illumination.[18] The cells are bipolarly flagellated in the stationary phase of growth, but most of the cells carry a bundle of flagella only at one end during the exponential growth phase. However, they swim both back and forth equally well in either phase of the growth.[18]

A clockwise or counterwise rotation of the flagellar bundle pushes or pulls the cell, respectively.[18] A spontaneous change of the sense of rotation of a flagellar bundle frequently reverses the swimming direction. The period of smooth swimming lasts 3 to 15 s, interrupted by a sudden stop or reversal. A slight reorientation of the cell axis is brought about by the Brownian motion during the period, resulting in a zigzag, random-walk pattern of a trajectory of the cells. It has been an issue whether an intracellular oscillator[19-21] or a stochastic mechanism of the flagellar motor switching[22,23] controls the period of the smooth swimming.

Modulation of the frequency of the spontaneous change in the swimming direction results in a net migration of a population of randomly swimming microorganisms.[24] In enteric bacteria such as *Escherichia coli* or *Salmonella typhimurium,* a sudden increase in the concentration of chemoattractants or a sudden decrease in that of chemorepellents leads to a suppression of the directional changes for 1 to 3 min.[25,26] In contrast, a decrease or an increase in the concentration of chemoattractants or chemorepellents, respectively, elicit erratic motion (tumbling) in these organisms. Halobacterial photoresponses occur in a similar fashion. In halobacteria, however, the modulation does not last long enough to affect the second or subsequent reversals after a stimulus. The cells show only a single, sudden reversal of the swimming direction or merely a stop response[18] when subjected to a repellent light ($\lambda < 520$ nm; a step-up photophobic response[27]) or after turning off an attractant light ($\lambda > 530$ nm; a step-down photophobic response[27]). Their behavior seems normal after the first reversal. Likewise, a photoattractant stimulus, either a sudden onset of the attractant light or a sudden removal of the repellent light, prolongs only a period of the smooth swimming for 15 to 30 s (a step-up or a step-down photoattractant response, respectively).[28]

Because a step-up or step-down photophobic response is always accompanied by a step-down or a step-up photoattractant response (a reversal suppression), respectively, the step-up and step-down responses to the repellent light is sometimes (not quite correctly) called negative phototaxis.

III. QUANTITATION OF THE MOVEMENT RESPONSES

A. POPULATION METHOD

A classical but still useful tool for quantitative analysis of the behavior of microorganisms is to measure changes in the population density resulting from migration of cells toward or away from chemicals or other environmental stimuli.[29] An earlier work of a successful and quantitative description of the chemotactic responses of the bacteria was essentially based on this methodology.[30] Recently, Wolff et al.[31] described a photomultiplier-linked microscope system for halobacterial phototaxis. In that system, changes in intensity of scattered light from cells are detected under dark-field illumination. In order to quantify the response, only the initial rate of phototactic accumulation is measured in order to avoid complicated effects of local changes in cell density.

B. INDIVIDUAL CELL TRACKING

In contrast to population techniques, methods of tracking individual cells potentially allow high time-resolution. The cells can be exposed to stimuli for a very short period of time. When averages of many cells are taken, the result is highly reproducible and quantitative. For example, temporal sensing[26] and the adaptation kinetics[25] of chemotactic responses in *S. typhimurium* were first measured by manual examination of individual trajectories recorded on a film. Because the error that accompanies this method is mainly statistical and depends on the sample size, automated analysis based on this methodology is beneficial.

A hardware configuration for a computerized cell tracking method usually consists of a microscope with a 32 to 40× objective lens, a 1/2- or 2/3-in. vidicon video camera (with a silicon-intensified target; charged coupled device, CCD; Newvicon; or other infrared (IR)-sensitive devices), a digitizer that converts video data to digital codes, a computer system, and, if necessary, a videotape recorder.

Three parameters have so far been utilized as a measure of the magnitude of the behavioral responses of halobacteria: percentages of reversing cells in a cell population,[8] averages of the reversals occurring in a cell population at a time point,[34] and latency periods between stimuli and the resultant reversals of a cell.[6]

1. Commercially Available System

The first description of a computerized image analysis system for individual cell tracking appeared in 1970,[32] 1 year before the development of a servo-controlled, three-dimensional tracking microscope for bacteria.[33] The multipurpose image analysis system for moving objects is now commercially available (ExpertVision system: Motion Analysis Systems Inc., Santa Rosa, CA). However, it took many years to be utilized by investigators in the field of bacterial taxis who had to develop their own user program from a software package provided by the manufacturer.[9,34,35] In this system, images of each object in consecutive frames of video data were first stored as a group of x- and y-coordinates (for economy of storage, only coordinates of the outline of objects are digitized and stored). Then the software calculates the position of the center-of-mass (centroid) of each object in each frame from the set of coordinates. The next stage of the analysis is to connect them in time-ordered fashion using a user-defined maximal possible displacement of the cells between two consecutive frames in order to find the trajectories. The file, a group of time series of x- and y-coordinates describing the centroids, is called a path file, with which the movement analysis begins. A set of command procedures is provided by the manufacturer for movement analyses: calculating velocities, orientation, acceleration, and rate of change of direction as a time series for each trajectory, along with statistical treatment either in the time domain or over a population of trajectories. Use of a computer system that has higher power than a usual personal system is recommended for statistical analyses of the behavior over many trajectories of the cells, since a floating-point computation of the centroids takes much time.

2. Utilization of a Video Digitizer

A variety of analog-to-digital converter systems (video digitizer) which freeze a video image and send the data to a personal computer are now commercially available. Utilizing one of them, we constructed an automated video-computer system for *H. halobium* almost 10 years ago.[8] Later, the system was improved to allow real-time data acquisition without videotape recording.[36] In our system, the video digitizer binarizes video images with an appropriate threshold. The binarized data of 256 $\times$ 256 pixels take 8 Kbyte in the main memory of a personal computer. The image of a halobacterial cell in a digitized video frame consists of 4 to 5 pixels, when a video camera with a 1/2-in. CCD target is used on a microscope with a 40$\times$ objective. Although a relatively large memory size is necessary for the storage of a digitized frame, storing the position of the objects in this way allows faster calculation of the displacement of each object between two successive frames.[8] The principle of the calculation using the correlation method[37] is illustrated in Figure 1. Only displacements of a cell between successive frames are used for detection of a reversal response.

A serious problem in automated analysis of moving objects is the so-called occlusion, where two moving objects come close together or shade each other.[38] Ambiguity due to occlusion is inevitable because sometimes we found it difficult to track occluded trajectories even by eyes. Therefore, our software was programmed so as to ignore the trajectory when more than two ambiguous local maxima appeared in the correlation matrix.

C. HOW TO DETECT THE REVERSAL RESPONSES

1. Commercially Available System

Sundberg et al.[9] reported a method of measuring the reversal frequency of *H. halobium* cells as a function of time by use of the ExpertVision system. To detect reversals of the cells, their method utilized a function called "rate-of-change-of-direction" that was calculated as a time series. The rate of change of direction represents the angle between the velocity vector of an object from the previous frame to the current one and that from the current one to the next. In order to assess whether or not the reversal occurred, they introduced a threshold value of 300 (°/s, i.e., 20°/frame to frame). When the function had a value above

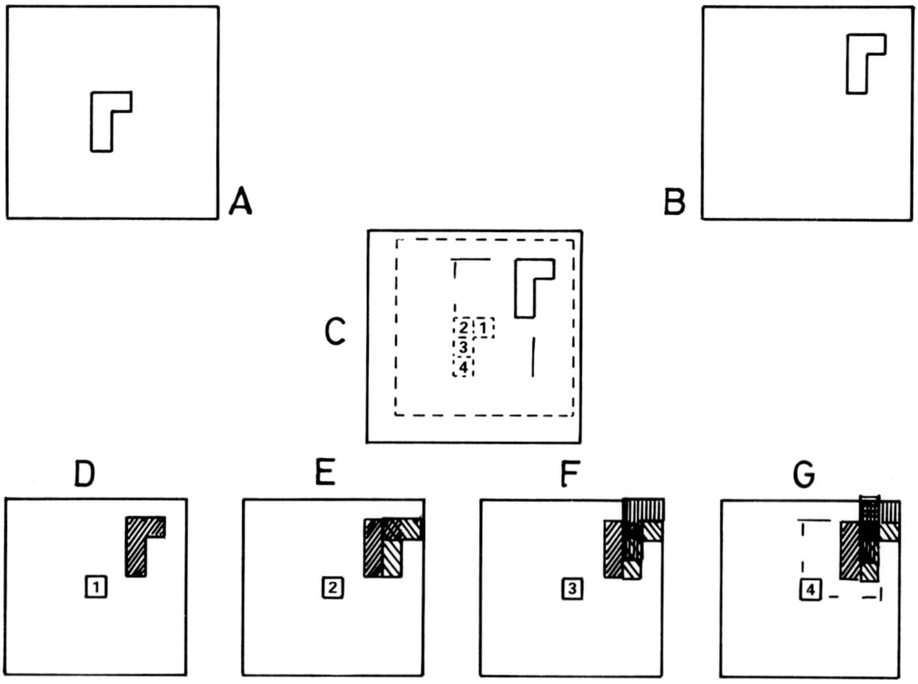

FIGURE 1. Schematic drawing of the method of calculating the displacement of a moving object between two successive frames. (A) The first frame, (B) the second frame. A grid of m × n pixels is made so that the center of the grid is located at one of those pixels which constitute an object in the first frame. Then, the grid is superimposed onto the second frame (C). Pixels within the grid are then queried if they have a value in the second frame (D). Thereafter, the grid is moved onto the next pixel of those which constitute the object in the first frame, and the query is repeated (E and F). After the summation of the grids with respect to all the pixels which constitute the object in the first frame, the resulting correlation matrix shows a maximum value at the element that represents the position of the displacement (G).

the threshold at a time point, the cell was assumed to show a reversal of swimming. Although this was the first successful quantitative description of a time-resolved behavior of halobacteria, the authors also reported that the efficiency of correct identification of actual reversals or that of smooth paths was 83 to 92%, compared with visual tracking.

Apparently, the low efficiency of tracking is due to false detections of reversal responses or smooth swimming behaviors. In order to improve the precision of the behavioral data, it seems therefore crucial to abandon those trajectories which show ambiguous behavior. In visual tracking by human eyes, we sometimes abandon these particular cell tracks unconsciously. Therefore, we have recently developed improved software for the detection of exact reversals.[39] The error due to false detections was considerably reduced when an appropriate buffer zone was introduced around the threshold value. The improved strategy[39] for the discrimination of the reversal response from smooth swimming is illustrated in Figure 2. The tracking efficiency in this method is higher than 95%.[39]

2. A Video Digitizer System

In contrast to the commercial motion-analyzer system, our small system[8] does not calculate the parameter that describes the swimming behavior. Instead, the position of the maximal value in a correlation matrix represents the two-dimensional displacement of the object from one frame to another. To simplify the algorithm and to reduce the size of the software written by assembler language, we divided two-dimensional directions of 360° into

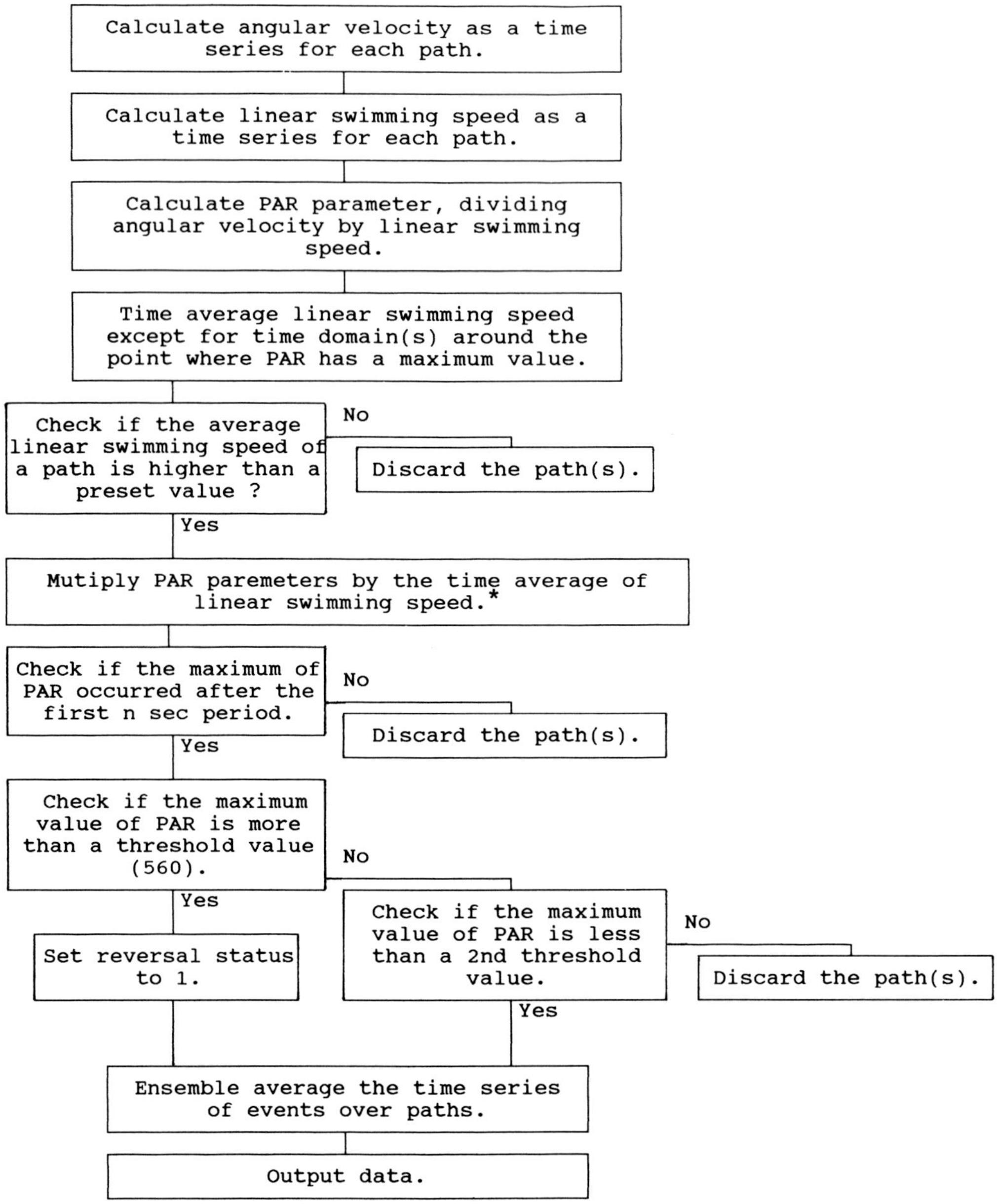

* This means that the angular velocities were divided by normalized linear speed.

FIGURE 2. Block diagram of the strategy for precise detection of the reversal or the stop response of halobacteria.[39]

eight subdivisions, i.e., the swimming directions of cells were quantitized. The smooth swimming of a cell is thus detected when the displacement of a cell from the current frame to the next frame falls in either the subdivision to which the previous displacement belongs or in the neighboring one.[8] An alternative method would be to calculate the scalar product of the two displacement vectors and to find the directional change by the sign of the product.[40]

D. PERFORMANCE OF THE TWO AUTOMATED SYSTEMS

A notable merit of using the ExpertVision system is the capability of measuring averages

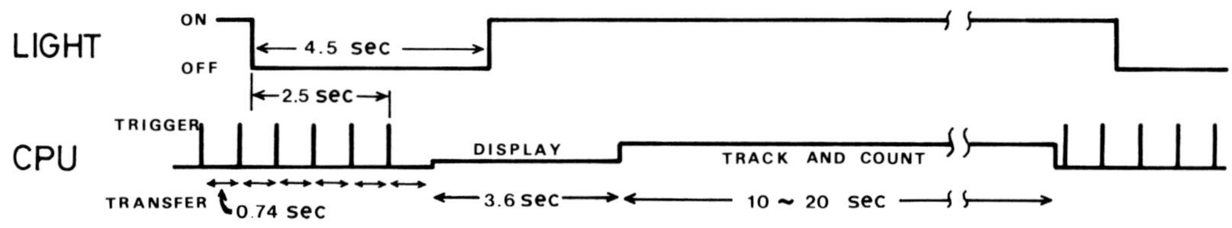

FIGURE 3. A timing chart of light stimuli for measurement of step-down photophobic response and acquisition and processing of digitized data.[8] In this case, the central processing unit (CPU) sends a signal for triggering the video digitizer six times at a frequency (frame rate) of $1.35\ s^{-1}$. Then the data are transferred to the main memory and analyzed by the CPU. Two frames of each sequence are frozen before the light stimulus so that the swimming direction before the stimulus is recognized.

of the reversal probability (reversal frequency) over a cell population for minutes as a function of time. It detects 30 to 100 trajectories of *H. halobium* per frame. As described above, a most time-consuming process is the computation of centroids. By using a Sun2/120 work station as a host computer, total processing time is about 1 min per digitized frame of video data. This means it takes about 20 min to process a sequence of video data digitized for 4 s at a frame rate of $5\ s^{-1}$. Although it is not so slow for real-time processing, a videotape recording of raw video data is necessary for increasing the sample size in order to reduce statistical errors. However, real-time analysis becomes possible by using a recently manufactured work station whose performance is greatly enhanced. Usually, a sample size of about 250 to 1000 trajectories is employed for the calculation of the reversal frequency.[9]

In contrast to the system described above, our video digitizer system[8] recognizes 8 to 30 cell trajectories per sequence of video data. One sequence consists of six consecutive frames digitized at 1 to 5 frames per second. The frame rate and the total number of frames per sequence can be increased by using a recent personal computer system. The original system[8] processes a sequence of data within 30 s. Therefore, real-time analysis of the behavioral response for 15 min usually yields a sample size of 150 to 300 cell trajectories. A timing chart describing data acquisition and the processing sequences is shown in Figure 3. It should be noted that shortening the interval between the acquisition of one sequence of data does not improve the performance of the total system because *H. halobium* needs an adaptation time of at least 20 to 30 s before the next stimulus.[22]

IV. SENSORY PHOTOPIGMENT AND THE AUTOMATED MEASUREMENT OF THE PHOTOBEHAVIOR

A. DISCOVERY OF SENSORY RHODOPSIN

In 1982, at about the same time as Tsuda et al.,[41-43] and Bogomolni and Spudich[12] reported that *H. halobium* contains a third retinal-protein pigment in addition to bR and hR. The strain they used[44] is deficient in the synthesis of both bR and hR, but showed full function of the photobehavioral responses, i.e., both positive phototaxis to yellow-orange light and negative phototaxis to blue-near-UV light. They proposed a hypothesis that the pigment mediates the two different photobehavioral responses using photochromism.[12,45] The basis of their reasoning was that the photoreaction scheme of the pigment involves an extraordinary long-lived ($t_{1/2} \approx 0.8$ s) intermediate species with the maximal absorption at about 370 nm.[13] Indeed, the absorption maximum of the intermediate matches the action peek of the photophobic response,[6] which was supposed to be mediated by a photosystem called PS370.

The action spectrum of PS370[6] sometimes showed a shoulder at around 500 nm.[7] This had been attributed to an antenna function of carotenoids since a strain rich in carotenoids showed the shoulder and cells grown with a carotenoid synthesis inhibitor did not.[7] However, S_{373} showed no absorption band around 450 nm, while the strain Bogomolni and Spudich

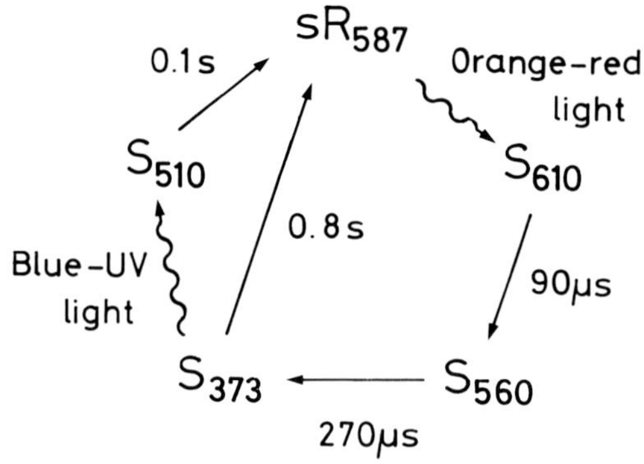

FIGURE 4. Photoreaction scheme of sensory rhodopsin.[12,45,53,54,72] Subscripts indicate the absorption maxima of the species. Wavy and other arrows represent photoprocesses and thermal reactions, respectively. The red-absorbing form sR_{587} was suggested to function as the receptor for the attractant light, and the blue-UV-absorbing species S_{373} was suggested to function as a repellent light receptor.[12] Under normal illumination conditions in the natural habitat of *Halobacterium halobium*, the slow thermal decay of S_{373} allows a photostationary state where almost half of the total amount of the pigment exists as S_{373}.[12] The isomeric form of the chromophore of sR_{587} and that of S_{373} is all-*trans* and 13-*cis*, respectively.[72]

used contained a large amount of carotenoids. This is not consistent with the hypothesis in view of the Grotthus-Draper law.

An alternative explanation[11] was that there were two or three retinal-protein pigments other than hR and bR. In this respect, action spectroscopic studies by visual tracking of the cells had been extensively done with the use of different strains[46-49] or substituting the chromophore by chemical analogs.[47,50] In addition, an interesting observation was reported by Hildebrand and Schimz,[51] that the PS370 was separated from the photoattractant system by inhibiting the protein synthesis with puromycin. Therefore, it had been of primary importance to obtain evidence on whether or not the S_{373} intermediate functions as a blue-UV-repellent light receptor.

B. AUTOMATED MEASUREMENT OF PHOTOTAXIS UNDER BACKGROUND ILLUMINATION

If, as shown in Figure 4, PS370 is part of the photochromic system of sensory rhodopsin, we can expect a notable effect of continuous illumination with blue-UV light on the photoattractant response of *H. halobium* to yellow-orange light as well as that of the yellow-orange light on the photophobic response to blue-UV light. It might be hinted in the early work[8] and was demonstrated[36,45] that the blue-UV response (PS370) is dependent on the intensity of yellow-orange background light. Furthermore, continuous background illumination with blue-UV light drastically changed the behavior of the cells to yellow-orange attractant light.[45,52] Our result was as follows. When monochromatic, continuous illumination at 410 nm was superimposed on the IR background light for observation, the fraction of the reversing cells measured upon subjecting the step-down stimulus of yellow-orange actinic light decreased with an increase in the intensity of the 410-nm background light. In contrast, the step-up response to yellow-orange light, which is usually less than the level of the spontaneous reversal, increased with increasing intensity of the 410-nm light. Finally, the step-up and the step-down responses were completely inverted at the highest intensity of the 410-nm background light. In other words, under sufficiently high blue-UV background illumination, the cells responded to the yellow-orange attractant light as if it were a repellent light[52] (Figure 5).

The result is easily explained by accepting the hypothesis proposed by Bogomolni and

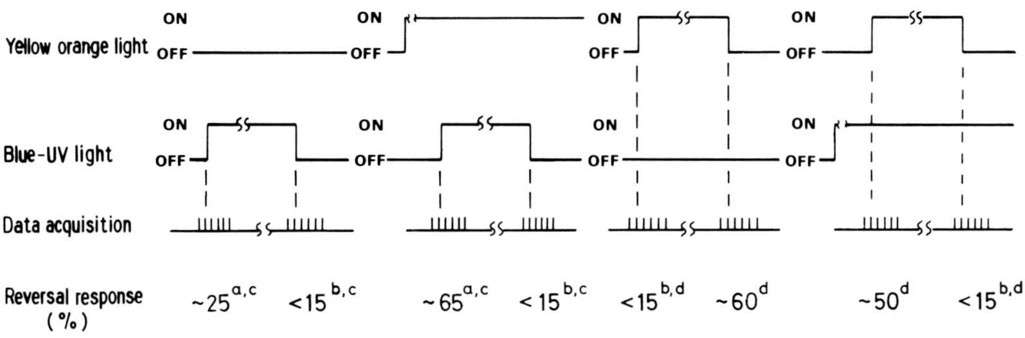

FIGURE 5. Blue-UV background illumination reverses the step-up and the step-down responses of *Halobacterium halobium* to yellow-orange light. The timing chart of data acquisition and the magnitude of the response measured at each timing are illustrated (a) at moderate intensity of the blue-UV actinic light and (b) less than a spontaneous reversal level. (c) Data from Reference 36; (d) data from Reference 52.

Spudich[12] if we pay attention to the fact that a repellent does not function as such until the receptor emerges. The photorepellent receptor S_{373} is produced almost instantaneously (with a rise time of 300 μs[53,54]) when yellow-orange light is applied to *H. halobium*. If the response is elicited by something that may occur after the photoreception by the photoreceptor molecule — this seems to always be the case since this is just an extension of the first law of photochemistry — the appearance of the photoreceptor molecule under continuous illumination is equivalent to the onset of light to the cell having the continuously existing photoreceptor pigment, S_{373}. On the other hand, it is difficult to explain this observation with the idea that PS370 is mediated by some other, unknown photopigment.[52]

Once S_{373} is concluded to be the photoreceptor molecule of the PS370, a further important consequence is deduced from this observation. This is related to the question of what in the photoreaction scheme illustrated in Figure 4 is responsible for the step-down photophobic response to blue-UV light. It might be any one of the processes denoted by arrows in Figure 4 or, alternatively, a change in the amount of a photointermediate state which can activate a certain enzyme, as in our visual transduction. As described above, however, the sequence before S_{373}, the thermal decay process of S_{373}, and the decrease in the amount of S_{373} turned out not to be responsible because such mechanisms could not generate the step-up photophobic response to yellow-orange light under a blue-UV background. An attractive hypothesis regarding the first step of the PS370 signal transduction is that the intermediate denoted by S_{510} in Figure 4 is the signaling state.

Furthermore, using the background effect, a difference action spectrum between step-up and step-down photophobic responses becomes measurable. Because the blue-UV background light causes a step-up photophobic response to the light absorbed by sR_{587}, the spectrum of this action can be subtracted from the action spectrum of the step-down photophobic response to yellow-orange light. If the latter is mediated by an unidentified photoreceptor, an action spectrum measured under blue-UV background light becomes a difference spectrum between sR_{587} and the unknown photoreceptor. On this basis, it was strongly suggested that the receptor for the step-down photophobic response is by sR_{587}.[52]

C. ACTION SPECTROSCOPY

Although action spectroscopy is a powerful tool in photobiology, its application to halobacterial phototaxis seemed sometimes unsuccessful and misleading, which might come from the difficulty in obtaining large amounts of data in a short time. Microorganisms may change their sensitivity to stimuli day by day. Also, the subjectivity inevitably accompanying visual tracking may make this task more difficult. Therefore, it is quite useful to adopt automated systems for action spectroscopy. The principle of action spectroscopy is described elsewhere.[55-57]

As a next step following the identification of S_{373} as the receptor pigment of PS370, we measured the action spectrum of the step-down photophobic response of a *H. halobium* mutant that lacks both bR and hR.[58] Our effort was partially successful, but also partially unsuccessful due to contamination by an unknown pigment. The successful part was the action spectrum beyond 580 nm[58] where the spectrum matched the absorption curve of sensory rhodopsin.[59] A slight problem was that the apparent standard error was relatively large (20 to 30% O.D.), presumably due to the long experimental period, during which time the cells might change their sensitivity. However, the shape of the curve must be much more accurate than that expected from the error bar because we measured all wavelength points as evenly as possible within 1 d.

The unsuccessful part was even more fruitful, because we detected by flash spectrophotometry the unknown pigment that distorted the action spectrum.[13] The flash spectrophotometric survey was done at around the wavelength where a mutant strain lacking the step-down photophobic response showed a maximal sensitivity of step-up photophobic response. Therefore, the new pigment became the first bacterial rhodopsin that was discovered by action spectroscopy. The pigment was named phoborhodopsin[60] (pR: also called sensory rhodopsin II[61]). The pigment is synthesized at the early exponential growth phase and the photophobic sensitivity gradually decreases at the stationary phase, whereas sensory rhodopsin is induced at the stationary phase.[60] A biological role of pR is therefore to avoid blue-green light or total sunlight which is harmful when oxygen is abundant in the medium. Taking this pigment into account, it seemed reasonable that Hildebrand and Schimz[61] observed a separation of the step-up photophobic system from the photoattractant system. Also, the spectral region where Dencher[7] observed a shoulder in the spectrum of the PS370 is indeed covered by this pigment.

D. PHOBORHODOPSIN OR SENSORY RHODOPSIN II

We have measured action spectra of two mutant strains which show no step-down photophobic response, but do show the step-up photophobic response.[62] The two spectra are indistinguishable and match the absorption spectrum of pR.[60] A notable feature is that it shows a shoulder at 460 nm along with the main peak at 485 nm. This unusual shape of the spectrum is strongly suggested to be a vibrational fine structure,[63] known in no visual rhodopsins nor in any other rhodopsin-like pigments thus far. In this respect, a mechanism of the unique color regulation in pR and of the spectral broadening in other bacterial rhodopsin was proposed.[63]

The chromophore of pR is all-*trans* retinal.[64] Isomerization of the 13-*cis* form is necessary to elicit the photophobic response because the response was inhibited by substitution of retinal by an analog whose 13-14 double bond isomerization was prohibited.[65] Also, the computerized behavioral analysis was utilized to identify the signal generating state(s) in the photoreaction scheme of pR.[66,67]

V. KINETIC ANALYSIS

Since kinetic properties of the photoreceptor molecules are measurable by flash spectrophotometry, measurement of the latency period of a behavioral response may provide information about the molecular events that occur in photoexcited halobacterial cells. Using an automated system with 67-ms resolution, Sundberg et al.[9] showed that at least 9.4 s of latency time is necessary to evoke the photophobic response of *H. halobium* after onset of intense 450-nm light. Interestingly, photophobic response to 400-nm light and step-down photophobic response to yellow-orange needed longer latency times.[9] Although this may reflect a different molecular process of signal amplification or transducing mechanism[68] between the photosystems, interpretation of the data awaits further work.

When a step-like increase in chemorepellent or decrease in attractant is subjected to *S. typhimurium* or *E. coli,* sudden flagellar motor reversal causes the cells to constantly change their swimming direction erratically, and thereafter the cells recover their original behavior slowly.[25,26] The latter process is called adaptation. Using the automated cell tracking method, McCain et al.[22] studied the adaptation process in halobacterial phototaxis. They observed occurrence of reversal in a cell population (reversal frequency) as a function of time for 90 s at the longest. When a step-up stimulus of 600-nm attractant light was delivered, the cells showed a drop in reversal frequency for about 30 s and recovered their prestimulus level frequency under the continuous illumination. The period of reversal suppression is presumably identical to those observed in a single cell-tracking experiment,[28] and this slow process probably involves methylation of membrane protein,[15,69,70] as in *E. coli* or *S. typhimurium.* In relation to the slow process, they observed an interesting effect of continuous irradiation with attractant light on halobacterial photobehavior: the cells showed an inverted response to a short pulse of 600-nm attractant light when it was delivered 4 to 8 s after the cell exhibited an intense step-down photophobic response to the turning off of the continuous irradiation with the attractant light. This may indicate that adaptation to attractant light lasts long, while that to repellent light terminates within the mean smooth swimming period of a cell.

VI. CONCLUDING REMARKS

Because this short review is mainly focused on the application of the computerized automated methods, some important or interesting observations have not been mentioned. These include the issue of the flagellar motor switch,[19-23,71] the hypothetical cellular oscillator that controls reversal probability,[19-21] methyl-accepting membrane proteins in halobacteria,[15] and the evolutionary relationship between the four halobacterial rhodopsins and our visual pigments.[63] Fortunately, however, our knowledge regarding these issues is rapidly increasing. The computerized automated technique for observation of bacterial behavior also might be of help for studies in this field in the near future.

Rapid progress in technology will certainly affect our technique and instrumentation. For example, the high-definition TV system seems to be a good candidate for the input data of our image analysis system in the next generation because a key to improving the precision of data in a statistical treatment is to collect as many samples (e.g., trajectories) as possible. Time-consuming measurements such as in action spectroscopy can easily be performed if it is introduced. Also, useful mutant strains[34] or chemicals that affect the behavioral response may readily be surveyed.

REFERENCES

1. **Oesterhelt, D. and Stoeckenius, W.,** Rhodopsin-like protein from the purple membrane on *Halobacterium halobium, Nature New Biol.,* 233, 149, 1971.
2. **Stoeckenius, W., Lozier, R. H., and Bogomolni, R. A.,** Bacteriorhodopsin and the purple membrane of halobacteria, *Biochim. Biophys. Acta,* 505, 215, 1979.
3. **Stoeckenius, W. and Bogomolni, R. A.,** Bacteriorhodopsin and related pigments of halobacteria, *Annu. Rev. Biochem.,* 52, 587, 1982.
4. **Khorana, H. G.,** Bacteriorhodopsin, a membrane protein that uses light to transfer protons, *J. Biol. Chem.,* 263, 7439, 1988.
5. **Sumper, M., Reitmeier, H., and Oesterhelt, D.,** Biosynthesis of the purple membrane of Halobacteria. *Angew. Chem. Int. Ed. Engl.,* 15, 187, 1976.

6. **Hildebrand, E. and Dencher, N.,** Two photosystems controlling behavioral responses of *Halobacterium halobium, Nature,* 257, 46, 1975.
7. **Dencher, N.,** Light-induced behavioral reactions of *Halobacterium halobium:* evidence for two rhodopsins acting as photopigments, in *Energetics and Structure of Halophilic Microorganisms,* Caplan, S. R. and Ginzburg, M., Eds., Elsevier/North-Holland, Amsterdam, 1978, 67.
8. **Takahashi, T. and Kobatake, Y.,** Computer-linked automated method for measurement of the reversal frequency in phototaxis of *Halobacterium halobium, Cell Struct. Funct.,* 7, 183, 1982.
9. **Sundberg, S. A., Alam, M., and Spudich, J. L.,** Excitation signal processing times in *Halobacterium halobium* phototaxis, *Biophys. J.,* 50, 895, 1986.
10. **Mukohata, Y., Matsuno-Yagi, A., and Kaji, Y.,** Light induced proton uptake and ATP synthesis by bacteriorhodopsin-depleted *Halobacterium halobium,* in *Saline Environment,* Morishita, H. and Masui, M., Eds., Academic Center Press Japan, Tokyo, 1980, 41.
11. **Dencher, N.,** The five retinal-protein pigments of halobacteria: bacteriorhodopsin, halorhodopsin, P 565, P 370, and slow-cycling rhodopsin, *Photochem. Photobiol.,* 38, 753, 1983.
12. **Bogomolni, R. A. and Spudich, J. L.,** Identification of a third rhodopsin-like pigment in phototactic *Halobacterium halobium, Proc. Natl. Acad. Sci. U.S.A.,* 79, 6250, 1982.
13. **Takahashi, T., Tomioka, H., Kamo, N., and Kobatake, Y.,** A photosystem other than PS370 also mediates the negative phototaxis of *Halobacterium halobium, FEMS Microbiol. Lett.,* 28, 161, 1985.
14. **Spudich, E. N. and Spudich, J. L.,** Biochemical characterization of halorhodopsin in native membrane, *J. Biol. Chem.,* 260, 1208, 1985.
15. **Spudich, E. N., Takahashi, T., and Spudich, J. L.,** Sensory rhodopsin I and II modulate a methylation/demethylation system in *Halobacterium halobium* phototaxis, *Proc. Natl. Acad. Sci. U.S.A.,* 86, 7746, 1989.
16. **Blanck, A. and Oesterhelt, D.,** The haloopsin gene. II. Sequence, primary structure of halorhodopsin and comparison with bacteriorhodopsin, *EMBO J.,* 6, 265, 1987.
17. **Schobert, B. and Lanyi, J. K.,** Halorhodopsin is a light-driven chloride pump, *J. Biol. Chem.,* 257, 10306, 1982.
18. **Alam, M. and Oesterhelt, D.,** Morphology, function and isolation of halobacterial flagella, *J. Mol. Biol.,* 176, 459, 1984.
19. **Schimz, A. and Hildebrand, E.,** Response regulation and sensory control in *Halobacterium halobium* based on an oscillator, *Nature,* 317, 641, 1985.
20. **Schimz, A. and Hildebrand, E.,** Entrainment and temperature dependence of the response oscillator in *Halobacterium halobium, J. Bacteriol.,* 166, 689, 1986.
21. **Hildebrand, E. and Schimz, A.,** Role of response oscillator in inverse responses of *Halobacterium halobium* to weak light stimuli, *J. Bacteriol.,* 169, 254, 1987.
22. **McCain, D., Amici, L., and Spudich, J. L.,** Kinetically resolved states of the *Halobacterium halobium* flagellar motor switch and modulation of the switch by sensory rhodopsin I, *J. Bacteriol.,* 169, 4750, 1987.
23. **Marwan, W. and Oesterhelt, D.,** Signal formation in the halobacterial photophobic response mediated by a fourth retinal protein (P480), *J. Mol. Biol.,* 195, 333, 1987.
24. **Berg, H. C. and Purcell, E. M.,** Physics of chemoreception, *Biophys. J.,* 20, 193, 1977.
25. **Spudich, J. L. and Koshland, D. E., Jr.,** Quantitation of the sensory response in bacterial chemotaxis, *Proc. Natl. Acad. Sci. U.S.A.,* 72, 710, 1975.
26. **Macnab, R. A. and Koshland, D. E., Jr.,** The gradient sensing mechanism in bacterial chemotaxis, *Proc. Natl. Acad. Sci. U.S.A.,* 69, 2509, 1972.
27. **Diehn, B., Feinleib, M., Haupt, W., Hildebrand, E., Lenci, F., and Nultsch, W.,** Terminology of behavioral responses of motile microorganisms, *Photochem. Photobiol.,* 26, 559, 1977.
28. **Spudich, J. L. and Stoeckenius, W.** Photosensory and chemosensory behavior of *Halobacterium halobium, Photobiochem. Photobiophys.,* 1, 45, 1979.
29. **Feinleib, M. E. H. and Curry, G. M.,** Methods for measuring phototaxis of cell populations and individual cells, *Physiol. Plant.,* 20, 1083, 1967.
30. **Adler, J.,** A method for measuring chemotaxis and use of the method to determine optimal condition for chemotaxis by *Escherichia coli, J. Gen. Microbiol.,* 74, 77, 1973.
31. **Wolff, E. K., Bogomolni, R. A., Scherrer, P., Hess, B., and Stoeckenius, W.,** Color discrimination in halobacteria: spectroscopic characterization of a second sensory receptor covering the blue-green region of the spectrum, *Proc. Natl. Acad. Sci. U.S.A.,* 83, 7272, 1986.
32. **Davenport, D., Culler, G. J., Graves, J. O. B., Forward, R. B., and Hand, W. B.,** The investigation of the behavior of microorganisms by computerized television, *IEEE Trans. Biomed. Eng.,* 17, 230, 1970.
33. **Berg, H. C.,** How to track bacteria, *Rev. Sci. Instrum.,* 42, 868, 1971.
34. **Sundberg, S. A., Bogomolni, R. A., and Spudich, J. L.,** Selection and properties of phototaxis-deficient mutants of *Halobacterium halobium, J. Bacteriol.,* 164, 282, 1985.
35. **Sager, B. M., Sekelsky, J. J., Matsumura, P., and Adler, J.,** Use of a computer to assay motility of bacteria, *Anal. Biochim.,* 173, 271, 1988.

36. **Takahashi, T., Watanabe, M., Kamo, N., and Kobatake, Y.,** Negative phototaxis from blue light and the role of the third rhodopsin-like pigment in *Halobacterium cutirubrum, Biophys. J.*, 48, 235, 1985.
37. **Smith, E. A. and Phillips, D. R.,** Automated cloud tracking using precisely aligned digital ATS pictures, *IEEE Trans. Comput.*, 21, 715, 1972.
38. **Yachida, M., Asada, M., and Tsuji, S.,** Automatic analysis of moving images, *IEEE Trans. Pattern Anal. Mach. Intelligence,* 3, 12, 1981.
39. **Takahashi, T., Yan, B., Amici, L., and Spudich, J. L.,** Improved method for detection of behavioral responses of *Halobacterium halobium* to light stimuli using computerized motion analysis, in preparation.
40. **Kondoh, H.,** personal communication, 1989.
41. **Tsuda, M., Hazemoto, N., Kondo, M., Kamo, N., Kobatake, Y., and Terayama, Y.,** Two photocycles in *Halobacterium halobium* that lack bacteriorhodopsin, *Biochem. Biophys. Res. Commun.*, 108, 970, 1982.
42. **Hazemoto, N., Terayama, Y, Kobatake, Y., and Tsuda, M.,** Photochemistry of two rhodopsinlike pigments in bacteriorhodopsin-free mutant of *Halobacterium halobium, Biophys. J.*, 44, 59, 1983.
43. **Hazemoto, N., Terayama, Y., Kobatake, Y., Tsuda, M., and Terayama, Y.,** Effect of salt on photocyde and ion-pumping of halorhodopsin and third rhodopsinlike pigment of *Halobacterium halobium, Biophys. J.*, 44, 59, 1983.
44. **Spudich, E. N. and Spudich, J. L.,** Control of transmembrane ion fluxes to select halorhodopsin-deficient and other energy-transduction mutants of *Halobacterium halobium, Proc. Natl. Acad. Sci. U.S.A.*, 79, 4208, 1982.
45. **Spudich, J. L. and Bogomolni, R. A.,** Mechanism of color discrimination by a bacterial sensory rhodopsin, *Nature,* 312, 509, 1984.
46. **Hildebrand, E. and Schimz, A.,** Photosensory behavior of a bacteriorhodopsin-deficient mutant, ET-15, on *Halobacterium halobium, Photochem. Photobiol.*, 37, 581, 1983.
47. **Schimz, A., Sperling, W., Ermann, P., Bestiman, H. J., and Hildebrand, E.,** Substitution of retinal by analogs in retinal pigments of *Halobacterium halobium:* contribution of bacteriorhodopsin and halorhodopsin to photosensory activity, *Photochem. Photobiol.*, 38, 417, 1983.
48. **Traulich, B., Hildebrand, E., Schimz, A., Wagner, G., and Lanyi, J. K.,** Halorhodopsin and photosensory behavior in *Halobacterium halobium* mutant strain L-33, *Photochem. Photobiol.*, 37, 577, 1983.
49. **Narurkar, V. and Spudich, J. L.,** Evidence that the light-driven sodium pump is a phototaxis receptor in *Halobacterium halobium, Biophys. J.*, 33, 218a, 1981.
50. **Schimz, A., Sperling, E., Hildebrand, E., and Köler-Hahn, D.,** Bacteriorhodopsin and the sensory pigment of the photosystem 565 in *Halobacterium halobium, Photochem. Photobiol.*, 36, 193, 1982.
51. **Hildebrand, E. and Schimz, A.,** Consecutive formation of sensory photosystems in growing *Halobacterium halobium, Photochem. Photobiol.*, 38, 593, 1983.
52. **Takahashi, T., Mochizuki, Y., Kamo, N., and Kobatake, Y.,** Evidence that the long-lifetime photointermediate of s-rhodopsin is a receptor for negative phototaxis in *Halobacterium halobium, Biochem. Biophys. Res. Commun.*, 127, 99, 1985.
53. **Ohtani, H., Kobayashi, T., and Tsuda, M.,** Photocycle of sensory rhodopsin: a new precursor of sR_{370}, *Photobiochem. Photobiophys.*, 13, 203, 1986.
54. **Bogomolni, R. A. and Spudich, J. L.,** The photochemical reaction of bacterial sensory rhodopsin. I. Flash photolysis study in the one microsecond to eight second time window, *Biophys. J.*, 52, 1071, 1987.
55. **Hartmann, K. M.,** Action spectroscopy, in *Biophysics,* Hoppe, W., Lohmann, W., Markl, H., and Ziegler, H., Eds., Springer-Verlag, Berlin, 1983, 115.
56. **Lipson, E. D.,** Action spectroscopy of photoresponses in microorganisms, in *Biophysics of Photoreceptors and Photomovements in Microorganisms,* NATO ASI Series, Lenci, F., Colombetti, G., Häder, D.-P., Song, P.-S., and Ghetti, F., Eds., in press.
57. **Takahashi, T.,** Spectroscopy of sensory rhodopsins, in *Biophysics of Photoreceptors and Photomovements in Microorganisms,* NATO ASI Series, Lenci, F., Colombetti, G., Häder, D.-P., Song, P.-S., and Ghetti, F., Eds., in press.
58. **Tomioka, H., Takahashi, T., Kamo, N., and Kobatake, Y.,** Action spectrum of the photoattractant response of *Halobacterium halobium, Biochim. Biophys. Acta,* 884, 578, 1986.
59. **Spudich, J. L. and Bogomolni, R. A.,** Spectroscopic discrimination of the three rhodopsinlike pigments in *Halobacterium halobium, Biophys. J.*, 43, 243, 1983.
60. **Tomioka, H., Takahashi, T., Kamo, N., and Kobatake, Y.,** Flash spectrophotometric identification of a fourth rhodopsin-like pigment in *Halobacterium halobium, Biochem. Biophys. Res. Commun.*, 139, 389, 1986.
61. **Spudich, E. N., Sundberg, S. A., Manor, D., and Spudich, J. L.,** Properties of a second sensory receptor protein in *Halobacterium halobium, Proteins Struct. Function Genet.*, 1, 239, 1986.
62. **Nakamori, Y., Takahashi, T., Tomioka, H., and Kobatake, Y.,** Action spectrum of the photorepellent system 480 in *Halobacterium halobium* mutant strain lacking photobehavioral responses mediated by sensory rhodopsin, in preparation.

63. **Takahashi, T., Yan, B., Mazur, P., Derguini, F., Nakanishi, K., and Spudich, J. L.,** Color regulation in the archaebacterial phototaxis receptor phoborhodopsin (sensory rhodopsin II), *Biochemistry,* 29, 8467, 1990.
64. **Takahashi, T., Tomioka, H., Nakamori, Y., Tsujimoto, K., Kamo, N., and Kobatake, Y.,** Phototaxis and second sensory pigment phoborhodopsin in *Halobacterium halobium, in Proc. Yamada Conference XXI,* Hara, T., Ed., Yamada Science Foundation, Osaka, 1988, 149.
65. **Yan, B., Takahashi, T., Johnson, R., Derguini, F., and Nakanishi, K.,** All-*trans*/13-*cis* isomerization of retinal is required for phototaxis signaling by sensory rhodopsins in *Halobacterium halobium, Biophys. J.,* 57, 807, 1990.
66. **Takahashi, T., Yan, B., Nakanishi, I., and Spudich, J. L.,** Sensitivity increase in photophobic response of *Halobacterium halobium* reconstituted with retinal analogs, in preparation.
67. **Yan, B., Takahashi, T., and Spudich, J. L.,** Identification of signaling states of a sensory receptor by modulation of lifetimes of stimulus-induced conformations: the case of archaebacterial senory rhodopsin II, *Biochemistry,* in press.
68. **Oesterhelt, D. and Marwan, W.,** Change in membrane potential is not a component of the photophobic transduction chain in *Halobacterium halobium, J. Bacteriol.,* 169, 3315, 1987.
69. **Schimz, A.,** Methylation of membrane protein is involved in chemosensory and photosensory behavior of *Halobacterium halobium, FEBS Lett.,* 125, 205, 1981.
70. **Alam, M., Lebert, M., Oesterhelt, D., and Hazelbauer, G. L.,** Methyl-accepting taxis protein in *Halobacterium halobium, EMBO J.,* 8, 631, 1989.
71. **Marwan, W., Schäfer, W., and Oesterhelt, D.,** Signal transduction in *Halobacterium* depends on fumarate, *EMBO J.,* 9, 355, 1990.
72. **Tsuda, M., Nelson, B., Chang, C. H., Govindjee, R., and Ebrey, T. G.,** Characterization of the chromophore of the third rhodopsinlike pigment of *Halobacterium halobium* and its photoproducts, *Biophys. J.,* 47, 721, 1985.

Chapter 17

SIMULTANEOUS TRACKING OF MULTIPLE SUBJECTS

David B. Dusenbery

TABLE OF CONTENTS

I. INTRODUCTION

Studies of the behavior of organisms are plagued by large variations in the behavior of individuals. Usually one is most interested in the average behavior of a population and much less interested in individual variation. In many instances, the information sought concerns relatively simple questions about the movements of the subjects. When this is the case and a suitable group of subjects can be viewed simultaneously by a camera, it is possible to have a computer automatically track the movements of the individuals and average together appropriate data about their behavior. The summary data can then be reported continuously within seconds of its occurrence. These data could be used directly or used to control some event in real time.

This chapter will focus on experience in my laboratory in developing and using systems for the simultaneous tracking of the locomotion of many individual subjects and real-time reporting of statistical data on their behavior. Although our experience with this technique is limited to use with nematodes (round worms), it should be applicable to a much broader range of organisms.

II. METHODS

A. STRATEGIES

1. Simple Recognition

In order to simultaneously track multiple subjects, it is important to identify their positions rapidly. Consequently, a simple strategy has been used. In our system, the location of each subject is determined by the location of the first pixel that is found to be above threshold. During each round of tracking, the video frame is examined and if the pixel at the previous position of a subject is still above threshold, no movement is recorded. If the pixel is now below threshold, a movement is recorded for the subject and a search is initiated.

During the search, adjacent pixels in an ever-widening circle are examined out to a limit of a radius of 9 pixels. The first pixel found to be above threshold is recorded as the new location of the subject. If no pixels above threshold are found within the radius of 9, the search is abandoned. This search area, the magnification of the camera lens, and the time between rounds of tracking are adjusted so that the fastest moving subjects are not likely to get out of the search area before a new position is defined. With this tracking strategy, it is actually the trailing edge of the subject that is tracked.

This crude method does not detect all movements. If a subject reverses direction of locomotion, it will appear to be stationary until its head passes the point where the tail was located when the change of direction occurred. A more precise detection of movement can be obtained by tracking fewer subjects and determining their centers, but for many purposes it is advantageous to track more subjects.

2. Visual Feedback

Early experience in developing this technique demonstrated that it is very useful to have a video screen that overlays the image from the camera with indications from the computer of where the system has recorded the position of each subject. This allows the user to determine if the system is seeing dirt or noise rather than appropriate subjects and if it is tracking the subjects properly. To indicate the positions, we use markers that are single pixels that differ greatly in intensity or color from the camera image.

B. HARDWARE

1. Computer

The personal computers that are now so widely available have more than sufficient capacity to perform the basic functions necessary for simultaneous tracking. Our first system[1] used a computer based on the Motorola 6809 microprocessor with 72 kbytes of memory and the OS-9 operating system. Our second-generation system[2] used an IBM XT computer. And our third-generation system[3] used an IBM AT computer. Much faster personal computers are currently available.

If a particular application requires that subjects be viewed in an environment that makes them difficult to distinguish from the background or other objects, it may be helpful and important to use a real-time image processor to enhance contrast before the tracking software searches for the subjects. Such processors are generally quite expensive, but are becoming increasingly available. If contrast is a potential problem, it would be wise to select a computer that can accommodate such a processor.

2. Camera Interface

The electronic equipment to interface the camera to the computer is the most specialized and possibly the most expensive hardware requirement. In assembling a system from scratch, the availability of an appropriate camera interface should be the first consideration. Although hardware to interface video cameras to personal computers is now widely available, when our investigations started in the early 1980s, the interface was built in the lab. This interface digitized the video in a single scan in 1/60 s into an array of 240 × 256 pixels with 1 bit of intensity. A second, slower commercial interface was used to generate markers on a video screen to indicate where the computer was tracking the subjects.

The interface used with the IBM computers is the PCVISION frame grabber from Imaging Technology, Inc. (600 West Commings Park, Woburn, MA 01801). It digitizes a standard RS-170 video signal into 6 bits of intensity and stores a video frame of 512 × 480 pixels in its own memory in 1/30 s. In addition to its greater resolution, it contains a color video output with a programmable look-up table for mapping bytes of video memory to any of 16 million colors. This output is connected to an RGB color monitor (Electrohome model ECM 1901). This system allows markers to be displayed in colors different from the camera input and also allows the use of color to distinguish between input intensities that are above and below threshold.

3. Camera

The requirements of the camera are relatively simple. Although color cameras are widely available, color is rarely necessary and the inexpensive closed-circuit television (CCTV) cameras used in surveillance are quite satisfactory. Even if color were useful in distinguishing the subjects from background, a color filter would probably be a better solution than a color camera. If high sensitivity is required in order to work in low light, very sensitive surveillance cameras are available, although they cost thousands rather than hundreds of dollars.

If the subjects are small, the camera may need to be fitted with a macro lens or a microscope. Maximum flexibility is provided by a camera that accepts C-mount lenses, which most surveillance cameras do, in contrast to home video cameras. For subjects on the order of 1 mm in size, an enlarger lens may be good because it is highly corrected for a flat field at appropriate working distances. We used a Computar f 1.9, 55-mm lens for small nematodes.[2]

4. Illumination

A key element in making this technique fast and easy is to have the subjects stand out in high contrast from the background so that a simple threshold intensity criterion reliably distinguishes them. The appropriate strategy depends on the subjects and their environment. People viewed from above against a concrete courtyard might have sufficient contrast, but it could probably be improved by using color filters to select the color of light that gives the best contrast. Black mice should be viewed against a white background and vice versa.

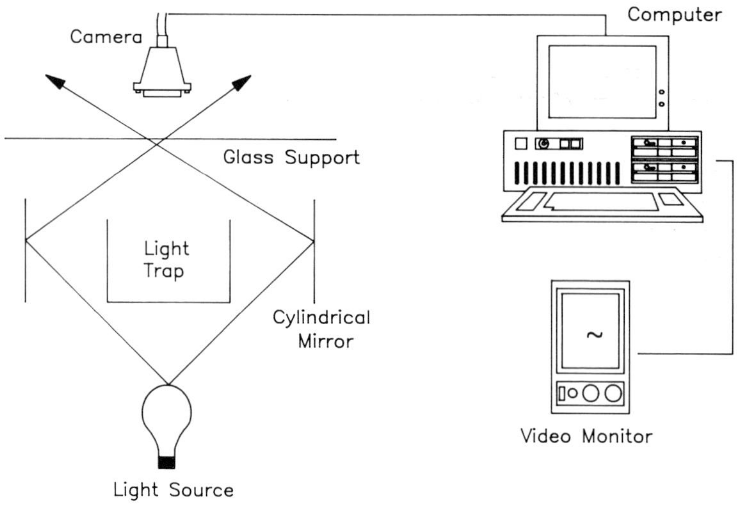

FIGURE 1. Computer tracking system with dark-field illumination. (From Dusenbery, D. B., *J. Nematol.*, 20, 351, 1988. With permission.)

In some circumstances it might be appropriate to attach retro-reflectors (e.g., glass beads as used on highway signs) to the subjects and illuminate them with a beam of light from the direction of the camera. For small organisms that scatter light, dark-field illumination is ideal. We have assembled such a system for use with nematodes ($\approx$1 mm long and translucent) (Figure 1).

C. SOFTWARE

1. Language

In principle, any computer language can be used as long as it can address the video memory. However, most applications will probably require a compiled language (unlike regular BASIC) in order to have sufficient speed to track multiple subjects. Our original system used BASIC09, a partially compiled and partially interpreted language, and some time-critical parts had to be written in assembly language. The later systems used Microsoft Pascal, and assembly-language modules were used only to provide access to I/O (input/output) ports not accessible from the higher-level language. Assembly language is probably not necessary in more recent versions of high-level languages.

2. Search Pattern

A key element in obtaining sufficient speed to make this system work is to avoid calculating the next nearest pixel to check during a search. Instead, a look-up table is generated that provides the relative positions of the pixels to be searched in order of increasing distance from the starting pixel. Thus, the search program merely has to work down the table until an appropriate pixel is located.

The generation of the search pattern starts with a table that contains pairs of positive integers (i, j) in order of increasing value of $i^2 + j^2$ up to but not including the pair 0, 9. Each pair of numbers is then used to generate four or eight sets of x- and y-offsets by assigning $x = \pm i$ and $y = \pm j$ in all combinations of sign and repeating with x and y reversed when $x \neq y$. The result is a table of x- and y-offsets that is in order of increasing distance from the origin. It includes every point inside a circle of radius 9.

Of course, different size areas of search patterns could be used, but 9 was considered to be a good compromise. In considering searching to larger distances, one should keep in mind that the time to search the whole area will increase with the square of the radius.

3. Measures of Behavior

In each tracking cycle, the search procedure provides the X- and Y-coordinates of each subject. These are compared against the previous values to determine the changed coordinates ΔX and ΔY, which are the raw data used to describe the behavior.

The most basic measure of behavior is the rate of locomotion, which is the distance moved divided by the time between tracking cycles. The distance would normally be calculated from $(\Delta X^2 + \Delta Y^2)^{1/2}$. However, calculating the squares and particularly the square root takes a lot of microprocessor time. We have avoided this delay by simply using the sum $(\Delta X + \Delta Y)$ for an approximation of the distance moved. Assuming a large number of subjects are moving in random directions, this simplification gives a constant overestimate of 27%, which is no problem for many applications.

In our initial interest in using this procedure to study chemotaxis, we wanted to measure the frequency with which the subjects changed direction of locomotion. However, a straightforward geometrical measure would have required far too much computer time. Consequently, a simple logical test was employed based on the current and previous signs of ΔX and ΔY and whether either of the current displacements was zero. A change in the direction of locomotion was identified when there was a change in direction along both axes or when there was a change in direction along one axis and no change in position along the other. This strategy avoided signaling a change in direction from a gradually curving path.

4. Finding Subjects

Some thought needs to be given to the procedure for identifying new subjects to track. Do you want the probability of a subject being tracked to be evenly distributed over the area viewed? Do you want to favor moving subjects or subjects moving toward the center? In our case, an even distribution was the main consideration. To obtain it, a position in the field of view is chosen at random and if that pixel is not above threshold, adjacent pixels are checked until a pixel above threshold is found or the time allowed for looking for new subjects is elapsed. If the image of the subjects includes several pixels, it is expedient not to check each adjacent pixel, but, rather, to skip over a few. The search is simplest if it proceeds linearly through the video memory without regard to how the memory maps to the two-dimensional field of view.

When a pixel with the appropriate intensity for a subject is found, the next step is to determine if it belongs to the image of a subject that is already being tracked. This is done by checking its position against the current positions of all the subjects being tracked. If it is within a certain distance of one of them, it is assumed to belong to the subject already being tracked and is not used. The distance used to make this decision is based on the size of the images of the subjects.

5. Collisions

When two subjects come together, there is no reliable way of determining which is which when they separate. In fact, with the procedures described so far, if both subjects are being tracked, when they separate, the computer may follow only one of the two images as two subjects. To avoid this problem, in each tracking cycle that determines new positions for each of the subjects being tracked, the position of each is compared against all the others and if any are identical, one of the two is abandoned. The choice as to which of the two to drop is arbitrary unless one has just registered a change of direction, in which case it is the one abandoned.

In most applications, it will be important to keep the density of the subjects sufficiently low that collisions occur infrequently. If there are so many subjects that collisions are frequent and many of the subjects are not tracked, the measure of the frequency of change of direction will be exaggerated by the system switching from one subject to another when they come together on paths of opposite direction and recording this as a change of direction when, in reality, neither subject changed direction.

III. APPLICATIONS

A. POSSIBLE RANGE OF APPLICATIONS

The technique of simultaneous tracking has a very wide range of applications. It should be applicable to bacteria which can be viewed by a video camera mounted on a compound microscope. Where studies of locomotion in two dimensions are sufficient, the approach would be straightforward. At the other extreme, simultaneous tracking of large animals should be possible if a camera can be positioned at an appropriate vantage point. For example it would appear straightforward to study the movement of people walking across a courtyard or automobiles with a camera looking down from a building.

In many cases, it would be desirable to track subjects in three dimensions. This will be much more difficult, and I am not aware that anyone has done simultaneous tracking of biological subjects in three dimensions. The simplest approach would be to view the volume with two cameras. The principle difficulty will probably be to identify corresponding images in the two cameras.

Simultaneous tracking of microscopic organisms in three dimensions faces additional problems because high resolution requires objective lenses of high numerical aperture, and there will be limits on how large the numerical aperture can be and still get the lenses sufficiently close to the subjects and at an angle to one another. One approach might be to use several cameras to examine the image from a single objective lens at different focal levels. Comparison of the intensity and spread of the image should provide a measure of the depth of the subject. A better approach would probably be to not try to resolve images of the subjects, but simply view them with two cameras in dark-field as bright spots smaller than the resolution of the optical system.

The possible range of application will be illustrated by presenting several examples of applications that have employed nematodes as subjects in the following sections.

B. STIMULI

1. Chemical Stimuli

The technique of simultaneous tracking was originally developed to study the response of nematodes to controlled changes in the concentrations of chemical stimuli. The previous technique which revealed the need for simultaneous tracking involved holding single nematodes by the tail with a suction pipet while various solutions were pumped past it and its shadow was projected on an array of light detectors wired to a multichannel recorder.[4] The chart was then examined by hand to identify behavior that normally leads to a change in the direction of locomotion. Testing for response to a chemical involved switching back and forth between two different concentrations of the chemical in the flowing solution and determining if behavior differed between the different exposures. This technique required many hours of recording to obtain statistically significant data for weak stimuli such as oxygen.[5]

This experience led to thinking about ways of obtaining data from many subjects simultaneously and ultimately to the development of the simultaneous tracking system. The nematodes are allowed to move around on the surface of a thin layer of agar across which a stream of hydrated air blows to carry volatile stimuli. Dark-field illumination provides a high-contrast image of the translucent nematodes. This technique provided the first method by which data on the behavior of a group of nematodes exposed to a controlled chemical stimulus could be obtained.[1] The original system could simultaneously track 25 animals with the position established once a second. Within a few minutes, statistically significant

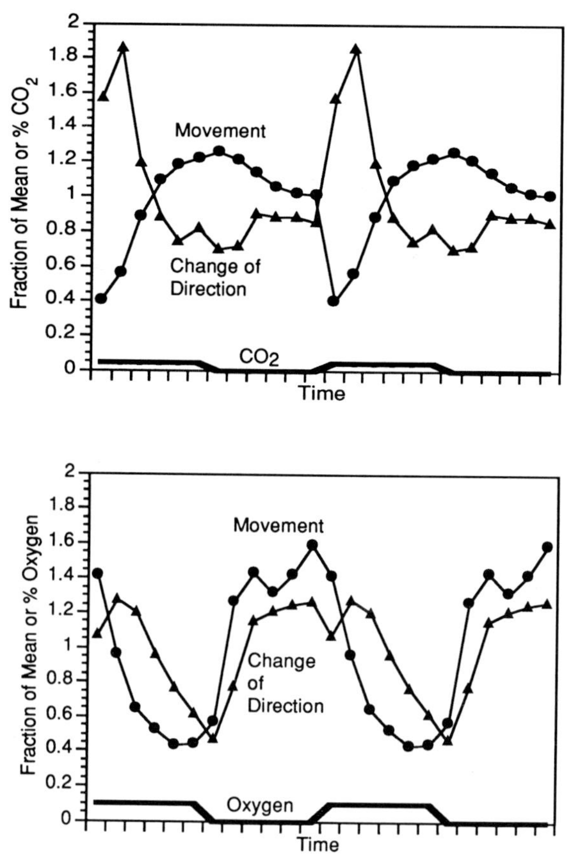

FIGURE 2. Behavioral responses to chemical stimulation. Average rates of locomotion and frequency of change of direction in response to cyclic changes in the concentrations of carbon dioxide or oxygen. In the upper graph, the concentration was changed back and forth between 0 and 4% carbon dioxide at 1-min intervals. In the lower graph, the oxygen concentration alternated between 0 and 10% on the same schedule. Shown are values averaged over 50 such stimulus cycles.[6]

data could be acquired that had previously required many hours to obtain with one nematode at a time.[6] The results (Figure 2) confirmed responses to oxygen and carbon dioxide and revealed very different patterns of adaptation, which is consistent with the suggestion that oxygen is sensed internally.[5]

Techniques were also developed to measure responses to spatial gradients. A parallel flow from two sources with different concentrations of the test chemical was established (Figure 3). Diffusion across the flow was sufficient to establish a nearly linear gradient in the vicinity of the nematodes. The software was modified to measure separately movements across and parallel to the flow. The results demonstrated clearly that the nematodes moved toward higher concentrations of carbon dioxide.[2]

2. Thermal Stimuli

Plant-parasitic nematodes migrate in very shallow temperature gradients[7] toward a preferred temperature.[8] They can respond to temperature gradients of 0.001°C/cm.[9] Questions about the mechanism of the response led to the use of simultaneous tracking experiments in which temperature was altered by shining an incandescent lamp on the agar containing

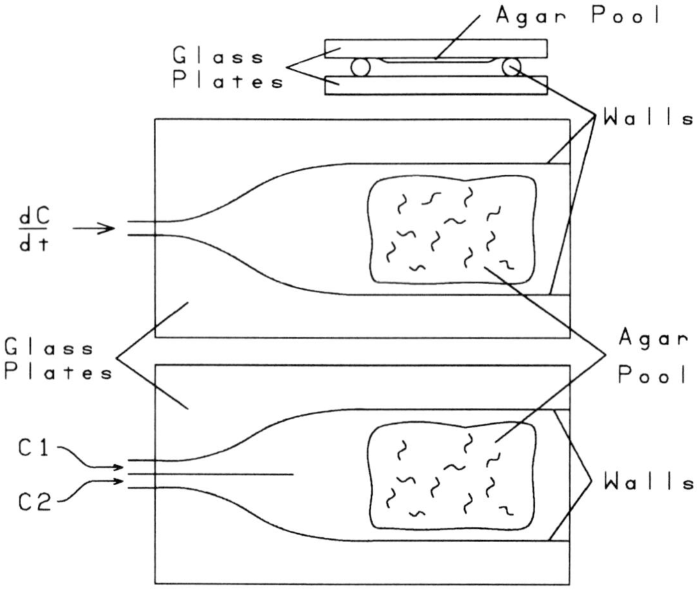

FIGURE 3. Diagram of the two types of flow chambers used with nematodes. The top figure is a cross-section of either chamber. The middle figure is the chamber used to generate a temporal, spatially uniform stimulus. The lower figure is the chamber used to generate a steady spatial gradient. (From Pline, M. and Dusenbery, D. B., *J. Chem. Ecol.*, 13, 873, 1987. With permission.)

the nematodes.[10] Turning the lamp on and off generated temperature changes that were estimated to be on the order of 10^{-4} °C/s. These experiments employed our second-generation system, which could simultaneously track 200 nematodes with positions established every 3 s.

As seen in Figure 4, behavior changed within about 10 s of the lamp being turned on or off. Consequently, the nematodes responded to a temperature change on the order of 0.001°C. Temperature changes away from the preferred temperature resulted in decreases in the rate of movement and increases in the rate of change of direction, whether the changes were toward warmer or cooler temperatures. These behavioral changes lasted about 30 s. Temperature changes toward the preferred temperature caused the response rates to change in the opposite directions. In this case, the behavioral changes persisted for several minutes. These results demonstrate that nematodes can respond to a purely temporal thermal stimulus in a manner consistent with efficient indirect orientation of klinokinesis. They further demonstrated very clearly different rates of adaptation for increases and decreases in stimulation, which has been suggested to lead to more efficient responses.[11]

C. BEHAVIORAL TOXICITY

Toxicity testing of environmental samples and synthetic chemicals is a problem of growing concern. Most such tests use survival of a period of exposure as a measure of toxicity. However, it is important to determine whether significant nonlethal effects occur at lower levels than those that are lethal. One of the most important areas of sublethal effects involves effects on the nervous system and behavior. Behavioral endpoints are desirable but, because of the large variability in behavior, difficult to carry out. Nematodes have been found useful in toxicity testing.[12] They are as sensitive as other organisms used in testing water toxicity[13] and seem to have good predictive values for human toxicity.[14] It seemed to us that simultaneous tracking offered an opportunity to obtain large amounts of behavioral information in a short time and relatively inexpensively.

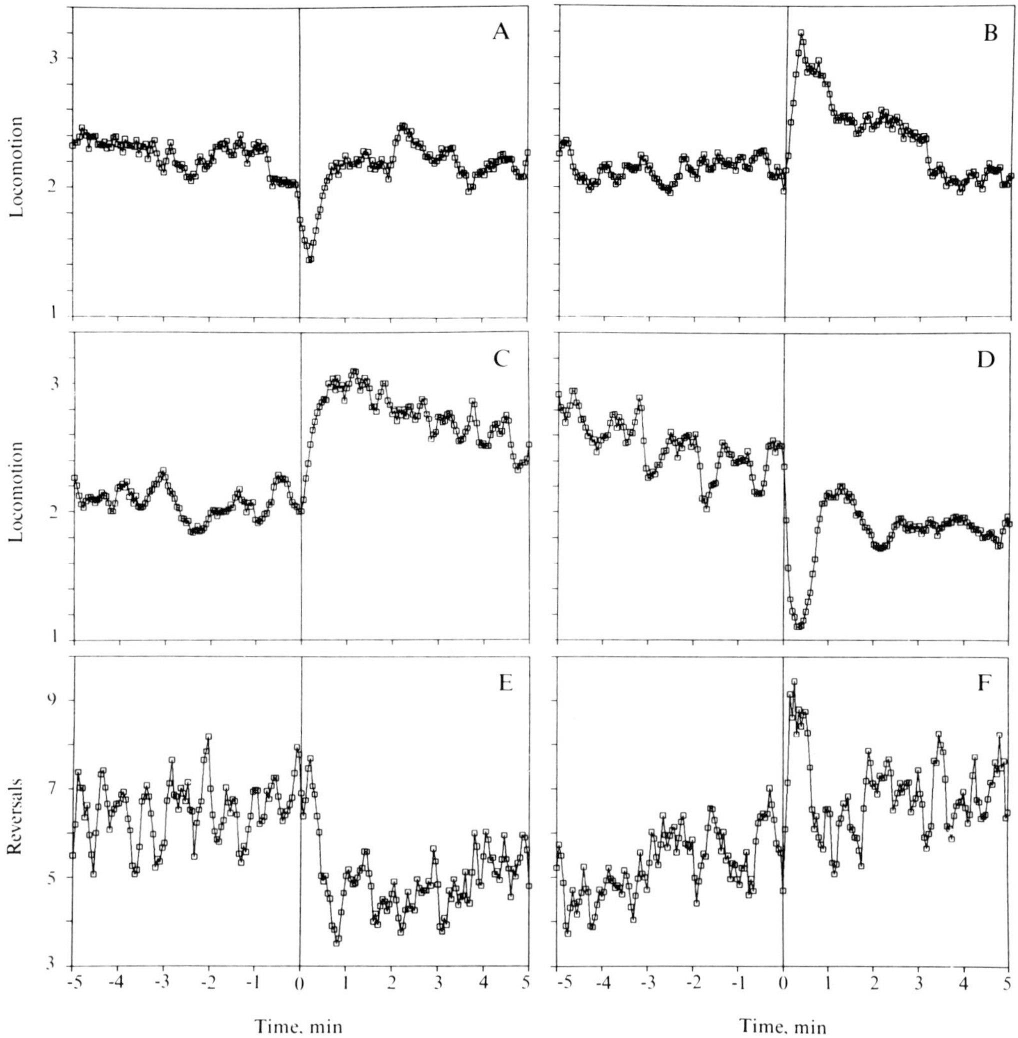

FIGURE 4. Changes in behavior of second-stage juveniles of *Meloidogyne incognita* caused by changes in temperature. The nematodes were grown at 23°C and consequently tended to move toward a "preferred" temperature of about 26°C.[8] Temperature changes were generated by turning a red-glowing incandescent lamp on (left-side plots) or off (right-side plots) at zero time. Experiments A and B were conducted at an ambient temperature of 29°C, and the others were conducted at 23°C.[10] Notice that a change in temperature toward the preferred temperature (B, C, E) resulted in an increase in the rate of locomotion and decrease in the frequency of changes of direction, and that these changes persisted for several minutes. In contrast, a change away from the preferred temperature (A, D, F) resulted in a decrease in locomotion and an increase in change of direction, and these changes in behavior lasted less than 1 min.

The nematodes were exposed to the test chemicals for 24 h and then transferred to a thin pool of agar under the video camera. We used our third-generation system, which could simultaneously track up to 300 subjects with positions established once a second. The average rate of locomotion was determined for animals exposed to the chemical being tested and to a parallel control group that was treated similarly except for being exposed to the chemical under test. The ratio of the rate of locomotion in the exposed animals to that of the control group was used as the basic indication of neurobehavioral toxicity. This parameter was determined for a number of concentrations of the chemical to develop a dose-response curve, which was then compared to the survival curve.

Initial testing with metallic ions[3,15] demonstrated behavioral effects at concentrations well below those that caused lethality for the neurotoxins lead and mercury (Figures 5b and c), but not for the nonneurotoxin copper (Figure 5a). Surprisingly, mercury produced hyperactivity in the nematodes, suggestive of its effects on humans.

D. GAS CHROMATOGRAPHY DETECTION

One of the strengths of simultaneously tracking is that statistically significant behavioral data can be obtained in a few seconds. This makes it attractive for use as a detector for gas chromatography. We have developed this application in order to search for volatile stimuli that plant-parasitic nematodes might use to locate the roots of host plants.

The setup is shown in Figure 6. A sample of gas was collected from around the roots and injected into the gas chromatograph. A temperature program was used to separate components of a wide range of volatility. As they eluted from the column, they passed through a thermal conductivity detector and then were carried in an air stream over nematodes on an agar slab. Data on the average distance moved and the number of changes of direction detected, and from the thermal conductivity detector, were collected by the computer and plotted together (Figure 7). These experiments indicated that carbon dioxide was the only volatile stimulus released by the roots, although the nematodes also respond to oxygen depletion.

IV. SUMMARY

Simultaneous tracking is a technique that can now be implemented at relatively low cost on personal computers. It provides a method of obtaining large amounts of behavioral data in short time periods. This permits automatic processing by the computer and presentation of results in nearly real time.

A wide range of applications has already been demonstrated. Future developments might include tracking in three dimensions or use of the behavioral data to control experimental parameters in real time.

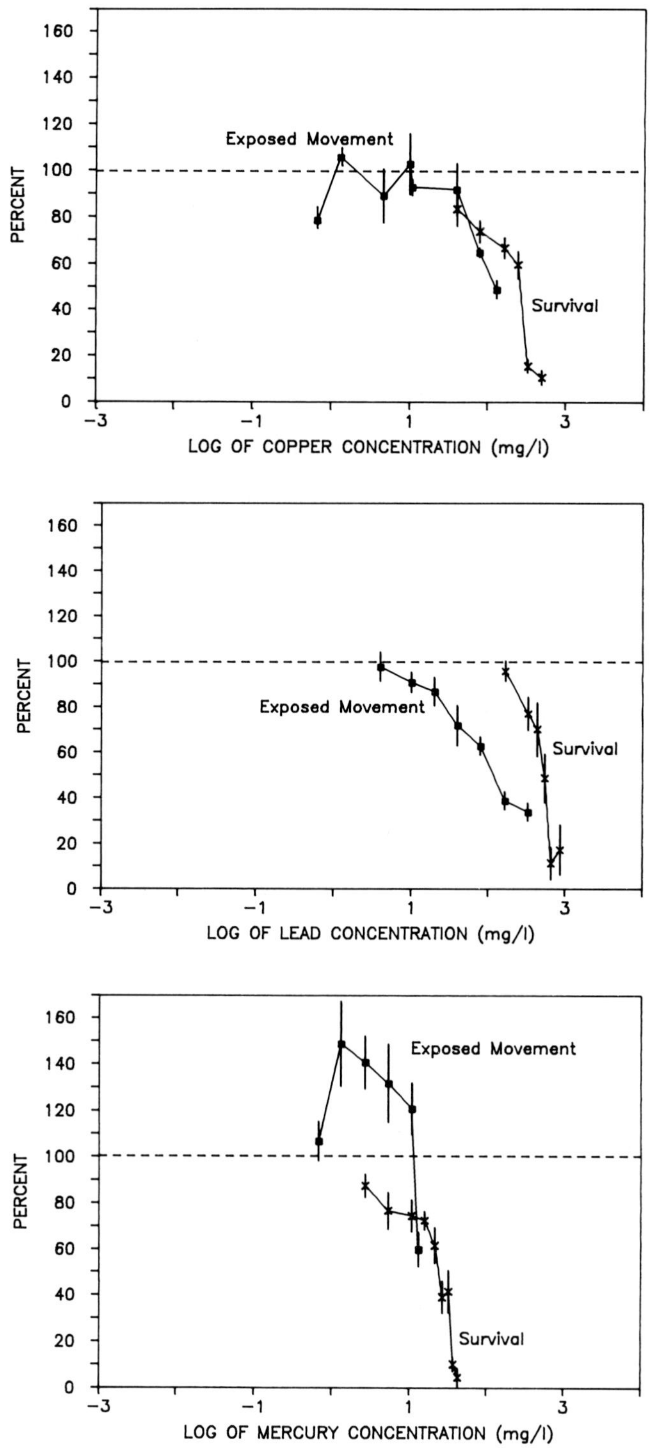

FIGURE 5. Rate of locomotion and survival of nematodes exposed to common salts of three different metals: copper, lead, and mercury. Values plotted are the percent survival or the average rate of locomotion of the nematodes exposed to the metal salt as a percent of the locomotion of nematodes from a parallel control that were not exposed. (From Williams, P. L., and Dusenbury, D. B., *Toxicol. Ind. Health,* 6, 425, 1990. With permission.)

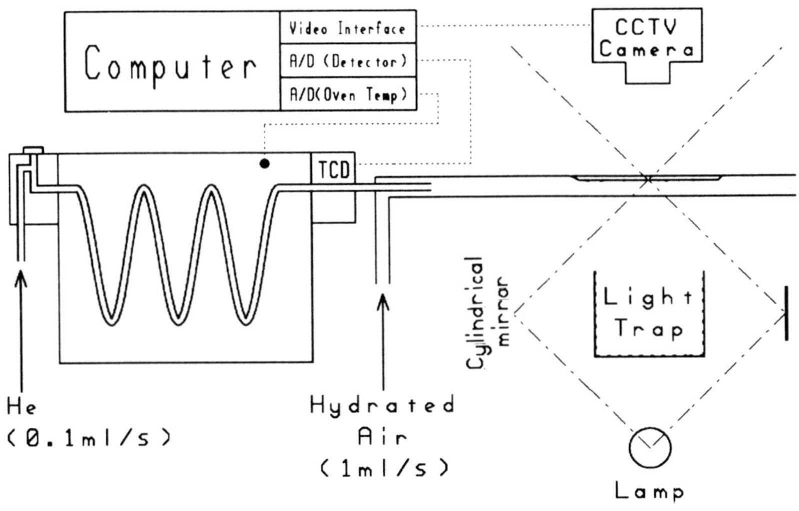

FIGURE 6. Diagram of system for using simultaneous tracking with a gas chromatograph.

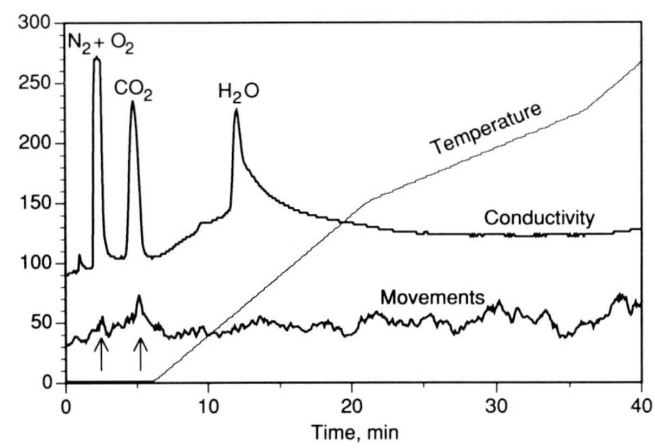

FIGURE 7. Data from using simultaneous tracking as a detector for gas chromatography. Vapor from plant roots was injected at time zero. The GC oven was held at 4°C for 7 min and then increased as indicated by the thin line. The upper thick line is a plot of the response of the thermal conductivity detector. It shows a small injection artifact and large peaks due to oxygen and nitrogen (which are not separated), carbon dioxide, and water. Other components are below detection. The lower thick line is a plot of the average rate of locomotion of a large number of nematodes. The only reproducible changes in locomotion are the two increases marked by arrows. Experiments with pure gases indicate that there are clear responses to oxygen and carbon dioxide.[2,16]

REFERENCES

1. **Dusenbery, D. B.,** Using a microcomputer and video camera to simultaneously track 25 animals, *Comput. Biol. Med.*, 15, 169, 1985.
2. **Pline, M. and Dusenbery, D. B.,** Responses of the plant-parasitic nematode *Meloidogyne incognita* to carbon dioxide determined by video camera-computer tracking, *J. Chem. Ecol.*, 13, 1617, 1987.

3. **Williams, P. L. and Dusenbery, D. B.,** A promising indicator of neurobehavioral toxicity using the nematode *Caenorhabditis elegans* and computer tracking, *Toxicol. Ind. Health,* 6, 425, 1990.
4. **Dusenbery, D. B.,** Responses of the nematode *Caenorhabditis elegans* to controlled chemical stimulation, *J. Comp. Physiol.,* 136, 327, 1980.
5. **Dusenbery, D. B.,** Appetitive response of the nematode *Caenorhabditis elegans* to oxygen, *J. Comp. Physiol.,* 136, 333, 1980.
6. **Dusenbery, D. B.,** Video camera-computer tracking of the nematode *Caenorhabditis elegans* to record behavioral responses, *J. Chem. Ecol.,* 11, 1239, 1985.
7. **El-Sherif, M. and Mai, W. F.,** Thermotactic response of some plant parasitic nematodes, *J. Nematol.,* 1, 43, 1969.
8. **Diez, J. A. and Dusenbery, D. B.,** Preferred temperature of *Meloidogyne incognita, J. Nematol.,* 21, 99, 1989.
9. **Pline, M., Diez, J. A., and Dusenbery, D. B.,** Extremely sensitive thermotaxis of the nematode *Meloidogyne incognita, J. Nematol.,* 20, 605, 1988.
10. **Dusenbery, D. B.,** Behavioral responses of *Meloidogyne incognita* to small temperature changes, *J. Nematol.,* 20, 351, 1988.
11. **Dusenbery, D. B.,** The value of asymmetric signal processing in klinokinesis, *Biol. Cybern.,* 61, 401, 1989.
12. **Samoiloff, M. R. and Bogaert, T.,** The use of nematodes in marine ecotoxicology, in *Ecotoxicological Testing for the Marine Environment,* Persoone, G., Jaspers, E., and Claus, C., Eds., State University of Ghent and Institute of Marine Scientific Research, Bredene, Belgium, 1984, 407.
13. **Williams, P. L. and Dusenbery, D. B.,** Aquatic toxicity testing using the nematode, *Caenorhabditis elegans, Environ. Toxicol. Chem.,* 9, 1285, 1990.
14. **Williams, P. L. and Dusenbery, D. B.,** Using the nematode *Caenorhabditis elegans* to predict mammalian acute lethality to metallic salts, *Toxicol. Ind. Health,* 4, 469, 1988.
15. **Williams, P. L. and Dusenbery, D. B.,** Screening test for neurotoxins using *Caenorhabditis elegans,* in *Model Systems in Neurotoxicology,* Shahar, A. and Goldberg, A. M., Eds., Alan R. Liss, New York, 1986, 163.
16. **McCallum, M. E. and Dusenbery, D. B.,** unpublished data, 1988.
17. **Pline, M. and Dusenbery, D. B.,** Responses of plant-parasitic nematode *Meloidogyne incognita* to carbon dioxide determined by video camera—computer tracking, *J. Chem. Ecol.,* 13, 873, 1987.

Chapter 18

IMAGE ANALYSIS OF MOTILE FLAGELLATES

Paolo Gualtieri and Primo Coltelli

TABLE OF CONTENTS

I. INTRODUCTION

The exact determination of the speed parameters of swimming microorganisms can be a very useful tool for the study of both behavioral and physiological aspects of motility. In this field or research, a complete and automatic analysis of the trajectories of microorganism motion under light or chemical stimuli could offer many advantages.

Usually, a wide variety of natural processes can be investigated by analyzing videotapes of microorganisms moving on a microscope slide. The article of Yashida et al.[1] provides a good review of the problems involved and of their possible solutions in dealing with video-recorded images. The procedure these authors apply consists of a preliminary coarse analysis in order to find a first location of the moving objects and of a successive fine analysis which exactly detects moving cells and measure quantitative parameters. The last step of this procedure is a control analysis of those objects which were misinterpreted during the tracking process. Other digital tracking microscopes, which determine speed parameters by reconstructing the entire movements of swimming microorganisms, use a videotape.[2,3] These systems detect the cells in the frames by means of a labeling procedure, and for each detected cell they reconstruct the trajectories by joining their baricenters in the successive frames. However, even if image analysis gives satisfactory results, the systems based on the video recording are time consuming because of the time necessary to advance the video recorder. As the number of reconstructed microorganism tracks should be high (greater than 500) to achieve accurate statistical results, the time necessary for the analysis is a fundamental constraint. Attempts to shorten the time of analysis are offered by the system of Häder and Lebert,[4] which extracts and follows one microorganism at a time in the microscope field, without using the videotape, or by the system described by Kondo et al.,[5] which stores successive video frames on a floppy disk and thins the recorded tracks; the process is completed by a tracing procedure which searches for suitable tracks.

The system we describe links a fast analysis with an exact track reconstruction. In fact, our system detects moving microorganisms, reduces the size of the acquired frames, stores them in a RAM board in real time (40 ms per frame, using the International Radio Consultative Committee [CCIR] standard), and then traces the trajectories of the moving objects by means of a fast algorithm. In order to detect moving cells, our method performs a temporal difference between successive frames; this temporal difference has already been applied by other authors for the detection of motion.[6-8] However, our system is much more simple, and the delay between two frames can be properly chosen in order to avoid the overlapping of the images of the same cell in successive frames.[9] The flagellate *Euglena gracilis* has been used as experimental subject.

II. HARDWARE CONFIGURATION

Our system consists of a Zeiss Axioplan microscope (Zeiss, Oberkochen, FRG), equipped with a black and white TV camera (saticon target, CCIR standard, Nical, Italy). The microorganisms can freely swim within a narrow layer of growth medium placed between a slide and a cover slip, in order to avoid out-of-focus movements.

The microscope field is acquired by the TV camera, and its video signal is digitized by an image system board (Imaging Technology, U.S.) which is plugged into the bus of a personal computer (IBM, U.S.).

The board consists of a frame memory, look-up tables (LUTs), and an A/D converter (Figure 1). In the 1024 × 1024 × 12-bit board, the operator can select a window of 640 × 512 × 8 bits. The size corresponds to a digitized image with equal vertical and horizontal spacing, according to the CCIR standard. The image can be reduced, panned horizontally, scrolled up and down, and stored in a selected position of the frame memory by means of

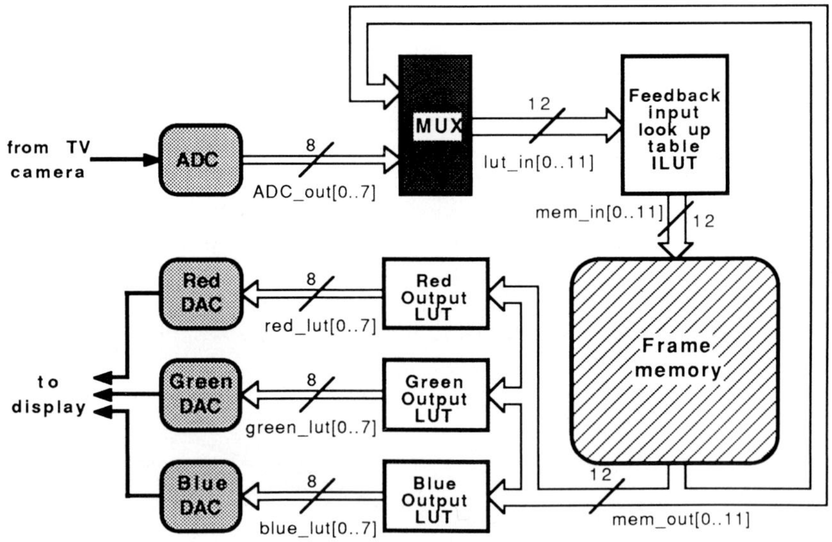

FIGURE 1. Configuration of the programmable image board.

the zoom, pan, and scroll hardware capabilities of the board. Up to 192 images can be stored in the frame memory. A black and white monitor (Nical, Italy) is used to display the signal output of the TV camera, while an RGB high-persistence monitor (Mitsubishi Model C-3479, Japan) is used to display the digital image.

The menu-driven software, whose commands are selected by means of a mouse, is written in the C language (Microsoft, U.S.).

III. OPERATION PROCEDURES

A. LOOK-UP-TABLE PROGRAMMING

Real-time detection of moving objects is performed by subtracting continuously each frame of the video image from the previous frame, with an experimentally determined delay. This operation is performed by programming the 12-bit input LUT of the board. This LUT, which is located between the digitization circuit and the frame memory, transforms the image before it is stored in the frame memory (Figure 1). Thanks to a feedback circuit between the frame memory and the input LUT, operations are made on combinations of stored and newly acquired data.

We programmed the input LUT in order to move the six most significant bits from the A/D converter (these bits represent the newly acquired data) to the six most significant bits of the frame memory; then the LUT subtracts the newly acquired data from the data previously stored in the six most significant bits of the frame memory (these bits represent the previously acquired data). The resulting six bits are then stored in the least significant bits of the frame memory (Figure 2). The difference image produced by these operations is always available in the lower six 1-bit planes of the frame memory, while the upper six 1-bit planes contain the most recent data of the analog-to-digital (A/D) converter, which are used as input for the next frame subtraction. When the images of a moving cell in two successive frames are partly overlapping, the subtraction operation gives a zero value for the overlapping region of the cell and for the background, a positive value for that part of the cell image which is present only in the newly acquired frame, and a negative value for that part of the cell image which is present only in the previously acquired frame. The input LUT is programmed to

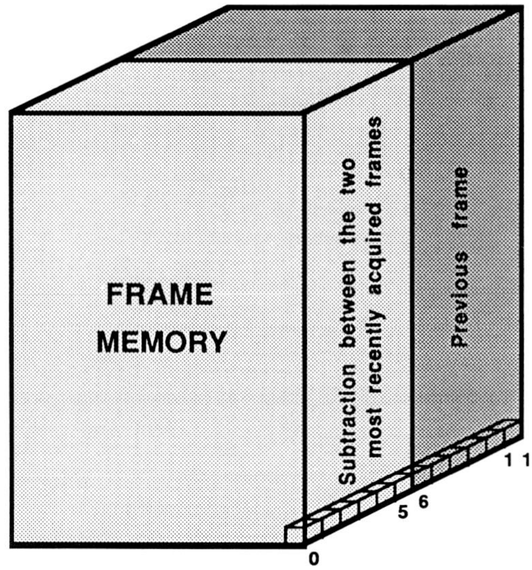

FIGURE 2. Frame memory partition between the image resulting from the subtraction operation and the stored image.

clip the negative values to zero. However, for a further simplification of this problem, we choose the delay between the two frames so that the two images do not overlap by testing the object area variation in the difference images. For the example shown in this chapter, a 480-ms delay was used. The output LUT displays the lower six 1-bit planes using a gray scale.

B. STORING MULTIPLE FRAMES

In order to store several difference images in the frame memory (in our examples, the difference images are 12), we reduce the spatial resolution of the image being acquired by means of the hardware zoom. The result of this operation is a 320 × 256-pixel image, whose size corresponds to half of the original image size. In order to store the reduced image into its proper position of the frame memory, the x- and y-coordinates of its origin are shifted by means of the hardware pan and scroll operations. In this way, we can store 12 images, by moving the coordinates of their origin toward the right and downward. At the end of this procedure, the frame memory is displayed as a patchwork of reduced images. Due to our hardware facilities, the zooming and storing operations are performed during the frame acquisition. The time necessary for the 12-image reduction depends only on the delay between the two frames on which the temporal difference has been performed; in our example, the acquisition time is 6 s. The 12-image reduction was selected because, during this period, the microorganisms swim across the whole microscope field. By using a greater reduction, more images could be stored, up to 192; a further reduction would be meaningless because of the very low spatial resolution achievable.

C. DETECTION OF MOVING OBJECTS

In order to identify moving cells and to extract features such as baricenter coordinates and areas, a segmentation and a labeling procedure are applied to each difference image. A method to recognize objects in the image is to calculate the gray-level histogram of the difference images. These histograms can easily be interpreted because they are bimodal, with a sharp dip between the peak corresponding to the objects and the peak corresponding to the background; in our case, a gray level equal to 5 isolates the cells from the background.

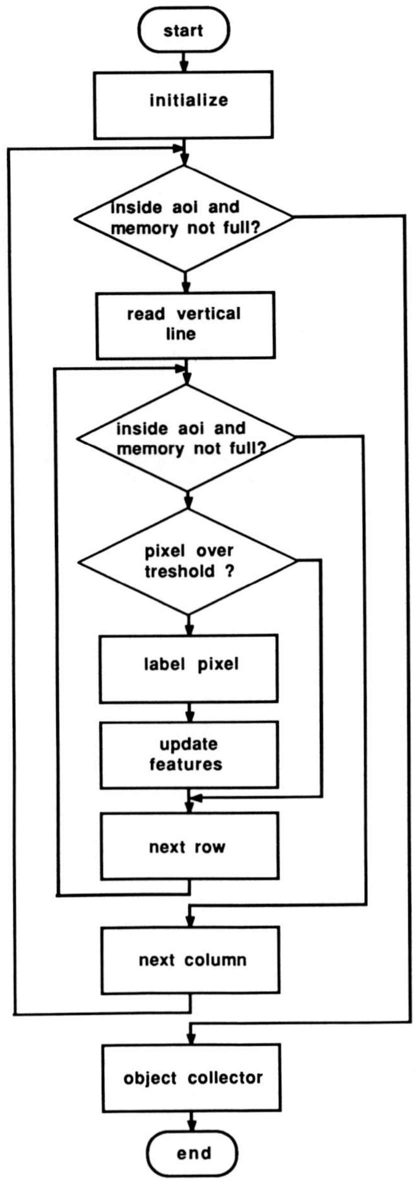

FIGURE 3. Flow chart of the labeling procedure.

Therefore, this value is used as the threshold value in the segmentation procedure.

The labeling algorithm performs a complete description of an image (Figures 3 and 4). Because the labeling process of connected regions is a typical recursive procedure, which requires backtracks at each point when regions are found to recombine, a second-pass analysis is necessary through the entire image to complete the process. Our algorithm combines in a single pass the labeling of connected regions with the generation of a data structure containing global features of all labeled regions. Therefore, backtrack from the connected points to previously labeled domains is avoided. Moreover, the speed of this algorithm depends on the handling of the random memory of a personal computer as if it were an associative memory, thus allowing parallel computation, since a single instruction can change the content of all the cells having the same content.[10]

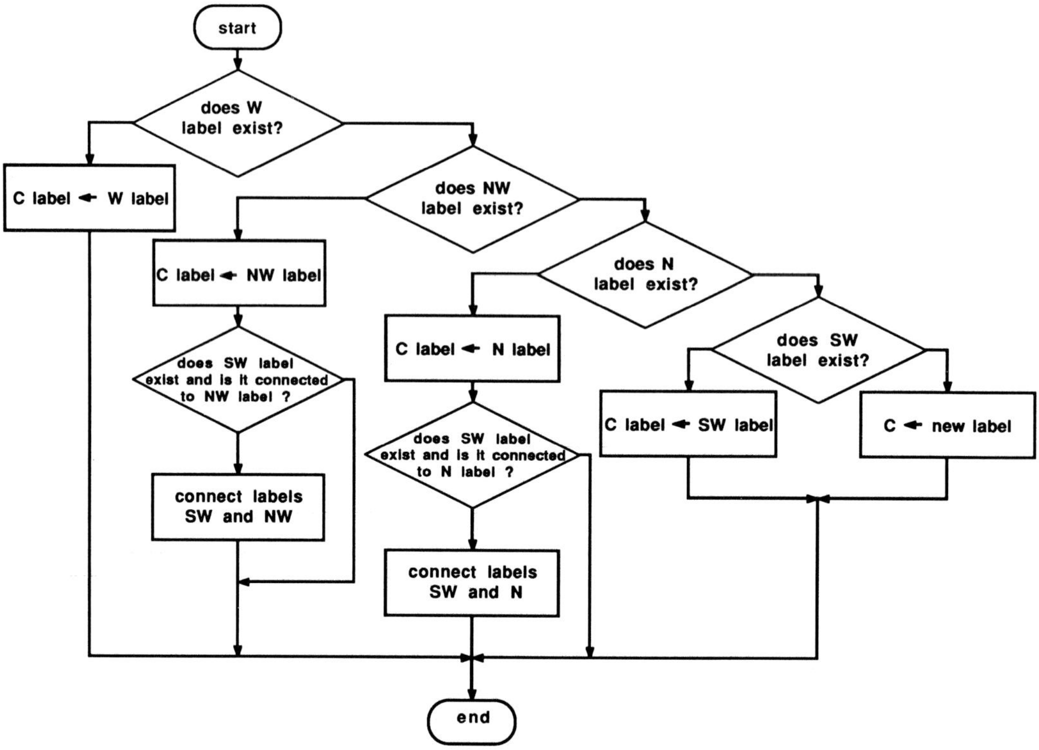

FIGURE 4. Flow chart of the procedure "label pixel".

D. TRACK RECONSTRUCTION

The data structure generated by the labeling algorithm has a list organization; for each difference image, the data structure is composed of a pointer, which points to a list of elements, and this list of elements. The number of elements of each list corresponds to the number of labelized cells in the corresponding difference image. The lists of the data structure are not of equal length because of the different number of cells present in each difference image. A list element consists of the cell baricenter coordinates, the cell area value, and a pointer to the successive elements of the list. The list pointers have a chronological order.

The algorithm of track reconstruction consists of two procedures which generate a new data structure, whose lists represent the reconstructed tracks. The first procedure of the reconstruction algorithm selects and removes the first element which is pointed to by the first valid pointer in the data structure generated by the labeling algorithm. This element represents the starting element of the first track list, i.e., the first point of the track which is pointed at by a header. Each header consists of the number of the starting difference image, the number of track elements, a pointer to the first element of the list, and a pointer to the header of the successive track.

Starting from the element selected by the first procedure, the second procedure scans the successive list of the data structure in order to find and remove the element whose baricenter lies in a preestablished distance range, and whose calculated motion vector has a preestablished angle value with respect to the forecast motion vector. The distance range can vary depending on the speed of the microorganisms. The last calculated motion vector is assumed to be the forecast motion vector. If this element does not exist, the track ends; if it exists, it represents the second element of the track list, i.e., the second point of the track. In the latter case, the second procedure repeats its operations on the successive lists iteratively until all the elements of the track being reconstructed are removed from the data

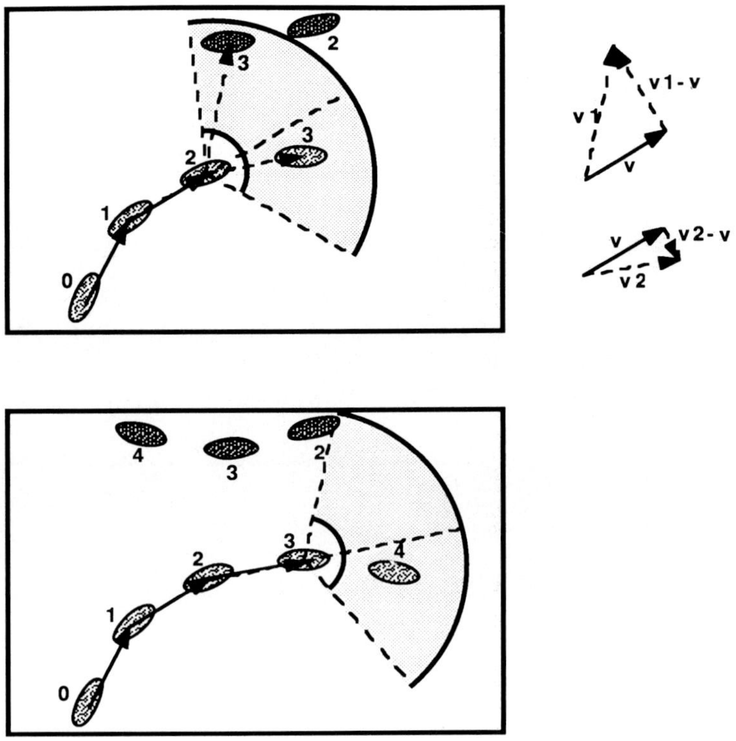

FIGURE 5. How a cell finds its successive position in the track, when more than one cell is present in the zone where the distance range and the angle range criteria are satisfied (in gray). The difference between each of the possible vectors and the last motion vector is calculated. The element whose baricenter has the smallest deviation will then be selected.

structure generated by the labeling procedure. It can happen that the second procedure encounters more than one element whose baricenter satisfies the selection criteria, i.e., more than one motion vector could be constructed; in this case, we choose the baricenter which allows us to construct the vector with the smallest deviation from the forecast motion vector. For this purpose, the difference between each of the possible vectors and the last motion vector is calculated. The element whose baricenter has the smallest deviation will then be selected (Figure 5).

When a track is completed, the first procedure selects a new starting element, and the second procedure uses this starting element to reconstruct another track. These operations are performed until all elements are removed from the data structure generated by the labeling procedure.

E. OCCLUSION PROBLEM AND BACKTRACKING OF MISINTERPRETED TRACKS

The reconstruction of the tracks of microorganisms freely swimming in a microscope slide can present the problem of overlapping cells, i.e., intersecting tracks. However, the peculiar characteristics of our experimental apparatus allow us to minimize this problem. The microscope has a low depth of focus, and the microorganisms swim within a narrow layer of medium between a slide and a cover slip. In this situation, a cell which turns along its major axis in the focused layer usually is not occluded by cells that swim in the out-of-focus layers, because the microscope does not focus on them. However, it can happen that two cells hide one another, but these rare events are left unsolved in the routine work since,

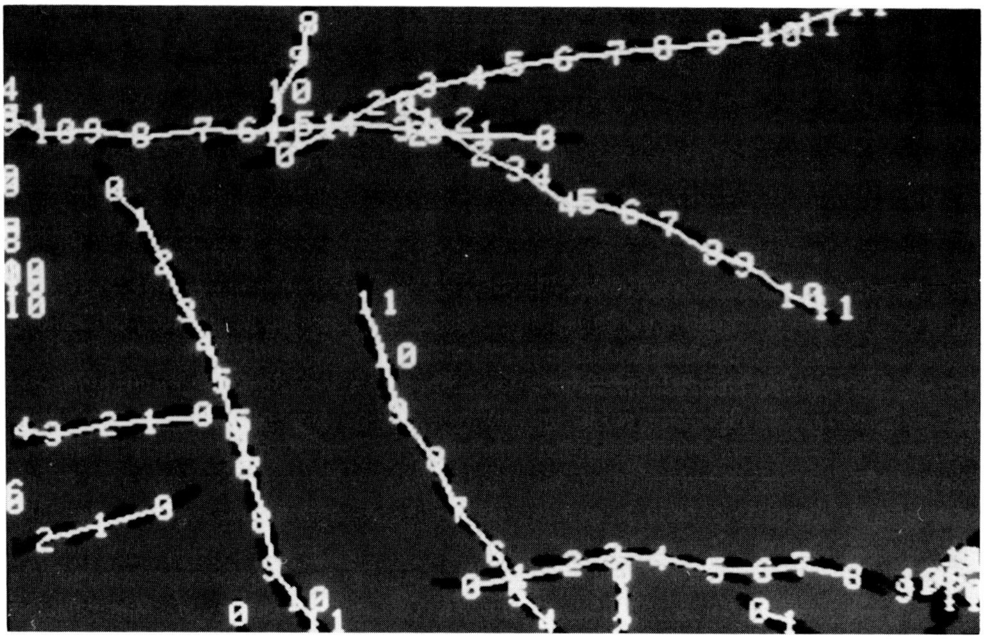

FIGURE 6. Zoom of the first difference image on which the reconstructed trajectories are shown. The cells are represented by the dark areas and are labeled with the number of the corresponding difference image superimposed on their baricenter. The reconstructed trajectories are the white lines.

for this kind of analysis, it is important to reconstruct more than 500 tracks rather than analyzing all tracks. In these cases, the labeling algorithm detects a cell with an area value about twice that of the average area value, but only one baricenter. Therefore, this baricenter is inserted in the track of one cell, while the other cell will have two tracks: the first track stops just before the intersection, and the second track starts just after the intersection. For the statistical analysis our samples demand, it is of no importance to link the two sections of the incomplete track. Therefore, in order to save time, in the routine work, we do not implement any procedure for a second-step analysis of incomplete tracks.

However, for a more precise analysis, we apply a procedure to solve these misinterpreted tracks. Because the data structure contains the positions of the cells, we can calculate the motion vectors in each difference image, and on the basis of this calculus, we can forecast the behavior of the cells. As the choice of the delay avoids the occurrence of the occlusion phenomena in more than one difference image, the procedure scans the headers of the tracks in order to find a track which ends in a difference image and a track which starts two difference images later. If this couple of tracks is found, the remaining tracks are then scanned in order to find a track element in the intermediate difference image which could join the two selected tracks. Therefore, its baricenter coordinates have to satisfy the distance and angle criteria, and its area value has to be about twice the average area value. When a track element with these characteristics is found, the two sections of the incomplete track can be joined.

Figure 6 is a zoom of the first difference image in which the reconstructed trajectories are shown for a 12-image sequence. In order to verify the accuracy of the reconstructed tracks, we apply a logical operation to the first stored frame and to all successive frames, which resembles a long time exposure with a photographic camera. The cells are represented by the dark areas and are labeled with the number of the corresponding difference image superimposed on their baricenter. The reconstructed tracks are the white lines. The computation time of the whole procedure depends on the number of the cells swimming in the microscope field. For the case shown in this chapter, it was 0.3 s per frame.

IV. CONCLUSIONS

This paper describes a digital system designed for the automatic, real-time detection and tracking of moving microorganisms in images acquired by means of a common TV camera mounted onto a microscope. The main characteristics of this system are

1. It performs a real-time gray-level difference between two successive frames in order to detect moving objects and to suppress stationary objects (subtraction procedure); the delay between two successive frames is chosen depending on the speed of the investigated microorganism.
2. It reduces the size of the images resulting from the subtraction procedure (difference images) and stores them in the frame memory; the result of these operations, all performed in real time, is a film of time sequences.
3. It performs an automatic labeling in order to recognize the moving microorganisms and to calculate their baricenter coordinates and their area in each difference image.
4. It calculates and plots the tracks of the cells moving in the microscope field, by joining with straight lines the baricenters of each cell in the time sequence.
5. It resolves ambiguous tracks using information such as cell positions and motion vectors obtained in previous frames.
6. It completes the analysis in only a few seconds.

ACKNOWLEDGMENT

This work has been partially supported by ''Progetto Finalizzato Sistemi Informatici e Calcolo Paralello'' of Consiglio Nazionale delle Ricerche (CNR).

REFERENCES

1. **Yashida, M., Asada, M., and Tsuji, S.,** Automatic analysis of moving images, *IEEE Trans. Pattern Anal. Mach. Intelligence,* 3 (1), 12, 1981.
2. **Gualtieri, P., Colombetti, G., and Lenci, F.,** Automatic analysis of the motion of microorganisms, *J. Microsc.,* 139, 57, 1985.
3. **Gualtieri, P., Ghetti, F., Passarelli, V., and Barsanti, L.,** Microorganism track reconstruction: an image processing approach, *Comput. Biol. Med.,* 18, 57, 1988.
4. **Häder, D. and Lebert, M.,** Real time computer-controlled tracking of motile microorganisms, *Photochem. Photobiol.,* 42, 5, 509, 1985.
5. **Kondo, T., Kubota, M., Aono, Y., and Watanabe, M.,** A computerized video system to automatically analyze movements of individual cells and its application to the study of circadian rhythms in phototaxis and motility in *Chlamydomonas reinhardtii, Protoplasma Suppl.,* 1, 185, 1988.
6. **Allen, R. D.,** New direction and refinements in video-enhanced microscopy applied to problems in cell motility, in *Advances in Microscopy,* Vol. 139, Cowden, R. R. and Harrison, F. W., Eds., R. Liss, New York, 1983, 3.
7. **Sugimoto, S. A., Matsuki, H., and Ichioka, Y.,** Implementation of tracking and extraction of moving objects in successive frames, *Appl. Opt.,* 25(6), 990, 1986.
8. **Jain, R., Militzer, D., and Nagel, H. H.,** Separating non-stationary from stationary scene components in a sequence of real word TV-images, in *Proc. 5th Int. Joint Conf. Artificial Intelligence,* IAAI, Kyoto, 1977, 612.
9. **Gualtieri, P. and Coltelli, P.,** A digital microscope for real time detection of moving microorganism, *Micron Microsc. Acta,* 20(2), 99, 1989.
10. **Coltelli, P. and Gualtieri, P.,** A procedure for the extraction of object features in microscope images, *J. Biomed. Comput.,* 25, 169, 1990.

Index

INDEX

A

B

C

D

E

F

G

H

I

K

L

M

N

Q

R

S

T

U

V

W

X

Y

Z

rendu le